APPLIED MATHEMATICS-I

By

Dr. J.K. TYAGI
Professor & Head
Deptt. of Applied Science
Maharaja Agarsain Instt. of Tech.
Ghaziabad (U.P.)

Dr. S.K. TYAGI
Senior Lecturer
Amity School of Engg. & Tech.
Amity University,
NOIDA (U.P.)

KHANNA BOOK PUBLISHING CO. (P) LTD.

DELHI

Applied Mathematics-I

By Dr. J.K. Tyagi
Dr. S.K. Tyagi

Laser Typeset by : **PLUS COMPUTERS**, 696/1, Madhav Puram, Meerut. Ph. 0-9897079785.

*Published by **Khanna Book Publishing Co. (P) Ltd.***
1694-95, Nai Sarak, Delhi - 110 006
Ph : 23256658; Fax : 23283211,
e-mail : Contact@khannaedu.com

Dedicated

To

OUR TEACHERS

Preface to the First Edition

This book is outcome of the experience of both the authors in teaching in Applied Mathematics for last few years in various institutions. The book is intended for the students of B.Tech. first semester Applied Mathematics course of Amity University Uttar Pradesh. Each chapter starts with comprehensive text written pragmatically in a very lucid manner and supported by a large number of illustrative examples to stimulate the interest of the students in the subject. The book comprises of a large number of solved and unsolved exercises for the better understanding of students. A huge number of unsolved problems are also added to each chapter for enough practice to be confident about the concept.

We express our profound and deep regards to our source of inspiration, respected Dr. Ashok K. Chauhan, Founder President, Ritnand Balved Education Foundation for his encouragement and blessing. We are also grateful to Prof. Balvinder Shukla, Director General, Amity School of Engineering and Technology, Noida for her generous support and motivation. The Director and Management of Maharaja Agarsain Institute of Technology, Ghaziabad are also thankfully acknowledged. Our sincere thanks are also due to Prof. A.L. Verma, HOD, Applied Science for his valuable guidance and to all our colleagus for their critical and constructive suggestions.

The authors are highly indebted to numerous authors, whose great and masterly work has been consulted to constuct this book.

We would like to extend our special thanks to our family members for their patience and constant support during the period to make this book a realty.

Last but not least, we owe our gratitude to M/s Khanna Publishers for their enthusiasm to publish this book.

August, 2009 **AUTHORS**

SYLLABUS

AMITY SCHOOL OF ENGINEERING & TECHNOLOGY

APPLIED MATHEMATICS-I (BTMT-10101)

UNIT-I : DIFFERENTIAL CALCULUS *14 Lectures*

Successive differentiation, Leibnitz's theorem (without proof), Expansion of Function of Several Variables, Asymptote & Curvature, Partial derivatives, Chain rule, Differentiation of Implicit functions, Total Differential, Jacobians, Maxima and Minima of two variables, Curve Tracing.

UNIT-II : INTEGRAL CALCULUS *12 Lectures*

Reduction formulae, Applications to length, Area and Volume, Gamma and Beta Integrals, Double Integrals, Change of Order of Integration, Triple Integrals, Applications to Area and Volume (Simple Applications).

Unit-III : ORDINARY DIFFERENTIAL EQUATIONS *10 Lectures*

Solution of Ordinary Differential Equations of First Order : Method of Separation of Variables, Homogeneous and Non-homogeneous Equations, Linear Differential Equations & Bernoulli's Equations, Exact Differential Equations, Linear Differential Equations of Higher Order, Complete Solution, Complementary Function, Particular Integrals, Variation of Parameters, Solution of Homogeneous Differential Equation—Cauchy's and Legendre's, Solution of Simple Simulateneous ODE.

Unit-IV : Vector Calculus *9 Lectures*

Scalar and Vector Field, Derivative of a Vector, Gradient, Directional Derivative, Divergence and Curl and their Physical Significance, Arc Length, Tangent, Directional Derivative, Evaluation of Line Integral, Green's Theorem in Plane (without proof), Representation of Surfaces, Tangent Plane and Surface Normal, Stoke's Theorem (without proof), Gauss Divergence Theorem (without proof).

CONTENTS

CHAPTER 1 : DIFFERENTIAL CALCULUS 1 - 122

CHAPTER 2 : INTEGRAL CALCULUS 123 - 244

CHAPTER 3 : ORDINARY DIFFERENTIAL EQUATIONS 245 - 363

1

Differential Calculus

1.1 ORDINARY DIFFERENTIATION

If $y = f(x)$ be a **function of only one variable** x, the differential coefficient $\frac{dy}{dx}$ of y, w.r.t x is called its **ordinary differential coefficient.**

But when $z = f(x, y)$ is a function of two independent variables, z is differentiated w.r.t. to one independent variable, x or y, keeping the other as constants, $\frac{\partial z}{\partial x}, \frac{\partial z}{\partial y}$ are called its **partial differentials coefficient** with respect to x and y respectively.

Here we consider the ordinary differentiation of function $y = f(x)$.

1.2 SUCCESSIVE DIFFERENTIATION

Differentiating a function $y = f(x)$, once, twice, thrice w.r.t. x, successively, is called its **successive differentiation**. The successive differential coefficients are denoted by

$$\frac{dy}{dx}, \frac{d^2y}{dx^2}, \ldots, \frac{d^ny}{dx^n}, \ldots \text{ or as } Dy, D^2y, \ldots \text{ or simplfy as } y_1, y_2, \ldots y_n.$$

Example 1.1 **If $y = e^{ax} \cos bx$, prove $y_2 - 2ay_1 + (a^2 + b^2)\, y = 0$.**

Solution : Differentiating and denoting the first derivative $\frac{dy}{dx}$ by y_1

$$y_1 = e^{ax} (a \cos bx - b \sin bx)$$

$$= ay - be^{ax} \sin bx$$

Differentiating again $y_2 = ay_1 - be^{ax} (a \sin bx + b \cos bx)$

$$= ay_1 - abe^{ax} \sin bx - b^2y$$

$$= ay_1 + a\{y_2 - ay\} - b^2y$$

$$= 2ay_1 - (a^2 + b^2)y$$

So $\quad y_2 - 2ay_1 + (a^2 + b^2)y = 0$

Example 1.2 **If $x = \sin t$, $y = \sin pt$, prove $(1 - x^2)y_2 - xy_1 + p^2y = 0$.**

Solution : We have $\quad pt = \sin^{-1} y$

or $\quad p \sin^{-1} x = \sin^{-1} y$

Differentiating w.r.t. x, both the sides

$$\frac{p}{\sqrt{1 - x^2}} = \frac{y_1}{\sqrt{1 - y^2}}$$

or $\quad (1 - x^2)y_1^2 = p^2(1 - y^2)$

Differentiating again

$$(1 - x^2)_2 y_1 y_2 - 2xy_1^2 = -p^2 2yy_1$$

or $\quad 2y_1 \{(1 - x^2)y_2 - xy_1 + p^2y\} = 0$, y_1 not always zero

so $\quad (1 - x^2)y_2 - xy_1 + p^2y = 0.$

Example 1.3 **Find $\dfrac{d^2y}{dx^2}$, when $x = a \cos^3 \theta$, $y = b \sin^3 \theta$.**

Solution :

$$\frac{dy}{dx} = \frac{\dfrac{dy}{d\theta}}{\dfrac{dx}{d\theta}} = \frac{-3b \sin^2 \theta \cos \theta}{3a \cos^2 \theta \sin \theta} = -\frac{b}{a} \tan \theta$$

So $\quad \dfrac{d^2y}{dx^2} = -\dfrac{b}{a} \dfrac{d}{dx} \tan \theta = -\dfrac{b}{a} \dfrac{d}{d\theta} \tan \theta \dfrac{d\theta}{dx}$

$$= -\frac{b}{a} \sec^2 \theta . \left(\frac{1}{-3a \cos^2 \theta \sin \theta}\right) = \frac{b}{3a^2} \sec^4 \theta \operatorname{cosec} \theta$$

Example 1.4 **If $ax^2 + 2hxy + by^2 = 1$, prove $\dfrac{d^2y}{dx^2} = \dfrac{h^2 - ab}{(hx + by)^3}$.**

Solution : Differentiating, the given equation, w.r.t. x,

$$ax + hy + (hx + by) \frac{dy}{dx} = 0$$

So, $\quad y_1 = \dfrac{dy}{dx} = -\dfrac{ax + hy}{hx + by}$

Differentiating again

$$y_2 = -\frac{(hx + by)(a + hy_1) - (ax + hy)(h + by_1)}{(hx + by)^2}$$

$$= -\frac{1}{(hx + by)^2} \{(ahx + aby - abx - h^2y) + (h^2x + hby - abx - hby)y_1\}$$

$$= \frac{h^2 - ab}{(hx + by)^2} \{y - xy_1\} = \frac{(h^2 - ab)(ax^2 + 2hxy + by^2)}{(hx + by)^3}$$

as $$ax^2 + 2hxy + by^2 = 1$$

Hence $$y_2 = (h^2 - ab)(hx + by)^{-3}.$$

Example 1.5 **If $x = a(\cos t + t \sin t)$, $y = a(\sin t - t \cos t)$; find $\dfrac{d^2y}{dx^2}$.**

Solution :

$$\frac{dy}{dx} = \frac{dy/dt}{dx/dt} = \frac{a(\cos t + t \sin t - \cos t)}{a(-\sin t + \sin t + t \cos t)} = \tan t$$

Again differentiating,

$$\frac{d^2y}{dx^2} = \frac{d}{dx}(\tan t) \cdot \frac{dt}{dx} = \frac{\sec^2 t}{at \cos t} = \frac{\sec^3 t}{at}.$$

Example 1.6 **If $p^2 = a^2 \cos^2\phi + b^2 \sin^2\phi$, show that**

$$p + \frac{d^2p}{d\phi} = \frac{a^2b^2}{p^3}.$$

Solution : We are given

$$p^2 = a^2 \cos^2\phi + b^2 \sin^2\phi \qquad ...(1)$$

Differentiating both sides of equation (1) w.r. to ϕ, we get

$$2p \frac{dp}{d\phi} = -2a^2 \cos\phi \sin\phi + 2b^2 \sin\phi \cos\phi$$

$$\Rightarrow \quad p \frac{dp}{d\phi} = -a^2 \cos\phi \sin\phi + b^2 \sin\phi \cos\phi = (b^2 - a^2) \sin\phi \cos\phi \qquad ...(2)$$

Differentiating both sides of equation (2) w.r. to ϕ, we get

$$p \frac{d^2p}{d\phi^2} + \left(\frac{dp}{d\phi}\right)^2 = (b^2 - a^2)(\cos^2\phi - \sin^2\phi)$$

$$= (b^2 \cos^2\phi + a^2 \sin^2\phi) - (a^2 \cos^2\phi + b^2 \sin^2\phi)$$

$$= (b^2 \cos^2\phi + a^2 \sin^2\phi) - p^2$$

So, $$p \frac{d^2p}{d\phi^2} + p^2 = (b^2 \cos^2\phi + a^2 \sin^2\phi) - \left(\frac{dp}{d\phi}\right)^2$$

$$= (b^2 \cos^2\phi + a^2 \sin^2\phi) - \frac{(b^2 - a^2)^2 \sin^2\phi \cos^2\phi}{p^2} \qquad \text{[From equation (2)]}$$

$$= \frac{1}{p^2}[p^2(b^2 \cos^2\phi + a^2 \sin^2\phi) - (b^2 - a^2)^2 \sin^2\phi \cos^2\phi]$$

$$= \frac{1}{p^2}[(a^2 \cos^2\phi + b^2 \sin^2\phi)(b^2 \cos^2\phi + a \sin^2\phi) - (b^2 - a^2)^2 \sin^2\phi \cos^2\phi]$$

$$= \frac{1}{p^2}[a^2b^2 \cos^4\phi + a^2b^2 \sin^4\phi + 2a^2b^2 \sin^2\phi \cos^2\phi]$$

$$= \frac{a^2b^2}{p^2}[\cos^2\phi + \sin^2\phi]^2 = \frac{a^2b^2}{p^2}$$

Dividing both sides by p.

$$\frac{d^2p}{d\phi^2} + p = \frac{a^2b^2}{p^3}$$

Example 1.7 **If** $y = \left[\log\left(\frac{x + \sqrt{x^2 - a^2}}{a}\right)\right]^2 + K \log(x + \sqrt{x^2 - a^2})$, **then prove that**

$$(x^2 - a^2)\frac{d^2y}{dx^2} + x\frac{dy}{dx} = 2a.$$

Solution : We are given

$$y = \left[\log\left(\frac{x + \sqrt{x^2 - a^2}}{a}\right)\right]^2 + K \log(x + \sqrt{x^2 - a^2}) \qquad ...(1)$$

On differentiating equation (1), w.r. to x, we get

$$\frac{dy}{dx} = 2 \log\left[\frac{x + \sqrt{x^2 - a^2}}{a}\right] \cdot \frac{a}{x + \sqrt{x^2 - a^2}} \cdot \frac{(x + \sqrt{x^2 - a^2})}{(\sqrt{x^2 - a^2})} + \frac{K}{(x + \sqrt{x^2 - a^2})} \cdot \frac{(x + \sqrt{x^2 - a^2})}{\sqrt{x^2 - a^2}}$$

or $$\frac{dy}{dx} = \frac{2a}{\sqrt{x^2 - a^2}} \log\left(\frac{x + \sqrt{x^2 - a^2}}{a}\right) + \frac{K}{\sqrt{x^2 - a^2}}$$

or $$\sqrt{x^2 - a^2}\frac{dy}{dx} = 2a \log\left(\frac{x + \sqrt{x^2 - a^2}}{a}\right) + K$$

Againg differentiating, we get

$$\sqrt{x^2 - a^2}\frac{d^2y}{dx^2} + \frac{dy}{dx} \cdot \frac{x}{\sqrt{x^2 - a^2}} = \frac{2a}{(x + \sqrt{x^2 - a^2})} \cdot \frac{x + \sqrt{x^2 - a^2}}{\sqrt{x^2 - a^2}}$$

or $$\sqrt{x^2 - a^2}\frac{d^2y}{dx^2} + \frac{x}{\sqrt{x^2 - a^2}}\frac{dy}{dx} = \frac{2a}{\sqrt{x^2 - a^2}}$$

or $$(\sqrt{x^2 - a^2})\frac{d^2y}{dx^2} + x\frac{dy}{dx} = 2a.$$

PROBLEM SET 1.1

1. If $y = e^{ax} \sin bx$, show that $y_2 - 2ay_1 + (a^2 + b^2)y = 0$.
2. If $y = a \cos(\log x) + b \sin(\log x)$, show that $x^2y_2 + xy_1 + y = 0$.
3. If $y = Ae^{-ax} \cos bx + c$, show that $y_2 + 2ay_1 + (a^2 + b^2)y = 0$.
4. If $y = \sin(a \sin^{-1} x)$, show that $(1 - x^2)y_2 - xy_1 + a^2y = 0$.
5. If $x^3 + y^3 = 3axy$, show that $y_2 = \dfrac{-2a^2xy}{(y^2 - ax)^3}$.

6. If $y = \sin(\sin x)$, prove that $y_2 + \tan x y_1 + y \cos^2 x = 0$.

7. If $y = \dfrac{ax+b}{cx+d}$, show that $2y_1y_3 = 3y_2^2$.

8. If $y = \tan^{-1} x$, prove that $(1+x^2)y_2 + 2xy_1 = 0$.

9. If $y = \dfrac{\log x}{x}$, show that $\dfrac{d^2y}{dx^2} = \dfrac{2\log x - 3}{x^3}$.

10. If $y = \log(\sin x)$ show that $y_3 = \dfrac{2\cos x}{\sin^3 x}$.

11. If $y = x + \tan x$, then prove that $\cos^2 x \dfrac{d^2y}{dx^2} - 2y + 2x = 0$.

12. If $y = \left(\dfrac{1}{x}\right)^x$, show that $y_2(1) = 0$.

13. If $y^3 + 3ax^2 + x^3 = 0$, then show that $\dfrac{d^2y}{dx^2} + \dfrac{2a^2x^2}{y^5} = 0$.

14. If $y = (\tan^{-1} x)^2$, show that $(1+x^2)^2 \dfrac{d^2y}{dx^2} + 2x(1+x^2)\dfrac{dy}{dx} - 2 = 0$.

15. If a $y = \sin(x+y)$, show that that $y\left(1+\left(\dfrac{dy}{dx}\right)\right)^3 = 0$.

16. If $y = e^{-x}(A\cos x + B\sin x)$, prove that $\dfrac{d^2y}{dx^2} + 2\dfrac{dy}{dx} + 2y = 0$.

1.3 *n*th DERIVATIVE

We find the *n*th derivatives $\dfrac{d^ny}{dx^n}$ of some functions.

(i) $\boldsymbol{y = e^{ax}}$.

The successive derivatives are $y_1 = ae^{ax}$, $y_2 = a^2e^{ax}$ ..., $y_n = a^ne^{ax}$.

(ii) $\boldsymbol{y = \sin x}$, The successive derivatives are given as

$$y_1 = \cos x = \sin\left(x + \frac{\pi}{2}\right)$$

$$y_2 = \cos\left(x + \frac{\pi}{2}\right) = \sin\left(x + \frac{2\pi}{2}\right)$$

... ...

So
$$\frac{d^n(\sin x)}{dx^n} = y_n = \sin\left(x + \frac{n\pi}{2}\right)$$

(iii) $\boldsymbol{y = \cos x}$. The successive derivatives are given as

$$y_1 = -\sin x = \cos\left(x + \frac{\pi}{2}\right)$$

$$y_2 = -\sin\left(x + \frac{\pi}{2}\right) = \cos\left(x + \frac{2\pi}{2}\right)$$

... ...

So $$D^n(\cos x) = \boldsymbol{y_n = \cos\left(x + \frac{n\pi}{2}\right)}$$

(iv) $\boldsymbol{y = \frac{1}{ax+b}}$, $y_1 = \frac{(-a)\,1\,!}{(ax+b)^2}$, $y_2 = \frac{(-a)^2\,2\,!}{(ax+b)^3}$ ... in general

$$\boldsymbol{y_n = (-1)^n a^n n\,!\,(ax+b)^{-n-1}}.$$

(v) $\boldsymbol{y = \log(ax+b)}$, $y_1 = a\left(\frac{1}{ax+b}\right)$

$$y_n = \frac{d^{n-1}}{dx^{n-1}}\left(\frac{dy}{dx}\right) = \frac{d^{n-1}}{dx^{n-1}}\left(\frac{a}{ax+b}\right).$$

Using the above result.

$$= a(-1)^n (n-1)\,!\,a^{n-1}(ax+b)^{n-1+1}$$

So $$D^n \log(ax+b) = y_n = (-1)^{n-1}(n-1)\,!\,a^n (ax+b)^n.$$

(vi) $\boldsymbol{y = x^n}$, $\boldsymbol{n}$ **= + ive integer**

$$y_1 = nx^{n-1} = \frac{n\,!}{(n-1)\,!}x^{n-1}$$

$$y_2 = n(n-1)x^{n-2} = \frac{n\,!}{(n-2)\,!}x^{n-2}$$

... ...

$$D^n(x^n) = y_n = \frac{n\,!}{(n-n)\,!}x^{n-n} = \frac{n\,!}{0\,!} = n\,!$$

$$D^{n+1}(x^n) = y_{n+1} = 0 = y_{n+2} \cdots$$

(vii) $$\boldsymbol{y = \tan^{-1}\frac{x}{a}}$$

$$\frac{dy}{dx} = \frac{1}{1 + \frac{x^2}{a^2}} \cdot \frac{1}{a} = \frac{a}{(x+ia)(x-ia)} = \frac{1}{2i}\left\{\frac{1}{x-ia} - \frac{1}{x+ia}\right\}$$

$$y_n = \frac{(-1)^{n-1}(n-1)\,!}{2i}\left\{\frac{1}{(x-ia)^n} - \frac{1}{(x+ia)^n}\right\}$$

Let $x = r\cos\alpha$, $a = r\sin\alpha$

$$= \frac{(-1)^{n-1}(n-1)\,!}{2i\,r^n}\{(\cos\alpha - i\sin\alpha)^{-n} - (\cos\alpha + i\sin\alpha)^{-n}\}$$

$$= \frac{(-1)^{n-1}(n-1)!}{2i\, r^n}\{\cos(-n\alpha) + i \sin n\alpha - \cos(-n\alpha) + i \sin n\alpha\}$$

$$= (-1)^{n-1}(n-1)!\, \frac{\sin n\alpha}{r^n},$$

where $r = (x^2 + a^2)^{1/2}$ and $\alpha = \tan^{-1}\frac{a}{x} = \cot^{-1}\frac{x}{a}$.

(viii) $y = e^{ax}\sin(bx + c)$

$$y_1 = e^{ax}.\cos(bx + c).b + \sin(bx + c).a.e^{ax}$$

$$= e^{ax}[a\sin(bx + c) + b\cos(bx + c)]$$

Let $a = r\cos\theta$ and $b = r\sin\theta$

$\Rightarrow$ $r = \sqrt{a^2 + b^2},\ \theta = \tan^{-1}\frac{b}{a}$

$\therefore$ $y_1 = e^{ax}[r\cos\theta\sin(bx + c) + r\sin\theta\cos(bx + c)]$

$$= re^{ax}\sin(bx + c + \theta)$$

In a similar manner we can obtain

$$y_2 = r^2 e^{ax}\sin(bx + c + 2\theta)$$

$$y_3 = r^3 e^{ax}\sin(bx + c + 3\theta)$$

....

$\therefore$ $y_n = r^n e^{ax}\sin(bx + c + n\theta)$

or $y_n = (a^2 + b^2)^{n/2} e^{ax}\sin\left(bx + c + n.\tan^{-1}\frac{b}{a}\right)$

(iv) $y = e^{ax}\cos(bx + c)$

$$y_1 = ae^{ax}\cos(bx + c) - be^{ax}\sin(bx + c)$$

$$= e^{ax}[a\cos bx + c) - b\sin(bx + c)]$$

Let $a = r\cos\theta,\ b = r\sin\theta$

$\Rightarrow$ $r = \sqrt{a^2 + b^2},\ \theta = \tan^{-1}\frac{b}{a}$

$\therefore$ $y_1 = re^{ax}[\cos\theta\cos(bx + c) - \sin\theta\sin(bx + c)]$

$$= e^{ax}[a\cos(bx + c) = b\sin(bx + c)]$$

Let $a = r\cos\theta,\ b = r\sin\theta$

$\Rightarrow$ $r = \sqrt{a^2 + b^2},\ \theta = \tan^{-1}\frac{b}{a}$

$\therefore$ $y_1 = re^{ax}[\cos\theta\cos(bx + c) - \sin\theta\sin(bx + c)]$

$$= re^{ax}[\cos(bx + c + \theta)]$$

In a similar manner, we can obtain,

$$y_2 = r^2 e^{ax} [\cos (bx + c + 2\theta)]$$

$$y_3 = r^3 e^{ax} [\cos (bx + c + 3\theta)]$$

....

$\therefore$ $$y_n = r^n e^{ax} [\cos (bx + c + n\theta)]$$

$\therefore$ $$y_n = r^n e^{ax} \cos (bx + c + n\theta)$$

or $$y_n = (a^2 + b^2)^{n/2} e^{ax} \cos\left(bx + c + n \tan^{-1} \frac{b}{a}\right)$$

Example 1.8 **Find the *n*th derivative of $y = \sin 3x \sin 2x$.**

Solution : We have $$y = \frac{1}{2} [\cos x - \cos 5x]$$

So $$y_n = \frac{1}{2}\left[\cos\left(x + \frac{n\pi}{2}\right) - 5^n \cos\left(5x + \frac{n\pi}{2}\right)\right]$$

Example 1.9 **Find the *n*th derivative of $y = \dfrac{x^3}{(x-1)(x-2)}$.**

Solution : As numerator is of higher degree than the denominator, we first divide it by the denominator before breaking it into partial fractions. We then have

$$y = x + 3 + \frac{7x-6}{(x-1)(x-2)} = x + 3 - \frac{1}{x+1} + \frac{8}{x-2}.$$

for $n = 1$, $$y_1 = 1 - \frac{-1}{(x-1)^2} + \frac{-8}{(x-2)^2}$$

for $n > 1$ $$y_n = 0 + (-1)^n n!\left\{-\frac{1}{(x-1)^{n+1}} + \frac{8}{(x-2)^{n+1}}\right\}$$

Example 1.10 **Find *n*th derivative of $y = \dfrac{x^2}{(x+2)(2x+3)}$.**

Solution : We have

$$y = \frac{x^2}{(x+2)(2x+3)}$$

or $$y = \frac{1}{2}\left[1 - \frac{8}{x+2} + \frac{9}{2x+3}\right]$$

$\therefore$ $$\frac{d^n y}{dx^n} = -\frac{(-1)^n . n!\, 8}{2(x+2)^{n+1}} + \frac{(-1)^n n!\, 2^n . 9}{2(2x+3)^{n+1}}$$

$$= \frac{(-1)^n n!}{2}\left[\frac{9 . 2^n}{(2x+3)^{n+1}} - \frac{8}{(x+2)^{n+1}}\right]$$

Example 1.11 **Find *n*th derivative of $y = \dfrac{x}{(x^2 + a^2)}$.**

Solution : We have

$$y = \frac{x}{(x^2 + a^2)}$$

or $$y = \frac{x}{(x+ai)(x-ai)} = \frac{1}{2}\left[\frac{1}{(x-ai)} + \frac{1}{(x+ai)}\right]$$

$$\therefore \quad \frac{d^n y}{dx^n} = \frac{1}{2}\left[\frac{d}{dx^n}\frac{1}{(x-ai)} + \frac{d^n}{dx^n}\left(\frac{1}{x+ai}\right)\right]$$

$$= \frac{1}{2}\left[\frac{(-1)^n\, n!}{(x-ai)^{n+1}} + \frac{(-1)^n\, n!}{(x+ai)^{n+1}}\right]$$

$$= \frac{(-1)^n}{2} n!\left[\frac{1}{(x-ai)^{n+1}} + \frac{1}{(x+ai)^{n+1}}\right]$$

Let $x = r\cos\theta,\ a = r\sin\theta$, so that $r = \sqrt{x^2+a^2},\ \theta = \tan^{-1}\left(\frac{a}{x}\right)$

$$\therefore \quad \frac{1}{(x-ai)^{n+1}} = \frac{1}{r^{n+1}}(\cos\theta - i\sin\theta)^{-(n+1)}$$

$$= \frac{1}{r^{n+1}}[\cos(n+1)\theta - i\sin(n+1)\theta]$$

and $$\frac{1}{(x+ai)^{n+1}} = \frac{1}{r^{n+1}}(\cos\theta + i\sin\theta)^{-(n+1)}$$

$$= \frac{1}{r^{n+1}}[\cos(n+1)\theta - i\sin(n+1)\theta)]$$

Hence, $$\frac{d^n y}{dx^n} = \frac{(-1)^n\, n!\cos(n+1)\theta}{r^{n+1}}$$

where, $r = \sqrt{x^2+a^2},\quad \theta = \tan^{-1}\left(\frac{a}{x}\right)$.

Example 1.12. Find the *n*th differential coefficient of $y = \cos^4 x$

Solution : As $$\cos^4 x = \left(\frac{1+\cos 2x}{2}\right)^2 = \frac{1}{4}\left\{1 + 2\cos 2x + \frac{1+\cos 4x}{2}\right\}$$

So $$y = \frac{3}{8} + \frac{1}{2}\cos 2x + \frac{1}{8}\cos 4x$$

Hence $$y_n = 0 + \frac{1}{2}\times 2^n\cos\left(2x + \frac{n\pi}{2}\right) + \frac{1}{8}4^n\cos\left(4x + \frac{n\pi}{2}\right).$$

Example 1.13 Find the *n*th derivative of $y = \tan^{-1}\dfrac{1+x}{1-x}$.

Solution : $$y = \tan^{-1}\frac{\tan\frac{\pi}{4} + \tan\theta}{1 - \tan\frac{\pi}{4}\tan\theta}$$

where $x = \tan\theta$

$$= \tan^{-1}\tan\left(\frac{\pi}{4} + \theta\right) = \frac{\pi}{4} + \tan^{-1} x.$$

So $\dfrac{dy}{dx} = 0 + \dfrac{1}{x^2+1} = \dfrac{1}{(x+i)(x-i)} = \dfrac{1}{2i}\left\{\dfrac{1}{x-i} - \dfrac{1}{x+i}\right\}$

$$D^n y = D^{n-1}\left(\frac{dy}{dx}\right) = \frac{1}{2i}(-1)^{n-1}(n-1)!\left\{\frac{1}{(x-1i)^n} - \frac{1}{(x+1i)^n}\right\}$$

Let $x = r\cos\alpha,\ 1 = r\sin\alpha$ so $r = (\sin\alpha)^{-1} = \operatorname{cosec}\alpha.$

$$y_n = \frac{(-1)^n(n-1)!}{2i}\frac{1}{r^n}\left\{\frac{1}{(\cos\alpha - i\sin\alpha)^n} - \frac{1}{(\cos\alpha + i\sin\alpha)^n}\right\}$$

$$= \frac{(-1)^n(n-1)!}{2i}\frac{1}{r^n}\{\cos n\alpha + i\sin n\alpha - (\cos n\alpha - i\sin n\alpha)$$

$$= (-1)^n(n-1)!\,\frac{1}{r^n}\sin n\alpha$$

$$= (-1)^n(n-1)!\,\sin^n\alpha\,\sin(n\cot^{-1}x),$$

where $\sin\alpha = \dfrac{1}{\sqrt{x^2+1}}$.

Example 1.14 **Find $\dfrac{d^n y}{dx^n}$, if $y = e^{ax}\cos^2 x\sin x$.**

Solution : We have $y = e^{ax}\cos^2 x\sin x$

$$\because \quad \cos^2 x\sin x = \frac{1}{2}(1+\cos 2x)\sin x$$

$$= \frac{1}{2}\sin x + \frac{1}{4}2\sin x\cos 2x$$

$$= \frac{1}{2}\sin x + \frac{1}{4}(\sin 3x - \sin x)$$

$$= \frac{1}{4}\sin x + \frac{1}{4}\sin 3x$$

$$\therefore \quad \frac{d^n y}{dx^n} = \frac{d^n}{dx^n}(e^{ax}\cos^2 x\sin x)$$

$$= \frac{1}{4}\frac{d^n}{dx^n}(e^{ax}\sin x) + \frac{1}{4}\frac{d^n}{dx^n}(e^{ax}\sin 3x)$$

$$= \frac{1}{4}(a^2+1)^{n/2}e^{ax}\sin\left(x + n\tan^{-1}\frac{1}{a}\right) + \frac{1}{4}(a^2+9)^{n/2}e^{ax}\sin\left(3x + \tan^{-1}\frac{3}{a}\right)$$

Example 1.15 **If $y = \sin ax + \cos ax$, show that**

$$y_n = a^n\sqrt{\{1+(-1)^n\sin 2ax\}}.$$

Solution : We have

$$y = \sin ax + \cos ax$$

Differentiating n times, we get

$$y_n = a^n\sin\left(ax + \frac{n\pi}{2}\right) + a^n\cos\left(ax + \frac{n\pi}{2}\right)$$

$$= a^n \sqrt{\left\{\sin\left(ax + \frac{n\pi}{2}\right) + \cos\left(ax + \frac{n\pi}{2}\right)\right\}^2}$$

$$= a^n \sqrt{\{1 + \sin(2ax + n\pi)}$$

$$= a^n \sqrt{1 + \sin 2ax \cos n\pi}$$

$$= a^n \sqrt{1 + (-1)^n \sin 2ax}$$

1.4 *n*th DERIVATIVE OF PRODUCT OF TWO FUNCTIONS (LEIBNITZ THEOREM)

If u and v be the two functions of x, then $\frac{d^n}{dx^n}(uv) = D^n(uv)$, has the following representation

$$D^n(uv) = D^n(u) \,.\, v + {}^nC_1 D^{n-1}(u)\, D(u) + {}^nC_2\, D^{n-2}(u)\, D^2(v) + \ldots$$

$$+ \ldots \text{ General term } {}^nC_r\, D^{n-r}(u)\, D^r(v) + \ldots + uD^n(v).$$

Proof is beyond the scope of syllabus.

Example 1.16 **Find the *n*th derivative of $y = x^2 e^{ax}$.**

Solution : Taking $u = e^{ax}$, $v = x^2$

$$D^n(e^{ax} x^2) = D^n(e^{ax}) \,.\, x^2 + {}^nC_1\, D^{n-1}(e^{ax})\, D(x^2) + {}^nC_2\, D^{n-2}(e^{ax})\, D^2(x^2)$$

$$+ \text{ other terms are all zero as } D^3(x^2) = 0 \ldots$$

$$= a^n e^{ax} x^2 + n a^{n-1} e^{ax} \,.\, 2x + \frac{n(n-1)}{2} a^{n-2} e^{ax} \,.\, 2$$

$$= e^{ax}(a^n x^2 + 2n\, a^{n-1} x + n(n-1)\, a^{n-2}).$$

Value of $(y_n)_0$, (the value of *n*th derivative of y at $x = 0$).

Example 1.17 **Find *n*th derivative of $y = x^2 e^x \cos x$.**

Solution : We have $y = x^2 e^x \cos x$

To find *n*th derivative, we take $e^x \cos x$ as a first factor and x^2 as the second.

$$\therefore \quad \frac{d^n y}{dx^n} = D^n(e^x \cos x \,.\, x^2)$$

$$= D^n(e^x \cos x) \,.\, x^2 + {}^nC_1\, D^{n-1}(e^x \cos x) \,.\, 2x + {}^nC_2\, D^{n-2}(e^x \cos x) \,.\, 2$$

$$= 2^{n/2} e^x \cos(x + n \tan^{-1} 1) \,.\, x^2 + n 2^{(n-1)/2} e^x \cos[x + (n-1)\tan^{-1} 1] \,.\, 2x$$

$$+ \frac{n(n-1)}{2} \,.\, 2^{(n-2)/2} e^x \cos[x + (n-2)\tan^{-1} 1)]\, 2$$

$$= 2^{(n-2)/2} e^x \left[2x^2 \cos\left(x + \frac{n\pi}{4}\right)\right] + 2^{3/2} \,.\, nx \,.\, \cos\left(x + (n-1)\frac{\pi}{4}\right)$$

$$+ \left[n(n-1) \cos\left(x + \frac{(n-2)\pi}{4}\right)\right]$$

Example 1.18 **If $\cos^{-1}\left(\frac{y}{b}\right) = \log\left(\frac{x}{n}\right)^n$, prove that**

$$x^2y_{n+2} + (2n+1)xy_{n+1} + 2n^2y_n = 0.$$

Solution : We are given

$$\cos^{-1}\left(\frac{y}{b}\right) = \log\left(\frac{x}{n}\right)^n$$

$$\Rightarrow \quad \cos^{-1}\left(\frac{y}{b}\right) = n\log\left(\frac{x}{n}\right)$$

Differentiating both sides w.r. to x; we get

$$\frac{-y_1}{\sqrt{b^2-y^2}} = \frac{n}{x}$$

or $$x^2y_1^2 = n^2(b^2-y^2)$$

Differentiating again w.r. to x, we get

$$2x^2y_1y_2 + 2xy_1^2 = -2n^2yy_1$$

$$\Rightarrow \quad x^2y_2 + xy_1 + n^2y = 0$$

Applying Leibnitz theorem, we get

$$x^2y_{n+2} + {}^nC_1 \cdot y_{n+1} \cdot 2x + {}^nC_2y_n \cdot 2 + xy_{n+1} + {}^nC_1y_n + n^2y_n = 0$$

or $$x^2y_{n+2} + (2n+1)xy_{n+1} + 2n^2y_n = 0.$$

Example 1.19 **If $y = (\sin^{-1}x)^2$, show that $(1-x^2)y_{n+2} - (2n+1)xy_{n+1} - x^2y_n = 0$ hence find $(y_n)_0$.**

Solution : We have $y = (\sin^{-1}x)^2$, on differentiation w.r.t. x

$$y_1 = \frac{2\sin^{-1}x}{\sqrt{1-x^2}} = 2\sqrt{\frac{y}{1-x^2}}$$

So $$(1-x^2)y_1^2 - 4y = 0 \quad \text{...(1)}$$

From above $y(0) = 0$,

$$y_1(0) = 0 \quad \text{...(2)}$$

Differentiating (1)

$$(1-x^2)2y_1y_2 - 2xy_1y_1 - 2\times 2y_1 = 0$$

or $$(1-x^2)y_2 - xy_1 - 2 = 0 \quad \text{...(3)}$$

We have $$(y_2)_0 - 0 - 2 = 0 \text{ or } (y_2)_0 = 2 \quad \text{...(4)}$$

Differentiating (3), n times

$$D^n\{(1-x^2)y_2\} - D^n(xy_1) - D^n(2) = 0$$

or $$(1-x^2)y_{n+2} + {}^nC_1(-2x)y_{n+1} - 2\,{}^nC_2\,y_n + 0 - xy_{n+1} - {}^nC_1 \cdot 1y_n + 0 = 0$$

or $$(1-x^2)y_{n+2} - (2n+1)xy_{n+1} - n^2y_n = 0 \quad \text{...(5)}$$

Putting $x = 0$,

$$(y_{n+2})_0 = n^2(y_n)_0 \quad \text{...(6)}$$

replacing $(n + 2)$ by n

$$(y_n)_0 = (n-2)^2 (y_{n-2})_0 \quad \text{...(7)}$$

Case (i) Let n = even ... ...

$$(y_4)_0 = 2^2(y_2)_0 \quad \text{...(8)}$$

$$(y_2)_0 = 2 \text{ from (4)} \quad \text{...(9)}$$

Multiplying (6), (7), ... (8), (9)

$$(y_{n+2})_0 (y_n)_0 \cdots (y_4)_0(y_2)_0 = n^2(n-2)^2 \ldots 4^2.\ 2^2.\ 2\ (y_n)_0(y_{n-2}) \cdots (y_2)_0$$

Hence, $$(y_{n+2})_0 = \{2^2.\ 4^2 \ldots n^2\}.\ 2.$$

Case (ii) When n = odd $(y_1)_0 = 0$ so $(y_3)_0 = 1^2(y_1)_0 = 0$

So $(y_n)_0 = 0$.

Example 1.20 **If $y^{1/m} + y^{-1/m} = 2x$, prove that**

$$(x^2 - 1)y_{n+2} + (2n + 1)x\, y_{n+1} + (n^2 - m^2)y_n = 0.$$

Solution : Let $y^{1/m} = z$ then $z + \dfrac{1}{z} - 2x = 0$

or $$z^2 - 2xz + 1 = 0 \text{ so } z = x \pm \sqrt{x^2 - 1}$$

considering only +sign.

So $$y = z^m = (x + \sqrt{x^2 - 1})^m$$

This now becomes the question solved in Ex. 1.23.

Example 1.21 **If $(x + y) = 1$, then prove that**

$$D^n (x^n y^n) = n!\, [y^n - ({}^nC_1)^2\, y^{n-1} .\, x + ({}^nC_2)^2\, y^{n-2}\, x^2 + \ldots (-1)^n x^n].$$

Solution : It is given that $x + y = 1 \Rightarrow y = 1 - x$

Let $x^n = u$ and $y^n = v$, then

$$u_n = n!,\ u_{n-1} = n!\, x,\ u_{n-2} = \frac{n!}{2} x^2 \ldots \qquad (u_n = D^n u)$$

$$\therefore \quad \frac{d^n}{dx^n}(x^n y^n) = D^n (uv)$$

$$= u_n v + {}^nC_1\, u_{n-1} .\, v_1 + {}^nC_2\, u_{n-2}\, v_2 + \ldots {}^nC_n\, v_n$$

$$= n! y^n + {}^nC_1\, n!\, x(-\, {}^nC_1\, y^{n-1}) + {}^nC_2 \left(\frac{n!}{2!} x^2\right) {}^nC_2\, 2!\, y^{n-2} + \ldots + x^n(-1)^n\, n!$$

$$= n!\, [y^n - ({}^nC_1)^2\, y^{n-1} + ({}^nC_2)^2\, y^{n-2}\, x^2 + \ldots + (-1)^{n^n}]$$

Example 1.22 **By finding in two different ways, the nth derivative of x^{2n}, prove that**

$$1 + \frac{n^2}{1^2} + \frac{n^2(n-1)^2}{1^2 .\ 2^2} + \frac{n^2(n-1)^2\,(n-2)^2}{1^2 .\ 2^2 .\ 3^2} + \ldots = \frac{(2n)!}{(n!)^2}.$$

Solution : We know that

$$D^n(x^{2n}) = 2n(2n-1)(2n-2)\ldots(2n-n+1)(x)^{2n-n}$$
$$= 2n(2n-1)(2n-2)\ldots(n+1)x^n$$
$$= \frac{(2n)!}{n!}x^n \quad \ldots(1)$$

By Leibnitz theorem, we get

$$D^n(x^{2n}) = D^n(x^n \,.\, x^n)$$
$$= D^n(x^n)\,.\,x^n + {}^nC_1 D^{n-1}(x^n)\,.\,D(x^n) + {}^nC_2 D^{n-2}(x^n)\,.\,D^2(x^n) + {}^nC_3 D^{n-3}(x^n)\,.\,D^3(x^n) + \ldots$$
$$= (n!)x^n + {}^nC_1 (n!\,x)n\,x^{n-1} + {}^nC_2\left(\frac{n!\,x^2}{2!}\right)n(n-1)x^{n-2} + {}^nC_3\frac{n!\,x^3}{3!}[n(n-1)(n-2)x^{n-3}] + \ldots$$
$$= n!\,x^n\left[1 + \frac{n^2}{1^2} + \frac{x^2(n-1)^2}{1^2\,.\,2^2} + \frac{x^2(n-1)^2(n-2)^2}{1^2\,.\,2^2\,.\,3^2} + \ldots\right] \quad \ldots(2)$$

Equating $D^n \,.\, x^{2n}$ as given by equation (1) and (2), we get

$$x^n n!\left[1 + \frac{n^2}{1^2} + \frac{n^2(n-1)^2}{1\,.\,2^2} + \frac{n^2(n-1)^2(n-2)^2}{1^2\,.\,2^2\,.\,3^2} + \ldots\right] = \frac{(2n)!\,x^n}{n!}$$

$$\Rightarrow \quad 1 + \frac{n^2}{1^2}\,\frac{n^2(n-1)^2}{1^2\,.\,2^2} + \frac{n^2(n-1)^2(n-2)^2}{1^2\,.\,2^2\,.\,3^2} + \ldots = \frac{(2n)!}{(n!)^2}$$

Example 1.23 **If $y = (x + \sqrt{x^2+1})^m$, prove**

(i) $(x^2+1)y_2 + xy_1 - m^2y = 0$,

(ii) $(y_{n+2})_0 = (m^2 - n^2)(y_n)_0$. Hence Find $(y_n)_0$.

Solution : Differentiating y,

$$\frac{dy}{dx} = m(x + \sqrt{x^2-1})^{m-1}\left\{1 + \frac{x}{\sqrt{x^2+1}}\right\} = \frac{my}{\sqrt{x^2+1}}$$

So $$(x^2+1){y_1}^2 - m^2y^2 = 0. \quad \ldots(1)$$

Differentiating (1)

$$2y_1(x^2+1)y_2 + 2x{y_1}^2 - m^2 2y_1 y = 0$$

or $$(x^2+1)y_2 + xy_1 - m^2y = 0 \quad \ldots(2)$$

From $y = (x + \sqrt{x^2+1})^m$, $y_1 = \dfrac{my}{\sqrt{1-x^2}}$ and (2)

$$y(0) = 1,\; y_1(0) = m \text{ and } (y_2)_0 = m^2y_0 = m^3 \quad \ldots(3)$$

Differentiating (2), n times

$$D^n\{(x^2+1)y_2\} + D^n(xy_1) - m^2 D^n y = 0$$

or $$(x^2+1)y_{n+2} + 2xny_{n+1} + \frac{2n(n-1)}{2} y_n + x\,y_{n+1} + ny_n - m^2 y_n = 0$$

so $$(x^2+1)y_{n+2} + (2n+1)xy_{n+1} + (n^2 - m^2)y_n = 0 \qquad \text{...(4)}$$

Putting $x = 0$

$$(y_{n+2})_0 = -(n^2 - m^2)(y_n)_0 \qquad \text{...(5)}$$

By repeated use of (5)

$$(y_n)_0 = \{-(n-2)^2 + m^2\}\{-(n-4)^2 + m^2\} \ldots (-2^2 + m^2)(y_0)_0$$

if n be even

$$(y_n) = \{-(n-2)^2 + m^2\}\{m^2 - (n-4)^2\} \ldots (m^2 - 2^2) \,.\, m^3.$$

and **if n is odd**

$$(y_n)_0 = \{m^2 - (n-2)^2\}\{m^2 - (n-4)^2\} \ldots (m^2 - 1^2)(y_1)_0$$

where $(y_1)_0 = m$.

Example 1.24 **If $y = (x^2 - 1)^n$, prove that**

$$\mathbf{(x^2-1)\,y_{n+2} + 2xy_{n+1} - (n+1)\,y_n = 0.}$$

Solution : Differentiating y, we get

$$\frac{dy}{dx} = n\,2x\,(x^2-1)^{n-1} = \frac{2nxy}{x^2-1}$$

or $$(x^2-1)y_1 - 2nxy = 0$$

Differentiating $(n+1)$ times

$$(x^2-1)y_{n+2} + (n+1)c_1\,2x\,y_{n+1} + \frac{(n+1)n}{2}2y_n + 0$$
$$- 2n\,x\,y_{n+1} - 2n\,(n+1)\,1\,.\,y_n + 0 = 0$$

Adding

$$(x^2-1)y_{n+2} + (2n+2-2n)\,y_{n+1} + \{1-2\}\,n(n+1)\,y_n = 0$$

or $$(x^2-1)y_{n+2} + 2x\,y_{n+1} - n(n+1)y_n = 0.$$

Example 1.25 **If $y = \cos(m\sin^{-1}x)$, show that**

$$\mathbf{(1-x^2)y_{n+2} - (2n+1)xy_{n+1} + (m^2-n^2)y_n = 0} \textbf{ and hence find } \mathbf{y_n(0).}$$

Solution : We have

$$y = \cos\,(m\sin^{-1}x) \qquad \text{...(1)}$$

$$y_1 = -\sin\,(m\sin^{-1}x)\cdot\frac{m}{\sqrt{1-x^2}} \qquad \text{...(2)}$$

$$\Rightarrow \qquad (1-x^2)y_1^2 = m^2\sin^2\,(m\sin^{-1}x)$$
$$= m^2\,[1 - \cos^2\,(m\sin^{-1}x)]$$
$$= m^2\,(1-y^2)$$

Differentiating we get

$$(1-x^2)\,2y_1y_2 - 2xy_1^2 = -\,2m^2yy_1$$

$$\Rightarrow \quad (1-x^2)y_2 - xy_1 + m^2y = 0 \qquad \text{...(3)}$$

Applying Leibnitz theorem, we get

$$(1-x^2)y_{n+2} + {}^nC_1\,(-2x)y_{n+1} + {}^nC_2\,(-2)y_n - xy_{n+1} - {}^nC_1y_n + m^2y_n = 0$$

$$\Rightarrow \quad (1-x^2)y_{n+2} - (2n+1)xy_{n+1} + (m^2-n^2)y_n = 0$$

Putting $x = 0$, we get

$$y_{n+2}(0) = (n^2-m^2)y_n(0) \qquad \text{...(4)}$$

From (1), (2) and (3), we get

$$y(0) = 1,\ y_1(0) = 0,\ y_2(0) = m^2,\ y(0) = m^2$$

Putting $n = 1, 2, 3, 4$ and so on in equation (4), we get

$$y_3(0) = (1^2-m^2)y_1(0) = 0$$

$$y_4(0) = (2^2-m^2)y_2(0) = m^2(2^2-m^2)$$

$$y_5(0) = (3^2-m^2)y_3(0) = 0$$

$$y_6(0) = (4^2-m^2)y_4(0) = 0$$

$$= m^2(2^2-m^2)\,(4^2-m^{2)}$$

In general we can write

$$y_n(0) = 0 \text{ if } n \text{ is odd.}$$

$$= m^2(2^2-m^2)\,(4^2-m^2)\ldots((n-2)^2-m^2), \text{ if } n \text{ is even.}$$

PROBLEM SET 1.2

1. Find the nth derivative of $\dfrac{x+1}{x^2-4}$.

Ans. $\dfrac{(-1)^n n!}{4}\left[\dfrac{3}{(x-2)^{n+1}} - \dfrac{1}{(x+2)^{n+1}}\right]$.

2. Find the nth derivative of $y = \dfrac{4x}{(x-1)^2\,(x+1)}$.

Ans. $(-1)^n n!\left[\dfrac{1}{(x-1)^{n+1}} - \dfrac{1}{(x+1)^{n+1}} + \dfrac{2(n+1)}{(x-1)^{n+2}}\right]$.

3. Find the nth derivative of $y = \dfrac{1}{(a^2-x^2)}$.

Ans. $\dfrac{1}{2a}(-1)^n\, n!\left[\dfrac{3}{(x+a)^{n+1}} - \dfrac{1}{(x-a)^{n+1}}\right]$.

4. Find the nth derivative of the following :

(i) $y = \tan^{-1} \dfrac{2x}{1 - x^2}$

Ans. $2(-1)^{n-1}(n-1)! \sin^n \theta \sin n\theta$, where $\theta = \cot^{-1} x$.

(ii) $y = \sin^{-1} \dfrac{2x}{1 + x^2}$

Ans. $(-1)^{n-1}(n-1)! \sin^n \theta \sin n\theta$, where $\theta = \cot^{-1} x$.

(iii) $y = \tan^{-1} \dfrac{\sqrt{1 - x^2} - 1}{x}$

Ans. $\dfrac{1}{2}(-1)^{n-1}(n-1)! \sin^n \theta \sin n\theta$, where $\theta = \cot^{-1} x$.

5. Find the nth derivative of the following :

(i) $\sin^3 x$

Ans. $\dfrac{3}{4} \sin\left(x + \dfrac{n\pi}{2}\right) - \left(\dfrac{3}{4}\right)^n \sin\left(3x + \dfrac{n\pi}{2}\right)$.

(ii) $\sin 5x \sin 3x$

Ans. $\dfrac{1}{2}\left[2^n \cos\left(2x + \dfrac{n\pi}{2}\right) - 8^n \cos\left(8x + \dfrac{n\pi}{2}\right)\right]$.

(ii) $e^x \sin^4 x$

Ans. $\dfrac{e^x}{8}[3 - 4^{n/2} \cos(2x + n \tan^{-1} 2) - 17^{n/2} \cos(4x + n \tan^{-1} 4)]$.

(iv) $e^x \cos x \cos 2x$

Ans. $\dfrac{e^x}{2}\left[(10)^{n/2} \cos(3x + n \tan^{-1} 3) - 2^{n/2} \cos\left(x + \dfrac{n\pi}{4}\right)\right]$.

6. Find nth derivating of the following functions :

(i) $x^n e^x$

Ans. $e^x\left[x^n + \dfrac{n^2}{1!}x^{n-1} + \dfrac{n^2(n-1)^2}{2!}x^{n-2} + \ldots \dfrac{n^2(n-1)^2 \ldots 1^2}{n!}\right]$

(ii) $x^3 \cos x$

Ans. $x^3 \cos\left(x + \dfrac{n\pi}{2}\right) + 3nx^2 \cos\left[\dfrac{x + (n-1)\pi}{2}\right] + 3n(n-1)x \cos\left[x + \dfrac{(n-2)\pi}{2}\right]$

$+ (n(n-1)(n-2) \cos\left[x + \dfrac{(n-3)\pi}{2}\right]$

7. Find the nth derivative of :

(i) $y = e^{ax}(bx + c)^3$.

[**Ans.** $y_3 = e^{ax}[a^n(bx + c)^3 + 3nba^{n-1}(bx + c)^2 a^{n-1}$

$+ 3b^2 n(n-1)a^{n-2}(bx + c) + n(n-1)(n-2)b^3 a^{n-3}]$

(ii) $y = e^x \log x$.

[Ans. $e^x\left[\log x + \frac{^nC_1}{x} - \frac{2\,.\,^nC_2}{x^2} + \ldots + \frac{(-1)^{n-1}(n-1)\,!}{x^n}\right]$

8. If $y = x^2e^x$. Show that

$$y_n = \frac{1}{2}n(n-1)y_2 - n(n-1)y_1 + \frac{1}{2}(n-1)(n-2)\,y.$$

9. If $y = e^{a\sin^{-1}x}$, prove that $(1-x^2)y_{n+2} - (2n+1)xy_{n+1} - (x^2+a^2)y_n = 0$. Hence find y_n at $x = 0$.

10. If $y = a\cos(\log x) + b\sin(\log x)$. Show :

(i) $x^2y_2 + xy_1 + y = 0$

(ii) $x^2y_{n+2} + (2n+1)xy_{n+1} + (n^2+1)y_n = 0$

11. If $y = \sin(m\sin^{-1}x)$, prove that

(i) $(1-x^2)y_2 - xy_1 + m^2y = 0$,

(ii) $(1-x^2)y_{n+2} - 2(n+1)xy_{n+1} + (m^2-n^2)y_n = 0$

Also find $(y_n)_0$.

12. Find $V_n = D^n\{x^n\log x\}$, show that $V_n - nV_{n-1} = (n-1)\,!$

Hence, show that $V_n = n\,!\left\{\log x + 1 + \frac{1}{2} + \ldots + \frac{1}{n}\right\}$.

13. If $y = \sin^{-1}x$, prove that $(1-x^2)y_{n+2} - (2n+1)xy_{n+1} - n^2y_n = 0$.

14. If $x = \sin t$, $y = \cos pt$, show that

(i) $(1-x^2)\,y_2 - xy_1 + p^2y = 0$

(ii) $(1-x^2)y_{n+2} - (2n+1)\,xy_{n+1} - (n^2-p^2)y_n = 0$

15. If $y = \sin\{2\log(x+1)\}$, prove that

(i) $(x+1)^2\,y_2 + (x+1)y_1 + 4y = 0$

(ii) $(x+1)^2y_{n+2} + (2n+1)xy_{n+1} + (n^2+4)y_n = 0$

16. If $x = \cosh\left(\frac{1}{m}\log y\right)$, prove that

$$(x^2-1)y_{n+2} + (2n+1)\,xy_{n+1} + (n^2-m^2)y_n = 0.$$

17. If $y = \tan^{-1}x$, prove that $(1-x^2)y_2 - 2xy_1 = 0$ and hence determine $y_n(0)$.

Ans. $\begin{cases} 0 & \text{if } n \text{ is even} \\ (-1)^{(n-1)/2}\,(n-1)! & \text{if } n \text{ is odd} \end{cases}$

1.5 EXPANSION OF FUNCTIONS (Single Variable)

Let $f(x)$ be a function which can be expanded in a series in powers of x and let the expansion be differentiable term by term any number of times

Then $$f(x) = f(0) + \frac{x}{1!}f'(0) + \frac{x^2}{2!}f''(0) + \ldots$$

Proof : Let $$f(x) = a_0 + a_1x + a_2x^2 + \ldots \qquad \ldots(1)$$

Putting $x = 0$ in (1),

$$f(0) = a_0 \qquad \ldots(2)$$

Differentiating (1)

$$f'(x) = 0 + a_1 + 2a_2x + 3a_3x^2 + \ldots \qquad \ldots(3)$$

Putting $x = 0$ in (3),

$$f'(0) = a_1 \qquad \ldots(4)$$

Differentiating (3)

$$f''(x) = 0 + 2!\,a_2 + 3.2a_3x + 4.3a_4x^2 + \ldots \qquad \ldots(5)$$

Putting $x = 0$ in (5)

$$f''(0) = 2!\,a_2 \qquad \ldots(6)$$

In this way, we get in general $a_r = \dfrac{f^r(0)}{r!}$.

Putting values of $a_0, a_1, a_2, \ldots$ in (1), we get

$$f(x) = f(0) + \frac{x}{1!}f'(0) + \frac{x^2}{2!}f''(0) + \ldots + \frac{x^n}{n!}f^n(0) + \ldots$$

which is known as **Maclaurin's series** expansion of $f(x)$.

1.6 SOME IMPORTANT SERIES EXPANSIONS

1. $e^{ax} = 1 + \dfrac{ax}{1!} + \dfrac{a^2x^2}{2!} + \ldots + \dfrac{a^nx^n}{n!} + \ldots$

2. (i) $\cos x = 1 - \dfrac{x^2}{2!} + \dfrac{x^4}{4!} + \ldots$ (terms alternately +, –)

   (ii) $\cosh x = 1 + \dfrac{x^2}{2!} + \dfrac{x^4}{4!} + \ldots$ (terms, all positive)

3. $\sin x = x - \dfrac{x^3}{3!} + \dfrac{x^5}{5!} + \ldots$

   $\sinh x = x + \dfrac{x^3}{3!} + \dfrac{x^5}{5!} + \ldots$

   $\tan x = x + \dfrac{x^3}{3} + \dfrac{2}{15}x^5 + \ldots$

4. $y = \tan^{-1} x = x - \dfrac{x^3}{3} + \dfrac{x^5}{5} - \dfrac{x^7}{7} + \ldots$ (known as Gregories series).

5. $f(x) = \sin^{-1} x = x + \frac{x^3}{3!} + \frac{9}{5!} x^5 + 3^2 5^2 \frac{x^7}{7!} + \ldots$

6. $\log(1 + x) = x - \frac{x^2}{2} + \frac{x^3}{3} - \frac{x^4}{4}$ .... valid when $|x| < 1$.

7. $\sec x = 1 + \frac{x^2}{2!} + \frac{5}{4!} x^4 + \frac{61}{6!} x^6 + \ldots$

8. $(a + x)^n$ when n is a positive integer

$$= a^n + {}^nc_1 x a^{n-1} + {}^nc_2 x^2 a^{n-2} + \ldots + {}^nc_r a^{n-r} x^r + \ldots + {}^nc_n x^n.$$

9. $(1 + x)^n$, when n is negative integer or a fraction with $|x| < 1$

$$= 1 + \frac{n}{1} x + \frac{n(n-1)}{1.2} x^2 + \ldots$$

$$(1 - x)^{-2} = 1 + 2x + 3x^2 + 4x^3 + \ldots + (n + 1)x^n + \ldots \quad |x| < 1$$

$$(1 - x)^{-1} = 1 + x + x^2 + \ldots + x^n + \ldots \quad |x| < 1$$

$$(1 - x)^{-3} = 1 + 3x + 6x^2 + \ldots + \frac{(n+1)(n+2)}{2} x^n + \ldots$$

1.7 TAYLOR'S SERIES EXPANSION

Let $f(a + h)$ be a function of h which can be expanded in powers of h, and let the expansion be differentiable w.r.t. h, any number of times term by term

then
$$f(a + h) = f(a) + \frac{h}{1!} f'(a) + \frac{h^2}{2!} f''(a) + \ldots \frac{h^n}{n!} f^n(a) + \ldots \quad \ldots(1)$$

Proof : Let
$$f(a + h) = a_0 + ha_1 + h^2 a_2 + \ldots \quad \ldots(2)$$

Putting $h = 0$ we get $f(a) = a_0$.

Differentiating (1), w.r.t. h

$$f'(a + a) = 1!\, a_1 + 2ha_2 + \ldots \quad \ldots(3)$$

Putting $h = 0$ in (3) $f'(a) = 1!\, a_1$, similarly we get $f''(a) = 2! a_2$ and so on. Putting values of a_0, a_1, a_2 in (2), we get the disired result (1).

When the expression of remainder after n terms is found it is given as

$$R_n = \frac{h^n}{n!} f^n(a + \theta h), \; 0 < \theta < 1,$$

the expansion with the term R_n is called Taylor's Theorem with Lagrange's form of remainder.

Example 1.26 Expand $y = e^{\sin x}$, by Maclarin's expansion, upto the terms containing x^4.

Solution :
$$y = e^{\sin x} \text{ so } y_1 = e^{\sin x} \cos x = y \cos x$$

$$y(0) = e^0 = 1,$$

$$y_1(0) = y(0) \cos 0 = 1$$

$$y_2 = y_1 \cos x - y \sin x$$

So $y_2(0) = y_1(0) - 0 = 1$

$$y_3 = y_2 \cos x - 2y_1 \sin x - y \cos x$$

So $y_3(0) = 1 - 0 - 1 = 0$

$$y_4 = y_3 \cos x - 3y_2 \sin x - 3y_1 \cos x + y \sin x$$

So $y_4(0) = 0 - 3 + 0$

$$\text{Hence } e^{\sin x} = y(0) + \frac{y_1(0)}{1\,!}x + \frac{y_2(0)}{2\,!}x^2 + \frac{y_3(0)}{3\,!}x^3 + \frac{y_4(0)}{4\,!}x^4 + \ldots$$

$$= 1 + \frac{x}{1\,!} + \frac{x^2}{2\,!} + 0x^3 - \frac{3x^4}{4\,!} + \ldots$$

Example 1.27 **Expand $y = \log(1 + \sin^2 x)$, in powers of x upto the term containing x^6.**

Solution : Instead of finding 6th derivative and its value $y_6(0)$, it would be easier to use expansions of $\log(1 + x)$ and that of $x = \sin^2 x$.

$$\sin^2 x = \left(x - \frac{x^3}{3\,!} + \frac{x^5}{5\,!} + \ldots\right)^2$$

$$= x^2 + \frac{x^6}{36} + \ldots - \frac{2xx^3}{6} + \frac{2xx^5}{120} + \ldots$$

$$\sin^2 x = x^2 - \frac{x^4}{3} + \frac{2}{45}x^6 + \ldots$$

Hence $\log(1 + \sin^2 x)$

$$= \left(x^2 - \frac{x^4}{3} + \frac{2}{45}x^6 + \ldots\right) - \frac{\left(x^2 - \frac{x^4}{3} + \ldots\right)^2}{2} + \frac{\left(x^2 - \frac{x^4}{3} + \ldots\right)^3}{3}$$

$$= x^2 - \frac{x^4}{3} + \frac{2}{45}x^6 + \ldots - \frac{1}{2}\left\{x^4 - \frac{2x^6}{3}\right\} + \frac{x^6}{3} + \ldots$$

$$= x^2 - \left(\frac{1}{2} + \frac{1}{3}\right)x^4 + \left(\frac{2}{45} + \frac{1}{3} + \frac{1}{3}\right)x^6 + \ldots$$

$$= x^2 - \frac{5}{6}x^4 + \frac{32}{45}x^6 + \ldots$$

Example 1.28 **Expand $y = e^{a \sin^{-1} x}$ in ascending powers of x, as far as terms containing x^4.**

Solution :

$$y = e^{a \sin^{-1} x}$$

So $$y(0) = e^0 = 1$$

$$y_1 = y \cdot \frac{a}{\sqrt{1 - x^2}} \quad \ldots(1)$$

So $$y_1(0) = a.$$

$$(1-x^2)y_1^{\,2} = a^2y^2 \qquad \ldots(2)$$

Differentiating (2)

$$(1-x^2)2y_1y_2 - 2xy_1^{\,2} - 2a^2yy_1 = 0$$

or $$(1-x^2)y_2 - xy_1 - a^2y = 0 \qquad \ldots(3)$$

Putting $x = 0$, $\quad (y_2)_0 - 0 = a^2(y)_0 = a^2$

Differentiating (3) n times

$$(1-x^2)y_{n+2} - 2nx\,y_{n+1} - n(n-1)y_n - xy_{n+1} - ny_n - a^2y_n = 0$$

or $$(1-x^2)y_{n+2} - (2n+1)\cdot x\cdot y_{n+1} - (n^2+a^2)y_n = 0$$

Putting $x = 0$,

$$(y_{n+2})_0 = (n^2+a^2)(y_n)_0 \qquad \ldots(4)$$

Putting $n = 1$,

$$(y_3)_0 = (1^2+a^2)(y_1)_0 = (1+a^2)a$$

Putting $n = 2$, in (4)

$$(y_4)_0 = (2^2+a^2)(y_2)_0 = (2^2+a^2)a^2.$$

Using above values, we have

$$y = e^{a\sin^{-1}x} = y(0) + \frac{x}{1\,!}y'(0) + \frac{x^2}{2\,!}y''(0) + \frac{x^3}{3\,!}y'''(0) + \frac{x^4}{4\,!}y^{\text{iv}}(0) + \ldots$$

$$= 1 + \frac{x}{1\,!}a + \frac{x^2}{2\,!}a^2 + \frac{x^3}{3\,!}(1+a^2)a + \frac{x^4}{4\,!}(4+a^2)a^2 + \ldots$$

Example 1.29 **Expand $f(x+h) = \log\sin(x+h)$ in power of h by Taylor's expansion.**

Solution : $f(x) = \log\sin x \quad f'(x) = \cot x, \quad f''(x) = -\operatorname{cosec}^2 x$

$f'''(x) = 2\operatorname{cosec}^2 x \cot x$ etc.

So $$f(x+h) = f(x) + hf'(x) + \frac{h^2}{2\,!}f''(x) + \ldots$$

$$= \log\sin x + h\cot x + \frac{h^2}{2}(-\operatorname{cosec}^2 x) + \frac{h^3}{3}\operatorname{cosec}^2\cot x + \ldots$$

Example 1.30 **Expand $f(x) + f\{2+(x-2)\} = x^4 - 3x^3$ in powers of $(x-2)$.**

Solution : $$f(x) = f(2) + \frac{(x-2)}{1\,!}f'(2) + \frac{(x-2)^2}{2\,!}f''(2) + \frac{(x-2)^3}{3\,!}f'''(2) + \frac{(x-2)^4}{4\,!}f^{\text{iv}}f(2) + 0$$

As
$f(x) = x^4 - 3x^3$ $\qquad f(2) = 2^4 - 24 = -8$

$f'(x) = 4x^3 - 9x^2$ $\qquad$ so $\qquad f'(2) = -4$

$f''(x) = 12x^2 - 18x$ $\qquad f''(2) = 12$

$f'''(x) = 24x - 18$ $\qquad f'''(2) = 30$

$f^{(\text{iv})}(x) = 24$ $\qquad f^{\text{iv}}(2) = 24$

So $$f(x) = -8 - \frac{(x-2)}{1\,!}(-4) + \frac{12(x-2)^2}{2} + \frac{30}{6}(x-2)^3 + \frac{24}{120}(x-2)^4.$$

Example 1.31 **Expand $\log_e x$ in powers of $(x - 1)$ and hence evaluate $\log_e (1.1)$ correct to 4 places of decimal.**

Solution :

$$f(x) = \log_e x, \quad f'(x) = \frac{1}{x}, \quad f''(x) = -\frac{1}{x^2}, \quad f'''(x) = \frac{2}{x^3}, \quad f^{(iv)}(1) = \frac{-6}{x^4}$$

so $f(1) = 0, \quad f'(1) = 1, \quad f''(1) = -1, \quad f'''(1) = 2 \quad f^{(iv)}(1) = -6.$

Hence $$f(1 + x - 1) = f(1) + \frac{(x-1)}{1!} f'(1) + \ldots$$

$$= \log x + (x - 1) - \frac{(x-1)^2}{2} + \frac{(x-1)^3}{3} - \frac{(x-1)^4}{4} + \ldots$$

For $x = 1.1$

$x - 1 = 0.1$

So $\log(x) = \log(1.1)$

$$= \log 1 + (0.1) - \frac{(0.1)^2}{2} + \frac{(0.1)^3}{3} - \frac{(0.1)^4}{4} + \ldots$$

$$= 0 + 0.1 - 0.005 + 0.0003 - 0.00002$$

$$= 0.0953.$$

Example 1.32 **Use Taylor's expansion to prove that**

$$\tan^{-1}(x + a) = \tan^{-1} x + (h \sin z) \sin \frac{z}{1!} - (h \sin z)^2 \sin \frac{2z}{2} + (h \sin z)^3 \frac{\sin 3z}{3} + \ldots$$

Solution :

$$\tan^{-1}(x + h) = \tan^{-1} x + \frac{h}{1!} f'(x) + \frac{h^2}{2!} f''(x) + \ldots$$

where $f(x) = \tan^{-1} x$

So $$f'(x) = \frac{1}{1 + x^2} = \frac{1}{(x+i)(x-i)} = \frac{1}{2i}\left[\frac{1}{x-i} - \frac{1}{x+i}\right]$$

Let $x = r \cos z,\ 1 = r \sin z$

$$f^n(x) = \frac{1}{2i}(-1)^{n-1}(n-1)!\left[\frac{1}{(x-i)^n} - \frac{1}{(x+i)^n}\right]$$

$$= \frac{1}{2i}\frac{(-1)^{n-1}(n-1)!}{r^n}\left[\frac{1}{(\cos z - i \sin z)^n} - \frac{1}{(\cos z + i \sin z)^n}\right]$$

$$= \frac{(-1)^{n-1}(n-1)!}{r^n}\frac{1}{2i}[\cos nz + i \sin nz - \cos nz + i \sin nz]$$

$$f^n(x) = (-1)^{n-1}(n-1)! \sin^n z \sin nz$$

$$\frac{f^n(x)}{n!} = \frac{(-1)^{n-1}}{n} \sin^n z \sin nz$$

Hence

$$f(x+h) = \tan^{-1}(x+h)$$

$$= \tan^{-1} x + h\frac{f'(x)}{1!} + \frac{h^2 f''(x)}{2!} + \ldots$$

$$= \tan^{-1} x + h\frac{(-\sin z \sin z)}{1} + h^2\frac{\sin^2 z \sin 2z}{2} - \frac{h^3 \sin^3 z \sin 3z}{3} + \ldots$$

where $z = \cot^{-1} x$

PROBLEM SET 1.3

1. Give the Maclaurin's expansion of the following function :

(i) $f(x) = \sin x$

Ans. $x - \frac{x^3}{3!} + \frac{x^5}{5!} + \ldots$

(ii) $f(x) = \sinh x$

Ans. $x + \frac{x^3}{3!} + \frac{x^5}{5!} + \ldots$

2. Prove the following :

(i) $e^{x \sin x} = 1 + x + \frac{x^2}{2} - \frac{x^4}{8} - \ldots$

(ii) $\log \sec x = \frac{x^2}{2} + \frac{x^4}{12} + \frac{x^6}{45} \ldots$

(iii) $\sec x = 1 + \frac{x^2}{2!} + \frac{5x^4}{4!} + \ldots$

(iv) $e^x \cos x = 1 + x - \frac{2x^3}{3!} - \frac{4x^4}{4!} + \ldots$

(v) $\log(1 + \sin x) = x - \frac{x^2}{2} + \frac{x^3}{6} - \frac{x^4}{12} + \ldots$

(vi) $\frac{1}{2} \log \frac{1+x}{1-x} = x + \frac{x^3}{3} + \frac{x^5}{5} + \ldots$

(vii) $\log(1 + e^x) = \log_e 2 + \frac{x}{2} + \frac{x^2}{8} - \frac{x^4}{192} + \ldots$

(viii) $\sin^{-1} \frac{2x}{1+x^2} = 2\left\{x - \frac{x^3}{3} + \frac{x^5}{5} - \frac{x^7}{7} + \ldots\right\}$

(ix) $\frac{x}{\sin x} = 1 + \frac{x^2}{6} + \frac{x^4}{40} + \ldots$

3. If $y = \sin (m \sin^{-1} x)$. Show $(1 - x^2)y_2 - xy_1 + m^2y = 0$. Hence expand $\sin m\theta$, in powers of $\sin \theta$

Ans. $\left[\dfrac{m \sin \theta}{1\,!} - \dfrac{m(m^2 - 1)}{3\,!} \sin^3 \theta \ldots\right]$

4. Expand e^x in powers of $(x - 1)$ upto 4 terms.

Ans. $e\left\{1 + \dfrac{x-1}{1\,!} + \dfrac{(x-1)^2}{2\,!} + \dfrac{(x-1)^3}{3\,!} + \ldots\right\}$

5. Expand $\sin x$ in power of $\left(x - \dfrac{\pi}{2}\right)$. Hence find the value of $\sin 91°$ correct upto 4 places of decimal.

Ans. $1 - \dfrac{\left(x - \dfrac{\pi}{2}\right)^2}{2\,!} + \dfrac{\left(x - \dfrac{\pi}{2}\right)^4}{4\,!} + \ldots;$ 0.9998.

6. Expand $\tan^{-1} x$ in powers of $(x - 1)$.

Ans. $\dfrac{\pi}{4} + \dfrac{1}{2}(x - 1) - \dfrac{1}{4}(x - 1)^2 + \dfrac{1}{12}(x - 1)^3 + \ldots$

7. Find Taylor's series expansion for $\log \cos x$ about the points $\dfrac{\pi}{3}$.

Ans. $\log \dfrac{1}{2} - \sqrt{3}\left(x - \dfrac{\pi}{3}\right) - 2\left(x - \dfrac{\pi}{3}\right)^2 - \dfrac{4\sqrt{3}}{3}\left(x - \dfrac{\pi}{3}\right)^3 + \ldots$

8. Give $\log_{10} 4 = 0.6021$. Calculate $\log_{10} 4.04$ and $\log_{10} 404$.

Ans. 0.6121; 2.6121.

9. Compute $\cos 32°$ using Taylor's series expansions for $\cos x$ about $x = 30°$.

Ans. 0.8482.

1.8 ASYMPTOTES

When a curve extends to infinity, the tangents at far away point on the curve, when it tends to a line at a finite distance from the origin is called an asymptote. So an asymptote is defined as a straight line at a finite distance from origin to which the tangent to the curve tends as the point of contact goes away and away from the origin. Now (x, y) may tend to infinity in three ways :

(i) both $x \to \infty$ and $y \to \infty$, in this case the asymptote will have the slope $m \neq 0$, $m \neq \infty$.

(ii) $x \to \infty$, $y \to$ a finite number a. The asymptote in this case will be parallel to x-axis.

(iii) $x \to b$, finite and $y \to \infty$, the asymptote will be parallel to y-axis.

Equation of tangent to the curve $f(x, y) = 0$, at (x, y) is

$$Y - y = \frac{dy}{dx}(X - x)$$

or
$$Y = \frac{dy}{dx} X + \left(y - x\frac{dy}{dx}\right)$$

If it tends to
$$Y = mX + c \text{ as } x \to \infty$$

then
$$m = \underset{x \to \infty}{\text{Lt}} \frac{dy}{dx}$$

and
$$c = \lim_{x \to \infty} \left(y - x\frac{dy}{dx}\right) = \text{finite}$$

So
$$\underset{x \to \infty}{\text{Lt}} \left(\frac{c}{x}\right) = 0 = \underset{x \to \infty}{\text{Lt}} \left(\frac{y}{x} - \frac{dy}{dx}\right)$$

So at infinity
$$\underset{x \to \infty}{\text{Lt}} \frac{dy}{dx} = \underset{x \to \infty}{\text{Lt}} \left(\frac{y}{x} \text{ for the curve}\right) = m$$

1.9 ASYMPTOTES OF ALGEBRAIC CURVE

Let $\{a_0x^3 + a_1x^2y + a_2xy^2 + a_3y^3\} + \{b_0x^2 + b_1xy + b_2y^2\} + \{c_1x + c_2y\} + k = 0$ be an algebraic cruve of degree 3. To find the m of its asymptotes we are to find $\lim\limits_{x \to \infty} \frac{y}{x}$. So we divide through out by x^3, 3 is the highest degree in the equation, and have

$$\left\{a_0 + a_1\left(\frac{y}{x}\right) + a_2\left(\frac{y}{x}\right)^2 + a_3\left(\frac{y}{x}\right)^3\right\} + \frac{1}{x}\left\{b_0 + b_1\left(\frac{y}{x}\right) + b_2\left(\frac{y}{x}\right)^2\right\} + \frac{1}{x^2}\left\{c_1 + c_2\left(\frac{y}{x}\right)\right\} + \frac{k}{x^3} = 0 \quad \ldots(1)$$

Taking limit as $x \to 0$, we get
$$a_0 + a_1m + a_2m^2 + a_3m^3 + 0 + 0 = 0$$

We denote it by
$$\phi_3(m) = 0 \quad \ldots(2)$$

It can be obtained by putting $x = 1$ and $y = m$ in the highest degree terms. Solving this let $m = m_1$, $m = m_2$, $m = m_3$ be the value of m. We consider $m \neq 0$, $m \neq \infty$ (meaning that the asymptotes are not parallel to the axes, which case will be taken up separately).

Writing (1) as
$$\phi_3\left(\frac{y}{x}\right) + \frac{1}{x}\phi_2\left(\frac{y}{x}\right) + \frac{1}{x^2}\phi_1\left(\frac{y}{x}\right) + \frac{k}{x^3} = 0$$

and differentiating, we get

$$\left\{\phi_3'\left(\frac{y}{x}\right) + \frac{1}{x}\phi_2'\left(\frac{y}{x}\right) + \frac{1}{x^2}\phi_1'\left(\frac{y}{x}\right)\right\}\left\{\frac{x\frac{dy}{dx} - y}{x^2}\right\} - \frac{1}{x^2}\phi_2\left(\frac{y}{x}\right) - \frac{2}{x^3}\phi_1\left(\frac{y}{x}\right) = 0 \quad \ldots(1)$$

Multiplying by x^2 and then taking limit as $x \to \infty$, we get

$$\phi_3'(m) \cdot (-c) - \phi_2(m) = 0 \qquad ...(2)$$

So for non parallel asymptotes

$$c = \frac{-\phi_2(m)}{\phi_3'(m)} \qquad ...(3)$$

In the general case equation (2) and (3) are written as

$$\phi_n(m) = 0 \text{ and } c = -\frac{\phi_{n-1}(m)}{\phi_n'(m)} = 0 \qquad ...(4)$$

These equations give as the values of m and the corresponding values of c and the asymptotes are thus found.

1.10 PARALLEL ASYMPTOTES

If equation (3) for some $m = m_1$ reduces to the form $0c + 0 = 0$, this will be case when $m = m_1$ is a repeated root of $\phi_n(m) = 0$ and is also a root of $\phi_{n-1}(m) = 0$. In that case the equation for determining the value of c is found by differentiating (3) once again, it gives the equation, after taking limits as $x \to \infty$,

$$\frac{1}{2}c^2 \phi_n''(m) + c\,\phi_{n-1}'(m) + \phi_{n-2}(m) = 0 \qquad ...(5)$$

Asymptotes parallel to x-axis. This will be the case when $m = 0$ instead of finding such asymptotes also in the routine way, they can easily be found merely by inspection by equating to zero, the coefficient of the highest power of x, provided it is not a constant. Thus for $x^2y - 3x^2 + 3xy + 2y + 5 = 0$ Asymptote parallel to x-axis is $y - 3 = 0$.

Asymptotes parallel to y-axis. In this case we have $m = \infty$, and x remains finite but $y \to \infty$. These asymptotes can easily be found by equating to zero the coefficient of the highest power of y, provided it is not a constant. Thus, in the above example the highest power of y is one and equating to zero the coefficient of y, we have $x^2 + 3y + 2 = 0$ or $x = -1$ and $x = -2$, as the asymptotes of the curve which are parallel to y-axis.

1.11 NUMBER OF ASYMPTOTES

If an equation is of degree n, theoretically $\phi_n(m) = 0$ can be resolved in n linear factors so there will be n values of m and then n values of c so in general there will be n asymptotes. If some of the values of m be imaginary, the number of real asymptotes will be less than n.

1.12 NUMBER OF POINTS OF INTERSECTION OF THE CURVE AND ALL ITS ASYMPTOTES

An asymptote will in general meet a curve of n th degree in n points. Two of them coincide at its points of contact. So it will cut the curve in $(n - 2)$ points. Thus, the n asymptotes cut the curve in $n(n - 2)$ points. If $f(x, y) = 0$ be the equation to the curve and $\phi(x, y) = 0$ be the combined equation of all its asymptotes then the equation $f(x, y) - \phi(x, y) = 0$ is satisfied at all the common points. So the common points lie on this curve. This curve is of smaller degree than the degree of $f(x, y) = 0$.

1.13 WORKING RULE FOR FINDING THE ASYMPTOTES

(i) Put $y = m$ and $x = 1$, in the highest degree terms and equating them to zero have the equation $\phi_n(m) = 0$. Let it give $m = m_1, m_2, \ldots m_n$.

(ii) For non-parallel asymptotes, or for the non-repeated values of m, the value of c is given by $c = -\dfrac{\phi_{n-1}(m)}{\phi_n'(m)}$.

(iii) If two values of m be equal, for these parallel asymptotes the value of c is found from the quadratic equation.

$$\frac{1}{2}c^2\phi_n''(m) + c\phi_{n-1}'(m) + \phi_{n-2}(m) = 0$$

(iv) If the coefficient of the highest power of x is not merely a constant, then equating it to zero, we have the asymptotes which are parallel to x-axis.

(v) If the coefficient of the highest power of y is not merely a constant, then equating it to zero, we get the asymptotes which are parallel to y-axis.

Example 1.33 **Find the asymptote parallel to co-ordinate axis of the curves :**

(i) $(x^2 + y^2)x - ay^2 = 0$

(ii) $x^2y^2 - 4(x^2 - y^2) = 0$

Solution : (i) To find parallel asymptotes to the curve $(x^2 + y^2)x - ay^2 = 0$. We have to equate the coefficient of highest powers of x only. The coefficient of highest power y^2 of y, is $x - a$. Hence the asymptote parallel to y-axis is $x - a = 0$. The coefficient of highest power x^3 of x is l, which is a constant, hence there is no asymptote parallel to x-axis.

(ii) To find asymptote parallel to x and y axis of the curve. $x^2y^2 - a^2(x^2 + y^2) = 0$, we equate the coefficient highest power of x and y respectively.

The coefficient of highest power of x^2 of x is $y^2 - a^2$. Hence the asymptote parallel to x-axis is

$$y^2 - a^2 = 0$$

$$\Rightarrow \quad (y + a)(y - a) = 0$$

$$\Rightarrow \quad y + a = 0, \; y - a = 0$$

In a similar manner the asymptotes parallel to y-axis are $x - a = 0$ and $x + a = 0$.

Example 1.34 **Find the asymptotes of the cubic curve**

$$\mathbf{2x^2 - x^2y + 2xy^2 + y^3 - 4x^2 + 8xy - 4x + 1 = 0.}$$

(GNDU 2004)

Solution : We have the ruve $2x^2 - x^2y + 2xy^2 + y^3 - 4x^2 + 8xy - 4x + 1 = 0$

We observe that, there is no parallel asymptote to the given curve.

Putting $x = 1$ and $y = m$ in the third and second degree terms, we get

$$\phi_3(m) = 2 - m - 2m^2 + m^3,\ \phi_2(m) = -4 + 8m$$

The slopes of asymptotes can be obtained by $\phi_3(m) = 0$, *i.e.*,

$$\phi_3 = 2 - m - 2m^2 + m^3 = 0$$

$$\Rightarrow \quad (m + 1)(m - 1)(m - 2) = 0$$

$$\Rightarrow \quad m = -1, 1, 2$$

Again c can be obtained as

$$c\phi_3'(m) + \phi_2(m) = 0$$

$$\Rightarrow \quad c(-1 - 4m + 3m^2) + (-4 + 8m) = 0$$

For $m = -1$, 1 and 2, we get c as 2, 2, -4 respectively.

Therefore, the asymptotes are

$$y = -x + 2,\ y = x + 2 \text{ and } y = 2x - 4.$$

Example 1.35 **Find the asymptotes of the curve**

$$x^3 - x^2y - xy^2 + y^3 + 2x^2 - 4y^2 + 2xy + x + y + 1 = 0. \qquad \textit{(Rohilkhand (2001)}$$

Solution : We observe that there is no parallel asymptote to the given curve

Putting $x = 1$ and $y = m$ in the third and second degree terms of the given curve, we get

$$\phi_3(m) = 1 - m - m^2 + m^3 = (1 - m) - m^2(1 - m)$$

$$= (1 - m)(1 - m^2)$$

$$= (1 - m)^2 (1 + m)$$

$$\phi_2(m) = 2 + 2m - 4m^2$$

The slopes of asymptotes are given by $\phi_3(m) = 0$, *i.e.*,

$$(1 - m)^2 (1 + m) = 0$$

$$\Rightarrow \quad m = 1, 1, -1$$

To determine c, we have

$$c\phi_3'(m) + \phi_2(m) = 0$$

$$\Rightarrow \quad c(-1 - 2m + 3m^2) + (2 + 2m - 3m^2) = 0 \qquad \text{...(1)}$$

For $m = -1$, we get $c = 1$, hence $y = -x + 1$ is the corresponding asymptote.

For $m = 1$, the equation (1) becomes $0, c + 0 = 0$, in this type of exceptional case, c is determined from the equation

$$\left(\frac{c^2}{2}\right) . \phi_3''(m) + c\phi_2'(m) + \phi_1(m) = 0$$

$$\Rightarrow \quad \frac{c^2}{2}(-2 + 6m) + c(2 - 8m) + 1(1 + m) = 0$$

For $m = 1$, this equation reduces to

$$2c^2 - 6c + 2 = 0$$

$$\Rightarrow \quad c = \frac{3 \pm \sqrt{5}}{2}$$

Hence, $y = x + \frac{3+\sqrt{5}}{2}$ and $y = x + \frac{3-\sqrt{5}}{2}$ are the two asymptotes corresponding to $m = 1$.

Example 1.36 **Find the asymptotes of $x^3 + 2x^2y - xy^2 - 2y^3 + xy - y^2 - 1 = 0$**

Solution : We are given the curve

$$x^3 + 2x^2y - xy^2 - 2y^3 + xy - y^2 - 1 = 0 \quad ...(1)$$

Let $y = mx + c$, the asymptotes of the given equation of curve.

Substituting $mx + c$ for y in equation (1), we get

$$x^3 - 2x^2(mx + c) - x(mx + c)^2 - 2(mx + c)^3 + x(m + c) - (mx + c)^2 - 1 = 0$$

$$\Rightarrow \quad x^3(1 + 2m - m - 2m^3) + x^2(2c - 2mc - 6m^2c + m - m^2) + \ldots = 0$$

To find m and c, we equate the coefficient of x^3 and x^2 to zero.

$$1 + 2m - m^2 - 2m^3 = 0 \quad ...(2)$$

and $$2c - 2mc - 6m^2c + m - m^2 = 0 \quad ...(3)$$

From equation (2)

$$1 - m^2 + 2m(1 - m^2) = 0 \Rightarrow (1 - m^2)(1 + 2m) =)$$

$$\Rightarrow \quad m = 1, -1, -\frac{1}{2}$$

From equation (3)

$$2c(1 - , - 3m^2) = m^2 - m$$

$$\Rightarrow \quad c = \frac{m^2 - m}{2(1 - m - 3m^3)}$$

From $m = 1$,

$$c = \frac{1 - 1}{2(1 - 1 - 3)} = 0$$

From $m = -1$,

$$c = \frac{1 + 1}{2(1 + 1 - 3)} = -1$$

From $m = -\frac{1}{2}$,

$$c = \frac{\frac{1}{4} + \frac{1}{2}}{2\left(1 + \frac{1}{2} - \frac{3}{4}\right)} = \frac{\frac{3}{4}}{\frac{3}{2}} = \frac{1}{2}$$

For the above pair of values of m and c, we get the asymptotes as

$$y = x, \; y = -x - 1 \text{ and } y = -\frac{1}{2}x + \frac{1}{2}.$$

Example 1.37 **Find the asymptotes of the curve**

$$(\mathbf{3x^3 + 2x^2y - 7xy^2 + 2y^3}) + (\mathbf{-14xy + 7y^2}) + (\mathbf{4x + 5y}) = \mathbf{0}.$$

Solution : Putting $x = 1$ and $y = m$, in the third degree terms, we get

$$\phi_3(m) = 3 + 2m - 3m^2 + 2m^3$$

The values of m for the asymptotes are the roots of $\phi_3(m) = 0$, which gives $m_1 = -\frac{1}{2}$, $m_2 = 1$, $m_3 = 3$.

For the corresponding values of c, we have the equation

$$c = -\frac{\phi_2(m)}{\phi_3'(m)} = -\frac{(-14m + 7m^2)}{6m^2 - 14m + 2}$$

For $m_1 = -\frac{1}{2}$, $c_1 = -\frac{5}{6}$, for $m_2 = 1$, $c_2 = -\frac{7}{6}$ and for $m_3 = 3$, $c_3 = -\frac{3}{2}$.

So the asymptotes are

$$y = -\frac{1}{2}x - \frac{5}{6}\,;\; y = 1x - \frac{7}{6}\,;\; y = 3x - \frac{3}{2}.$$

Example 1.38 **Find the asymptotes of the curve**

$$(x^3 + 3x^2y - 4y^3) + (-x + y) + 3 = 0.$$

Solution : Putting $x = 1$ and $y = m$, in the third degree terms, we get $\phi_3(m) = 1 + 3m - 4m^3$. The m's of asymptotes are the root of $\phi_3(m) = 0$ or $1 + 3m - 4m^3 = 0$ or $(x - 1)(2m + 1)^2 = 0$ we have $m_1 = 1$, $m_2 = m_3 = -\frac{1}{2}$. So for $m_1 = 1$ we have a non-parallel asymptote for which the value of c is given by$c = -\frac{\phi_2(m)}{\phi_3'(m)} = 0$, as the second degree terms in the equation are absent.

Hence the asymptote is

$$y = 1x + 0 \qquad \ldots(1)$$

For repeated roots $m = -\frac{1}{2} - \frac{1}{2}$, the value of c is found from

$$\frac{1}{2}c^2\,\phi_3''(m) + c\,\phi_2'(m) + \phi_1(m) = 0$$

or $$\frac{1}{2}c^2\{-24m\} + 0 + (-1 + m) = 0, \quad \text{for } m = -\frac{1}{2}$$

$$\frac{24}{4}c^2 - \frac{3}{2} = 0 \text{ so } c = \pm\frac{1}{2}$$

So the three asymptotes are

$$y = x_1, \quad y = -\frac{1}{2}x + \frac{1}{2}, \; y = -\frac{1}{2}x - \frac{1}{2}.$$

Example 1.39 **Find the asymptotes of the curve** $\frac{a^2}{x^2} - \frac{b^2}{y^2} = 1.$

Solution : The equation to the curve can be written as $x^2y^2 - a^2y^2 + b^2x^2 = 0$ equating to the zero the coefficient of highest power of x, we get $y^2 + b^2 = 0$. So the asymptotes parallel to the x-axis are imaginary.

The asymptotes parallel to y-axis are given by $x^2 - a^2 = 0$. So the curve has only two real asymptotes $x = \pm a$.

Example 1.40 **Find the asymptotes of $x^2(x-y)^2 - a^2(x^2+y^2) = 0$.**

Solution : The equation can be simplied to

$$x^4 - 2x^3y + x^2y^2 - a^2x^2 - a^2y^2 = 0$$

Equating to zero the coefficient of the highest power of y we have $x^2 - a^2 = 0$ or $x = \pm a$ are asymptotes parallel to y-axis.

Putting $x = 1$, $y = m$ in the highest degree terms, the equation $\phi_4(m) = 0$ is only of degree 2 given as $(1-m)^2 = 0$

So there are two parallel asymptotes for which $m = 1, 1$. The equation for determining the values of c is

$$\frac{1}{2}c^2\,\phi_4''(m) + c\,\phi_3'(m) + \phi_2(m) = 0$$

$$\phi_4''(m) = 2,\ \phi_3(m) = 0 = \phi_3'(m),\ \phi_2(m) = -a^2m^2 - a^2$$

So $\quad \frac{1}{2}c^2 \times 2 + 0 - a^2(1^2+1) = 0$

So $\quad c = \pm a\sqrt{2}$

The four asymptotes are $x = \pm a$, $y = 1x \pm a\sqrt{2}$.

Example 1.41 **Find the asymptotes of the curve $x^2y - xy^2 + xy + y^2 + x - y = 0$, and show that cut the curve again in three points which lie on the line $x + y = 0$.**

Solution : The coefficient of the highest power of x is y.

So $y = 0$ is the asymptote parallel to x-axis.

The highest power of y in the equation is 2, and the coefficient of y^2 is $-x+1$, so $-x+1 = 0$ is the asymptote parallel to y-axis.

The equation is of degree 3, so it may have one more asymptote. Putting $x = 1$, $y = m$ in highest (3rd) degree terms we get

$$\phi_3(m) = m - m^2.$$

So the m's of asymptotes are given by $m - m^2 = 0$ or $m = 0$ and $m = 1$. The asymptotes for $m = 0$ (one which is parallel to x-axis has already been found).

Now $\phi_2(m) = m + m^2$

So $\quad c = -\frac{\phi_2(m)}{\phi_3'(m)} = -\left(\frac{m^2+m}{1-2m}\right)_{\text{at } m=1} = \frac{-2}{-1} = 2$

So the asymptotes are $(1-x) = 0$, $y = 0$, $y = x + 2$

The combined equation of the asymptotes is

$$y(1-x)(2+x-y) = 0$$

or $\quad \phi(x, y) \equiv -x^2y + xy^2 - xy - y^2 + 2y = 0 \quad \ldots(1)$

Where the equation of the curve is

$$F(x, y) \equiv x^2y - xy^2 + xy + y^2 + x - y = 0 \quad \ldots(2)$$

Additing (1) and (2) we get $x + y = 0$ as the curve on which the asymptotes cut the curve.

Example 1.42 **Find the asymptotes of the curve**

$$y^2(x-2a)-x^3+a^3=0.$$

Solution : Coefficient of highest power of y is $x-2a$ so $x-2a=0$ in the asymptote parallel to the y-axis.

Rearranging the terms $y^2x-x^3-2ay^2+a^3=0$.

Putting $y=m$ and $x=1$, in the highest degree (3rd) terms $\phi_3(m)=m^2-1$.

So values of m's of asymptotes are the roots of $\phi_3(m)=m^2-1=0$, so $m=\pm 1$.

$$\phi_2(m)=-2am^2,\ c=-\frac{\phi_2(m)}{\phi_3'(m)}=\frac{2am^2}{2m}=am$$

Hence, the asymptotes are $y=1x+a$, $y=-x-a$ and $x=2a$.

Example 1.43 **Find the asymptotes of the curve :**

$$(x-y+2)(2x-3y+4)(4x-5y+6)+5x-6y+7=0.$$

Solution : For the given curve :

$(x-y+2)(2x-3y+4)(4x-5y+6)+5x-6y+7=0$, the asymptote parallel to $x-y+2=0$ is

$$x-y+2+\lim\frac{5x-6y+7}{(2x-3y+4)(4x-5y+6)}=0 \quad \text{as} \quad x\to\infty;\ \frac{y}{x}\to 1$$

$$\Rightarrow \qquad x-y+2+\lim\left[\frac{5-6\frac{y}{x}+\frac{7}{x}}{\left(2-\frac{3y}{x}+\frac{y}{x}\right)\left(4-\frac{5y}{x}+\frac{6}{x}\right)x}\right]=0$$

$$\Rightarrow \qquad x-y+2=0 \qquad (\because \text{ the limit is being zero as } x\to\infty)$$

In a similar manner we get the other asymptotes as

$$2x-3y+4=0 \text{ and } 4x-5y+6=0.$$

1.14 ASYMPTOTES OF POLAR CURVES

For curves going the infinity the polar equation to the curve will be of the form $\frac{1}{r}=f(\theta)$.

If α be a root of the equation $f(\theta)=0$, the corresponding asymptotes has the equation

$$r\sin(\theta-\alpha)=\frac{1}{f'(\alpha)}$$

Example 1.44 **Find the asymptotes of the curve $\frac{a}{r}=\sin 2\theta$.**

Solution : $f(\theta)=\sin 2\theta=0$ when $2\theta=n\pi$ or $\theta=\frac{n\pi}{2}$, $n\in N$

The curve has infinite number of asymptotes.

$$f'(\theta)=\frac{2\cos 2\theta}{a}$$

$$f'\left(\frac{n\pi}{2}\right) = \frac{2\cos(n\pi)}{a} = (-1)^n \frac{2}{a}.$$

So the asymptotes are

$$r\sin\left(\theta - \frac{n\pi}{2}\right) = (-1)^n \frac{a}{2}$$

for $n = 0$, $r\sin\theta = \frac{a}{2}$, for $n = 1$, $r\cos\theta = \pm\frac{a}{2}$.

Example 1.45 **Find the asymptotes of the spiral $\frac{1}{r} = \frac{1}{a}\theta$.**

Solution : $f(\theta) = \frac{1}{a}\theta$.

So $f(\theta) = 0$ gives $\theta = 0 = \alpha$.

$$f'(\theta) = \frac{1}{a}$$

Equation of asymptote is

$$r\sin(\theta - \alpha) = \frac{1}{f'(\theta)}$$

or $$r\sin(\theta - 0) = a$$

or $$r\sin\theta = \alpha$$

Example 1.46 **Find the asymptotes of the curve $r = a\tan\theta$.**

Solution : Equation in the form $\frac{1}{r} = f(\theta)$ is

$$\frac{1}{r} = \frac{1}{a}\cot\theta.$$

$$f(\theta) = \cot\theta = 0 \Rightarrow \theta = n\pi + \frac{\pi}{2}$$

$$f'(\theta) = -\frac{\operatorname{cosec}^2\theta}{a} = -\operatorname{cosec}^2\left(n\pi + \frac{\pi}{2}\right) = -\frac{1}{a}$$

So the asymptotes are

$$r\sin(\theta - \alpha) = \frac{1}{f'(\theta)}$$

$$r\sin\left(\theta - n\pi - \frac{\pi}{2}\right) = -a$$

for $n = 0$, $$r\cos\theta = a.$$

Example 1.47 **Find the asymptotes of the curve :**
$x^2y - xy^2 + xy + y^2 + x - y = 0$ and show that they cut the curve again in three points which lie on the line $x + y = 0$.

Solution : We can easily obtain the asymptotes of the given curve as

$$y = 0, x = 1, x - y + 2 = 0$$

The joint equation of the asymptotes is

$$y(x-1)(x-y+2) = 0$$

i.e., $x^2y - xys + xy + y^2 - 2y = 0$

The equation of the curve can be written as

$$x^2y - xy^2 + xy + y^2 - 2y + (x + y) = 0$$

Hence $F_3 = x^2y - xy^2 + xy + y^2 - 2y$

$F_1 = x + y$

Hence, the point of intersection lie on the line

$F_1 = x + y = 0.$

Example 1.48 **Find the equation of the cubic which has the same asymptotes as the curve.**

$$\mathbf{x^3 - 6x^2y - 11xy^2 - 6y^3 + x + y + 1 = 0}$$

and which passes through the points (0, 0), (1, 0) and (0, 1).

Solution : For the given curve

$$x^3 - 6x^2y - 11xy^2 - 6y^3 + x + y + 1 = 0$$

We write $F_3 = x^3 - 6x^2y + 11xy^2 - 6y^3$

$= (x - y)(x - 2y)(x - 3y)$

and $F_1 = x + y + 1$

The equation of the curve can be written in the form. $F_3 + F_1 = 0$, where F_3 has non-repeated linear factors. Thus, $F_3 = 0$ is the joint equation of the asymptotes of the cubic.

The general equation of the cubic can be of the form

$$F_3 + ax + by + c = 0$$

$\Rightarrow$ $x^3 - 6x^2y + 11xy^2 - 6y^3 + ax + by + c = 0$

where $ax + by + c$ is the general linear expression since it passes through the points (0, 0), (1, 0) and (0, 1), we get

$c = 0$

$1 + a = 0 \Rightarrow a = -1$

$-6 + 6 = 0 \Rightarrow 6 = 6$

Thus, the required curve (cubic) is

$$x^3 - 6x^2y + 11xy^2 - 6y^3 - x + 6y = 0$$

PROBLEM SET 1.4

Find the asymptotes of the following curves :

1. (i) $y^2x - a^2(x - a) = 0$

Ans. $x = 0, y = \pm a$

(ii) $x^2y - 3x^2 = 5xy + 6y + 2 = 0$

Ans. $x = 2, x = 3, y = 3$

(iii) $y = \dfrac{x}{s^2 - 1}$

Ans. $x = \pm 1, y = 0$

2. $4x^3 - 3xy^2 - y^3 + 2x^2 - xy - y^2 = 1$

Ans. $y = x,\ y = -2x,\ y = -2x - 1$

3. $4x^3 - x^2y - 4xy^2 + y^3 + 3x^2 + 2xy - y^2 + 10 = 0$

Ans. $y = 4x + \dfrac{1}{3},\ y = x + \dfrac{2}{3},\ y = -x$

4. $(x + y)^2(x + 2y + 2) - x - 9y + 5 = 0$

Ans. $1 + 2y + 2 = 0,\ x + y = \pm 2\sqrt{2}$

5. $2x^3 + 3x^2y - 3xy^2 - 2y^3 + 3x^2 - 3y^2 + y - 3 = 0$

Ans. $y = x,\ y = -2x - 1,\ 2y = -x - 1$

6. $y^3 - 6xy^2 + 11x^2y - 6x^3 + x + y + 10 = 0$

Ans. $y = x,\ y = -2x,\ y = 3x$

7. $2x(y - 3)^2 - 3y(x - 1)^2 = 0$

Ans. $y = 0,\ x = 0,\ 2y = 3x + 6$

8. $x^2y^2 - xy^2 - x^2y + x + y + 1 = 0$

Ans. $y = 0,\ y = 1,\ x = 0,\ x = 1.$

9. $\dfrac{a^2}{x^2} + \dfrac{b^2}{y^2} = 1$

Ans. $y = \pm b,\ x = \pm a$

10. $x^2y^2 - a^2(x^2 + y^2) - a^3(x + y) + a^4 = 0.$

Ans. $x = \pm a,\ y = \pm a$

11. $x(y - x)^2 = x(y - x) + 2$

Ans. $x = 0, y = x, y = x + 1$

12. $x^2(x - y)^2 + a^2(x^2 - y^2) = a^2xy$

Ans. $y = x \pm a, x = \pm a$

13. $y^3 - x^2y + 2xy^2 - y + 1 = 0$

Ans. $y = , y + x = \pm 1$

14. $xy^2 = (x + y)^2$

Ans. $x = 1$

15. $(x + y)^2\ (x^2 + xy + y^2) = a^2x^2 + a^2(y - x)$

Ans. $x + y \pm a = 0$

16. $y^3 + 3y^2x - x^2y - 3x^2 + y^2 - 2xy + 3x^2 + 4y + 5 = 0$

Ans. $4(x - x) + 1 = 0,\ 2(y + x) = 3,\ 4(y + 3x) + 9 = 0$

17. $(x - y)^2\ (x - 2y)\ (x - 3y) - 2a(x^3 - y^3) - 2a^2(x - 2y)\ (x + y) = 0$

Ans. $x = y + a,\ x = 2a,\ 2y = x + 14a,\ 3y = x - 13a$

18. Show that the asymptotes of the curve $x^2y^2 = a^2(x^2 + y^2)$ form a square of side $2a$.

19. Find the asymptotes of $x^3 + y^3 - 3axy = 0$

Ans. $x + y + a = 0$

20. $y^2(x - 2a) + x^2(y - b) + 2x + 3y + 5 = 0,$

Ans. $x = 2a,\ y = b,\ x + y + 2a + b = 0$

Find the asymptotes of the following polar curves :

21. (i) $r = a\ (\sec\theta + \tan\theta)$

Ans. $r\cos\theta = 0,\ r\cos\theta = 2a.$

(ii) $r\sin\theta = 2\cos 2\theta$

Ans. $r\sin\theta = 2$

(iii) $r\sin n\theta = a$

Ans. $r\sin\left(\theta - \frac{m\pi}{n}\right) = \frac{a}{n}\cos m\pi$

22. Find the rectilinear asymptotes of the following curves

(i) $\frac{2}{r} = 1 + r\sin\theta,$

Ans. $r\sin\theta\left(\theta \pm \frac{\pi}{6}\right) = \frac{2}{\sqrt{3}}$

(ii) $\frac{2\theta}{r} = \sin\theta$

Ans. $r\sin\theta = 2n\pi,\ n \in N$

(iii) $r = a\coth\frac{\theta}{2}$

Ans. $r\sin\theta = 2a.$

23. Show that the eight points of intersections of the curve $x^4 - 5x^2y^2 + 4y^4 + x^2 - y^2 + x + y + 1 = 0$ and its asymptotes lie on a rectangular hyperbola.

24. Show that the asymptotes of the curve $x^3 - 2y^3 + 2x^2y - xy^2 + xy - y^2 + 1 = 0$. Cut the curve in three points which lie on the line $x - y + 1 = 0$.

Show that the following curves have no asymptotes :

25. $x^3 + y^4 = a^2(x^2 - y^2)$

26. $y^2 = x(x + 1)^2$

27. $a^4y^2 = x^5y\ (2a - x)$

28. Find the asymptotes of the curve $x(y - 3)^2 = 4y(x - 1)^3$.

Ans. $x = 0,\ y = 0,\ 2y = 4x + 33,\ 4x + 2y = 15$

29. Find the asymptotes of $x(y^2 - 3by + 2b^2) = y^3 - 36x^2 + b^3$.

Ans. $y - x = 0$

30. Find the equation of the cubic which has the same asymptotes as the curve :

$$x^3 - 6x^2y + 11xy^2 - 6y^3 + x + y + 1 = 0$$

and which touches the axis of y at origin and passes through the point (3, 2).

Ans. $x^3 - 6x^2y + 11xy^3 - 6y^2 - x = 0$

31. Find the asymptotes of the curve :

$$(2x - 3y + 1)^2 (x + y) - 8x + 2y - 9 = 0$$

and show that they intersect the curve again in three points, which lie on a straight line. Obtain the equation of the line.

Ans. $x + y = 0,\ 2x - 3y - 1 = 0,\ 2x - 3y + 3 = 0,\ 4x - 6y + 9 = 0$

1.15 CURVATURE

On looking at the curves C_1, C_2, C_3, we notice that C_1 bends more sharply than C_2. We express it by saying that C_1 has greater curvature than C_2 and the straight line C_3 has no curvature. A measure of curvature k at a point P on a curve in defined as $k = \dfrac{d\psi}{ds}$, where s is the arc length of P from some standard points A and ψ is the angle which the tangents to the curve at P makes with the positive direction of x-axis. The reciprocal of curvature k is called the radius of curvature. Radius of curvature is also defined as follows:

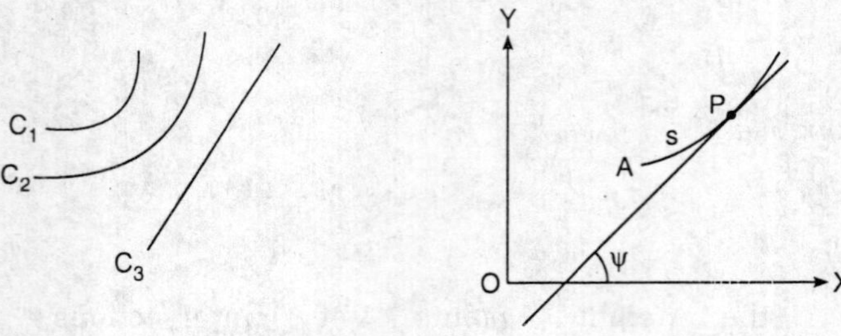

Fig. 1.1

Radius of curvature. Let $y = f(x)$ be a curve and P and Q be two neighbouring points (x, y) and $(x + \delta x, y + \delta y)$ with arc $PQ = \delta s$. Let the tangents at P and Q make angles ψ and $\psi + \delta\psi$ with x axis and let the normals at P and Q intersects at N. So $\angle PNQ = \delta\psi$. As $Q \to P$, the points N tends to a definite point C.

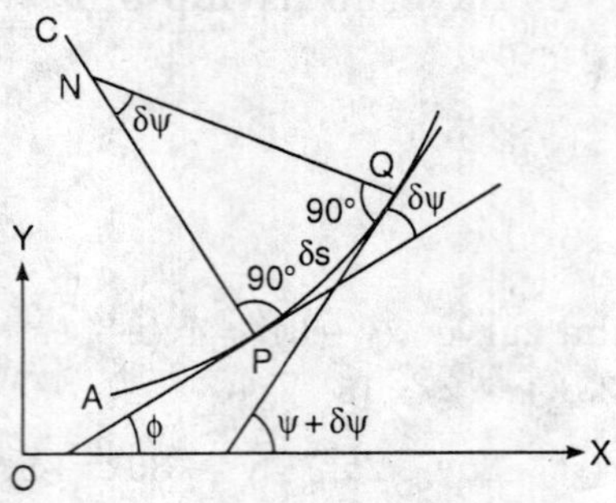

Fig. 1.2

1.16 CENTRE OF CURVATURE THE RADIUS OF CURVATURE AT *P*

So as $\frac{\text{arc}}{\text{radius}}$ = angle, $\frac{\delta s}{PN} = \delta\psi$,

Radius of curvatrue $= \rho = \lim\limits_{P \to Q} PN = \text{Lt}\,\frac{ds}{d\psi}$.

Circle of curvature at *P*. The circle *C* and radius ρ touches the curve at *P* and is called the circle of curvature at *P*.

1.17 FORMULAE FOR RADIUS OF CURVATURE ρ

(i) Equation to a curve given in intrinsic form as $s = f(\psi)$

$$\rho = \frac{ds}{d\psi}$$

(ii) Equation gives in cartesian form as $f(x, y) = 0$

$$\rho = \frac{\left[1 + \left(\frac{dy}{dx}\right)^2\right]^{3/2}}{\frac{d^2y}{dx^2}} = \frac{[\dot{x}^2 + \dot{y}^2]^{3/2}}{\dot{x}\ddot{y} - \dot{y}\ddot{x}} \quad \text{(parametric formula for } \rho)$$

If $\frac{dy}{dx} = \infty$, we use the formula

$$\rho = \frac{\left\{1 + \left(\frac{dx}{dy}\right)^2\right\}^{3/2}}{\frac{d^2x}{dy^2}}$$

When the equation is expressed in the parameter form as $x = \phi(t)$, $y = \psi(t)$, where t is a parameter. Here $\dot{x}$ and $\dot{y}$ etc are derivatives with respect to the parameter t.

(iii) When equation is given in the pedal form

$$r = f(p)$$

$$\rho = r\frac{dr}{dp}.$$

(iv) When the equation of the curve is given in polar form as

$$r = f(\theta)$$

$$\rho = \frac{\left\{r^2 + \left(\frac{dr}{d\theta}\right)^2\right\}^{3/2}}{r^2 + 2\left(\frac{dr}{d\theta}\right)^2 - r\frac{d^2r}{d\theta^2}}$$

Pedal Equation. p, r relation for a curve is called its pedal equation.

Example 1.49 **Find the radius of curvature at the points (s, ψ) on the following curves :**

(i) $s = c \tan \psi$, (Catenary)

(ii) $s = 4a \sin \psi$ (Cycloid)

(iii) $s = c \log \sec \psi$

(iv) $s = 8a \sin^2 \frac{\psi}{6}$ (Cardioid).

Solution : (i) $\rho = \frac{ds}{d\psi} = c \cdot \sec^2 \psi$

(ii) $\rho = \frac{ds}{d\psi} = 4a \cos \psi$

(iii) $\rho = \frac{ds}{d\psi} = \frac{c}{\sec \psi} \sec \psi \tan \psi = c \tan \psi,$

(iv) $s = 4a\left(1 - \cos \frac{\psi}{3}\right)$

So $\rho = \frac{ds}{d\psi} = \frac{4a}{3} \sin \frac{\psi}{3}.$

Example 1.50 **Prove that for the ellipse $\frac{x^2}{a^2} + \frac{y^2}{b^2} = 1$, $\rho = \frac{a^2b^2}{p^3}$, where p is the length of perpendicular from (0, 0) on the tangents at $P(x, y)$.**

Solution : Writing the equation in parametric form as

$$x = a \cos \phi,\ y = b \sin \phi,$$

$$\dot{x} = -a \sin \phi,\ \ddot{x} = -a \cos \phi,\ \dot{y} = b \cos \phi,\ \ddot{y} = -b \sin \phi,$$

$$\rho = \frac{[\dot{x}^2 + \dot{y}^2]^{3/2}}{\dot{x}\ddot{y} - \dot{y}\ddot{x}} = \frac{(a^2 \sin^2 \phi + b^2 \cos^2 \phi)^{3/2}}{ab\,\{(-\sin \phi)^2 - \cos \phi\,(-\cos \phi)\}}$$

$$= \frac{1}{ab}(a^2 \sin^2 \phi + b^2 \cos^2 \phi)^{3/2} \qquad \ldots(1)$$

Equation to tangent at $P(a \cos \phi, b \sin \phi)$ is

$$y - y' = \frac{dy}{dx}(x - x')$$

or $$y - b \sin \phi = -\frac{b \cos \phi}{a \sin \phi}(x - a \cos \phi)$$

or $$bx \cos \phi + ay \sin \phi + ab = 0$$

Length of perpendicular on it from $(0, 0) = p = \frac{0 + ab}{\sqrt{a^2 \sin^2 \phi + b^2 \cos^2 \phi}}$

So $$(a^2 \sin^2 \phi + b^2 \cos^2 \phi)^{-3/2} = \left(\frac{p}{ab}\right)^3 \qquad \ldots(2)$$

Hence from (1)

$$\rho = \frac{1}{ab}\left(\frac{p}{ab}\right)^{-3} = \frac{a^2b^2}{p^3}.$$

Example 1.51 **Prove that the radius of curvature of cycloid $x = a(\theta + \sin\theta)$, $y = a(1 - \cos\theta)$ is $4a\cos\frac{\theta}{2}$.**

Solution : We have $\dot{x} = a(1 + \cos\theta)$, $\dot{y} = a\sin\theta$

So $$\dot{x}^2 + \dot{y}^2 = (2a\cos^2\theta/2)^2 + \left(2a\sin\frac{\theta}{2}\cos\frac{\theta}{2}\right)^2$$

$$= 4a^2\cos^2\frac{\theta}{2}$$

and $$\dot{x}\ddot{y} - \dot{y}\ddot{x} = a^2\{(1+\cos\theta)\cos\theta - \sin\theta(-\sin\theta)\} = a^2(1+\cos\theta)$$

Hence $$\rho = \frac{(\dot{x}^2 + \dot{y}^2)^{3/2}}{\dot{x}\ddot{y} - \dot{y}\ddot{x}} = \frac{\left(2a\cos\frac{\theta}{2}\right)^3}{2a^2\cos^2\frac{\theta}{2}} = 4a\cos\frac{\theta}{2}$$

Example 1.52 **Find the radius of the curvature of $\left(\frac{3a}{2}, \frac{3a}{2}\right)$ of the Folium**

$$x^3 + y^3 = 3axy. \quad \text{...(1)}$$

Solution : Differentiating (1)

$$x^2 + y^2\frac{dy}{dx} - a\left(x\frac{dy}{dx} + y\right) = 0 \quad \text{...(2)}$$

$\left(\frac{dy}{dx}\right)$ at $\left(\frac{3a}{2}, \frac{3a}{2}\right)$ is

$$\left(\frac{dy}{dx}\right) = \frac{ay - x^2}{y^2 - ax} = \frac{1 - \frac{3}{2}}{\frac{3}{2} - 1} = -1$$

$$2x + 2y\left(\frac{dy}{dx}\right)^2 + y^2\frac{d^2y}{dx^2} - a\left(x\frac{d^2y}{dx^2} + 2\frac{dy}{dx}\right) = 0$$

at $x = \frac{3a}{2}$, $y = \frac{3a}{2}$, $\frac{dy}{dx} = -1$,

$$\frac{d^2y}{dx^2}(y^2 - ax) = 2\left\{a\frac{dy}{dx} - y\left(\frac{dy}{dx}\right)^2 - x\right\}$$

$$= 2\left\{-a - \frac{3a}{2}\times 1 - \frac{3a}{2}\right\} = -8a$$

So $$\frac{d^2y}{dx^2} = \frac{-8a}{\frac{3a^2}{2}\left(\frac{3}{2} - 1\right)} = \frac{32}{3a}.$$

Hence $$\rho = \frac{\{1 + (-1)^2\}^{3/2}}{-\frac{32}{3a}} = \frac{3a \times 2\sqrt{2}}{-32} = \frac{3a\sqrt{2}}{-16}.$$

Example 1.53 **Show that the radius of curvature at any point P of the astroid $x^{2/3} + y^{2/3} = a^{2/3}$, is three times the length of the perpendicular from the origin upon the tangents at P.** *(Madras 2003)*

Solution : Writing the equation in the parametric form as

$$x = a\cos^3\theta,\; y = a\sin^3\theta$$

$$\dot{x} = -3a\cos^2\theta\sin\theta$$

$$\dot{y} = 3a\sin^2\theta\cos\theta$$

$$\ddot{x} = 3a(2\cos\theta\sin^2\theta - \cos^3\theta)$$

$$\ddot{y} = 3a(-\sin^3\theta + 2\sin\theta\cos^2\theta)$$

So $(\dot{x}^2 + \dot{y}^2) = (3a\sin\theta\cos\theta)^2(\cos^2\theta + \sin^2\theta) = (3a\sin\theta\cos\theta)^2$

and
$$(\dot{x}\ddot{y} - \dot{y}\ddot{x}) = 9a^2\{-\cos^2\theta\sin\theta\sin\theta(2\cos^2\theta - \sin^2\theta) - \sin^2\theta\cos^2\theta(2\sin^2\theta - \cos^2\theta)\}$$

$$= 9a^2\sin^2\theta\cos^2\theta\{-2\cos^2\theta + \sin^2\theta - 2\sin^2\theta + \cos^2\theta\}$$

$$= -9a^2\sin^2\theta\cos^2\theta$$

Hence
$$\rho = \frac{(\dot{x}^2 + \dot{y}^2)^{3/2}}{|\dot{x}\ddot{y} - \dot{y}\ddot{x}|} = \frac{(3a\sin\theta\cos\theta)^3}{9a^2\sin^2\theta\cos^2\theta} = 3a\sin\theta\cos\theta \qquad \text{...(1)}$$

Tangent at P is

$$y - a\sin^3\theta = -\frac{3a\sin^2\theta\cos\theta}{3a\cos^2\theta\sin\theta}(x - a\cos^3\theta)$$

or $y\cos\theta - a\sin^3\theta\cos\theta + x\sin\theta - a\sin\theta\cos^3\theta = 0$

or $x\sin\theta + y\cos\theta - a\sin\theta\cos\theta = 0$

$$p = \frac{0 - (-a\sin\theta\cos\theta)}{\sqrt{\sin^2\theta + \cos^2\theta}} = a\sin\theta\cos\theta \qquad \text{...(2)}$$

From (1) and (2)

$$\rho = 3p.$$

Example 1.54 **Show that for the rectangular hyperbola $xy = c^2$,**

$$\rho = \frac{(x^2 + y^2)^{3/2}}{2c^2}.$$

(Madras 2000 PT)

Solution : Writing the equation in the parametric form as $x = ct$, $y = \frac{c}{t}$

$$\dot{x} = c, \qquad \dot{y} = -\frac{c}{t^2}$$

$$\ddot{x} = 0 \qquad \ddot{y} = \frac{2c}{t^3}.$$

$$\rho = \frac{(\dot{x}^2 + \dot{y}^2)^{3/2}}{\dot{x}\ddot{y} - \dot{y}\ddot{x}} = \frac{c^3\left(1 + \frac{1}{t^4}\right)^2}{c \cdot \frac{2c}{t^3}}$$

or $$\rho = \frac{\left\{c^2 t^2 + \dfrac{c^2}{t^2}\right\}^{3/2}}{2c^2} \text{ or } \frac{(x^2+y^2)^{3/2}}{2c^2}$$

Example 1.55 **Find the radius of curvature at (r, θ) on each of the following curves**

(i) $r^n = a^n \cos n\theta$

(ii) $r = a(1 - \cos\theta)$

(iii) $r\,\theta^2 = a$

Find also its pedal equation.

Solution : (i) $n \log r = n \log a + \log \cos n\theta$ differentiating w.r.t. θ

$$\frac{n}{r}\frac{dr}{d\theta} = -n \tan n\theta$$

$$-\frac{1}{r^2}\left(\frac{dr}{d\theta}\right)^2 + \frac{1}{r}\frac{d^2r}{d\theta^2} = -n \sec^2 n\,\theta$$

So $$\rho = \frac{\left[r^2 + \left(\dfrac{dr}{d\theta}\right)^2\right]^{3/2}}{r^2 + 2\left(\dfrac{dr}{d\theta}\right)^2 - r\dfrac{d^2r}{d\theta^2}}$$

$$= \frac{(r^2 + r^2\tan^2 n\theta)^{3/2}}{r^2\{1 + 2\tan^2 n\theta - \tan^2 n\theta + n\sec^2 n\theta\}}$$

$$= \frac{r^3 \sec^3 n\theta}{r^2 (n+1)\sec^2 n\theta} = \frac{r \sec n\theta}{(n+1)}$$

(ii) $r = a(1 - \cos\theta) = 2a \sin^2 \dfrac{\theta}{2}$

$$\log r = \log 2a + 2 \log \sin \frac{\theta}{2}$$

$$\frac{1}{r}\frac{dr}{d\theta} = 2 \times \frac{1}{2} \cot \frac{\theta}{2}$$

So $$\frac{dr}{d\theta} = r \cot \frac{\theta}{2}$$

$$\frac{d^2r}{d\theta^2} = \frac{dr}{d\theta}\cot\frac{\theta}{2} - \frac{1}{2}\operatorname{cosec}^2\frac{\theta}{2}\cdot r$$

$$= r\left\{\cot^2\frac{\theta}{2} - \frac{\operatorname{cosec}^2\dfrac{\theta}{2}}{2}\right\}$$

So $$r^2 + 2\left(\frac{dr}{d\theta}\right)^2 - r\frac{d^2r}{d\theta^2} = r^2\left[1 + 2\cot^2\frac{\theta}{2} - \cot^2\frac{\theta}{2} + \frac{1}{2}\operatorname{cosec}^2\frac{\theta}{2}\right]$$

$$= \frac{3}{2}r^2\operatorname{cosec}^2\frac{\theta}{2}$$

$$\rho = \frac{r^2\left[1 + \left(\frac{1}{r}\frac{dr}{d\theta}\right)^2\right]^{3/2}}{\frac{3}{2}r^2\operatorname{cosec}^2\frac{\theta}{2}} = \frac{2}{3}r\operatorname{cosec}\frac{\theta}{2} = \frac{2}{3}\sqrt{2ar}.$$

Pedal Equation : As $\tan\phi = r\frac{d\theta}{dr} = \tan\frac{\theta}{2}$ so $\phi = \frac{\theta}{2}$. $p = r\sin\phi$ so $p = r\sin\frac{\theta}{2}$ or $p^2 = \frac{r^3}{2a}$

(iii) $r\theta^2 = a$ so $r = \frac{a}{\theta^2}$, $\frac{dr}{d\theta} = \frac{-2a}{\theta^3} = -\frac{2r}{\theta}$, $\frac{d^2r}{d\theta^2} = \frac{6a}{\theta^4} = \frac{6r}{\theta^2}$

$$\rho = \frac{\left\{r^2 + \left(\frac{dr}{d\theta}\right)^2\right\}^{3/2}}{r^2 + 2\left(\frac{dr}{d\theta}\right)^2 - r\frac{d^2r}{d\theta^2}} = \frac{r^3\left(1 + \frac{4}{\theta^2}\right)^{3/2}}{r^2\left\{1 + 2\times\frac{4}{\theta^2} - \frac{6}{\theta^2}\right\}}$$

$$= \frac{r}{\theta}\frac{(\theta^2+4)^{3/2}}{(\theta^2+2)} = \frac{a\,(\theta^2+4)^{3/2}}{\theta^3(\theta^2+2)}.$$

Example 1.56 **Find the radius of curvature for the following curves :**

(i) parabola $p^2 = ar$, with pole at focus.

(ii) $\frac{1}{p^2} = \frac{1}{a^2} + \frac{1}{b^2} - \frac{r^2}{a^2b^2}$, (ellipse)

(iii) $r^3 = a^2p$, (lemniscate)

Solution : (i) $p^2 = ar.$

So $$\rho = r\frac{dr}{dp} = r\frac{2p}{a}$$

(ii) $$-\frac{2}{p^3}\frac{dp}{dr} = -2\frac{r}{a^2b^2}.$$

So $$\rho = r\frac{dr}{dp} = \frac{a^2b^2}{p^3}$$

(iii) $$3r^2\frac{dr}{dp} = a^2.$$

So $$\rho = r\frac{dr}{dp} = r\frac{a^2}{3r^2} = \frac{a^2}{3r}.$$

1.18 RADIUS OF CURVATURE AT ORIGIN

Let the equation $y = f(x)$, when expanded about origin be

$$y = f(0) + \frac{x}{1!} f'(0) + \frac{x^2}{2!} f''(0) + \ldots$$

If curve passes through origin then $f(0) = 0$

So $$y = 0 + \frac{x f'(0)}{1!} + \frac{x^2}{2!} f''(0) + \ldots = px + \frac{qx^2}{2!} + \ldots$$

$$(\rho)_{\text{at origin}} = \frac{(1 + (f'(0))^2)^{3/2}}{f''(0)} = \frac{(1 + p^2)^{3/2}}{q}.$$

Example 1.57 **Find the radius of curvature at origin at (0, 0), for the curve**

$$y = 3x + 3x^2 + 5x^3 + \ldots$$

Solution :
$$y = f(0) + \frac{x}{1!} f'(0) + \frac{x^2}{2!} f''(0) + \ldots$$

About origin $f'(0) = 3$, $\frac{f''(0)}{2} = 3$ (by comparison)

So $$\rho_{\text{at origin}} = \frac{(1 + 3^2)^{3/2}}{6} = \frac{5\sqrt{5}}{6}.$$

1.19 NEWTON METHODS FOR CURVATURE AT ORIGIN

Radius of curvature when x axis is tangent $\left(\frac{dy}{dx} = 0\right)$

The equation to the curve $y = f(x)$, will have the expansion

$$y = f(0) + \frac{x}{1!} f'(0) + \frac{x^2}{2!} f''(0) + \ldots = 0 + px + \frac{qx^2}{2} + \ldots$$

Here p also $= 0$ as $\left(\frac{dy}{dx}\right) = 0$ at origin.

When x axis is tangent so

$$\underset{x \to 0}{\text{Lt}} \frac{2y}{x^2} = q$$

But $$\rho = \frac{(1 + p)^{3/2}}{q} = \frac{(1 + 0)^{3/2}}{q} = \underset{x \to 0}{\text{Lt}} \left(\frac{x^2}{2y}\right),$$

when x axis is tangent at (0, 0).

If the equation to the curve is not given in the form $y = f(x)$ but involves other powers of y also. The lowest degree terms in the equation to curve equated to zero gives $y = 0$. Then x axis is tangent at origin and ρ is given by $\underset{x \to 0}{\text{Lt}} \frac{x^2}{2y} = \rho$. Similarly if the curve passes through (0, 0) and y axis is the tangent at (0, 0) then $\rho = \underset{x \to 0}{\text{Lt}} \left(\frac{y^2}{2x}\right)$.

Then $$\rho = \underset{y \to 0}{\text{Lt}} \frac{y^2}{2x}.$$

Example 1.58 **Find the radius of curvature at the origin for the curve**

$$y = x^2 - y^3 - x^3 \quad \text{...(1)}$$

Solution : Since the lowest degree term equated to zero gives the tangent at origin to be $y = 0$

So $$\rho = \underset{x \to 0}{\text{Lt}} \frac{x^2}{2y}$$

Dividing (1) by y and taking limit as $x \to 0$, $y \to 0$, $\frac{x^2}{2y} \to \rho$.

$$1 = \left[\frac{x^2}{y} - y^2 - \frac{x^2}{y}.x\right]$$

$$= 2\rho - 0 - 0 \qquad \text{so} \qquad \rho = \frac{1}{2}$$

Example 1.59 **Find the radius of curvature for $y = x + 3x^2 - x^3$ at (0, 0).**

Solution : Curve passes through (0, 0). Comparing with

$$y = f(0) + \frac{x}{1!} f'(0) + \frac{x^2}{2!} f''(0) \ldots$$

$$f(0) = 0, \quad f'(0) = 1, \quad \frac{f''(0)}{2} = 3$$

$$(\rho)_{\text{at }(0,\,0)} = \frac{\{1 + f'(0)\}^{3/2}}{f''(0)}$$

$$= \frac{(1+1)^{3/2}}{6} = \frac{\sqrt{2}}{3}.$$

Example 1.60 **Find the radius of curvature at origin to**

$$a_1x + a_2y + b_1x^2 + b_2xy + b_3y^2 = 0.$$

Solution : Let $$y = px + \frac{qx^2}{2}$$

where $p = f'(0)$, $q = f''(0)$

So, $$a_1x + a_2\left(px + \frac{qx^2}{2}\right) + b_1x^2 + b_2x\left(px + \frac{qx^2}{2}\right) + b_3\left(px + \frac{qx^2}{2}\right)^2$$

$$= x(a_1 + a_2p) + x^2\left\{\frac{a^2q}{2} + b_1 + b_2p + b_3p^2\right\} + \text{terms of } x^3$$

Equating to zero the coefficients of various power of x

$$a_1 + a_2p = 0$$

So $$p = -\frac{a_1}{a_2}$$

$$a^2\frac{q}{2}+b_1+b_2\left(\frac{-a}{a_2}\right)+b_3\frac{a_1^2}{a_2^2}=0$$

So $$q=\frac{2}{a_2{}^3}(a_1a_2b_2-a_1{}^2b_3-a_2{}^2b_1)$$

Putting values of p and q in $\rho=\frac{(1+p^2)^{3/2}}{q}$, we get ρ.

Example 1.61 **Find curvature at (0, c) in the catenary $y=c\cosh\frac{x}{c}$.**

Solution : $y=c\left\{1+\frac{x^2}{2c^2}+\frac{x^4}{4!\,c^4}+\ldots\right\}$.

Tangent at $(0, c)$ is $y-c=0$.

$$\underset{x\to\infty}{\text{Lt}}\frac{(y-c)}{x^2}=\underset{x\to\infty}{\text{Lt}}\left\{\frac{1}{2c}+\frac{x^2}{4!\,c^3}\right\}=\frac{1}{2c}$$

So $$\underset{x\to\infty}{\text{Lt}}\frac{x^2}{2(y-c)}=\rho=c.$$

PROBLEM SET 1.5

1. Find the radius of curvature at any point $(at^2, 2at)$ on the parabola $y^2=4ax$.

Ans. $2a(1+t^2)^{3/2}$.

2. Prove that the radius of curvature of the catenary $y=c\cosh\frac{x}{c}$ at any point (x, y) is $\frac{y^2}{c}$.

3. Find the radius of curvature at $\left(\frac{a}{4},\frac{a}{4}\right)$ on the curve $\sqrt{x}+\sqrt{y}=\sqrt{a}$.

Ans. $\frac{a}{\sqrt{2}}$.

4. Show that the radius of curvature at $(a\cos^3\theta, a\sin^3\theta)$ on $x^{2/3}+y^{2/3}=a^{2/3}$ is $3a\sin\theta\cos\theta$.

5. Find the radius of curvature at any point on $x=a\cos\theta$, $y=b\sin\theta$.

Ans. $\frac{(a^2\sin^2\theta+b^2\cos^2\theta)^{3/2}}{ab}$

6. Show that the radius of curvature at an end of the major axis of ellipse $\frac{x^2}{a^2}+\frac{y^2}{b^2}=1$ is equal to the semi latus rectum.

7. Show that the radius of curvature ρ at P on the ellipse $\frac{x^2}{a^2} + \frac{y^2}{b^2} = 1$ is $\frac{CD^3}{ab}$ where CD is the semi diameter conjugate to CP.

8. If ρ_1 and ρ_2 be the radii of curvature at the ends of a focal chord of $y^2 = 4ax$, then show $\rho_1^{-2/3} + \rho_2^{-2/3} = (2a)^{-2/3}$.

9. Show that the radius of curvature at $(a, 0)$ on $y^2 = a^2\left(\frac{a-x}{x}\right)$ is $\frac{a}{2}$.

10. Find the radius of curvature at any point on $x = a(\cos t + t \sin t)$, $y = a(\sin t - t \cos t)$.
Ans. at

11. Find the radius of curvature at the origin for
(i) $x^3 + y^3 - 2x^2 + 6y = 0$
Ans. $\frac{3}{2}$
(ii) $2x^4 + 3y^4 + 4x^2y + xy - y^2 + 2x = 0$
Ans. 1
(iii) $y^2 = x^2 \frac{(a+x)}{a-x}$
Ans. $a\sqrt{2}$.

12. Show that the radius of curvature at $x = \frac{\pi}{2}$ of the curve $y = 4 \sin x - \sin 2x$ is $\rho = \frac{5\sqrt{5}}{4}$.

13. For the cycloid $s = 4a \sin \psi$, show that $\rho^2 + s^2 = 16a^2$.

14. For a catenary $s = c \tan \psi$, show that $\rho c = c^2 + s^2$

15. Find the radius of curvature in $x^3 + y^3 = 3axy$, where it intersects the line $y = x$.
Ans. $\frac{3\sqrt{2}a}{16}$.

16. Find the radius of curvature at (r, θ) on $r(1 + \cos\theta) = a$
Ans. $\rho = 2r\sqrt{\frac{2r}{a}}$.

17. Find radius of curvature at (r, θ) on $r = a(1 - \cos\theta)$
Ans. $\rho = \frac{2}{3}\sqrt{2ar}$.

1.20 CHORDS OF CURVATURE PARALLEL TO THE AXES

Let $P(x, y)$ be a point on the curve $y = f(x)$. Let PQA be the circle of curvature at P with centre C and diameter $PQ = 2\rho$. Let the tangent to the curve make an angle ψ with a line drawn parallel to x-axis. If lines are drawn through P and Q drawn parallel to X and Y

axis respectively which meet at A on the circle, PA and QA are called the chords parallel to the axes their lengths are given by

$$PA = 2\rho \sin \psi$$

and

$$QA = 2\rho \cos \psi.$$

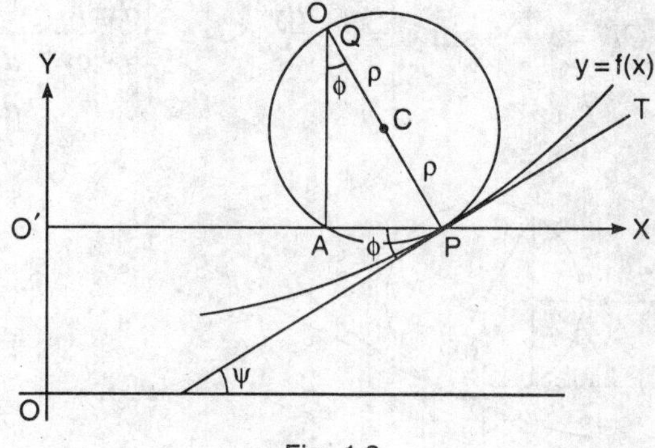

Fig. 1.3

1.21 CHORD THROUGH ORIGIN OR POLE

If the points P on the curve be joining to O, the intersects PA on this line O lying in the circle of curvature at P is called the chord through the origin. It length is

$$PA = 2\rho \sin \phi$$

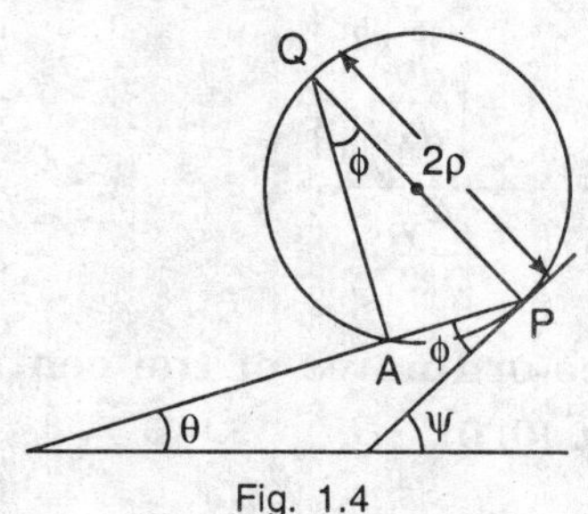

Fig. 1.4

Example 1.62 **Show that for the curve $y = a \log \sec \dfrac{x}{a}$, the chord of curvature parallel to y axis is of constant length.**

Solution :

$$\tan \psi = \frac{dy}{dx} = \frac{a \dfrac{1}{a} \sec \dfrac{x}{a} \tan \dfrac{x}{a}}{\sec \dfrac{x}{a}} = \tan \frac{x}{a}$$

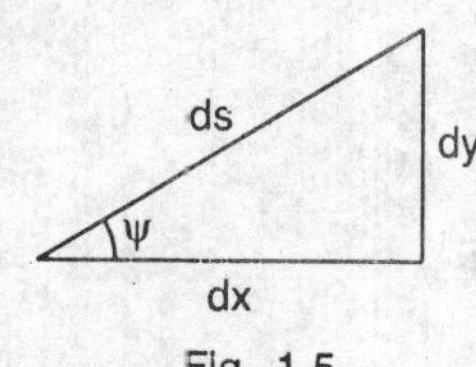

Fig. 1.5

So $\qquad \psi = \dfrac{x}{a}$ or $\dfrac{dx}{d\psi} = a$

Chord parallel to y axis $= 2\rho \cos \psi$

$$= 2 \frac{ds}{d\psi} \frac{dx}{ds} = 2 \frac{dx}{d\psi} = 2a.$$

Centre of Curvature. Let $y = f(x)$ be a curve and PQA be the circle of curvature at P. Its centre C is called the centre of curvature. The co-ordinates of C denoted by $(\bar{x}, \bar{y})$ are

$$\bar{x} = ON = x - \rho \sin \psi = x - \frac{dy}{ds}\rho = x - \frac{dy}{dx}\frac{\left\{1 + \left(\frac{dy}{dx}\right)^2\right\}}{\frac{d^2y}{dx^2}}$$

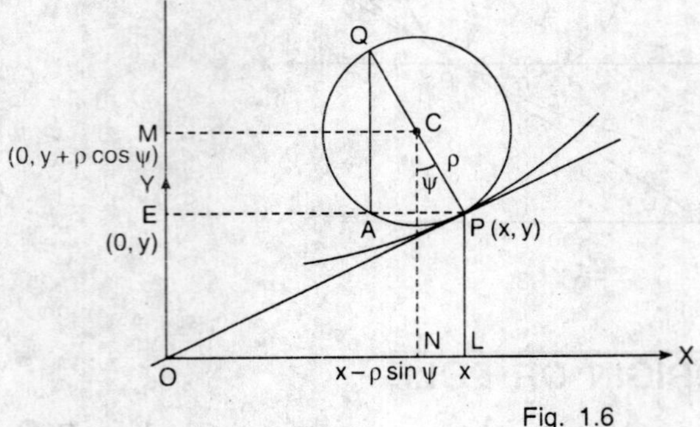

Fig. 1.6

and $$\bar{y} = OM = OE + EM = y + \rho \cos \psi = y + \rho \frac{dx}{ds}$$

$$= y + \frac{\left\{1 + \left(\frac{dy}{dx}\right)^2\right\}}{\frac{d^2y}{dx^2}}$$

Example 1.63 **Find ρ and the co-ordinates of the centre of curvature on the curve $y = 2x + 5x^2 + x^3$ at (0, 0).**

Solution :

$$\frac{dy}{dx} = 2 + 10x + 3x^2 \text{ at } (0, 0)\ \frac{dy}{dx} = 2$$

$$\frac{d^2y}{dx^2} = 10 + 6x \qquad \frac{d^2y}{dx^2} = 10$$

$$\rho = \frac{\left\{1 + \left(\frac{dy}{dx}\right)^2\right\}^{3/2}}{\frac{d^2y}{dx^2}} = \frac{5\sqrt{5}}{10} = \frac{\sqrt{5}}{2}.$$

$$\bar{x} = x - \frac{dy}{dx}\frac{\left\{1 + \left(\frac{dy}{dx}\right)^2\right\}}{\frac{d^2y}{dx^2}} = 0 - 2\left\{\frac{1+4}{10}\right\} = -1$$

$$\bar{y} = y + \frac{\left\{1 + \left(\frac{dy}{dx}\right)^2\right\}}{\frac{d^2y}{dx^2}} = 0 + \frac{5}{10} = \frac{1}{2}.$$

Example 1.64 **Find the circle of curvature at $P\left(\frac{a}{4}, \frac{a}{4}\right)$ of the curve $\sqrt{x} + \sqrt{y} = \sqrt{a}$.**

Solution : The circle of curvature has the equation

$$(x - \bar{x})^2 + (y - \bar{y})^2 = \rho^2$$

where $(\bar{x}, \bar{y})$ are the co-ordinates of centre of curvature.

We have $\sqrt{y} = \sqrt{a} - \sqrt{x}$

So $y = a + x - 2\sqrt{ax}$

$$\frac{dy}{dx} = 1 - \frac{2\sqrt{a}}{2\sqrt{x}} = 1 - \sqrt{\frac{a}{x}}$$

$$\left(\frac{dy}{dx}\right)_{\text{at } P} = +1 - \sqrt{4} = -1$$

$$\frac{d^2y}{dx^2} = \frac{1}{2}\frac{\sqrt{a}}{x\sqrt{x}}$$

$$\left(\frac{d^2y}{dx^2}\right)_{\text{at } P} = \frac{2\sqrt{4}}{a} = \frac{4}{a}$$

So $$\rho = \frac{\{1 + (-1)^2\}^{3/2}}{\frac{4}{a}} = a\,\frac{2\sqrt{2}}{4} = \frac{a}{\sqrt{2}}$$

$$\bar{x} = x - \frac{dy}{dx}\,\frac{\left\{1 + \left(\frac{dy}{dx}\right)^2\right\}}{\frac{d^2y}{dx^2}} = \frac{a}{4} - (-1)\,\frac{1+1}{4}\,a = \frac{3a}{4}$$

$$\bar{y} = y + \frac{1 + \left(\frac{dy}{dx}\right)^2}{\frac{d^2y}{dx^2}} = \frac{a}{4} + \frac{2a}{4} = \frac{3a}{4}$$

Hence the circle is $\left(x - \frac{3a}{4}\right)^2 + \left(y - \frac{3a}{4}\right)^2 = \frac{a^2}{2}$

or $$x^2 + y^2 - \frac{3}{2}a\,(x + y) + \frac{5}{8}a^2 = 0.$$

PROBLEM SET 1.6

1. Find the centre of curvature of the curve $xy = c^2$ at (x, y)

   Ans. $\left(\frac{3x}{2} + \frac{c^4}{2x^3}, \frac{3}{2}y + \frac{x^3}{2c^2}\right)$

2. Prove that the centre of curvature at (x, y) on $x^{2/3} + y^{2/3}$ is

   Ans. $(x - 3x^{1/3}y^{2/3}, y + 3x^{2/3}y^{1/3})$

3. Prove that the centre of curvature at (x, y) on the catenary $y = c \cosh \frac{x}{c}$ is $\left(x - \frac{y}{c}\sqrt{y^2 - c^2}, 2y\right)$.

4. Show that the chord of curvature through the focus of a parabola $p^2 = ar$ is of length $4r$.

5. Show that the chord of curvature parallel to y axis for the catenary $y = c \cosh \frac{x}{c}$ is double its ordinate and the chord parallel to x-axis of length $c \sinh \frac{2x}{c}$.

1.22 ENVELOPE AND EVOLUTE

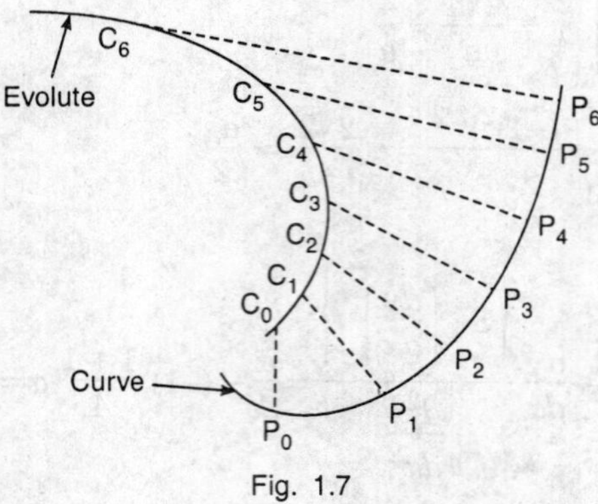

Fig. 1.7

Envelope. If $f(x, y, a) = 0$ be the equation of a family of curves, the locus of the points of intersection of two consecutive members of the family is called the envelope of the family. It is obtained by eliminating a between the equations $f(x, y, a) = 0$ and $\frac{\partial f}{\partial a} = 0$.

Example 1.65 **Find the family of lines $y = mx + \frac{a}{m}$.**

Solution :

$$f(x, y, m) = \left(y - mx - \frac{a}{m}\right) \qquad ...(1)$$

$$\frac{\partial f}{\partial m} = -x + \frac{a}{m^2} \qquad ...(2)$$

So $m = \sqrt{\frac{a}{x}}$

Eliminating m, $y = \sqrt{\frac{a}{x}}\, x + a\sqrt{\frac{x}{a}} = 2\sqrt{ax}$

Hence the parabola $y^2 = 4ax$ is the envelope of the family (1) which represents its tangents.

Evolute. The locus of the centre of curvature for the curve $y = f(x)$, is called its evolute and the curve is called the involute of the evolute.

1.23 EVOLUTE OF A CURVE IS THE ENVELOPE OF ITS NORMALS

Example 1.66 **Find the co-ordinates of the centre of curvature at any point $\left(\frac{a}{m^2}, \frac{2a}{m}\right)$ on the parabola $y^2 = 4ax$ and hence find its evolute.**

Solution : $\frac{dx}{dm} = -\frac{2a}{m^3}$, $\frac{dy}{dm} = -\frac{2a}{m^2}$ so $\frac{dy}{dx} = m$

$$\frac{d^2y}{dx^2} = \frac{d}{dm}\left(\frac{dy}{dx}\right)\cdot\frac{dm}{dx} = 1\cdot\left(-\frac{m^3}{2a}\right).$$

$$\bar{x} = x - \frac{dy}{dx}\frac{\left\{1 + \left(\frac{dy}{dx}\right)^2\right\}}{\frac{d^2y}{dx^2}} = \frac{a}{m^2} - \frac{m\{1 + m^2\}}{-\frac{m^3}{2a}} = 2a + \frac{3a}{m^2}.$$

$$\bar{y} = y + \frac{\left\{1 + \left(\frac{dy}{dx}\right)^2\right\}}{\frac{d^2y}{dx^2}} = \frac{2a}{m} + \frac{1 + m^2}{-\frac{m^3}{2a}} = -\frac{2a}{m^3}.$$

So $(\bar{x} - 2a)^3 = 27\frac{a^3}{m^6}$

and $(\bar{y})^2 = \frac{4a^2}{m^6}$

Eliminating m^6, $4(\bar{x} - 2a)^3 = 27a(\bar{y})^2$.

Hence the locus is $27ay^2 = 4(x - 2a)^3$.

Example 1.67 **Find the envelope of the family of lines $y = mx + \sqrt{1 + m^2}$, m being the parameter.** ***(Madras 1990S)***

Solution : $f(x, y, m) = (y - mx - \sqrt{1 + m^2}) = 0$

$$\frac{\partial f}{\partial m} = -x - \frac{m}{\sqrt{1+m^2}} = 0$$

or $$x = -\frac{m}{\sqrt{1-m^2}}$$

So $$(1+m^2)x^2 = m^2$$

or $$m^2 = \frac{x^2}{1-x^2}$$

or $$m = -\frac{x}{\sqrt{1-x^2}}.$$

Hence the envelope is

$$y = x\frac{(-x)}{\sqrt{1-x^2}} + \sqrt{1+\frac{x^2}{1-x^2}}$$

$$= \frac{-x^2+1}{\sqrt{1-x^2}} = \sqrt{1-x^2}$$

or $$y^2 = 1-x^2$$

or $$x^2+y^2 = 1.$$

Example 1.68 **Show that the envelope of the family of lines $x \sec\alpha + y \operatorname{cosec}\alpha = l$ is the astroid $x^{2/3}+y^{2/3} = l^{2/3}$.**

Solution :
$$f(x, y, l) = x\sec\alpha + y\operatorname{cosec}\alpha - l = 0$$

$$\frac{\partial f}{\partial \alpha} = x\sec\alpha\tan\alpha - y\operatorname{cosec}\alpha\cot\alpha = 0$$

So $$\tan^3\alpha = \frac{y}{x}$$

Fig. 1.8

Eliminating α $$\sqrt{x^{2/3}+y^{2/3}}\left(\frac{x}{x^{1/3}}+\frac{y}{y^{1/3}}\right) = l$$

Hence $$\{x^{2/3}+y^{2/3}\}^{3/2} = l$$

or $$x^{2/3}+y^{2/3} = l^{2/3}.$$

Example 1.69 **Find the envelope of $\frac{x^2}{a^2}+\frac{y^2}{b^2} = 1$ for which $a+b = c$.** *(Madras 2000 S)*

Solution : We have taking the total differentials of the given equations, regarding a and b as variables.

$$\left.\begin{aligned}\frac{x^2}{a^3}da + \frac{y^3}{b^3}d\,b &= 0\\ da + d\,b &= 0\end{aligned}\right\}$$

$$\Rightarrow \quad \begin{vmatrix}\frac{x^2}{a^3} & \frac{y^3}{b^3}\\ 1 & 1\end{vmatrix} = 0$$

Hence $$\frac{a^3}{x^2} = \frac{b^3}{y^2}$$

or $$\frac{a}{\frac{x^2}{a^2}} = \frac{b}{\frac{y^2}{b^2}} = \frac{a+b}{\frac{x^2}{a^2}+\frac{y^2}{b^2}} = \frac{c}{1}$$

Hence $a^3 = cx^2$, $b^3 = cy^2$

Putting values of a and b in $a + b = c$

$$(x^{2/3} + y^{2/3})\, c^{1/3} = c$$

or $$x^{2/3} + y^{2/3} = c^{2/3}$$ is the required envelope.

Example 1.70 **Find the envelope of he lines $\frac{x}{a} + \frac{y}{b} = 1$, where the parameters a and b are connected by $a^n + b^n = c^n$.**

Solution : Taking total differntials with a, b as variables

$$\left.\begin{aligned} \frac{x}{a^2}\, da + \frac{y}{b^2}\, db = 0 \\ a^{n-1}\, da + b^{n-1}\, db = 0 \end{aligned}\right\} \Rightarrow \begin{vmatrix} \frac{x}{a^2} & \frac{y}{b^2} \\ a^{n-1} & b^{n-1} \end{vmatrix} = 0$$

or $$\frac{xb^{n-1}}{a^2} = \frac{ya^{n-1}}{b^2} \text{ so } \frac{\frac{x}{a}}{a^n} = \frac{\frac{y}{b}}{b^n} = \frac{1}{c^n}$$

Hence $a^{n+1} = xc^n$, $b^{n+1} = yc^n$

from $$a^n + b^n = c^n$$

$$(xc^n)^{\frac{n}{n+1}} + (yc^n)^{\frac{n}{n+1}} = c^n$$

or $$x^{\frac{n}{n+1}} + y^{\frac{n}{n+1}} = c^{\frac{n}{n+1}}$$

Different cases

(i) Condition $a' + b' = c'$ so $n = 1$, envelope is $x^{1/2} + y^{1/2} = c^{1/2}$.

(ii) Condition $a^2 + b^2 = c^2$, so $n = 2$, envelope is $x^{2/3} + y^{2/3} = c^{2/3}$.

Example 1.71 **Find the envelope of a system of concentric and coaxial ellipses of constant area.** ***(Tirupati 98 S)***

Solution : The ellipse are $\frac{x^2}{a^2} + \frac{y^2}{b^2} = 1$, the parameters a and b are connected by $\pi ab = \pi c^2$.

Finding the total differntials with a, b as variables

$$\left.\begin{aligned} \frac{x^2}{a^3}\, da + \frac{y^2}{b^3}\, db = 0 \\ b\, da + a\, db = 0 \end{aligned}\right\} \Rightarrow \begin{vmatrix} \frac{x^2}{a^3} & \frac{y^2}{b^3} \\ b & a \end{vmatrix} = 0$$

$$\text{or} \quad \frac{\frac{x^2}{a^2}}{1} = \frac{\frac{y^2}{b^2}}{1} = \frac{\frac{x^2}{a^2}+\frac{y^2}{b^2}}{2} = \frac{1}{2}$$

So $\quad a = \sqrt{2}x,\ b = \sqrt{2}y$ as $c^2 = ab$

So $\quad c^2 = 2xy.$

Example 1.72 **Find the evolute of the parabola $y^2 = 4ax$.** *(Madras 96)*

Solution : We know evolute is the envelope of the normals. So we find the envelope of $y = mx - 2am - am^3$ which is normal to the parabola at $(am^2, -2am)$.

$$\frac{\partial f}{\partial m} = x - 2a - 3am^2 = 0$$

So $$m = \left(\frac{x-2a}{3a}\right)^{1/2}$$

Hence the envelope is

$$y = \sqrt{\frac{x-2a}{3a}}\left\{x - 2a - a\left(\frac{x-2a}{3a}\right)\right\}$$

On simplifying we get $27ay^2 = 4(x-2a)^3$ as equation of evolute.

PROBLEM SET 1.7

1. Find the envelope of the family of lines $y = mx + am^2$, m being the parametric.
Ans. $x^2 + 4ay = 0$ *(Madras 98 S)*

2. Find the envelope of the family of lines $y = mx - 2am - am^3$. *(Madurai 98 S)*
Ans. $27ay^2 = 4(x-2a)^3$

3. Find the envelope of the family of lines $y = mx + \sqrt{a^2m^2 + b^2}$ *(Madras 96)*
Ans. $\frac{x^2}{a^2} + \frac{y^2}{b^2} = 1$

4. Find the envelope of the family of straights lines $\frac{x}{a}\cos\alpha + \frac{y}{b}\sin\alpha = 1$, α being the parametric and a, b are constants. *(Coimbatore 90)*
Ans. $\frac{x^2}{a^2} + \frac{y^2}{b^2} = 1$

5. Find the envelope of the straight line $\frac{x}{a} + \frac{y}{b} = 1$, having a, b are parameters which are related by
(i) $ab = c^2$, (ii) $a^2 + b^2 = c^2$ (iii) $x + y = c$ *(JNTU 98)*
Ans. (i) $4xy = c^2$, (ii) $x^{2/3} + y^{2/3} = a^{2/3}$, (iii) $\sqrt{x} + \sqrt{y} = \sqrt{c}$

6. Find the envelope of the projectile path $y = x \tan\alpha - \frac{1}{2} g \frac{x^2}{u^2 \cos^2\alpha}$, α being the parameter. ***(Madras 98 S)***

Ans. $y = \frac{u^2}{2g} - \frac{g}{2u^2}x^2$

7. Find the evolute of the ellipse $\frac{x^2}{a^2} + \frac{y^2}{b^2} = 1$

Ans. $(ax)^{2/3} + (by)^{2/3} = (a^2 - b^2)^{2/3}$.

8. Find the evolute of the hyperbola $\frac{x^2}{a^2} - \frac{y^2}{b^2} = 1$.

Ans. $(ax)^{2/3} - (by)^{2/3} = (a^2 + b^2)^{2/3}$

9. Find the evolute of the parabola $y^2 = 4ax$. ***(Osmania 95)***

Ans. $27ay^2 = 4(x - 2a)^3$.

Partial Differentiation

1.24 CONTINUITY

In the relation $y = f(x)$, y is a function of the single independent variable x, provided for every value of x, in its domain, $f(x)$ has a single and definite value. If $f(x)$ be a quotient, such as $f(x) = \frac{x^2 - a^2}{x - a}$, it assumes the indeterminate form $\frac{0}{0}$ at $x = a$, so **there the value of $f(x)$ does not exist.** So in order to have an idea about the value to which $f(x)$ tends at $x = a$, leaving the **exact point $x = a$, $f(x)$ is investigated in a close neighbourhood of $x = a$, by taking $x = a + h$ and allowing h to tend to zero, ultimately, after cancelling terms as $\frac{h}{h}$, since $\lim\limits_{h \to 0} \frac{h}{h} = 1$, through the value $\frac{0}{0}$ is not one.** What we thus get is called the limit of $f(x)$ as x tends to a. Though the value of $f(x)$ as such points does not exist as at $x = a$, its limit exists. At such points, the value of $f(x)$ is generally given separately. Thus $f(x)$, in the above example is fully defined as

$$f(x) = \frac{x^2 - a^2}{x - a},\ x \neq a$$

and $f(x) = k$ also denoted as $f(a) = k$, at $x = a$.

While taking the limit, if x tends to a, from the left then the limit is denoted as $\lim\limits_{h \to 0\, h > 0} f(a - h) = l_1$ (say) and **is called the left hand limit** and when $x \to a$, from the right then it is written as $\lim\limits_{h \to 0\, h > 0} f(a + h) = l_2$ and is called the **right hand limit at $x = a$.**

1.24.1 The Function $f(x)$ is said to be Continuous At $x = a$

A function $f(x)$ when the

$$\text{left} \lim_{x \to a^-} f(x) = \text{right} \lim_{x \to a^+} f(x) = \text{Value } f(a).$$

or

$$\lim_{h \to 0} f(a - h) = \lim_{h \to 0} f(a + h) = f(a).$$

1.25 FUNCTIONS OF TWO INDEPENDENT VARIABLES

If $z = f(x, y)$ be a function of two independent variables x and y, (x, y) may now be regarded as the coordinates of a point P in a plane, whose neighbourhood is not an interval of line as $a - h \leq x \leq a + h$ but a circle with centre $P(x, y)$ and radius, a small number δ. For taking the limit as $Q(x + h, y + k)$ tends to P, where $h^2 + k^2 \leq \delta^2, \delta \to 0$, **$Q$ may tend to P along so many paths.**

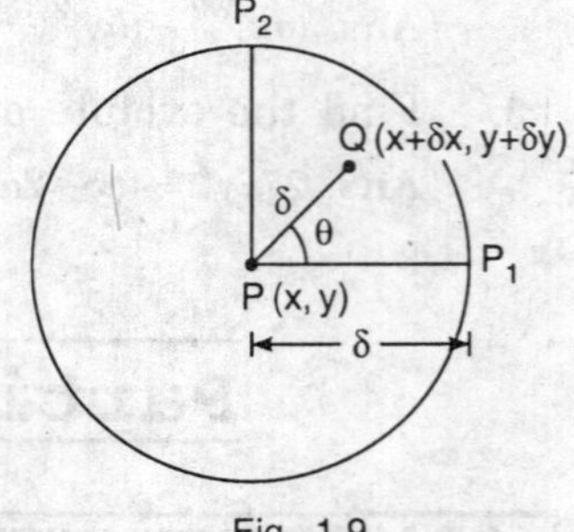

Fig. 1.9

If Q tend to P along $P_1 P$, we take $k = 0$ initially and allow h to tend to zero. While along the path $P_2 P$, h is made zero in the beginning and then k is made to tend to zero. Along QP, a line with slope $m = \tan \theta$, we let $k = h \tan \theta$ and then take the limits as $h \to 0$. **If the limit comes out to be unique $= l$ (say) in all cases, irrespective of the path along which Q tends to P, then the limit is said to exist at P,** and we write

$$\lim_{\delta x \to 0,\, \delta y \to 0} f(x + \delta x, y + \delta y) = l$$

So when the limit exists

$$\lim_{\delta x \to 0} \left| \lim_{\delta y \to 0} f(x + \delta x, y + \delta y) \right| = \lim_{\delta y \to 0} \left| \lim_{\delta x \to 0} f(x + \delta x, y + \delta y) \right| = l \text{ (unique)}$$

1.26 CONTINUITY OF $f(x, y)$ AT (a, b)

In case the function assume the form $\frac{0}{0}$ at $x = a, y = b$, the value of $f(x, y)$ at it is given separately. Denoting it by $f(a, b)$, $f(x, y)$ is said to be continuous at $P(a, b)$ if

$$\lim_{x \to a,\, y \to b} f(x, y) = f(a, b)$$

1.27 THEOREMS ON LIMITS AND CONTINUITY

As for the case of a single variable, the following results hold for functions of two variables also.

If $\lim_{x \to a, y \to b} f(x, y) = l$, and $\lim_{x \to a\, y \to b} g(x, y) = m$

then (i) $\lim_{x \to a, y \to b} \{f(x, y) \pm g(x, y)\} = l \pm m$

(ii) $\lim\limits_{x \to a, y \to b} \{f(x, y) \cdot g(x, y)\} = l \cdot m$

(iii) $\lim\limits_{x \to a, y \to b} \{f(x, y)/g(x, y)\} = \dfrac{l}{m}$

provided $m \neq 0$.

If $f(x, y)$ and $g(x, y)$ be continuous at $(x = a, y = b)$ then $f(x, y) \pm g(x, y)$; $f(x, y) \cdot g(x, y)$ and $f(x, y)/g(x, y), g(x, y) \neq 0$, are also continuous.

Example 1.73 **Show that the following functions are continuous at (a, b).**

(i) $$f(x, y) = \frac{x^2 - a^2}{x - a} \cdot \frac{y^3 - b^3}{y^2 - b^2}, (x, y) \neq (a, b)$$
$$= 3ab, \text{ at } (x, y) = (a, b)$$

(ii) $f(x, y) = (x + b)(y + a)$, **for all** (x, y)

Solution : (i) Function is continuous at (a, b) if

$$\lim_{\substack{x \to a \\ y \to b}} f(x, y) = \lim_{y \to b} \left\{ \lim_{x \to a} f(x, y) \right\} = \lim_{x \to a} \left\{ \lim_{y \to b} f(x, y) \right\}$$

$$\lim_{y \to b} \left\{ \lim_{x \to a} f(x, y) \right\} = \lim_{y \to b} \left\{ \lim_{x \to a} \left\{ \frac{(x - a)}{x - a} (x + a) \right\} \lim \frac{(y - b)(y^2 + by + b^2)}{(y - b)(y + b)} \right\}$$

$$= \lim_{y \to b} \left\{ 2a \frac{(y - b)(y^2 + by + b^2)}{(y - b)(y + b)} \right\}$$

$$= 2a \cdot \frac{(b^2 + b^2 + b^2)}{2b} = 3ab = f(a, b) \text{ given}$$

$$\lim_{x \to a} \left\{ \lim_{y \to b} f(x, y) \right\} = \lim_{x \to a} \left\{ \frac{x - a}{x - a} (x + a) \cdot \frac{3b^2}{2b} \right\}$$

$$= 2a \cdot \frac{3b}{2} = 3(a \cdot b) = f(a \cdot b)$$

Hence f is continuous at (a, b).

(ii) $$\lim_{y \to b} \left\{ \lim_{x \to a} f(x, y) \right\} = \lim_{y \to b} \left\{ \lim_{h \to 0} (a + h + b)(y + a) \right\}$$

$$= \lim_{k \to 0} [(a + b)\{b + k + a\}] = (a + b)(a + b) \qquad ...(1)$$

Value $$f(a, b) = (a + b)(b + a) \qquad ...(2)$$

$$\lim_{x \to a} \left\{ \lim_{y \to b} f(x + y) \right\} = \lim_{x \to a} \left\{ \lim_{k \to 0} (x + b)[b + k + a] \right\} = \lim_{x \to a} (x + b)(a + b)$$

$$= \lim_{h \to 0} \{a + h + b\}(a + b) = (a + b)^2 \qquad ...(3)$$

By (1), (2), (3), the three values are each $= (a + b)^2$. So $f(x, y)$ is continuous at (a, b).

Example 1.74 **Show that the function $f(x, y) = \dfrac{x-y}{x+y}$, is not continuous at (0, 0).**

Solution :

$$\lim_{x \to 0}\left\{\lim_{y \to 0} f(x, y)\right\} = \lim_{x \to 0}\left\{\frac{x-0}{x+0}\right\} = 1$$

$$\lim_{y \to 0}\left\{\lim_{x \to 0} f(x, y)\right\} = \lim_{y \to 0}\left\{\frac{0-y}{0+y}\right\} = -1$$

Since the two limits are not equal, the function is not continuous at origin.

Example 1.75 **Find the limit of the function $f(x, y) = \dfrac{x-y}{x+y}$, as $(x, y) \to (0, 0)$, along the line $y = mx$ and hence show that the function is not continuous at origin.**

Solution : Putting $y = mx$ in $f(x, y)$.

$$f(x, y) = \frac{x(1-m)}{x(1+m)} \quad \lim_{x \to 0}$$

$$\{f(x, y)\} = \frac{1-m}{1+m}$$

As the limit will be different for different values of m, it is therefore not unique. So the function is not continuous at the origin (0, 0).

Example 1.76 **Find the following limits**

(i) $\displaystyle\lim_{\substack{x \to a \\ y \to b}} \frac{x^2 y + y^2 x}{x^2 + y^2 + 1}$ (ii) $\displaystyle\lim_{\substack{x \to \infty \\ y \to b}} \frac{xy + 5}{2x^2 + 3y^2}$

Solution :

$$\text{Limit} = \lim_{x \to a}\left\{\lim_{y \to b} \frac{yx^2 + xy^2}{x^2 + y^2 + 1}\right\}$$

$$= \lim_{x \to a}\left\{\frac{bx^2 + b^2x}{x^2 + b^2 + 1}\right\} = \frac{ab(a+b)}{a^2 + b^2 + 1}$$

$$\lim_{y \to b}\left\{\lim_{x \to a} \frac{yx^2 + xy^2}{x^2 + y^2 + 1}\right\} = \lim_{y \to b}\left\{\frac{ya^2 + ay^2}{a^2 + y^2 + 1}\right\} = \frac{ab(a+b)}{a^2 + b^2 + 1}.$$

As the two limits are same, the limit exists and is $= \dfrac{ab(a+b)}{a^2 + b^2 + 1}$

(ii) $$\lim_{x \to \infty}\left\{\lim_{y \to b} \frac{xy + 5}{2x^2 + 3y^2}\right\} = \lim_{x \to \infty}\left\{\frac{5x + 5}{2x^2 + 3b^2}\right\}$$

$$= \frac{x}{x^2}\left(\frac{5 + 5/x}{2 + 3b^2/x^2}\right) = 0$$

$$= \lim_{y \to b}\left\{\left\{\lim_{x \to \infty} \frac{x(y + 5/x)}{x^2\,(2 + 3\,y^2/x^2)}\right\}\right\} = \lim_{y \to b} \quad \{0\} = 0.$$

As the two limits are same, so the limit exists and is = 0.

Example 1.77 **Show that the following limits do not exist**

(i) $\lim\limits_{\substack{x \to 0 \\ y \to b}} \dfrac{xy}{x^2 + y^2}$

(ii) $\lim\limits_{\substack{x \to a \\ y \to b}} \dfrac{x(y - b)}{y(x - a)}$

Solution : (i) $$\lim_{x \to 0} \left\{ \lim_{y \to 0} \frac{xy}{x^2 + y^2} \right\} = \lim_{x \to 0} \left\{ \frac{0}{x^2} \right\} = 0$$

or $$\lim_{y \to 0} \left\{ \lim_{x \to 0} \frac{xy}{x^2 + y^2} \right\} = \lim_{y \to 0} \left\{ \frac{0}{y^2} \right\} = 0$$

But along $y = mx$

$$\lim_{x \to 0} f(x, y) = \lim_{x \to 0} \left\{ \frac{mx^2}{x^2 (1 + m^2)} \right\} = \lim_{x \to 0} \frac{m}{1 + m^2}$$

Since the limit is dependent on m. So it is not unique. Hence the limit does not exist.

(ii) $$\lim_{y \to b} \left\{ \lim_{x \to a} f(x) \right\} = \lim_{y \to b} \left\{ \lim_{h \to 0} \frac{a + h}{h} \frac{y - b}{b} \right\} = \infty$$

$$\lim_{x \to a} \left\{ \lim_{y \to b} f(x) \right\} = \lim_{x \to a} \left\{ \frac{0x}{b(x - a)} \right\} = 0$$

Since the two limits are different. So the limit does not exist.

PROBLEM SET 1.8

1. Evaluate

(i) $\lim\limits_{\substack{x \to 1 \\ y \to 2}} \dfrac{x^2 y}{x^2 + y^2 + 1}$

(ii) $\lim\limits_{\substack{x \to 0 \\ y \to 0}} \dfrac{x^2 y^2}{x^2 + y^2}$

(iii) $\lim\limits_{\substack{x \to \infty \\ y \to 2}} \dfrac{xy + 1}{x^2 + 2y^2}$

Ans. (i) $\frac{1}{3}$; (ii) 0; (iii) 0

2. Evaluate the limit

$$\lim_{\substack{x \to 1 \\ y \to 3}} \frac{(x-2)(y-3)}{y(x-1)}$$

Ans. Limit does not exist

3. Evaluate the limit

$$f(x, y) = \frac{xy}{x^2 + y^2}, (x, y) \neq 0 = 0 \text{ at } (0, 0)$$

Ans. Limit does not exist.

4. Show that the function $f(x, y) = x^2 + 2y + y^2$

$(x, y) \neq (1, 2)$

$f(1, 2) = 0$

Though the limit $x \to 1$, $y \to 2$ exists, the function is not continuous at (1, 2).

Limit $= 9 \neq$ value so discontinuous at (1, 2).

In what follows, the functions taken for differentiation will be continuous along with their partial differential coefficients ?

1.28 PARTIAL DERIVATIVES

When $z = (x, y)$ is a functions of two independent variables so while differentiating z, w.r.t. x, y is treated as constant and the differential coefficient is denoted by $\frac{\partial f}{\partial x}$ or p

So
$$\frac{\partial f}{\partial x} = \lim_{\delta x \to 0} \frac{f(x + \delta x, y) - f(x, y)}{\delta x} \equiv \boldsymbol{p}$$

and similarly
$$\frac{\partial f}{\partial y} = \lim_{\delta y \to 0} \frac{f(x, y + \delta y) - f(x, y)}{\delta y} \equiv \boldsymbol{q}$$

1.29 NOTATIONS

The higher order derivatives are denoted as

$$\frac{\partial^2 f}{\partial x^2} = f_{xx} \equiv r, \quad \frac{\partial^3 f}{\partial x^3} = f_{xxx}$$

Assuming f and its partial derivative to be continuous we have

$$\frac{\partial}{\partial x}\left(\frac{\partial f}{\partial y}\right) = \frac{\partial}{\partial y}\left(\frac{\partial f}{\partial x}\right)$$

We denote $f_{xy} = f_{xy} \equiv s$, $f_{yy} = \frac{\partial^2 f}{\partial y^2} = t$.

1.30 GEOMETRICAL MEANING

If $f(x, y, z) = 0$ be a three dimensional surface, the partial derivatives $\frac{\partial f}{\partial x}, \frac{\partial f}{\partial y}, \frac{\partial f}{\partial z}$ are proportional to the direction cosines of the normal at $P(x, y, z)$ and the tangent plane at P has the equation

$$(X - x)\frac{\partial f}{\partial x} + (Y - y)\frac{\partial f}{\partial y} + (Z - z)\frac{\partial f}{\partial z} = 0.$$

Example 1.78 **If $u = e^{xyz}$, show that $\frac{\partial^3 u}{\partial x\, \partial y\, \partial z} = (1 + 3xyz + x^2y^2z^2)\, e^{xyz}$.**

Solution : $u = e^{xyz}$

So $$\frac{\partial u}{\partial x} = yze^{xyz} = yzu, \frac{\partial u}{\partial y} = zxu, \frac{\partial u}{\partial z} = xyu$$

$$\frac{\partial^2 u}{\partial y\, \partial z} = \frac{\partial}{\partial y}\left(\frac{\partial u}{\partial z}\right) = \frac{\partial}{\partial y}(xyu)$$

$$= x\left\{u + y\frac{\partial u}{\partial y}\right\} = x(1 + xyz)\, u$$

$$\frac{\partial^3 u}{\partial x\, \partial y\, \partial z} = \frac{\partial}{\partial x}(x + x^2yz)\, u$$

$$= (1 + 2xyz)\, u + (x + x^2 yz)\frac{\partial u}{\partial x}$$

$$= (1 + 2xyz)\, u + (x + x^2 yz)\, yzu$$

$$= (1 + 3xyz + x^2y^2z^2)\, u.$$

Example 1.79 **If $u = x^2 \tan^{-1}\frac{y}{x} - y^2 \tan^{-1}\frac{x}{y}$, show that $\frac{\partial^2 u}{\partial x\, \partial y} = \frac{x^2 - y^2}{x^2 + y^2}$.**

Solution : If $\tan\theta = \frac{y}{x}$, $\cot\theta = \frac{x}{y}$

So $\theta = \tan^{-1}\frac{y}{x} = \cot^{-1}\frac{x}{y}$

or $\tan^{-1}\frac{x}{y} = \cot^{-1}\frac{y}{x}$

Fig. 1.10

$$u = x^2 \tan^{-1}\frac{y}{x} - y^2 \tan^{-1}\frac{x}{y},$$

$$= x^2 \tan^{-1}\frac{y}{x} - y^2 \cot^{-1}\frac{y}{x}$$

$$\frac{\partial u}{\partial x} = 2x \tan^{-1}\frac{y}{x} + x^2 \frac{1}{1 + \frac{y^2}{x^2}}\left(\frac{-y}{x^2}\right) - \frac{y^2(-1)}{1 + \frac{y^2}{x^2}}\left(\frac{-y}{x^2}\right)$$

$$= 2x \tan^{-1}\frac{y}{x} - \frac{y(x^2 + y^2)}{(x^2 + y^2)} = 2x \tan^{-1}\frac{y}{x} - y$$

Differentiating partially w.r.t. y

$$\frac{\partial^2 u}{\partial y\,\partial x} = 2x\,\frac{1}{1+y^2/x^2}\left(\frac{1}{x}\right) - 1$$

$$= \frac{2x^2}{x^2+y^2} - 1 = \frac{x^2-y^2}{x^2+y^2}$$

Example 1.80 **If $u = (x^2+y^2+z^2)^{-\frac{1}{2}}$, prove that $\frac{\partial^2 u}{\partial x^2} + \frac{\partial^2 u}{\partial y^2} + \frac{\partial^2 u}{\partial z^2} = 0$.**

Solution :

$$\frac{\partial u}{\partial x} = \frac{-\frac{1}{2}\cdot 2x}{(x^2+y^2+z^2)^{-3/2}} = -xu^3 \qquad ...(1)$$

$$\frac{\partial u}{\partial y} = -yu^3 \qquad ...(2)$$

$$\frac{\partial u}{\partial z} = -zu^3 \qquad ...(3)$$

$$\frac{\partial^2 u}{\partial x^2} = -\left(1\,u^3 + 3xu^2\,\frac{\partial u}{\partial x}\right) = -(u^3 - 3x^2u^5) \qquad \text{Using (1)}$$

Similarly from (2) and (3)

$$\frac{\partial^2 u}{\partial y^2} = -(u^3 - 3y^2u^5)$$

$$\frac{\partial^2 u}{\partial z^2} = -(u^3 - 3z^2u^5)$$

So $$\frac{\partial^2 u}{\partial x^2} + \frac{\partial^2 u}{\partial y^2} + \frac{\partial^2 u}{\partial z^2} = -\ \{3u^3 - 3\,(x^2+y^2+z^2)\,u^5\}$$

$$= -\ \{3u^3 - 3\,u^{-2}\,u^5\} = 0. \text{ Hence Proved.}$$

Example 1.81 **If $u = \log(x^3+y^3+z^3-3xyz)$. Show that**

(i) $u_x + u_y + u_z = \frac{3}{x+y+z}$

(ii) $\left(\frac{\partial}{\partial x} + \frac{\partial}{\partial y} + \frac{\partial}{\partial z}\right)^2 u = \frac{-9}{(x+y+z)^2}$

(iii) $\frac{\partial^2 u}{\partial x^2} + \frac{\partial^2 u}{\partial y^2} + \frac{\partial^2 u}{\partial z^2} = \frac{-3}{x+y+z}$.

Solution : We know $x^3+y^3+z^3-3xyz = (x+y+z)\,(x+\omega y+\omega^2 z)\,(x+\omega^2 y+\omega z)$

where ω is the imaginary cube root of unity with $\omega^3 = 1$ and $1+\omega+\omega^2 = 0$.

So $$u = \log(x+y+z) + \log(x+\omega y+\omega^2 z) + \log(x+\omega^2 y+\omega z)$$

Differentiating

$$\frac{\partial u}{\partial x} = \frac{1}{x+y+z} + \frac{1}{x+\omega y+\omega^2 z} + \frac{1}{x+\omega^2 y+\omega z} \qquad ...(1)$$

$$\frac{\partial u}{\partial y}=\frac{1}{x+y+z}+\frac{\omega}{x+\omega y+\omega^2 z}+\frac{\omega^2}{x+\omega^2 y+\omega z} \quad ...(2)$$

$$\frac{\partial u}{\partial z}=\frac{1}{x+y+z}+\frac{\omega^2}{x+\omega y+\omega^2 z}+\frac{\omega}{x+\omega^2 y+\omega z} \quad ...(3)$$

Adding (1), (2), (3),

$$\frac{\partial u}{\partial x}+\frac{\partial u}{\partial y}+\frac{\partial u}{\partial z}=\frac{3}{x+y+z}+0+0 \text{ as } 1+\omega+\omega^2=0 \quad ...(4)$$

Let $\frac{\partial u}{\partial x}+\frac{\partial u}{\partial y}+\frac{\partial u}{\partial z}=\upsilon=\frac{3}{x+y+z}$

So $\frac{\partial \upsilon}{\partial x}=\frac{\partial \upsilon}{\partial y}=\frac{\partial \upsilon}{\partial z}=\frac{-3}{(x+y+z)^2}$ and $\left(\frac{\partial}{\partial x}+\frac{\partial}{\partial y}+\frac{\partial}{\partial z}\right)\left(\frac{\partial u}{\partial x}+\frac{\partial u}{\partial y}+\frac{\partial u}{\partial z}\right)$

$$=\left(\frac{\partial}{\partial x}+\frac{\partial}{\partial y}+\frac{\partial}{\partial z}\right)\upsilon=\frac{-3+(-3)+(-3)}{(x+y+z)^2}=\frac{-9}{(x+y+z)^2} \quad ...(5)$$

Differentiating (1), (2), (3), partially w.r.t. x, y, z, respectively.

$$\frac{\partial^2 u}{\partial x^2}=\frac{-1}{(x+y+z)^2}+\frac{-1}{(x+\omega y+\omega^2 y)^2}+\frac{-1}{(x+\omega^2 y+\omega z)^2} \quad ...(6)$$

$$\frac{\partial^2 u}{\partial y^2}=\frac{-1}{(x+y+z)^2}+\frac{-\omega\cdot\omega}{(x+\omega y+\omega^2 z)^2}+\frac{-\omega^2\cdot\omega^2}{(x+\omega^2 y+\omega z)^2} \quad ...(7)$$

$$\frac{\partial^2 u}{\partial z^2}=\frac{-1}{(x+y+z)^2}+\frac{-\omega^2\cdot\omega^2}{(x+\omega y+\omega^2 z)^2}+\frac{-\omega\cdot\omega}{(x+\omega^2 y+\omega z)^2} \quad ...(8)$$

Adding (6), (7), (8),

$$\frac{\partial^2 u}{\partial x^2}+\frac{\partial^2 u}{\partial y^2}+\frac{\partial^2 u}{\partial z^2}=\frac{-3-3-3}{(x+y+z)^2}+0+0. \text{ Hence proved}$$

Example 1.82 **If $x^x y^y z^z = c$, show that at $x = y = z$, $\frac{\partial^2 z}{\partial y\,\partial x}=-(x\log_e ex)^{-1}$.**

Solution : On taking logarithm, we have

$$x\log x+y\log y+z\log z=\log c$$

Taking z to be the dependent and x and y as independent variable

$$z\log z=\log c-x\log x-y\log y$$

So differentiating w.r.t. x

$$\left(\frac{z}{z}+\log z\right)\frac{\partial z}{\partial x}=-\left(\frac{x}{x}+\log x\right) \quad ...(1)$$

So $\left(\frac{\partial z}{\partial x}\right)_{\text{at } x=y=z}=-1$

Similarly

$$(1+\log z)\frac{\partial z}{\partial y}=-(1+\log y) \quad ...(2)$$

and $\left(\frac{\partial z}{\partial y}\right)_{x=y=z}=-1$

Differentiating (1) partially w.r.t. y.

$$(1 + \log z)\frac{\partial^2 z}{\partial y\,\partial x} + \left(0 + \frac{1}{z}\frac{\partial z}{\partial y}\right)\frac{\partial z}{\partial y} = 0$$

at $x = y = z$

$$(1 + \log x)\frac{\partial^2 z}{\partial y\,\partial x} + \frac{1}{x}(-1)(-1) = 0$$

So $$\frac{\partial^2 z}{\partial y\,\partial x} = \frac{-1}{x(\log_e e + \log_e x)} = -(x \log_e ex)^{-1}$$

Example 1.83 **If $\theta = t^n e^{-\frac{r^2}{4t}}$, what value of n will make $\frac{1}{r^2}\frac{\partial}{\partial r}\left(r^2\frac{\partial\theta}{\partial r}\right) = \frac{\partial\theta}{\partial t}$?**

Solution : Taking logarithm,

$$\log\theta = n\log t - \frac{r^2}{4t} \qquad ...(1)$$

(Taking logarithm is not necessary but it makes calculations easier).
Differentiating (1) w.r.t. r partially

$$\frac{1}{\theta}\frac{\partial\theta}{\partial r} = -\frac{2r}{4t}$$

So $$r^2\frac{\partial\theta}{\partial r} = -\frac{1}{2t}r^3\theta$$

Differentiating it partially w.r.t. r

$$\frac{\partial}{\partial r}\left(r^2\frac{\partial\theta}{\partial r}\right) = -\frac{1}{2t}\left\{3r^2\theta + r^3\frac{\partial\theta}{\partial r}\right\}$$

So $$\frac{1}{r^2}\frac{\partial}{\partial r}\left(r^2\frac{\partial\theta}{\partial r}\right) = -\frac{1}{2t}\left[3\theta + r\left(-\frac{1}{2t}r\right)\theta\right] = -\frac{1}{2t}\left(3 - \frac{r^2}{2t}\right)\theta \qquad ...(2)$$

Differentiating (1) partially w.r.t. t

$$\frac{1}{\theta}\frac{\partial\theta}{\partial t} = \left(\frac{n}{t} + \frac{r^2}{4t^2}\right) \qquad ...(3)$$

(2) and (3) are equal when $n = -\frac{3}{2}$.

Example 1.84 **If $\frac{x^2}{a^2+u} + \frac{y^2}{b^2+u} + \frac{z^2}{c^2+u} = 1$, prove $u_x^2 + u_y^2 + u_z^2 = 2(x\,u_x + y\,u_y + z\,u_z)$.**

Solution : In $\frac{\partial u}{\partial x}$, the u is regarded as the dependent and x, y, z as the independent variables.

Differentiating (1), partially w.r.t. x

$$\frac{2x}{a^2+u} + \left(-\frac{\partial u}{\partial x}\right)\left\{\frac{x^2}{(a^2+u)^2} + \frac{y^2}{(b^2+u)^2} + \frac{z^2}{(c^2+u)^2}\right\} = 0 \qquad ...(1)$$

Writing $$\frac{x^2}{(a^2+u)^2}+\frac{y^2}{(b^2+u)^2}+\frac{z^2}{(c^2+u)^2}=f \qquad ...(2)$$

From (1)

$$\frac{2x}{a^2+u}=u_x f \qquad ...(3)$$

Similarly, we may have

$$\frac{2y}{b^2+u}=u_y f \qquad ...(4)$$

and $$\frac{2z}{c^2+u}=u_z f \qquad ...(5)$$

Multiplying (3), (4), (5) by x, y, z respectively and adding

$$2\left\{\frac{x^2}{a^2+u}+\frac{y^2}{b^2+u}+\frac{z^2}{c^2+u}\right\}=2 \text{ (using given relation)}$$

$$=f(x\,u_x+y\,u_y+z\,u_z) \qquad ...(6)$$

Squaring, each of (3), (4), (5) and adding, we get

$$4\left\{\frac{x^2}{(a^2+u)^2}+\frac{y^2}{(b^2+u)^2}+\frac{z^2}{(c^2+u)^2}\right\}=4f=f^2\left\{u_x^2+u_y^2+u_z^2\right\} \qquad ...(7)$$

So $$f(u_x^2+u_y^2+u_z^2)=4 \qquad ...(8)$$

Dividing (8) by (6)

$$\frac{u_x^2+u_y^2+u_z^2}{(x\,u_x+y\,u_y+z\,u_z)}=\frac{4}{2}=2. \text{ Hence proved.}$$

Example 1.85 **If $z(x+y)=x^2+y^2$, show that $(z_x-z_y)^2=4(1-z_x-z_y)$.**

Solution : Differentiating the given relation partially, with z as the dependent variable

We have $$(x+y)\frac{\partial z}{\partial x}+z=2x \qquad ...(1)$$

and $$(x+y)\frac{\partial z}{\partial y}+z=2y \qquad ...(2)$$

so on subtraction, $(x+y)(z_x-z_y)=2(x-y)$

or $$(z_x-z_y)^2=4\left(\frac{x-y}{x+y}\right)^2 \qquad ...(3)$$

Hence $$4[1-(z_x+z_y)]=4\left[1-\frac{2(x+y)}{x+y}+\frac{2z}{x+y}\right]$$

$$=4\left[-1+2\frac{x^2+y^2}{(x+y)^2}\right]=4\left(\frac{x-y}{x+y}\right)^2 \qquad ...(4)$$

By (3) and (4), the result is proved.

Example 1.86 **If $u = f(r)$ be a function of r only, prove that**

$$\frac{\partial^2 u}{\partial x^2} + \frac{\partial^2 u}{\partial y^2} = f''(r) + \frac{1}{r} f'(r), \text{ where } r^2 = x^2 + y^2.$$

Solution : We know $r^2 = x^2 + y^2$ so $r\frac{\partial r}{\partial x} = x$, or $\frac{\partial r}{\partial x} = \frac{x}{r}, \frac{\partial r}{\partial y} = \frac{y}{r}$

Now $\frac{\partial u}{\partial x} = \frac{\partial u}{\partial r}\frac{\partial' r}{\partial x} = \frac{x}{r} f'(r)$, as $\frac{\partial u}{\partial r} = f'(r)$ because u is a function of r only. ..(1)

$$\frac{\partial u}{\partial x} = \frac{x}{r} f'(r) \text{ as } \frac{\partial u}{\partial r} = f'(r) \text{ because } u \text{ is a function of } r \text{ only.}$$

Similarly $\frac{\partial u}{\partial y} = \frac{y}{r} f'(r)$ ...(2)

Differentiating (1), partially w.r.t. x

$$\frac{\partial^2 u}{\partial x^2} = \frac{x}{r}\frac{x}{r} f''(r) + f'(r)\left\{\frac{1}{r} - \frac{x}{r^2}\frac{\partial r}{\partial x}\right\}$$

$$= \frac{x^2}{r^2} f''(r) + f'(r)\left\{\frac{1}{r} - \frac{x^2}{r^3}\right\} \quad ..(3)$$

Similarly $\frac{\partial^2 u}{\partial y^2} = \frac{y^2}{r^2} f''(r) + f'(r)\left\{\frac{1}{r} - \frac{y^2}{r^3}\right\}$ ...(4)

Adding (3) and (4)

$$\frac{\partial^2 u}{\partial x^2} + \frac{\partial^2 u}{\partial y^2} = \frac{x^2 + y^2}{r^2} f''(r) + f'(r)\left\{\frac{2}{r} - \frac{x^2 + y^2}{rr^2}\right\}$$

$$= f''(r) + \frac{1}{r} f'(r).$$

Hence proved.

Example 1.87 **If $u = lx + my, v = mx - ly$, then show that**

(i) $\left(\frac{\partial u}{\partial x}\right)_y \left(\frac{\partial x}{\partial u}\right)_\upsilon = \frac{l^2}{l^2 + m^2}$

(ii) $\left(\frac{\partial y}{\partial \upsilon}\right)_x \left(\frac{\partial \upsilon}{\partial y}\right)_u = \frac{l^2 + m^2}{l^2}.$

Solution : From $u = lx + my$, from $\upsilon = mx - ly$

$$\left(\frac{\partial u}{\partial x}\right)_y = l \quad ...(1)$$

$$\left(\frac{\partial v}{\partial y}\right)_x = m \quad ...(2)$$

From given relation we can get x, y expressed as

$$x = \frac{lu + m\upsilon}{l^2 + m^2} \quad ...(3)$$

and $$y = \frac{u - l\upsilon}{l^2 + m^2} \qquad \text{...(4)}$$

So $$\left(\frac{\partial x}{\partial u}\right)_\upsilon = \frac{l}{l^2 + m^2} \qquad \text{...(5)}$$

Using (1) and (5)

$$\left(\frac{\partial u}{\partial x}\right)_y \left(\frac{\partial x}{\partial u}\right)_\upsilon = l \cdot \frac{l}{l^2 + m^2}$$

So (i) is proved.

Writing $\upsilon = mx - ly$ as $y = \dfrac{mx - \upsilon}{l}$

$$\left(\frac{\partial y}{\partial \upsilon}\right)_x = -\frac{1}{l}$$

Writing (4),

$$y = \frac{mu - l\upsilon}{l^2 + m^2} \text{ as}$$

$$\upsilon = \frac{-(l^2 + m^2)\, y + mu}{l}$$

So $$\left(\frac{\partial \upsilon}{\partial y}\right)_u = -\frac{l^2 + m^2}{l}$$

Hence $$\left(\frac{\partial y}{\partial \upsilon}\right)_x \cdot \left(\frac{\partial \upsilon}{\partial y}\right)_u = \left(-\frac{1}{l}\right)\left(-\frac{l^2 + m^2}{l}\right) \qquad \text{...(ii)}$$

Hence (ii) is proved.

PROBLEM SET 1.9

1. If $u = \tan^{-1} \dfrac{2xy}{x^2 - y^2}$, prove that $u_{xx} + u_{yy} = 0$

2. If $u = \log(x^2 + y^2) + \tan^{-1} \dfrac{y}{x}$, prove that $u_{xx} + u_{yy} = 0$

3. If $z = x^3 + y^3 - 3axy$, verify $\dfrac{\partial^2 z}{\partial y\, \partial x} = \dfrac{\partial^2 z}{\partial x\, \partial y}$

4. If $u = y^2 \tan^{-1} \dfrac{x}{y} - x^2 \tan^{-1} \dfrac{y}{x}$, show that $\dfrac{\partial^2 u}{\partial x\, \partial y} = \dfrac{y^2 - x^2}{y^2 + x^2}$.

5. If $z = f(x + ct) + \phi(x - ct)$, show that $\dfrac{\partial^2 z}{\partial t^2} = c^2 \dfrac{\partial^2 z}{\partial x^2}$.

6. If $u = x^y$, show that $\dfrac{\partial^3 u}{\partial x^2\, \partial y} = \dfrac{\partial^3 u}{\partial x\, \partial y\, \partial z}$, each equal in
$$\{x^{y-2}(2y - 1) + y(y - 1)\, x^{y-2} \log x\}.$$

7. If $u = e^{x - at} \cos (x - at)$, show that $a^2 \dfrac{\partial^2 u}{\partial x^2} = \dfrac{\partial^2 u}{\partial t^2}$.

8. If $u = e^x (x \cos y - y \sin y)$, show that $u_{xx} + u_{yy} = 0$.

9. If $u = \tan^{-1} \dfrac{xy}{\sqrt{1 + x^2 + y^2}}$, show that $\dfrac{\partial^2 u}{\partial x \, \partial y} = (1 + x^2 + y^2)^{-3/2}$.

10. If $u = \log (\tan x + \tan y)$, prove that $\sin 2x \, u_x + \sin 2y \, u_y = 2$.

11. If $z = xf (x + y) + yg (x + y)$, show that $z_{xx} - 2 z_{xy} + z_{yy} = 0$.

12. If $z = f (x + ay) + \phi (x - ay)$. Show that $z_{yy} = a^2 z_{xx}$.

13. If $x^2 + y^2 + z^2 = \dfrac{1}{u^2}$, show that $u_{xx} + u_{yy} + u_{zz} = 0$.

14. If $x = r \cos \theta, y = r \sin \theta$, show that

(i) $\dfrac{\partial^2 r}{\partial x^2} + \dfrac{\partial^2 r}{\partial y^2} = \dfrac{1}{r} \left\{ \left(\dfrac{\partial r}{\partial x} \right)^2 + \left(\dfrac{\partial r}{\partial y} \right)^2 \right\}$

(ii) $\dfrac{\partial^2 r}{\partial x^2} \dfrac{\partial^2 r}{\partial y^2} = \left| \dfrac{\partial^2 r}{\partial x \, \partial y} \right|^2$.

15. If $z = \tan (y + ax) - (y - ax)^{3/2}$ show that $\dfrac{\partial^2 z}{\partial x^2} = a^2 \dfrac{\partial^2 z}{\partial y^2}$

16. If $\upsilon = \dfrac{1}{\sqrt{t}} e^{-\frac{x^2}{4a^2 t}}$, show that $\dfrac{\partial \upsilon}{\partial t} = a^2 \dfrac{\partial^2 \upsilon}{\partial x^2}$.

17. If $\upsilon = \log \sqrt{x^2 + y^2 + z^2}$, show that $(x^2 + y^2 + z^2) (\upsilon_{xx} + \upsilon_{yy} + \upsilon_{zz}) = 1$.

18. If $f(x, y) = (1 - 2xy + y^2)^{-1/2}$, show that $\dfrac{\partial}{\partial x} \left\{ (1 - x^2) \dfrac{\partial f}{\partial x} \right\} + \dfrac{\partial}{\partial y} \left\{ y^2 \dfrac{\partial f}{\partial y} \right\} = 0$.

If $z = e^{ax + by} f(ax - by)$, show $b \, z_x + a \, z_y = 2abz$.

19. If $V = r^n (3 \cos^2 \theta - 1)$, show that $\dfrac{\partial}{\partial r} \left(r^2 \dfrac{\partial V}{\partial r} \right) + \dfrac{1}{\sin \theta} \dfrac{\partial}{\partial \theta} \left(\sin \theta \dfrac{\partial V}{\partial \theta} \right) = 0$,

when $n = 2$ or -3.

20. If $V = r^m$, where $r^2 = x^2 + y^2 + z^2$, show that $V_{xx} + V_{yy} + V_{zz} = m (m + 1) r^{m - 2}$.

21. If $u = (1 - 2xy + y^2)^{-1/2}$, show that $x \dfrac{\partial u}{\partial x} - y \dfrac{\partial u}{\partial y} = y^2 u^3$.

1.31 HOMOGENEOUS FUNCTION

Let $u = f(x, y, z)$ be a function of independent variables x, y, z. It is called a homogeneous function of degree n, if **the sum of powers of x, y, z, in each term is same (say) n.** It can then be expressed as $x^n \phi\left(\frac{y}{x}, \frac{z}{x}\right)$.

Thus $u(x, y) = a_0 x^3 + a_1 x^2 y + a_2 xy^2 + a_3 y^3$ is a homogeneous function of degree 3 and can be expressed as $= x^3 \left\{a_0 + a_1 \frac{y}{x} + a_2 \left(\frac{y}{x}\right)^2 + a_3 \left(\frac{y}{x}\right)^3\right\} = x^3 f\left(\frac{y}{x}\right)$

If $u = \frac{x + y}{\sqrt{x} + \sqrt{y}}$ then $u = x^{\frac{1}{2}} \left(\frac{1 + \frac{y}{x}}{1 + \sqrt{\frac{y}{x}}}\right)$ is a homogeneous function of degree $\frac{1}{2}$.

$z = \sin\left(\frac{y}{x} + \frac{x}{y}\right) = x^0 \sin\left\{\frac{y}{x} + \left(\frac{y}{x}\right)^{-1}\right\}$ is a homogeneous function of degree zero.

1.32 EULER'S THEOREM ON HOMOGENEOUS FUNCTIONS

Swiss mathematician Leonhard Euler (1707-1783) gave the theorem which states, "If $u(x, y)$ be a homogeneous function of degree n in x and y, then

$$x \frac{\partial u}{\partial x} + y \frac{\partial u}{\partial y} = nu.$$

Proof : The proof is very simple. Calculating the terms on L.H.S. and adding, the theorem is proved.

As u is a homogeneous function of degree n, so u can be expressed as

$$u = x^n f\left(\frac{y}{x}\right) \quad ...(1)$$

Differentiating u, partially w.r.t. x and y

$$\frac{\partial u}{\partial x} = n x^{n-1} f\left(\frac{y}{x}\right) - \frac{y}{x^2} x^n f'\left(\frac{y}{x}\right) \quad ...(2)$$

and
$$\frac{\partial u}{\partial y} = 0 + \frac{1}{x} x^n f'\left(\frac{y}{x}\right) \quad ...(3)$$

Multiplying (2) by x, (3) by y and adding, we get

$$\boldsymbol{x \frac{\partial u}{\partial x} + y \frac{\partial u}{\partial y}} = n x^{1+n-1} f'\left(\frac{y}{x}\right) + 0 = \boldsymbol{nu} \quad ...(4)$$

Hence proved.

1.33 RESULTS FROM EULER'S THEOREM

If u be a homogeneous function of degree n, then

$$x\frac{\partial^2 u}{\partial x^2}+y\frac{\partial^2 u}{\partial y\,\partial x}=(n-1)\frac{\partial u}{\partial x} \quad \text{...(1)}$$

$$x\frac{\partial^2 u}{\partial y\,\partial x}+y\frac{\partial^2 u}{\partial y^2}=(n-1)\frac{\partial u}{\partial y} \quad \text{...(2)}$$

$$x^2\frac{\partial^2 u}{\partial x^2}+2xy\frac{\partial^2 u}{\partial x\,\partial y}+y^2\frac{\partial^2 u}{\partial y^2}=n(n-1)\,u \quad \text{...(3)}$$

Proof : By Euler's theorem

$$x\frac{\partial u}{\partial x}+y\frac{\partial u}{\partial y}=nu \quad \text{...(4)}$$

Differentiating (4) partially w.r.t. x

$$x\frac{\partial^2 u}{\partial x^2}+\frac{\partial u}{\partial x}+y\frac{\partial^2 u}{\partial x\,\partial y}=n\frac{\partial u}{\partial x}$$

So
$$x\frac{\partial^2 u}{\partial x^2}+y\frac{\partial^2 u}{\partial y\,\partial x}=(n-1)\frac{\partial u}{\partial x} \quad \text{...(5)}$$

So (1) is Proved.

Differentiating (5), partially w.r.t. y,

$$x\frac{\partial^2 u}{\partial y\,\partial x}+y\frac{\partial^2 u}{\partial y^2}+\frac{\partial u}{\partial y}=n\frac{\partial u}{\partial y}$$

So
$$x\frac{\partial^2 u}{\partial y\,\partial x}+y\frac{\partial^2 u}{\partial y^2}=(n-1)\frac{\partial u}{\partial y} \quad \text{...(6)}$$

Multiplying (5) by x and (6) by y and adding, we get

$$x^2\frac{\partial^2 u}{\partial x^2}+2xy\frac{\partial^2 u}{\partial y\,\partial x}+y^2\frac{\partial^2 u}{\partial y^2}=(n-1)\left\{x\frac{\partial u}{\partial x}+y\frac{\partial u}{\partial y}\right\}$$

$$=(n-1)\cdot nu$$

So (3) is proved.

Example 1.88 Verify the Euler's theorem for

$$\boldsymbol{u=\sin^{-1}\frac{x}{y}+\tan^{-1}\frac{y}{x}}.$$

Solution : It is given that

$$u=\sin^{-1}\frac{x}{y}+\tan^{-1}\frac{y}{x} \quad \text{...(1)}$$

$\therefore$
$$=\operatorname{cosec}^{-1}\frac{y}{x}+\tan^{-1}\frac{y}{n}=x^{\circ}f\left(\frac{y}{x}\right)$$

$\therefore$ $u(x,y)$, u is a homogeneous function of degree 0 in x and y.

$\therefore$ By Euler's theorem, we can write

$$x\frac{\partial u}{\partial x}+y\frac{\partial u}{\partial y}=0\times u=0 \quad \text{...(2)}$$

Now $$\frac{\partial u}{\partial x} = \frac{1}{\sqrt{1-\frac{x^2}{y^2}}} \cdot \frac{1}{y} + \frac{1}{1+\frac{y^2}{x^2}}\left(\frac{-y}{x^2}\right)$$

$$= \frac{1}{\sqrt{y^2 - x^2}} - \frac{y}{(x^2+y^2)}$$

$$x\frac{\partial u}{\partial x} = \frac{x}{\sqrt{y^2-x^2}} - \frac{xy}{x^2+y^2}$$

$$\frac{\partial u}{\partial y} = \frac{1}{\sqrt{1-\frac{x^2}{y^2}}}\left(-\frac{x}{y^2}\right) + \frac{1}{\left(1+\frac{y^2}{x^2}\right)} \cdot \frac{1}{x}$$

$$= -\frac{x}{y\sqrt{y^2-x^2}} + \frac{x}{x^2+y^2}$$

$$y\frac{\partial u}{\partial y} = -\frac{x}{\sqrt{y^2-x^2}} + \frac{xy}{x^2+y^2}$$

So $$x\frac{\partial u}{\partial x} + y\frac{\partial u}{\partial y} = 0$$

Hence, Euler's theorem is verified.

Example 1.89 **(i) If $u = (x^{1/4} + y^{1/4})(x^{1/5} + y^{1/5})$, apply Euler's theorem to find the value of $x\frac{\partial u}{\partial x} + y\frac{\partial u}{\partial y}$**

(ii) If $z = x^4y^2 \sin^{-1}\left(\frac{x}{y}\right) + \log x - \log y$ show that

$$x\frac{\partial z}{\partial x} + y\frac{\partial z}{\partial y} = 6x^4y^2 \sin^{-1}\left(\frac{x}{y}\right)$$

Solution : (i) It is given that

$$u(x, y) = (x^{1/4} + y^{1/4})(x^{1/5} + y^{1/5})$$

$$u(\lambda x, \lambda y) = \lambda^{1/4}(x^{1/4} + y^{1/4})\,\lambda^{1/5}(x^{1/5} + y^{1/5})$$

$$= \lambda^{9/20}(x^{1/4} + y^{1/4})(x^{1/5} + y^{1/5}) = \lambda^{9/20}\,u(x, y)$$

$\therefore$ $u(x, y)$ is a homogeneous function of degree $\frac{9}{20}$.

Hence by Euler's theorem

$$x\frac{\partial u}{\partial x} + y\frac{\partial u}{\partial y} = \frac{9}{20}u$$

(ii) For given function $z = x^4y^2 \sin^{-1}\left(\frac{x}{y}\right) + \log x - \log y$

Let $u(x, y) = x^4y^2 \sin^{-1}\left(\frac{x}{y}\right)$ and

$$v(x, y) = \log x - \log y = \log\left(\frac{x}{y}\right)$$

It is clear that $u(x, y)$ is a homogeneous function of degree 6 and $v(x, y)$ of degree 0.

$\therefore$ By Euler's theorem

$$x\frac{\partial u}{\partial x} + y\frac{\partial u}{\partial y} = 6u \quad \text{...(1)}$$

and
$$x\frac{\partial v}{\partial x} + y\frac{\partial v}{\partial y} = 0 \quad \text{...(2)}$$

Adding (1) and (2), we get

$$x\frac{\partial u}{\partial x} + x\frac{\partial v}{\partial x} + y\frac{\partial u}{\partial y} + y\frac{\partial v}{\partial y}$$

$$\Rightarrow \quad x\frac{\partial z}{\partial x} + y\frac{\partial z}{\partial y} = 6x^4y^2 \sin^{-1}\left(\frac{x}{y}\right)$$

Example 1.90 **Prove Euler's theorem, when $u = u\,(x, y, z)$ is a homogeneous function of three independent variables and is of degree n.**

Proof. u is a homogeneous function of degree n of three variables, so we may write u as function of degree n of three independent variable x, y, z

$$u = x^n f\left(\frac{y}{x}, \frac{z}{x}\right) \quad \text{...(1)}$$

Let
$$\frac{y}{x} = \xi, \quad \frac{z}{x} = \eta \quad \text{...(2)}$$

So
$$u = x^n f(\xi, n) \quad \text{...(3)}$$

So
$$\frac{\partial u}{\partial x} = n\,x^{n-1} f + x^n \left\{\frac{\partial f}{\partial \xi}\frac{\partial \xi}{\partial x} + \frac{\partial f}{\partial \eta}\frac{\partial \eta}{\partial x}\right\}$$

$$= n\,x^{n-1} f + x^n \left\{-\frac{y}{x^2}\frac{\partial f}{\partial \xi} - \frac{z}{x^2}\frac{\partial f}{\partial \eta}\right\} \quad \text{...(4)}$$

$$\frac{\partial u}{\partial y} = 0 + x^n \left\{\frac{1}{x}\frac{\partial f}{\partial \xi} + 0 \text{ as } \eta \text{ does not involve } y\right\} \quad \text{...(5)}$$

$$\frac{\partial u}{\partial z} = 0 + x^n \left\{0 + \frac{1}{x}\frac{\partial f}{\partial \eta}\right\} \quad \text{...(6)}$$

Multiplying (4) by x, (5) by y and (6) by z and adding, we get

$$x\frac{\partial u}{\partial x} + y\frac{\partial u}{\partial y} + z\frac{\partial u}{\partial z} = n\,x^n f + 0 \cdot \frac{\partial f}{\partial \xi} + 0\,\frac{\partial f}{\partial \eta}$$

$$= nu.$$

Example 1.91 **If $u = \tan^{-1}\dfrac{x^3 + y^3}{x - y}$, prove that $x\dfrac{\partial u}{\partial x} + y\dfrac{\partial u}{\partial y} = \sin 2u$.**

Solution :
$$\tan u = x^2 \left\{\frac{1 + (y/x)^3}{1 - y/x}\right\} = x^2 f\left(\frac{y}{x}\right)$$

So $\tan u$ is a homogeneous function of degree $n = 2$, using Euler's theorem

$$x \frac{\partial \tan u}{\partial x} + y \frac{\partial \tan u}{\partial y} = n \tan u = 2 \tan u$$

or
$$\sec^2 u \left(x \frac{\partial u}{\partial x} + y \frac{\partial u}{\partial y} \right) = 2 \frac{\sin u}{\cos u}$$

Hence
$$x \frac{\partial u}{\partial x} + y \frac{\partial u}{\partial y} = 2 \sin u \cos u = \sin 2u.$$

Example 1.92 **Verify Euler's Theorem for $u = \tan^{-1} \dfrac{x+y}{\sqrt{x} + \sqrt{y}}$.**

Solution :
$$\tan u = x^{1/2} \left(\frac{1 + y/x}{1 + \sqrt{y/x}} \right) = x^{1/2} f\left(\frac{y}{x}\right)$$

So Euler's Theorem gives

$$\sec^2 u \left(x \frac{\partial u}{\partial x} + y \frac{\partial u}{\partial y} \right) = \frac{1}{2} \tan u.$$

or
$$x \frac{\partial u}{\partial x} + y \frac{\partial u}{\partial y} = \frac{1}{2} \sin u \cos u \qquad ...(1)$$

Now we are to obtain this result by actual calculation on L.H.S. from

$$\tan u = \frac{x+y}{\sqrt{x} + \sqrt{y}} = \frac{\text{Numerator}}{\text{Denominator}} = \frac{N}{D} \text{ (say)}$$

$$\sec^2 u \frac{\partial u}{\partial x} = \frac{1}{D^2} \left\{ (\sqrt{x} + \sqrt{y})\, 1 - (x+y) \frac{1}{2\sqrt{x}} \right\} \qquad ...(2)$$

and
$$\sec^2 u \frac{\partial u}{\partial y} = \frac{1}{D^2} \left\{ \sqrt{x} + \sqrt{y} - (x+y) \frac{1}{2\sqrt{y}} \right\} \qquad ...(3)$$

Multiplying (2) by x and (3) by y and adding

$$\sec^2 u \left(x \frac{\partial u}{\partial x} + y \frac{\partial u}{\partial y} \right) = \frac{1}{D^2} \left\{ (\sqrt{x} + \sqrt{y}) \left(x + y - \frac{(x+y)}{2} (\sqrt{x} + \sqrt{y}) \right\} \right.$$

$$= \left(1 - \frac{1}{2} \right) \frac{ND}{DD} = \frac{1}{2} \frac{N}{D} = \frac{1}{2} \tan u$$

Hence
$$x \frac{\partial u}{\partial x} + y \frac{\partial u}{\partial y} = \frac{1}{2} \sin u \cos u.$$

This is same as (1). So the theorem is verified to be true.

Example 1.93 **If $z = x f\left(\dfrac{y}{x}\right) + g\left(\dfrac{y}{x}\right)$, show that $x^2 \dfrac{\partial^2 z}{\partial x^2} + 2xy \dfrac{\partial^2 z}{\partial x \partial y} + y^2 \dfrac{\partial^2 z}{\partial y^2} = 0.$**

Solution : Let
$$z = u + v \qquad ...(1)$$

where $u = xf\left(\dfrac{y}{x}\right)$ is a homogeneous function of degree 1 and $v = g\left(\dfrac{y}{x}\right)$ is of degree zero.

So by Euler's theorem

So $$x\frac{\partial u}{\partial x}+y\frac{\partial u}{\partial y}=1\,.\,u \quad ...(2)$$

From (2)

$$x\frac{\partial}{\partial x}\left\{x\frac{\partial u}{\partial x}+y\frac{\partial u}{\partial y}\right\}+y\frac{\partial}{\partial y}\left\{x\frac{\partial u}{\partial x}+y\frac{\partial u}{\partial y}\right\}=x\frac{\partial u}{\partial x}+y\frac{\partial u}{\partial y}$$

or $$x^2\frac{\partial^2 u}{\partial x^2}+x\frac{\partial u}{\partial x}+2xy\frac{\partial^2 u}{\partial y\,\partial x}+y^2\frac{\partial^2 u}{\partial y^2}+y\frac{\partial u}{\partial y}=x\frac{\partial u}{\partial x}+y\frac{\partial u}{\partial y}$$

so $$x^2\frac{\partial^2 u}{\partial x^2}+2xy\frac{\partial^2 u}{\partial y\,\partial x}+y^2\frac{\partial^2 u}{\partial y^2}=0 \quad ...(3)$$

Also as υ is homogeneous and 0, so,

$$x\frac{\partial \upsilon}{\partial x}+y\frac{\partial \upsilon}{\partial y}=0\,\upsilon \quad ...(4)$$

so $$x\frac{\partial}{\partial x}\left(x\frac{\partial \upsilon}{\partial x}+y\frac{\partial \upsilon}{\partial y}\right)+y\frac{\partial}{\partial y}\left(x\frac{\partial \upsilon}{\partial x}+y\frac{\partial \upsilon}{\partial y}\right)=x\times 0+y\times 0=0$$

or $$x^2\frac{\partial^2 \upsilon}{\partial x^2}+2xy\frac{\partial^2 \upsilon}{\partial y\,\partial x}+y^2\frac{\partial^2 \upsilon}{\partial y^2}=-\left(x\frac{\partial \upsilon}{\partial x}+y\frac{\partial \upsilon}{\partial y}\right)=0 \quad ...(5)$$

Adding (3) and (5)

$$\left(x^2\frac{\partial^2}{\partial x^2}+2xy\frac{\partial^2}{\partial x\,\partial y}+y^2\frac{\partial^2}{\partial y^2}\right)(u+\upsilon)=0 \quad ...(6)$$

As $u+\upsilon=z$. So the result is proved.

Example 1.94 **If $u=\sin^{-1}\dfrac{x+2y+3z}{x^8+y^8+z^8}$, find the value of $x\,u_x+y\,u_y+z\,u_z$.**

Solution : From given relation, we have

$$\sin u=x^{-7}\left\{\frac{1+2\dfrac{y}{x}+3\dfrac{z}{x}}{1+\left(\dfrac{y}{x}\right)^8+\left(\dfrac{z}{x}\right)^8}\right\}$$

So $\sin u$ is a homogeneous function of degree –7.

So By Euler's theorem

$$\left(x\frac{\partial}{\partial x}+y\frac{\partial}{\partial y}+z\frac{\partial}{\partial z}\right)\sin u=(-7)\sin u$$

or $$x\frac{\partial u}{\partial x}+y\frac{\partial u}{\partial y}+z\frac{\partial u}{\partial z}=-7\frac{\sin u}{\cos u}=-7\tan u.$$

Example 1.95 **If $z = x^n f_1\left(\frac{y}{x}\right) + y^{-n} f_2\left(\frac{x}{y}\right)$, prove that**

$$x^2 z_{xx} + 2xy + y^2 z_{yy} + x z_x + y z_y = n^2 z.$$

Solution : Let $x^n f_1\left(\frac{y}{x}\right) = u$ and $x^n \left\{\left(\frac{x}{y}\right)^n f_2\left(\frac{x}{y}\right)\right\} = \upsilon$

and u and υ are both homogeneous functions of degree n in x, and y, $z = u + \upsilon$ is therefore, a homogeneous function of degree n. So by Euler's theorem

$$x \frac{\partial z}{\partial x} + y \frac{\partial z}{\partial y} = nu$$

and
$$x \frac{\partial}{\partial x}\left\{x \frac{\partial z}{\partial x} + y \frac{\partial z}{\partial y}\right\} + y \frac{\partial}{\partial y}\left\{x \frac{\partial z}{\partial x} + y \frac{\partial z}{\partial y}\right\} = n\left\{x \frac{\partial u}{\partial x} + y \frac{\partial u}{\partial y}\right\},$$

or
$$x^2 \frac{\partial^2 z}{\partial x^2} + x \frac{\partial z}{\partial x} + 2xy \frac{\partial^2 z}{\partial x\, \partial y} + y^2 \frac{\partial^2 z}{\partial y^2} + y \frac{\partial z}{\partial y} = n \cdot nu$$

Hence proved.

Example 1.96 **If $u = \operatorname{cosec}^{-1}\left\{\frac{x^{\frac{1}{2}} + y^{\frac{1}{2}}}{x^{1/3} + y^{1/3}}\right\}^{\frac{1}{2}}$, prove that**

$$x^2 \frac{\partial^2 u}{\partial x^2} + 2xy \frac{\partial^2 u}{\partial x\, \partial y} + y^2 \frac{\partial^2 u}{\partial y^2} = \frac{\tan u}{144}(13 + \tan^2 u)$$

Solution : We have
$$\operatorname{cosec}^2 u = \frac{x^{\frac{1}{2}} + y^{\frac{1}{2}}}{x^{1/3} + y^{1/3}}$$

So
$$\sin^2 u = \frac{x^{1/3} + y^{1/3}}{x^{1/2} + y^{1/2}} = x^{-\frac{1}{6}} f\left(\frac{y}{x}\right).$$

Thus $\sin^2 u$ is a homogeneous function of x and y of degree $-1/6$ so by Euler's Theorem

$$\left(x \frac{\partial}{\partial x} + y \frac{\partial}{\partial y}\right) \sin^2 u = -\frac{1}{6} \sin^2 u$$

or
$$2 \sin u \cos u \left\{x \frac{\partial u}{\partial x} + y \frac{\partial u}{\partial y}\right\} = -\frac{1}{6} \sin^2 u$$

or
$$x \frac{\partial u}{\partial x} + y \frac{\partial u}{\partial y} = -\frac{1}{12} \tan u$$

so
$$\left(x \frac{\partial}{\partial x} + y \frac{\partial}{\partial y}\right)\left(x \frac{\partial u}{\partial x} + y \frac{\partial u}{\partial y}\right) = -\frac{1}{12}\left(x \frac{\partial}{\partial x} + y \frac{\partial}{\partial y}\right) \tan u$$

or $$x^2 \frac{\partial^2 u}{\partial x^2} + 2xy \frac{\partial^2 u}{\partial x\, \partial y} + \frac{\partial^2 u}{\partial y^2} + \left(x \frac{\partial u}{\partial x} + y \frac{\partial u}{\partial y}\right) = -\frac{1}{12} \sec^2 u \cdot \left(-\frac{1}{12} \tan u\right)$$

So $$\Sigma x^2 \frac{\partial^2 u}{\partial x^2} = \frac{1}{144} \sec^2 u \tan u - \left(-\frac{1}{12} \tan u\right)$$

$$= \frac{1}{144} \tan u \,|1 + \tan^2 u + 12| = \text{R.H.S.}$$

PROBLEM SET 1.10

1. If $z = (x + y)\, \phi\left(\frac{y}{x}\right)$, where $\phi(x, y)$ is an arbitrary function, show that $x \frac{\partial z}{\partial x} + y \frac{\partial z}{\partial y} = z$.

2. Show that $x \frac{\partial u}{\partial x} + y \frac{\partial u}{\partial y} = 2u \log u$, where $\log u = \frac{x^3 + y^3}{3x + 4y}$.

3. Verify Euler's theorem for the functions :

(i) $f(x, y) = 3x^2yz + 5xy^2z + 4z^4$

(ii) $f(x, y) = ax^2 + 2hxy + by^2$

(iii) $u = (x^{1/2} + y^{1/2})\, (x^n + y^n)$

(iv) $u = \frac{x^2\, (x^2 - y^2)^3}{(x^2 + y^2)^3}$

4. If $u = \tan^{-1} \frac{x^3 + y^3}{x - y}$ prove that $x^2 \frac{\partial^2 u}{\partial x^2} + 2xy \frac{\partial^2 u}{\partial x\, \partial y} + y^2 \frac{\partial^2 u}{\partial y^2} = \sin 4u - \sin 2u$.

Hint $\left[x \frac{\partial u}{\partial x} + y \frac{\partial u}{\partial y} = \frac{2 \tan u}{\sec^2 u} = \sin 2u\right.$

$$\left(x \frac{\partial}{\partial x} + y \frac{\partial}{\partial y}\right)\left\{x \frac{\partial u}{\partial x} + y \frac{\partial u}{\partial y}\right\} = \left\{\Sigma x^2 \frac{\partial^2 u}{\partial x^2}\right\} + \sin 2u$$

$$= \left(x \frac{\partial}{\partial x} + y \frac{\partial}{\partial y}\right) \sin 2u = 2 \cos 2u \cdot \sin 2u.$$

5. If $u = \log\left(\frac{x^4 + y^4}{x + y}\right)$, show that $x \frac{\partial u}{\partial x} + y \frac{\partial u}{\partial y} = 3$.

6. If $u = \sin^{-1} \frac{x + 2y + 3z}{\sqrt{x^8 + y^8 + z^8}}$, show that $\Sigma \left(x \frac{\partial u}{\partial x}\right) = -3 \tan u$.

7. If $u = \sin^{-1} \frac{x}{y} + \tan^{-1} \frac{y}{x}$, then find $x \frac{\partial u}{\partial x} + y \frac{\partial u}{\partial y}$.

Ans. 0.

8. If $u = \sin^{-1} \frac{x + y}{\sqrt{x} + \sqrt{y}}$, prove that $x^2 \frac{\partial^2 u}{\partial x^2} + 2xy \frac{\partial^2 u}{\partial x\, \partial y} + y^2 \frac{\partial^2 u}{\partial y^2} = \frac{-\sin u \cos 2u}{4 \cos^3 u}$

9. If $u = \cos^{-1} \dfrac{x + y}{\sqrt{x} + \sqrt{y}}$, show that $x \dfrac{\partial u}{\partial x} + y \dfrac{\partial u}{\partial y} = -\dfrac{1}{2} \cos u$.

10. If $u = \sin^{-1} \dfrac{x^2 + y^2}{x + y}$, prove that $x \dfrac{\partial u}{\partial x} + y \dfrac{\partial u}{\partial y} = \tan u$.

11. If $\sin u = \dfrac{x^2 y^2}{x + y}$, show that $x \dfrac{\partial u}{\partial x} + y \dfrac{\partial u}{\partial y} = 3 \tan u$.

12. If $e^u = \dfrac{x^5 + y^5 + z^5}{x^2 + y^2 + z^2}$, show that $x\, u_x + y\, u_y + z\, u_z = 3u$, also show that

$$x\, u_x + y\, u_y + z\, u_z = 2 \tan u$$

$$\sin u = \frac{x^3 + y^3 + z^3}{ax + by + cz}.$$

13. If $u = \tan^{-1} \dfrac{y^2}{x}$, prove that $x^2 u_{xx} + 2xy\, u_{xy} + y^2 u_{yy} = -\sin^2 u \sin 2u$.

14. If $f(x, y) = \dfrac{1}{x^2} + \dfrac{1}{xy} + \dfrac{\log x - \log y}{x^2 + y^2}$, prove that $x \dfrac{\partial f}{\partial x} + y \dfrac{\partial f}{\partial y} + 2f = 0$.

15. If $u = f\left(\dfrac{y}{x}\right) + \sqrt{x^2 + y^2}$, find the value of $x^2 \dfrac{\partial^2 u}{\partial x^2} + 2xy \dfrac{\partial^2 u}{\partial x\, \partial y} + y^2 \dfrac{\partial^2 u}{\partial y^2}$.

[**Ans.** = 0]

16. If $f(x, y) = \dfrac{1}{x^2} + \dfrac{1}{xy} + \dfrac{\log x - \log y}{x^2 + y^2}$, show that $x \dfrac{\partial u}{\partial x} + y \dfrac{\partial u}{\partial y} + 2f(x, y) = 0$.

17. If $u = \log\left(\dfrac{x^5 + y^5 + z^5}{x + y + z}\right)$, show tht $x \dfrac{\partial u}{\partial x} + y \dfrac{\partial u}{\partial y} + z \dfrac{\partial u}{\partial z} = 3$.

18. If $u = (x^2 + y^2)^{1/3}$, show that $x^2 \dfrac{\partial^2 u}{\partial x^2} + 2xy \dfrac{\partial^2 u}{\partial x \partial y} + y^2 \dfrac{\partial^2 u}{\partial y^2} = -\dfrac{2u}{9}$.

1.34 TOTAL DIFFERENTIAL

Let $u = u(x, y)$ be a function of two independent variables x and y.

Then $\quad u + \delta u = u(x + \delta x, y + \delta y)$

So $\quad \delta u = u(x + \delta x, y + \delta y) - u(x, y)$, adding and subtracting $u(x, y + \delta y)$

$$= \left\{\frac{u(x + \delta x, y + \delta y) - u(x, y + \delta y)}{\delta x}\right\} \delta x + \left\{\frac{u(x, y + \delta y) - u(x, y)}{\delta y}\right\} \delta y$$

Taking limit as $\delta x, \delta y \to 0$ and denoting the limits of $\delta u, \delta x, \delta y$ by du, dx, dy respectively, So Total Differential of $u = du$ is given by

$$\boldsymbol{du = \frac{\partial u}{\partial x}\, dx + \frac{\partial u}{\partial y}\, dy} \qquad \ldots(1)$$

If u be a function of single variable t, then it gives

$$\frac{du}{dt} = \frac{\partial u}{\partial x}\frac{\partial x}{\partial t} + \frac{\partial u}{\partial y}\frac{dy}{dt} \quad ...(2)$$

If u be a function of two variables t_1 and t_2 then from (1)

$$\frac{\partial u}{\partial t_1} = \frac{\partial u}{\partial x}\frac{\partial x}{\partial t_1} + \frac{\partial u}{\partial y}\frac{\partial y}{\partial t_1} \quad ...(3)$$

and

$$\frac{\partial u}{\partial t_2} = \frac{\partial u}{\partial x}\frac{\partial x}{\partial t_2} + \frac{\partial u}{\partial y}\frac{\partial y}{\partial t_2} \quad ...(4)$$

1.35 DERIVATIVE FOR AN IMPLICIT FUNCTION

Let $f(x, y) = 0$ be an implicit function. By total differential formula

$$0 = df = \frac{\partial f}{\partial x}dx + \frac{\partial f}{\partial y}dy.$$

So

$$\frac{dy}{dx} = -\frac{\frac{\partial f}{\partial x}}{\frac{\partial f}{\partial y}}.$$

Example. If $ax^2 + 2hxy + by^2 + 2gx + 2fy + c = f(x, y) = 0$

then $\frac{\partial f}{\partial x} = 2\,(ax + hy + g), \frac{\partial f}{\partial y} = 2\,(hx + by + f)$

and $\frac{dy}{dx} = -\frac{f_x}{f_y} = -\left(\frac{ax + hy + g}{hx + by + f}\right).$

1.36 THE SECOND ORDER DERIVATIVE d^2y/dx^2 FOR AN IMPLICIT FUNCTION $f(x, y) = 0$

Solution : We have

$$0 = f(x, y) \quad ...(1)$$

So by total differential formula

$$0 = df = \frac{\partial f}{\partial x}dx + \frac{\partial f}{\partial y}dy$$

Hence

$$\frac{dy}{dx} = -\frac{f_x}{f_y} \quad ...(2)$$

Applying total differential formula to (2)

$$d\left(\frac{dy}{dx}\right) = \frac{\partial}{\partial x}\left(-\frac{f_x}{f_y}\right)dx + \frac{\partial}{\partial y}\left(-\frac{f_x}{f_y}\right)dy$$

so

$$\frac{d^2y}{dx^2} = \frac{\partial}{\partial x}\left(-\frac{f_x}{f_y}\right)\frac{dx}{dx} + \frac{\partial}{\partial y}\left(-\frac{f_x}{f_y}\right)\frac{dy}{dx}$$

$$= \frac{1}{f_y^2}\left[(-f_y f_{xx} + f_x f_{xy}) + (-f_y f_{yx} + f_x f_{yy})\left(-\frac{f_x}{f_y}\right)\right]$$

$$= \frac{1}{f_y^3}[-f_y^2 f_{xx} + f_x f_y f_{xy} + f_x f_y f_{yx} - f_x^2 f_{yy}]$$

$$= -\frac{1}{q^3}[q^2 r - 2pqs + p^2 t]$$

where $f_x = p, f_y = q, f_{xx} = r, f_{xy} = s, f_{yy} = t.$

1.37 CHANGE OF VARIABLE

Example 1.97 **If $u = f(y-z, z-x, x-y)$, prove that $\frac{\partial u}{\partial x} + \frac{\partial u}{\partial y} + \frac{\partial u}{\partial z} = 0.$**

Solution : Let $X = y - z, Y = z - x, Z = x - y,$

So that
$$\frac{\partial X}{\partial x} = 0, \frac{\partial X}{\partial y} = 1, \frac{\partial X}{\partial z} = -1;$$

$$\frac{\partial Y}{\partial x} = -1, \frac{\partial Y}{\partial y} = 0, \frac{\partial Y}{\partial z} = 1;$$

$$\frac{\partial Z}{\partial x} = 1, \frac{\partial Z}{\partial y} = -1, \frac{\partial Z}{\partial z} = 0.$$

Now
$$\frac{\partial u}{\partial x} = \frac{\partial X}{\partial x}\frac{\partial u}{\partial X} + \frac{\partial Y}{\partial x}\frac{\partial u}{\partial Y} + \frac{\partial Z}{\partial x}\frac{\partial u}{\partial Z}$$

$$= 0\frac{\partial u}{\partial X} + (-1)\frac{\partial u}{\partial Y} + (1)\frac{\partial u}{\partial Z}$$

Similarly
$$\frac{\partial u}{\partial y} = 1\frac{\partial u}{\partial X} + 0 + -1\frac{\partial u}{\partial Z}$$

$$\frac{\partial u}{\partial z} = -1\frac{\partial u}{\partial X} + 1\frac{\partial u}{\partial Y} + 0$$

Adding
$$\frac{\partial u}{\partial x} + \frac{\partial u}{\partial y} + \frac{\partial u}{\partial z} = 0 + 0 + 0 = 0$$

Example 1.98 **If $z = f(x, y)$ and $x = e^u \cos \upsilon, y = e^u \sin \upsilon$, show that**

$$e^{2u}\frac{\partial z}{\partial y} = x\frac{\partial z}{\partial \upsilon} + y\frac{\partial z}{\partial u}$$

Solution : We have $z = f(z, y)$ also $= f(u, \upsilon)$

So
$$dz = \frac{\partial z}{\partial x}dx + \frac{\partial z}{\partial y}dy$$

or
$$dz = \frac{\partial z}{\partial u}du + \frac{\partial z}{\partial \upsilon}d\upsilon \qquad \text{...(2)}$$

From (2)
$$\frac{\partial z}{\partial y} = \frac{\partial z}{\partial u}\cdot\frac{\partial u}{\partial y} + \frac{\partial z}{\partial \upsilon}\frac{\partial \upsilon}{\partial y} \qquad \text{...(3)}$$

Now $x = e^u \cos \upsilon, y = e^u \sin \upsilon$, gives $e^{2u} = x^2 + y^2, \upsilon = \tan^{-1}\frac{y}{x}$

So $$2e^{2u}\frac{\partial u}{\partial y} = 2y \qquad ...(5)$$

and $$\frac{\partial \upsilon}{\partial y} = \frac{1}{1+\frac{y^2}{x^2}}\left(\frac{1}{x}\right) = \frac{x}{x^2+y^2} = \frac{x}{e^{2u}} \qquad ...(6)$$

Hence using values of $\frac{\partial u}{\partial y}, \frac{\partial \upsilon}{\partial y}$ from (5) and (6)

From (3)

$$\frac{\partial z}{\partial y} = \frac{\partial z}{\partial u}\left(\frac{y}{e^{2u}}\right) + \frac{\partial z}{\partial \upsilon}\left(\frac{x}{e^{2u}}\right)$$

Hence $$e^{2u}\frac{\partial z}{\partial y} = y\frac{\partial z}{\partial u} + x\frac{\partial z}{\partial \upsilon}$$

Example 1.99 **Given $u = \sin^{-1}(x-y), x = 3t, y = 4t^3$. Find $\frac{du}{dt}$.**

Solution : By total differential formula

$$du = \frac{\partial u}{\partial x}dx + \frac{\partial u}{\partial y}dy$$

So $$\frac{du}{dt} = \frac{\partial u}{\partial x}\frac{dx}{dt} + \frac{\partial u}{\partial y}\frac{dy}{dt} = \frac{1}{\sqrt{1-(x-y)^2}}[1(3) - 1(12t^2)]$$

$$= \frac{3(1-4t^2)}{\sqrt{1-(3t-4t^3)^2}} = \frac{3(1-4t^2)}{\sqrt{(1-3t+4t^3)(1+3t-4t^3)}}$$

$$= \frac{3(1-4t^2)}{\sqrt{(1+t)(1-4t^2)(1-t)(1-4t^2)}} = \frac{3}{\sqrt{1-t^2}}$$

Example 1.100 **If $x^2+y^2+z^2-2xyz = 1$, show that**

$$\frac{dx}{\sqrt{1-x^2}} + \frac{dy}{\sqrt{1-y^2}} + \frac{dz}{\sqrt{1-z^2}} = 0.$$

Solution : Regarding it as $f(x, y, z) = 1$, by the formula of total differential

$$df = \frac{\partial f}{\partial x}dx + \frac{\partial f}{\partial y}dy + \frac{\partial f}{\partial z}dz = 0 \qquad ...(1)$$

Now $$\frac{\partial f}{\partial x} = 2(x-yz), \quad \frac{\partial f}{\partial y} = 2(y-zx), \quad \frac{\partial f}{\partial z} = 2(z-xy) \qquad ...(2)$$

Now the given relation can be written as

$$(x-yz)^2 = 1-y^2-z^2+y^2z^2 = (1-y^2)(1-z^2)$$

So $$x-yz = \sqrt{(1-y^2)(1-z^2)} \qquad ...(3)$$

Similarly $$(y-zx) = \sqrt{(1-z^2)(1-x^2)} \qquad ...(4)$$

$$(z-xy) = \sqrt{(1-x^2)(1-y^2)} \qquad ..(5)$$

Hence using (2), (3), (4), (5) from (1) we have

$$\sqrt{(1-y^2)(1-z^2)}\,dx + \sqrt{(1-z^2)(1-x^2)}\,dy + \sqrt{(1-x^2)(1-y^2)}\,dz = 0$$

dividing through out by $\sqrt{(1-x^2)(1-y^2)(1-z^2)}$, we get the required result.

Example 1.101 **If $\phi(y, z) = 0$, $f(x, y) = 0$, show that**

$$\frac{\partial f}{\partial y}\frac{\partial \phi}{\partial z}\frac{dz}{dx} = \frac{\partial f}{\partial x}\frac{\partial \phi}{\partial y}$$

Solution : By total differential formula applied to $\phi(y, z) = 0$ and $f(x, y) = 0$, we get

$$d\phi = 0 = \frac{\partial \phi}{\partial y} dy + \frac{\partial \phi}{\partial z} dz$$

$$df = 0 = \frac{\partial f}{\partial x} dx + \frac{\partial f}{\partial y} dy$$

So $$-\frac{\partial \phi}{\partial y} dy = \frac{\partial \phi}{\partial z} dz \quad \text{...(1)}$$

$$-\frac{\partial f}{\partial y} dy = \frac{\partial f}{\partial x} dx \quad \text{...(2)}$$

Dividing (1) by (2)

$$\frac{\phi_y}{f_y} = \frac{\phi_z}{f_x}\frac{dz}{dx}$$

So $$\phi_y f_x = f_y \phi_z \frac{dz}{dx}$$

Hence proved

Example 1.102 **If $x^y + y^x = c$, find $\frac{dy}{dx}$.**

Solution : Let $x^y = \xi$, $y^x = \eta$

(**Note :** In sum, we can not take log)

So we have $\xi + \eta = c$

Hence $d\xi + d\eta = 0$ ...(1)

Now $\xi = x^y$, By total differential formula

$$d\xi = \frac{\partial \xi}{\partial x} dx + \frac{\partial \xi}{\partial y} dy = yx^{y-1} dx + (\log x)\, x^y\, dy \quad \text{...(2)}$$

$$\eta = y^x$$

So $$dn = \frac{d\eta}{\partial x} dx + \frac{\partial \eta}{\partial y} dy = (\log y)\, y^x\, dx + x\, y^{x-1}\, dy \quad \text{...(3)}$$

Putting values of $d\xi$, $d\eta$ in (1)

$$y\, x^{y-1}\, dx + x^y (\log x)\, dy + (\log y)\, y^x\, dx + x\, y^{x-1}\, dy = 0$$

So $$(x^y \log x + x\, y^{x-1})\frac{dy}{dx} = -\{y\, x^{y-1} + y^x \log y\}.$$

Hence we get $\frac{dy}{dx}$.

Example 1.103 **If $u = f\left(\frac{x}{y}, \frac{y}{z}, \frac{z}{x}\right)$, show that $x\frac{\partial u}{\partial x} + y\frac{\partial u}{\partial y} + z\frac{\partial u}{\partial z} = 0$.**

Solution : Let $$\frac{x}{y} = Z, \frac{y}{z} = X, \frac{z}{x} = Y \quad \text{...(1)}$$

then $$\frac{\partial X}{\partial x} = 0, \quad \frac{\partial X}{\partial y} = \frac{1}{z}, \quad \frac{\partial X}{\partial z} = -\frac{y}{z^2} \quad \text{...(2)}$$

as $$Y = \frac{z}{x}, \quad \frac{\partial Y}{\partial x} = -\frac{z}{x^2}, \quad \frac{\partial Y}{\partial y} = 0, \quad \frac{\partial Y}{\partial z} = \frac{1}{x} \quad ...(3)$$

as $$z = \frac{x}{y}, \quad \frac{\partial Z}{\partial x} = \frac{1}{y}, \quad \frac{\partial Z}{\partial y} = -\frac{x}{y^2}, \quad \frac{\partial Z}{\partial z} = 0 \quad ...(4)$$

Given that by total differential,

$$u = u(X, Y, Z) \quad ...(5)$$

So $$du = \frac{\partial u}{\partial X} dX + \frac{\partial u}{\partial Y} dY + \frac{\partial u}{\partial Z} dZ \quad ...(6)$$

Hence $$\frac{\partial u}{\partial x} = \frac{\partial u}{\partial X}\frac{\partial X}{\partial x} + \frac{\partial u}{\partial Y}\frac{\partial Y}{\partial x} + \frac{\partial u}{\partial Z}\frac{\partial Z}{\partial x} \quad ...(7)$$

$$\frac{\partial u}{\partial y} = \frac{\partial u}{\partial X}\frac{\partial X}{\partial y} + \frac{\partial u}{\partial Y}\frac{\partial Y}{\partial y} + \frac{\partial u}{\partial Z}\frac{\partial Z}{\partial y} \quad ...(8)$$

$$\frac{\partial u}{\partial z} = \frac{\partial u}{\partial X}\frac{\partial X}{\partial z} + \frac{\partial u}{\partial Y}\frac{\partial Y}{\partial z} + \frac{\partial u}{\partial Z}\frac{\partial Z}{\partial z} \quad ...(9)$$

Multiply (7) by x, (8) by y, (9) by z and adding

$$x\frac{\partial u}{\partial x} + y\frac{\partial u}{\partial y} + z\frac{\partial u}{\partial z} = \Sigma \frac{\partial u}{\partial X}\left\{x\frac{\partial X}{\partial x} + y\frac{\partial X}{\partial y} + z\frac{\partial X}{\partial z}\right\}$$

$$= \Sigma \frac{\partial u}{\partial X}\left\{x \times 0 + y\left(\frac{1}{z}\right) + z\left(-\frac{y}{z^2}\right)\right\}$$

$$= \Sigma\, 0 \frac{\partial u}{\partial X} = 0$$

Hence proved.

Example 1.104 **If $\phi(x, y, z) = 0$, show $\left(\frac{\partial y}{\partial z}\right)_x \left(\frac{\partial z}{\partial x}\right)_y \left(\frac{\partial x}{\partial y}\right)_z = -1$.**

As $\phi = 0$, $$d\phi = 0 = \frac{\partial \phi}{\partial x} dx + \frac{\partial \phi}{\partial y} dy + \frac{\partial \phi}{\partial z} dz = 0$$

Solution : If x be kept constant $dx = 0,\ 0 = \phi_y\, dy + \phi_z\, dz$

So $$\left(\frac{dy}{dz}\right)_{x = \text{const}} = -\frac{\phi_z}{\phi_y},$$ as y is function of more than one variable ...(1)

$$\left(\frac{dy}{dz}\right)_{z = \text{const}} = \left(\frac{\partial y}{\partial z}\right)_x = -\frac{\phi_z}{\phi_y} \quad ...(2)$$

Similarly

$$\left(\frac{\partial z}{\partial x}\right)_y = -\frac{\phi_x}{\phi_z} \quad ...(3)$$

and $$\left(\frac{\partial x}{\partial y}\right)_z = -\frac{\phi_y}{\phi_x} \quad ...(4)$$

Multiply (2), (3) and (4) to get the result.

Example 1.105 **If u be a function of x and y and $x = \xi \cos \alpha - \eta \sin \alpha$ and $y = \xi \sin \alpha + \eta \cos \alpha$**

show $$\frac{\partial^2 u}{\partial x^2} + \frac{\partial^2 u}{\partial y^2} = \frac{\partial^2 u}{\partial x^2} + \frac{\partial^2 u}{\partial \eta^2}$$

(Raj 2002)

Solution : We have $u = u(x, y)$ also $= u\ (\xi, n)$

So $$\frac{\partial u}{\partial x} = \frac{\partial u}{\partial \xi}\frac{\partial \xi}{\partial x} + \frac{\partial u}{\partial \eta}\frac{\partial \eta}{\partial x} \quad ...(1)$$

and $$\frac{\partial u}{\partial y} = \frac{\partial u}{\partial \xi}\frac{\partial \xi}{\partial y} + \frac{\partial u}{\partial \eta}\frac{\partial \eta}{\partial y} \quad ...(2)$$

From the given relation, we can have

$$\xi = x \cos \alpha + y \sin \alpha,\ \eta = -x \sin \alpha + y \cos \alpha$$

So $$\frac{\partial \xi}{\partial x} = \cos \alpha,\ \frac{\partial \xi}{\partial y} = \sin \alpha$$

$$\frac{\partial \eta}{\partial x} = -\sin \alpha,\ \frac{\partial \eta}{\partial y} = \cos \alpha$$

Hence from (1) and (2) we have the values of the operators $\frac{\partial}{\partial x}$ and $\frac{\partial}{\partial y}$ is given by

$$\frac{\partial}{\partial x} = \cos \alpha \frac{\partial}{\partial \xi} - \sin \alpha \frac{\partial}{\partial \eta}$$

and $$\frac{\partial}{\partial y} = \sin \alpha \frac{\partial}{\partial \xi} + \cos \alpha \frac{\partial}{\partial \eta}$$

So $$\frac{\partial^2 u}{\partial x^2} + \frac{\partial^2 u}{\partial y^2}$$

$$= \left(\cos \alpha \frac{\partial}{\partial \xi} - \sin \alpha \frac{\partial}{\partial \eta}\right)\left(\cos \alpha \frac{\partial u}{\partial \xi} - \sin \alpha \frac{\partial u}{\partial \eta}\right) + \left(\sin \alpha \frac{\partial}{\partial \xi} + \cos \alpha \frac{\partial}{\partial \eta}\right)^2 u$$

$$= \cos^2 \alpha \frac{\partial^2 u}{\partial \xi^2} - 2 \sin \alpha \cos \alpha \frac{\partial^2 u}{\partial \xi\, \partial \eta} + \sin^2 \alpha \frac{\partial^2 u}{\partial \eta^2} + \sin^2 \alpha \frac{\partial^2 u}{\partial \xi^2}$$

$$+ 2 \sin \alpha \cos \alpha \frac{\partial^2 u}{\partial \xi\, \partial \eta} + \cos^2 \alpha \frac{\partial^2 u}{\partial \eta^2}$$

$$= \frac{\partial^2 u}{\partial \xi^2} + \frac{\partial^2 u}{\partial \eta^2}.$$

Example 1.106 **Transform the equation $\frac{\partial^2 u}{\partial x^2} + \frac{\partial^2 u}{\partial y^2} = 0$, to polar coordinates.**

(Delhi 1997)

Solution : Relation connecting cartesian and polar coordinates are

$$x = r \cos \theta,\ y = r \sin \theta$$

$$r^2 = x^2 + y^2,\ \theta = \tan^{-1} \frac{y}{x}$$

So $$\frac{\partial r}{\partial x} = \frac{x}{r} = \cos \theta,\ \frac{\partial r}{\partial y} = \frac{y}{r} = \sin \theta,$$

$$\frac{\partial \theta}{\partial x} = \frac{1}{1 + \frac{y^2}{x^2}}\left(-\frac{y}{x^2}\right) = \frac{-\sin\theta}{r} \text{ and } \frac{\partial \theta}{\partial y} = \frac{\cos\theta}{r} \quad ...(1)$$

Now $\quad du = \dfrac{\partial u}{\partial r}dr + \dfrac{\partial u}{\partial \theta}d\theta$

So $$\frac{\partial u}{\partial x} = \frac{\partial r}{\partial x}\frac{\partial u}{\partial r} + \frac{\partial \theta}{\partial x}\frac{\partial u}{\partial \theta} = \left(\cos\theta\frac{\partial}{\partial r} - \frac{\sin\theta}{r}\frac{\partial}{\partial \theta}\right)u \quad ...(2)$$

and $$\frac{\partial u}{\partial y} = \frac{\partial r}{\partial y}\frac{\partial u}{\partial r} + \frac{\partial \theta}{\partial y}\frac{\partial u}{\partial \theta} = \left(\sin\theta\frac{\partial}{\partial r} + \frac{\cos\theta}{r}\frac{\partial}{\partial \theta}\right)u \quad ...(3)$$

Now $$\frac{\partial^2 u}{\partial x^2} = \frac{\partial}{\partial x}\left(\frac{\partial u}{\partial x}\right) = \left(\cos\theta\frac{\partial}{\partial r} - \frac{\sin\theta}{r}\frac{\partial}{\partial \theta}\right)\left(\cos\theta\frac{\partial u}{\partial r} - \frac{\sin\theta}{r}\frac{\partial u}{\partial \theta}\right)$$

$$= \cos^2\theta\frac{\partial^2 u}{\partial r^2} + \frac{1}{r^2}\sin\theta\cos\theta\frac{\partial u}{\partial \theta} - \frac{\sin\theta\cos\theta}{r}\frac{\partial^2 u}{\partial r\,\partial \theta}$$

$$+ \frac{\sin^2\theta}{r}\frac{\partial u}{\partial r} - \frac{\sin\theta\cos\theta}{r}\frac{\partial^2 u}{\partial \theta\,\partial r} - \frac{\sin\theta\cos\theta}{r^2}\frac{\partial u}{\partial \theta} + \frac{\sin^2\theta}{r^2}\frac{\partial^2 u}{\partial \theta^2} \quad ...(4)$$

Similarly

$$\frac{\partial^2 u}{\partial y^2} = \sin^2\theta\frac{\partial^2 u}{\partial r^2} - \frac{1}{r^2}\sin\theta\cos\theta\frac{\partial u}{\partial \theta} + \frac{\sin\theta\cos\theta}{r}\frac{\partial^2 u}{\partial r\,\partial \theta}$$

$$+ \frac{\cos^2\theta}{r}\frac{\partial u}{\partial r} + \frac{\sin\theta\cos\theta}{r}\frac{\partial^2 u}{\partial \theta\,\partial r} + \frac{\sin\theta\cos\theta}{r^2}\frac{\partial u}{\partial \theta} + \frac{\cos^2\theta}{r^2}\frac{\partial^2 u}{\partial \theta^2} \quad ...(5)$$

Adding (4) and (5)

$$\frac{\partial^2 u}{\partial x^2} + \frac{\partial^2 u}{\partial y^2} = \frac{\partial^2 u}{\partial r^2} + \frac{1}{r}\frac{\partial u}{\partial r} + \frac{1}{r^2}\frac{\partial^2 u}{\partial \theta^2} = 0.$$

PROBLEM SET 1.11

1. If $u = \sin(x^2 + y^2)$ where $a^2x^2 + b^2y^2 = c^2$, find $\dfrac{du}{dx}$.

Ans. $\left[2\left(\dfrac{b^2 - a^2}{b^2}\right)x\cos(x^2 + y^2)\right]$

2. If $u = x^2 + y^2 + z^2$ and $x = e^{2t}, y = e^{2t}\cos 3t, z = e^{2t}\sin 3t$. Find $\dfrac{du}{dt}$ as a total derivative and verify the result by direct substitution.

Ans. $8e^{4t}$.

3. If $f(x, y) = 0$, show that $\left(\dfrac{\partial f}{\partial y}\right)^3\dfrac{d^2y}{dx^2} = [2pqs - q^2r - p^2t]$,

where $p = f_x, q = f_y, r = f_{xx}, s = f_{xy}, t = f_{yy}$.

4. Given $u = \sin \frac{x}{y}$, $x = e^t$, $y = t^2$. Find $\frac{du}{dt}$ as a function of t.

Ans. $\frac{du}{dt} = \frac{t-2}{t^3} e^t \cos\left(\frac{e^t}{t^2}\right)$.

5. If $u = f(r, s)$ where $r = x + at$, $s = y + bt$, show that

$$\frac{\partial u}{\partial t} = a \frac{\partial u}{\partial x} + b \frac{\partial u}{\partial y}.$$

6. If $z = \log (u^2 + \upsilon)$, $u = e^{x^2+y^2}$, $\upsilon = x^2 + y$, find $\frac{\partial z}{\partial x}, \frac{\partial z}{\partial y}$.

Ans. $\frac{2x(2u^2+1)}{u^2+\upsilon}, \frac{4yu^2+1}{u^2+\upsilon}$

7. If $u = f(2x - 3y, 3y - 4z, 4z - 2x)$, prove that

$$\frac{1}{2} u_x + \frac{1}{3} u_y + \frac{1}{4} u_z = 0.$$

8. If $u = x^2 - y^2$, $\upsilon = 2xy$ and $f(x, y) = \phi(u, \upsilon)$. Show that

$$\frac{\partial^2 f}{\partial x^2} + \frac{\partial^2 f}{\partial y^2} = 4(x^2 + y^2)\left(\frac{\partial^2 \phi}{\partial u^2} + \frac{\partial^2 \phi}{\partial \upsilon^2}\right).$$

9. If $u = u(x, y)$ and $x = e^r \cos \theta$, $y = e^r \sin \theta$, show that

$$\frac{\partial^2 u}{\partial x^2} + \frac{\partial^2 u}{\partial y^2} = e^{-2r}\left(\frac{\partial^2 u}{\partial r^2} + \frac{\partial^2 u}{\partial \theta^2}\right).$$

1.38 JACOBIAN

Let $u = u(x, y)$, $\upsilon = \upsilon(x, y)$ be two functions of two independent variables x and y, then the functional determinant

$$\begin{vmatrix} \frac{\partial u}{\partial x} & \frac{\partial u}{\partial y} \\ \frac{\partial \upsilon}{\partial x} & \frac{\partial \upsilon}{\partial y} \end{vmatrix} \quad \text{also denoted as} = \frac{\partial(u, \upsilon)}{\partial(x, y)} \text{ or } J\left(\frac{u, \upsilon}{x, y}\right)$$

is called the **Jacobian of u, υ, with respect to x, y.**

For three functions u, υ, w of x, y, z, the Jacobian of u, v, w, *w.r.t.*, x, y, z is written as

$$\frac{\partial(u, \upsilon, w)}{\partial(x, y, z)} \equiv \begin{vmatrix} \frac{\partial u}{\partial x} & \frac{\partial u}{\partial y} & \frac{\partial u}{\partial z} \\ \frac{\partial \upsilon}{\partial x} & \frac{\partial \upsilon}{\partial y} & \frac{\partial \upsilon}{\partial z} \\ \frac{\partial w}{\partial x} & \frac{\partial w}{\partial y} & \frac{\partial w}{\partial z} \end{vmatrix}$$

Likewise we can define the Jacobian of n functions $u_1, u_2 \dots u_n$ of n variables $x_1, x_2 \dots x_n$.

1.39 PROPERTIES OF JACOBIAN

If $J = \dfrac{\partial(u, \upsilon)}{\partial(x, y)}$ and J' is its inverse defined as $J' = \dfrac{\partial(x, y)}{\partial(u, \upsilon)}$ then $J \cdot J' = 1$.

Proof : We have $du = \dfrac{\partial u}{\partial x} dx + \dfrac{\partial u}{\partial y} dy$

So
$$\frac{\partial u}{\partial u} = 1 = \frac{\partial u}{\partial x}\frac{\partial x}{\partial u} + \frac{\partial u}{\partial y}\frac{\partial y}{\partial u}$$

$$\frac{\partial u}{\partial \upsilon} = 0 = \frac{\partial u}{\partial x}\frac{\partial x}{\partial \upsilon} + \frac{\partial u}{\partial y}\frac{\partial y}{\partial \upsilon}$$

Similarly
$$\frac{\partial \upsilon}{\partial u} = 0 = \frac{\partial \upsilon}{\partial x}\frac{\partial x}{\partial u} + \frac{\partial \upsilon}{\partial y}\frac{\partial y}{\partial u}$$

and
$$\frac{\partial \upsilon}{\partial \upsilon} = 1 = \frac{\partial \upsilon}{\partial x}\frac{\partial x}{\partial \upsilon} + \frac{\partial \upsilon}{\partial y}\frac{\partial y}{\partial \upsilon}$$

So
$$\begin{vmatrix} \dfrac{\partial u}{\partial u} & \dfrac{\partial u}{\partial \upsilon} \\ \dfrac{\partial \upsilon}{\partial u} & \dfrac{\partial \upsilon}{\partial \upsilon} \end{vmatrix} = \begin{vmatrix} 1 & 0 \\ 0 & 1 \end{vmatrix} = 1$$

Putting values of partial derivatives

$$\text{L.H.S.} = \begin{vmatrix} \dfrac{\partial u}{\partial x}\dfrac{\partial x}{\partial u} + \dfrac{\partial u}{\partial y}\dfrac{\partial y}{\partial y} & \dfrac{\partial u}{\partial x}\dfrac{\partial x}{\partial \upsilon} + \dfrac{\partial u}{\partial y}\dfrac{\partial y}{\partial \upsilon} \\ \dfrac{\partial \upsilon}{\partial x}\dfrac{\partial x}{\partial u} + \dfrac{\partial \upsilon}{\partial y}\dfrac{\partial y}{\partial u} & \dfrac{\partial \upsilon}{\partial x}\dfrac{\partial x}{\partial \upsilon} + \dfrac{\partial \upsilon}{\partial y}\dfrac{\partial y}{\partial \upsilon} \end{vmatrix}$$

expressing it as product of two determinants

$$= \begin{vmatrix} \dfrac{\partial x}{\partial u} & \dfrac{\partial x}{\partial \upsilon} \\ \dfrac{\partial y}{\partial u} & \dfrac{\partial y}{\partial \upsilon} \end{vmatrix} \times \begin{vmatrix} \dfrac{\partial u}{\partial x} & \dfrac{\partial u}{\partial \upsilon} \\ \dfrac{\partial \upsilon}{\partial x} & \dfrac{\partial \upsilon}{\partial y} \end{vmatrix} = JJ' = 1$$

II. If u and υ are functions of ξ and η and ξ, η are functions of x and y then
$$\frac{\partial(u, \upsilon)}{\partial(x, y)} = \frac{\partial(u, \upsilon)}{\partial(\xi, \eta)} \cdot \frac{\partial(\xi, \eta)}{\partial(x, y)}.$$

Example 1.107 **In polar coordinates $x = r\cos\theta, y = r\sin\theta$, show that $\dfrac{\partial(x, y)}{\partial(r, \theta)} = r$.**

Solution : We have $x = r\cos\theta$, $y = r\sin\theta$

So
$$\frac{\partial x}{\partial r} = \cos\theta,\ \frac{\partial x}{\partial \theta} = -r\sin\theta,\ \frac{\partial y}{\partial r} = \sin\theta,\ \frac{\partial y}{\partial \theta} = r\cos\theta$$

So
$$\frac{\partial(x, y)}{\partial(r, \theta)} = \begin{vmatrix} \dfrac{\partial x}{\partial r} & \dfrac{\partial x}{\partial \theta} \\ \dfrac{\partial y}{\partial r} & \dfrac{\partial y}{\partial \theta} \end{vmatrix} = \begin{vmatrix} \cos\theta & -r\sin\theta \\ \sin\theta & r\cos\theta \end{vmatrix}$$

$$= r(\cos^2\theta + \sin^2\theta) = r$$

Example 1.108 **For change from Cartesian to cylindrical polar coordinates, defined by $x = \rho \cos \phi$, $y = \rho \sin \phi$, $z = z$, show $\dfrac{\partial(x, y, z)}{\partial(\rho, \phi, z)} = \rho$.**

Solution : We have $\dfrac{\partial x}{\partial \rho} = \cos \phi$, $\dfrac{\partial x}{\partial \phi} = -\rho \sin \phi$, $\dfrac{\partial x}{\partial z} = 0$,

$\dfrac{\partial y}{\partial \rho} = \sin \phi$, $\dfrac{\partial y}{\partial \phi} = \rho \cos \phi$, $\dfrac{\partial y}{\partial z} = 0$,

$\dfrac{\partial z}{\partial x} = 0$, $\dfrac{\partial z}{\partial y} = 0$, $\dfrac{\partial z}{\partial z} = 1$

So
$$\frac{\partial(x, y, z)}{\partial(\rho, \phi, z)} = \begin{vmatrix} \cos \phi & -\rho \sin \phi & 0 \\ \sin \phi & r \cos \phi & 0 \\ 0 & 0 & 1 \end{vmatrix} = \rho\,(\cos^2 \phi + \sin^2 \phi) = \rho.$$

Example 1.109 **For change from Cartesian coordinates to spherical polar coordinates, defined by $x = r \sin \theta \cos \phi$, $y = r \sin \theta \sin \phi$, $z = r \cos \theta$.**

Show that $\dfrac{\partial(x, y, z)}{\partial(r, \theta, \phi)} = r^2 \sin \theta$.

Solution : We have $\dfrac{\partial x}{\partial r} = \sin \theta \cos \phi$, $\dfrac{\partial x}{\partial \theta} = r \cos \theta \cos \phi$, $\dfrac{\partial x}{\partial \phi} = -r \sin \theta \sin \phi$

$\dfrac{\partial y}{\partial r} = \sin \theta \sin \phi$, $\dfrac{\partial y}{\partial \theta} = r \cos \theta \sin \phi$, $\dfrac{\partial y}{\partial \phi} = r \sin \theta \cos \phi$

$\dfrac{\partial z}{\partial r} = \cos \theta$, $\dfrac{\partial z}{\partial \theta} = -r \sin \theta$, $\dfrac{\partial z}{\partial \phi} = 0$

So
$$\frac{\partial(x, y, z)}{\partial(r, \theta, \phi)} = \begin{vmatrix} \sin \theta \cos \phi & r \cos \theta \cos \phi & -r \sin \theta \sin \phi \\ \sin \theta \sin \phi & r \cos \theta \sin \phi & r \sin \theta \cos \phi \\ \cos \theta & -r \sin \theta & 0 \end{vmatrix}$$

= taking r common from C_2 and $r \sin \theta$ from C_3

$$J = r^2 \sin \theta \begin{vmatrix} \sin \theta \cos \phi & \cos \theta \cos \phi & -\sin \phi \\ \sin \theta \sin \phi & \cos \theta \sin \phi & \cos \phi \\ \cos \theta & -\sin \theta & 0 \end{vmatrix}$$

Expanding in terms of elements of third row

$= r^2 \sin \theta\,[\cos \theta\,(\cos \theta\,(\cos^2 \phi + \sin^2 \phi)) + \sin \theta \sin \theta\,(\cos^2 \phi + \sin^2 \phi) + 0]$

$= r^2 \sin \theta$

Example 1.110 **If $u = x + y + z$, $u\upsilon = y + z$, $u\upsilon w = z$, show that $\dfrac{\partial(x, y, z)}{\partial(u, \upsilon, w)} = u^2\upsilon$.**

Solution : We have $z = u\upsilon w$, $y = u\upsilon - u\upsilon w$, $x = u - u\upsilon$

So
$$J = \begin{vmatrix} \dfrac{\partial x}{\partial u} & \dfrac{\partial x}{\partial \upsilon} & \dfrac{\partial x}{\partial w} \\ \dfrac{\partial y}{\partial u} & \dfrac{\partial y}{\partial \upsilon} & \dfrac{\partial y}{\partial w} \\ \dfrac{\partial z}{\partial u} & \dfrac{\partial z}{\partial \upsilon} & \dfrac{\partial z}{\partial w} \end{vmatrix} = \begin{vmatrix} 1 - \upsilon & -u & 0 \\ \upsilon - \upsilon w & u - uw & -u\upsilon \\ \upsilon w & wu & u\upsilon \end{vmatrix}$$

Taking u common from C_2 and $u\upsilon$ from C_3

$$J = u^2\upsilon \begin{vmatrix} 1-\upsilon & -1 & 0 \\ \upsilon(1-w) & 1-w & -1 \\ \upsilon w & w & 1 \end{vmatrix}, \text{ Expanding}$$

$$= u^2\upsilon \left\{(1-\upsilon)\begin{vmatrix} 1-w & -1 \\ w & 1 \end{vmatrix} + \upsilon \begin{vmatrix} 1-w & -1 \\ w & 1 \end{vmatrix} + 0\right\}$$

$$= u^2\upsilon \,|(1-\upsilon)(1-w+w) + \upsilon(1+w+w)| = u^2\upsilon$$

Example 1.111 **If $x = a \cosh \xi \cos \eta$, $y = a \sinh \xi \sin \eta$, show**

$$\frac{\partial(x, y)}{\partial(\xi, \eta)} = \frac{a^2}{2}(\cosh 2\xi - \cos 2\eta)$$

Solution : $\frac{\partial X}{\partial \xi} = a \sinh \xi \cos \eta$, $\frac{\partial x}{\partial \eta} = -a \cosh \xi \sin \eta$,

$\frac{\partial y}{\partial \xi} = a \cosh \xi \sin \eta$, $\frac{\partial y}{\partial \eta} = a \sinh \xi \cos \eta$

So $$J = \begin{vmatrix} \frac{\partial x}{\partial \xi} & \frac{\partial x}{\partial \eta} \\ \frac{\partial y}{\partial \xi} & \frac{\partial y}{\partial \eta} \end{vmatrix} = a^2 \begin{vmatrix} \sinh \xi \cos \eta & -\cosh \xi \sin \eta \\ \cosh \xi \sin \eta & \sinh \xi \cos \eta \end{vmatrix}$$

$$= a^2 |\sinh^2 \xi \cos^2 \eta + \cosh^2 \xi \sin^2 \eta|, \text{ we know } \cosh^2 \xi - \sinh^2 \xi = 1$$

$$\cosh^2 \xi + \sinh^2 \xi = \cos 2\xi$$

$$= a^2 |(-1 + \cosh^2 \xi)\cos^2 \eta + \cosh^2 \xi \sin^2 \eta| = a^2 |\cos^2 \eta + \cosh^2 \xi|$$

$$= a^2 \left\{-\frac{1 + \cos 2\eta}{2} + \frac{\cosh 2\xi + 1}{2}\right\}$$

$$= \frac{a^2}{2}(\cosh 2\xi - \cos \eta)$$

PROBLEM SET 1.12

1. If $r^2 = x^2 + y^2$, $\theta = \tan^{-1}\frac{y}{x}$, show $\frac{\partial(r, \theta)}{\partial(x, y)} = \frac{1}{r}$.

2. If $u = x^2 - 2y$, $\upsilon = x + y$, prove $\frac{\partial(u, \upsilon)}{\partial(x, y)} = 2(x + 1)$.

3. If $x = u(1 - \upsilon)$, $y = u\,\upsilon$, prove $JJ' = 1$.

4. For change from Cartesian to polar and polar to Cartesian defined by $x = r \cos \theta$, $y = r \sin 0$, $r^2 = x^2 + y^2$, $\theta = \tan^{-1}\frac{y}{x}$, show $\frac{\partial(r, \theta)}{\partial(x, y)} \cdot \partial(x, y)\partial(r, \theta) = 1$.

5. If $u = \frac{x + y}{1 - xy}$, $\upsilon = \tan^{-1} x + \tan^{-1} y$, find $\frac{\partial(u, \upsilon)}{\partial(x, y)}$.

Ans. 0.

6. If $x = a \cosh \xi \cos \eta$, $y = a \sinh \xi \sin \eta$, show $\dfrac{\partial(x, y)}{\partial(\xi, \eta)} = \dfrac{1}{2} a^2 (\cosh 2\xi - \cos 2\eta)$.

7. If $u = \dfrac{y^2}{2x}$, $\upsilon = \dfrac{x^2 + y^2}{2x}$, find $\dfrac{\partial(u, \upsilon)}{\partial(x, y)}$.

Ans. $(- y/2x)$

8. If $u = \dfrac{yz}{x}$, $\upsilon = \dfrac{zx}{y}$, $w = \dfrac{xy}{z}$, show that $\dfrac{\partial(u, \upsilon, w)}{\partial(\xi, y, z)} = 4$.

9. If $u = x^2 - 2y$, $\upsilon = x + y + z$, $w = x - 2y + 3z$, find $\dfrac{\partial(u, \upsilon, w)}{\partial(x, y, z)}$.

Ans. $(10x + 4)$

10. If $u = x + y + z$, $u\upsilon = y + z$, $u\upsilon w = z$, show that $\dfrac{\partial(x, y, z)}{\partial(u, \upsilon, w)} = u^2\upsilon$.

11. If $u = \dfrac{x}{y - z}$, $\upsilon = \dfrac{y}{z - x}$, $w = \dfrac{z}{x - y}$, show that $\dfrac{\partial(u, \upsilon, w)}{\partial(x, y, z)} = 0$

12. If $u = xyz$, $\upsilon = xy + yz$, $w = x + y + z$, show that $\dfrac{\partial(u, \upsilon, w)}{\partial(x, y, z)} = (x - y)(y - z)(z - x)$.

1.40 TAYLOR'S SERIES EXPANSION FOR FUNCTIONS OF TWO VARIABLES

We know, the Taylor's series expansion of $f(x + h)$ in powers of h an infinite series is

$$f(x + h) = f(x) + \frac{h}{1!} f'(x) + \frac{h^2}{2!} f''(x) + \ldots$$

When it is expressed as sum of first n terms plus R_n, the remainder after n terms, is known as Taylor's Theorem.

If f be a function of two variables and expanding w.r.t. one keeping the other as constant, the expansion may then be written as

$$f(x + h, y + k) = f(x, y + k) + \frac{h}{1!} \frac{\partial}{\partial x} f(x, y + k) + \frac{h^2}{2!} \frac{\partial^2}{\partial x^2} f(x, y + k) + \ldots$$

Now expanding the terms in the above expansion keeping x as constant,

$$\text{1st term R.H.S.} = f(x, y) + k \frac{\partial f(x, y)}{\partial y} + \frac{k^2}{2!} \frac{\partial^2 f(x, y)}{\partial y^2} + \ldots$$

$$\text{Second term R.H.S.} = h \frac{\partial}{\partial x} \left\{ f(x, y) + k \frac{\partial f(x, y)}{\partial y} + \ldots \right\}$$

$$= h \frac{\partial f}{\partial x} + hk \frac{\partial^2 f}{\partial x \, \partial y} + \ldots$$

$$\text{3rd term} = \frac{h^2}{2!} \frac{\partial^2}{\partial x^2} \left\{ f(x, y) + k \frac{\partial f}{\partial y} + \ldots \right\}$$

$$= \frac{h^2}{2!} \frac{\partial^2 f}{\partial x^2} + \ldots$$

Adding the terms, R.H.S. equal to

$$f(x, y) + \frac{1}{1!}\left\{h\frac{\partial}{\partial x} + k\frac{\partial}{\partial y}\right\} f + \frac{1}{2!}\left\{h\frac{\partial}{\partial x} + k\frac{\partial}{\partial y}\right\}^2 f + \frac{1}{3!}\left\{h\frac{\partial}{\partial x} + k\frac{\partial}{\partial y}\right\}^3 f + \ldots$$

where $\left(h\frac{\partial}{\partial x} + k\frac{\partial}{\partial y}\right)^2 f$ stands for $\left(h^2\frac{\partial^2}{\partial x^2} + 2hk\frac{\partial^2}{\partial x\,\partial y} + k^2\frac{\partial^2}{\partial y^2}\right) f$

Thus when the function $f(a + h, b + k)$ is expanded about (a, b),

$$\boldsymbol{f(a + h, b + k) = f(a, b) + \left(h\frac{\partial f}{\partial a} + k\frac{\partial f}{\partial b}\right) + \frac{1}{2!}\left(h\frac{\partial}{\partial a} + k\frac{\partial}{\partial b}\right)^2 f(a, b) + \ldots}$$

If the function $f(x, y)$ is expanded about (a, b), we write $x = a + (x - a), y = a + (y - a)$ So $f(x, y) = f\{a + (x - a), b + (y - b)\}$, now the expansion will be in powers of $x - a$ and $y - b$ given as

$$f(x, y) = f(a, b) + \left\{(x - a)\frac{\partial f}{\partial a}(a, b) + (y - b)\frac{\partial f}{\partial b}(a, b)\right\} + \frac{1}{2!}\left((x - a)\frac{\partial}{\partial a} + (y - b)\frac{\partial}{\partial b}\right)^2 f(a, b) + \ldots$$

1.41 MACLAURIN'S EXPANSION

Expansion involving $\frac{\partial f}{\partial x}, \frac{\partial f}{\partial y}$ etc. are calculated about (0, 0), is called the Maclaurin's expansion. So taking $a = b = 0$ in the above result

$$f(x, y) = f(0, 0) + \left\{x\frac{\partial f}{\partial x} + y\frac{\partial f}{\partial y}\right\}_{\text{at }(0,0)} + \frac{1}{2!}\left\{x\frac{\partial}{\partial x} + y\frac{\partial}{\partial y}\right\}^2_{\text{at }(0,0)} f(x, y) + \ldots$$

Example 1.112 **Expand $f(x, y) = e^x \log(1 + y)$ in powers of x and y upto terms of third degree.**

Solution : We have

$$\frac{\partial f}{\partial x} = \frac{\partial^2 f}{\partial x^2} = \frac{\partial^3 f}{\partial x^3} = e^x \log(1 + y)$$

Values of these, at

$$x = 0, y = 0 \text{ arc} = e^0 \log(1 + 0) = 0 \quad \ldots(1)$$

So $f_x(0, 0) = f_{xx}(0, 0) = f_{xxx}(0, 0) = 0$ and $f(0, 0) = 0$

$$\frac{\partial f}{\partial y} = \frac{e^x}{1 + y} \text{ so } f_y(0, 0) = 1$$

$$\frac{\partial^2 f}{\partial y^2} = -\frac{e^x}{(1 + y)^2}, \quad f_{yy}(0, 0) = -1$$

$$\frac{\partial^3 f}{\partial y^3} = \frac{2e^x}{(1 + y)^3}, \quad f_{yyy}(0, 0) = 2$$

$$f_{xy} = \frac{e^x}{1 + y} \text{ so } f_{xy}(0, 0) = 1$$

$$f_{xxy} = \frac{e^x}{1+y} \text{ so } f_{xxy}(0, 0) = 1$$

$$f_{xyy} = -\frac{e^x}{(1+y)^2} \text{ so } f_{xyy} = -1$$

Hence the expansion is

$$f(x, y) = 0 + \frac{x \times 0 + y \times 1}{1!} + \frac{x^2 \times 0 + 2xy\,(1) + y^2\,(-1)}{2!}$$

$$+ \frac{1}{3!}\,\{x^3 \times 0 + 3 \times 1x^2y - 3 \times 1\,xy^2 + 2y^3\} + \ldots$$

or $$f(x, y) = y + \frac{1}{2}(2xy - y^2) + \frac{1}{3!}(3x^2y - 3xy^2 + 2y^3) + \ldots$$

$$= y + xy - \frac{y^2}{2} + \frac{1}{2}(x^2y - xy^2) + \frac{1}{3}y^3 + \ldots$$

Example 1.113 **Expand $f(x, y) = e^x \sin y$, in powers of x and y as far as terms of third degree.**

Solution : The function is to be expanded at (0, 0).

We have

$$f_x = f_{xx} = f_{xxx} = e^x \sin y$$

$$f(0, 0) = e^0 \sin 0 = 0$$

So $$f_x(0, 0) = f_{xx}(0, 0)$$

$$f_{xxx}(0, 0) = e^0 \sin 0 = 0$$

$$f_y = e^x \cos y,\ f_{yy} = e^x(-\sin y),\ f_{yyy} = -e^x \cos y$$

So $$f_y(0, 0) = 1,\ f_{yy}(0, 0) = 0,\ f_{yyy}(0, 0) = -1$$

$$f_{xy} = e^x \cos y = f_{xxy}$$

$$f_{xyy} = -e^x \sin y$$

So $$f_{xy}(0, 0) = f_{xxy}(0, 0) = 1$$

$$f_{xyy}(0, 0) = 0$$

Hence from

$$f(x, y) = f(0, 0) + (x f_x + y f_y)_{(0,0)} + \frac{1}{2!}\left\{x^2 f_{xx} + 2xy\,f_{xy} + y^2 f_{yy}\right\}_{(0,0)}$$

$$+ \frac{1}{3!}\left\{x^3 f_{xxx} + 3x^2y\,f_{xxy} + 3xy^2\,f_{xyy} + y^3 f_{yyy}\right\}_{(0,0)} +$$

$$= 0 + \{0 + y\} + \frac{1}{2!}\{0 + 2xy\,(1) + 0\} + \frac{1}{3!}\{0 + 3x^2y + 0 - y^3\} \ldots$$

$$= y + xy + \frac{x^2y}{2} - \frac{y^3}{6} + \ldots$$

Example 1.114 **Expand $f(x, y) = x^2y + 3y - 2$, in powers of $(x - 1, y + 2)$ using Taylor's theorem.**

Solution : $$f(x, y) = x^2y + 3y - 2,\ f(1, -2) = -2 - 6 - 2 = -10$$

Differentiating,

$$f_x = 2xy$$

So $\quad f_x(1, -2) = 2(-2) = -4,$

$$f_{xx} = 2y$$

$$f_{xx}(1, -2) = -4,$$

$$f_{xxx} = 0$$

$$f_{xxx}(1, -2) = 0$$

$$f(x, y) = f\{1 + (x-1), -2 + y + 2\}$$

$$= f(1, -2) + \{(x-1) f_x + (y+2) f_y\}_{(1,-2)}$$

$$+ \frac{1}{2!}\left\{(x-1)^2 f_{xx} + 2(x-1)(y+2) f_{xy} + (y+2)^2 f_{yy}\right\}_{(1,-2)}$$

$$+ \frac{1}{3!}\left\{(x-1)^3 f_{xxx} + 3(x-1)^2 (y+2) f_{xxy}\right.$$

$$\left. + 3(x-1)(y+2)^2 f_{xyy} + (y+2)^3 f_{yy}\right\}_{(1,-2)} + \ldots$$

$$= -10 + \{-4(x-1) + 4(y+2)\} + \frac{1}{2}\{-4(x-1)^2 + 4(x-1)(y+2) + 0\}$$

$$+ \frac{1}{6}\{0 + 3(x-1)^2(y+2) \times 2 + 0 + 0\}$$

$$= [-10 - 4(x-1) + 4(y+2) - 2(x-1)^2 + 2(x-1)(y+2) + (x-1)^2(y+2)]$$

Example 1.115 **Expand $f(x, y) = \tan^{-1}\left(\frac{y}{x}\right)$ in powers of $(x-1)$, $(y-1)$ upto third degree terms. Hence compute $f(1{\cdot}1, {\cdot}9)$ approximately.**

Solution : $f(x, y) = \tan^{-1}\left(\frac{y}{x}\right)$, $f(1, 1) = \pi/4$.

So $\quad f_x = -\frac{y}{x^2 + y^2}, \quad f_{xx} = \frac{2xy}{(x^2 + y^2)^2}, \; f_{xxx} = \frac{2y}{(x^2 + y^2)^2} - \frac{8x^2 y}{(x^2 + y^2)^3}$

$$f_x(1, 1) = -\frac{1}{2}, \quad f_{xx}(1, 1) = \frac{1}{2}, \quad f_{xxx}(1, 1) = \left(\frac{1}{2} - 1\right) = -\frac{1}{2}$$

$$f_y = \frac{x}{x^2 + y^2}, \quad f_{yy} = \frac{-2xy}{(x^2 + y^2)^2}, \; f_{yyy} = \left\{\frac{-2x}{(x^2 + y^2)^2} + \frac{8xy^2}{(x^2 + y^2)^3}\right\}$$

$$f_y(1, 1) = \frac{1}{2}, \quad f_{yy}(1, 1) = -\frac{1}{2}, \; f_{yyy}(1, 1) = \left\{-\frac{1}{2} + 1\right\} = \frac{1}{2}.$$

$$f_{xy} = -\frac{1}{x^2 + y^2} + \frac{2y^2}{(x^2 + y^2)^2}, \qquad f_{xxy} = \frac{2x}{(x^2 + y^2)^2} - \frac{8xy^2}{(x^2 + y^2)^3}$$

$$f_{xy}(1, 1) = -\frac{1}{2} + \frac{1}{2} = 0, \qquad f_{xxy}(1, 1) = \frac{1}{2} - 1 = -\frac{1}{2}$$

$$f_{xyy} = \frac{-2y}{(x^2 + y^2)^2} + \frac{8x^2 y}{(x^2 + y^2)^3},$$

So $\quad f_{xyy}(1, 1) = -\frac{1}{2} + \frac{2}{2} = \frac{1}{2}.$

So $\quad f(x, y) = f\{1 + (x-1), 1 + (y-1)\}$

$$= f(1, 1) + \{(x-1) f_x + (y-1) f_y\}_{(1, 1)}$$

$$+ \frac{1}{!2}\{(x-1)^2 f_{xx} + 2(x-1)(y-1) f_{xy} + (y-1)^2 f_{yy}\}$$

$$= \frac{1}{3!}\{(x-1)^3 f_{xxx} + 3(x-1)^2 (y-1) f_{xxy} + 3(x-1)(y-1)^2 f_{xyy} + (y-1)^3 f_{yyy}\} + \ldots$$

$$= \frac{\pi}{4} + \left\{-\frac{x-1}{2} + \frac{y-1}{2}\right\} + \frac{1}{2}\left\{\frac{(x-1)^2}{2} + 2(x-1)(y-1)\,0 + \frac{(y-1)^2}{-2}\right\}$$

$$+ \frac{1}{6}\left\{\frac{(x-1)^3}{-2} + 3(x-1)^2(y-1)\left(-\frac{1}{2}\right) + 3(x-1)(y-1)^2\frac{1}{2} + \frac{(y-1)^3}{2}\right\} + \ldots$$

$$= \frac{\pi}{4} - \frac{1}{2}(x-1) + \frac{1}{2}(y-1) + \frac{1}{4}(x-1)^2 - \frac{1}{4}(y-1)^2$$

$$- \frac{1}{12}(x-1)^3 - \frac{1}{4}(x-1)^2(y-1) + \frac{1}{4}(x-1)(y-1)^2 + \frac{1}{12}(y-1)^3 + \ldots$$

Now

$$f(1{\cdot}1, 0{\cdot}9) = f(1 + 0{\cdot}1, 1 - 0{\cdot}1)$$

So $x - 1 = 0{\cdot}1, y - 1 = -0{\cdot}1$

$$= \frac{\pi}{4} - \frac{1}{2}(0{\cdot}1) + \frac{1}{2}(-0{\cdot}1) + \frac{1}{4}0{\cdot}01 - \frac{1}{4}0{\cdot}01$$

$$- \frac{1}{12}0{\cdot}001 + \frac{1}{4}0{\cdot}001 + \frac{1}{4}0{\cdot}001 - \frac{0{\cdot}001}{12}$$

$$= \frac{\pi}{4} - 0{\cdot}1 + \frac{1}{12}0{\cdot}001(-1 + 3 + 3 - 1)$$

$$= \frac{3{\cdot}14}{4} - 0{\cdot}1 + \frac{1}{3}0{\cdot}001$$

$$= 0{\cdot}7854 - 0{\cdot}0967$$

$$= 0{\cdot}6887.$$

PROBLEM SET 1.13

1. Expand $f(x + h, y + k) = \dfrac{(x+h)(y+k)}{(x+h)+(y+k)}$

in powers of h and k upto second degree terms.

Ans. $\dfrac{xy}{x+y} + \dfrac{y^2}{(x+y)^2}h + \dfrac{x^2}{(x+y)^2}k + \dfrac{(-y^2h^2 + 2xyhk - x^2k^2)}{(x+y)^3} + \ldots$

2. Expand $e^y \log(1 + x)$ in powers of x and y, upto second degree terms.

Ans. $1 + x + \dfrac{1}{2!}(x^2 - y^2) + \dfrac{1}{3!}(x^3 - 3xy^2) + \ldots$

3. Expand $f(x, y) = e^x \sin y$ upto third degree terms.

Ans. $\left\{y + xy + \frac{1}{2}x^2y - \frac{1}{6}y^3\right\}$

4. Expand $f(x, y) = xy^2 + 3x - 2$ in powers of $x + 2$ and $y - 1$, using Taylor's Theorem.

Ans. $-10 - 4(y-1) + 4(x+2) - 2(y-1)^2 + 2(y-1)(x+2) + (y-1)^2(x+2)$.

5. Expand $f(x, y) = \sin x \sin y$ as far as the third degree terms.

Ans. xy.

6. Show that $e^y \log(1 + x) = x + xy - \frac{x^2}{2}$ approximately.

7. Expand $\sin[(x + h)(y + k)]$ by Taylor's theorem in powers of h and k.

Ans. $\sin xy + (hy + kx)\cos xy + hk\cos xy - \frac{1}{2}(hy + kx)^2 \sin xy + \ldots$

8. Expand $e^{ax} \sin by$ in powers of x and y as far as the terms of third degree.

Ans. $by + abxy + \frac{1}{3!}(3a^2bx^2y - b^3y^3) + \ldots$

9. Expand $f(x, y) = e^x \cos y$ at $\left(1, \frac{\pi}{4}\right)$.

Ans. $\frac{e}{\sqrt{2}}\left\{1 + (x-1) - \left(y - \frac{\pi}{4}\right) + \frac{(x-1)^2}{2} - (x-1)\left(y - \frac{\pi}{4}\right) - \frac{1}{2}\left(y - \frac{\pi}{4}\right)^2 \ldots\right\}$

10. Expand $f(x, y) = e^x \cos y$ in powers of x and y upto the third degree terms.

Ans. $1 + x + \frac{1}{2!}(x^2 - y^2) + \frac{1}{3!}(x^3 - 3xy^2) + \ldots$

11. Expand e^{xy} at $(1, 1)$ upto second degree terms.

Ans. $e\,\{1 + (x-1) + (y-1)\} + \frac{1}{2!}\{(x-1)^2 + 4(x-1)(y-1) + (y-1)^2\}$

12. Expand $f(x, y) = \sqrt{1 + x + y^2}$ at $(1, 0)$ upto second degree terms.

Ans. $\sqrt{2}\left\{1 + \frac{x-1}{4} - \frac{(x-1)^2}{32} + \frac{y^2}{4} + \ldots\right\}$

1.42 MAXIMA AND MINIMA OF FUNCTIONS OF TWO VARIABLES

Let $z = f(x, y)$ be a function of two variables, representing a surface in a three daimensional space. Let on it $P(x, y, z)$ and $Q(x + \delta x, y + \delta y, z + \delta z)$ be two neighbouring points. If N and M be the feet of perpendiculars from P and Q on the XY plane $(z = 0)$, then M may be any point within a circle with centre N and radius δ where $(\delta x)^2 + (\delta y)^2 < \delta^2$.

P is a maxima if $QM - PN < 0$ for all points in the neighbourhood of N.

P is a minima if $PN < QM$ or $QM - PN > 0$ for all Q in the neighbourhood. The maximas or minimas are called the extreme points of the function.

Now

$$QM - PN = \delta z$$
$$= f(x + \delta x, y + \delta y) - f(x, y)$$
$$= \left(\delta x \frac{\partial f}{\partial x} + \delta y \frac{\partial f}{\partial y}\right) + \frac{1}{2!}\left((\delta x)^2 \frac{\partial^2 f}{\partial x^2} + 2\delta x\, \delta y \frac{\partial^2 f}{\partial x\, \partial y} + (\delta y)^2 \frac{\partial^2 f}{\partial y^2}\right) + \ldots$$

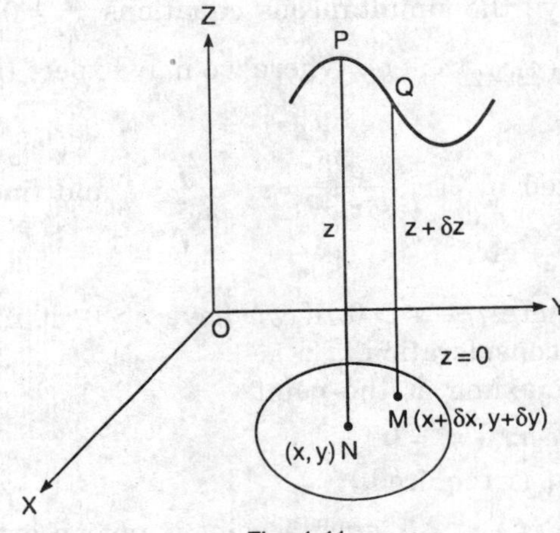

Fig. 1.11

When $\delta x, \delta y \to 0$, the sign of the first term on R.H.S., govern the sign of the whole expression. This term has single powers of δx and δy. So it changes sign as the sign of δx or δy or both are changed. But for maxima or minima $QM - PN$ must have an invariable sign. Hence the necessary conditions for a maxima or a minima at $P(x, y)$ are :

(i) $$\frac{\partial f}{\partial x} = 0, \frac{\partial f}{\partial y} = 0 \qquad ...(1)$$

Denoting $\frac{\partial^2 f}{\partial x^2} = r, \frac{\partial^2 f}{\partial x\, \partial y} = s, \frac{\partial^2 f}{\partial y^2} = t,$

The governing term on R.H.S.

$$= \frac{1}{2}\left\{(\delta x)^2 r + 2\, \delta x\, \delta y\, s + (\delta y)^2\, t\right\}$$
$$= \frac{1}{2r}\left\{(r\, \delta x + s\, \delta y)^2 + (rt - s^2)\, (\delta y)^2\right\} \qquad ...(2)$$

This expression will be positive definite if $rt - s^2 > 0$.

In that case the sign of this term will be governed by the sign of r. Hence the sufficient conditions for maxima or minima at $P(x, y, z)$ are

(i) $rt - s^2 > 0$ and

(ii) Maxima if $r < 0$

(iii) Minima if $r > 0$.

If $rt - s^2$ be $= 0$, then (2) reduces to

$\frac{1}{2r}(r\,\delta x + s\,\delta y)^2$, which becomes zero for some values of δx and δy. So no conclusion can be drawn and further investigation will be unnecessary.

1.43 WORKING RULE FOR SEARCHING THE MAXIMA OR MINIMA OF $f(x, y)$

(i) Find $\frac{\partial f}{\partial x}, \frac{\partial f}{\partial y}$ and solve the simultaneous equations $\frac{\partial f}{\partial x} = 0, \frac{\partial f}{\partial y} = 0$. It will give a set of points as $(x_1, y_1), (x_2, y_2)$ ... etc. Where we may expect the maxima or the minima to occur.

(ii) Calculate $\frac{\partial^2 f}{\partial x^2}$ denoted by $= \boldsymbol{r}$, $\frac{\partial^2 f}{\partial x\,\partial y} = \boldsymbol{s}$, $\frac{\partial^2 f}{\partial y^2} = \boldsymbol{t}$ and find $(\boldsymbol{rt - s^2})$ for each set of values $(x_1, y_1), (x_2, y_2)$ etc.

(iii) For those points where $rt - s^2 > 0$, if we have $r > 0$, it implies there is a minima at the point under consideration.

If $r < 0$, then there is a maxima at the point.

If at some point, we have $\boldsymbol{rt - s^2 = 0}$
then further investigation is required.

If at some point, we have $\boldsymbol{rt - s^2 < 0}$, such a point is not an extreme point, but a saddle point.

Example 1.116 **Show that the function $z = axy - x^2y - xy^2$, has a maxima at $(x, y) = \left(\frac{a}{3}, \frac{a}{3}\right)$.**

Solution : Differentiating z, partially w.r.t. x and y,

We have
$$\frac{\partial z}{\partial x} = ay - 2xy - y^2$$
$$\frac{\partial z}{\partial y} = ax - x^2 - 2xy.$$

At extreme point $\frac{\partial z}{\partial x} = 0, \frac{\partial z}{\partial y} = 0$

So $y(a - 2x - y) = 0$

and $x(a - x - 2y) = 0$

So either $y = 0$ or $2x + y = a$ and $x = 0$ or $x + 2y = a$

So we have the following set of points

$$(x_1, y_1) = (0, 0), (x_2, y_2) = (a, 0), (x_3, y_3) = (0, a)\ (x_4, y_4) = \left(\frac{a}{3}, \frac{a}{3}\right)$$

Further differentiating, we have

$$\frac{\partial^2 z}{\partial x^2} = -2y, \quad \frac{\partial^2 z}{\partial x\,\partial y} = a - 2x - 2y, \quad \frac{\partial^2 z}{\partial y^2} = -2x$$

at $\quad x = y = \frac{a}{3},\ r = -\frac{2a}{3},\ s = \frac{a}{3},\ t = -\frac{2a}{3}.$

We have $rt - s^2 = \left(\frac{4}{9} - \frac{1}{9}\right)a^2 > 0$ and $r < 0$

Hence there is a maxima at $\left(\frac{a}{3}, \frac{a}{3}\right)$.

Example 1.117 Find the extreme points of the function $z = x^3 + y^3 - 3axy$.

Solution : At the extreme points $\frac{\partial z}{\partial x} = 0,\ \frac{\partial z}{\partial y} = 0$, so we have $x^2 - ay = 0, y^2 - ax = 0.$

Hence $x = y = a$, or $x = y = 0$

Now $\quad r = \frac{\partial^2 z}{\partial x^2} = 2x,\ \frac{\partial^2 z}{\partial x\, \partial y} = -a,\ \frac{\partial^2 z}{\partial y^2} = 2y$

At (0, 0), $\quad rt - s^2 = 0$

At (a, a), $\quad rt - s^2 = (2a)(2a) - (-a)^2 > 0$

and $\quad r = 2a > 0$

So there is a minima at (a, a).

Example 1.118 Find the extreme points of the function $z = xy + \frac{a^3}{x} + \frac{b^3}{y}$.

Solution : At the extreme points $\frac{\partial z}{\partial x} = 0, \frac{\partial z}{\partial y} = 0$, so we have

$$y - \frac{a^3}{x^2} = 0,\ x - \frac{b^3}{y^2} = 0 \quad \text{or} \quad xy = \frac{a^3}{x^2}\frac{b^3}{y^2}$$

So $\quad xy = ab$ and $x = \frac{a^2}{b}, y = \frac{b^2}{a}.$

Now $\quad r = \frac{\partial^2 z}{\partial x^2} = \frac{2a^3}{x^3},\ s = \frac{\partial^2 z}{\partial x\, \partial y} = 1,\ t = \frac{\partial^2 z}{\partial y^2} = \frac{2b^3}{y^3}$

So $\quad (rt - s^2)$ at $\left(\frac{a^2}{b}, \frac{b^2}{a}\right) = \frac{4a^3b^3}{(xy)^3} - 1 = 4 - 1 = 3 > 0$

and $\quad r = \frac{2a^3}{x^3} = 2\frac{b^3}{a^3} > 0.$

Hence there is minima at $\left(\frac{a^2}{b}, \frac{b^2}{a}\right)$ with $z_{\min} = ab + ab + ab = 3ab.$

Example 1.119 Examine the function $f(x, y) = x^4 + y^4 + 2x^2 + 4xy - 2y^2$, for extreme points and find the values at them.

Solution : It extreme points $\frac{\partial f}{\partial x} = 0, \frac{\partial f}{\partial y} = 0$

or $\quad 4x^3 + 4y - 4x = 0$ and $4y^3 + 4x - 4y = 0$

Adding $4(x^3 + y^3) = 0$ so $x = -y$, other values are imaginary.

Hence the points are

(i) $\quad (x_1, y_1) = (\sqrt{2}, -\sqrt{2}),\ (x_2, y_2) = (-\sqrt{2}, \sqrt{2})$, (iii) (0, 0)

Now

$r = f_{xx} = 12x^2 - 4$	at $P_1(0, 0)$	at $P_2(\sqrt{2}, -\sqrt{2})$	$P_3(-\sqrt{2}, \sqrt{2})$
$s = f_{xy} = 4$	$r = -4$	$r = 20$	$r = 20$
$t = f_{yy} = 12y^2 - 4$	$s = 4$	$s = 4$	$s = 4$
	$t = -4$	$t = 20$	$t = 20$
	$rt - s^2 = 0$	$rt - s^2 > 0$	$rt - s^2 > 0$
	So no conclusion	$r > 0$	$r > 0$
		So there is minima at P_2	Minima at P_3

So there are minima at P_2 and P_3

For investigating further at (0, 0) where $rt - s^2 = 0$

We have $$f(x, y) = 0$$

For points along x-axis $x = 0, y = k, k \to 0$.

$$f(x, y) = k^4 - 2k^2 > 0$$

For points along y-axis $x = h, y = 0, h \to 0$

$$f(x, y) = h^4 - 2h^2 < 0$$

At points (h, k) near origin taking $h = k$

$$f = 2h^4 - 2h^2 + 4h^2 - 2k^2 > 0$$

So near the point (0, 0) the function is both positive and negative in its neighbourhood. Hence (0, 0) is not an extreme point.

Example 1.120 **For a plane ΔABC, find the maximum value of $Z = \cos A \cos B \cos C$.**

Solution : As $A + B + C = \pi$, $C = \pi - A - B$

So $$Z = -\cos A \cos B \cos (A + B).$$

$$\frac{\partial Z}{\partial A} = \cos B \{\sin A \cos (A + B) + \cos A \sin (A + B)\} = 0, \text{ at extreme points}$$

So $\cos B = 0$ or $\sin (2A + B) = 0$,

either $B = \pi/2$ or $2A + B = \pi$ (as $A \geq 0, B \geq 0$).

Similarly

$$\frac{\partial Z}{\partial B} = \cos A \sin (A + 2B) = 0, \text{ when either}$$

$$A = \frac{\pi}{2} \text{ or } A + 2B = \pi$$

Hence the point is $A = B = \frac{\pi}{3}$, where

$$r = -1,\ s = -\frac{1}{2},\ t = -1 \text{ and } rt - s^2 = \frac{3}{4} > 0 \text{ as } r < 0.$$

Hence Z is maximum for $A = B = C = \pi/3$, with $Z_{max} = \frac{1}{8}$.

1.44 LAGRANGE'S METHOD OF UNDETERMINED MULTIPLIERS

Let $u = f(x, y, z)$ be a function of variables (x, y, z), which are constrained to satisfy the condition

$$\phi(x, y, z) = 0$$

u is stationary, where $du = 0$, so we have

$$du = 0 = \frac{\partial f}{\partial x} dx + \frac{\partial f}{\partial y} dy + \frac{\partial f}{\partial z} dz = 0 \qquad ...(1)$$

and
$$\frac{\partial \phi}{\partial x} dx + \frac{\partial \phi}{\partial y} dy + \frac{\partial \phi}{\partial z} dz = 0 \qquad ...(2)$$

Multiplying (2) by λ and adding

$$(f_x + \lambda\, \phi_x)\, dx + (f_y + \lambda\, \phi_y)\, dy + (f_z + \lambda\, \phi_z)\, dz = 0$$

As dx, dy, dz may vary independently. We must have

$$f_x + \lambda\, \phi_x = 0, \quad f_y + \lambda\, \phi_y = 0, \quad f_z + \lambda\, \phi_z = 0 \qquad ...(3)$$

These equations are obtained also by considering the function

$$U = f(x, y, z) + \lambda\, \phi(x, y, z) = 0,$$

now without any constraint. So for extreme points of

$$U, \frac{\partial U}{\partial x} = 0\ \frac{\partial U}{\partial y} = 0, \frac{\partial U}{\partial z} = 0$$

gives the same equation as in (3). Solving these equations, we get values of x, y, z and a equation in U such that the extreme values are the roots of this equation when only the maximum or minimum points are required, it is not essential to do further consideration for determining their nature.

Example 1.121 **Find the stationary values of $r^2 = x^2 + y^2$, subject to the condition $ax^2 + 2hxy + by^2 = 1$.**

Solution : We defined the Lagrangian function L by

$$L(x, y, z) = (x^2 + y^2) + \lambda\, (ax^2 + 2hxy + by^2 - 1) = 0$$

When L is stationary

$$\frac{\partial L}{\partial x} = 0, \frac{\partial L}{\partial y} = 0$$

So
$$2\,|x + \lambda\,(ax + hy)| = 0 \qquad ...(2)$$

$$2\,|y + \lambda\,(hx + by)| = 0 \qquad ...(3)$$

Multiplying (2) by x, (3) by y and adding, we get

$$x^2 + y^2 + \lambda\,(ax^2 + 2hxy + by^2) = 0$$

or
$$u^2 + \lambda = 0 \qquad ...(4)$$

Eliminating x and y from

$$x(1 + \lambda a) + h\lambda y = 0$$

$$x\, h\lambda + (1 + \lambda b)\, y = 0$$

$$\begin{vmatrix} 1 + \lambda & h\lambda \\ h\lambda & 1 + b\lambda \end{vmatrix} = 0$$

or
$$1 + (a + b)\,\lambda + \lambda^2\,(ab - h^2) = 0 \qquad ...(5)$$

Hence the maximum or minimum values of u are roots of

$$u^4 (ab - h^2) - u^2 (a + b) + 1 = 0$$

Example 1.122 Find the volume of the greatest rectangular parallelopiped, that can be inscribed in the ellipsoid

$$\frac{x^2}{a^2} + \frac{y^2}{b^2} + \frac{z^2}{c^2} = 1.$$

Solution : If $2x$, $2y$, $2z$ be the sides of the parallelopiped, then its volume

$$V = 8xyz \qquad \text{...(1)}$$

We defined the Lagrangian function L as

$$L(x, y, z) = 8xyz + \lambda\left(\frac{x^2}{a^2} + \frac{y^2}{b^2} + \frac{z^2}{c^2} - 1\right)$$

For L to be extreme,

$$\frac{\partial L}{\partial x} = 0, \frac{\partial L}{\partial y} = 0, \frac{\partial L}{\partial z} = 0$$

So

$$4yz + \frac{\lambda x}{a^2} = 0 \qquad \text{...(2)}$$

$$4zx + \frac{\lambda y}{b^2} = 0 \qquad \text{...(3)}$$

$$4xy + \frac{\lambda z}{c^2} = 0 \qquad \text{...(4)}$$

Multiplying (2) by x, (3) by y, (4) by z and adding

$$12xyz + \lambda = 0 \text{ or } \lambda = -\frac{3}{2} V \qquad \text{...(5)}$$

Further from (2), (3), (4), we get

$$\frac{-4xyz}{\lambda} = \frac{\frac{x^2}{a^2}}{1} = \frac{\frac{y^2}{b^2}}{1} = \frac{\frac{z^2}{c^2}}{1} = \frac{\frac{x^2}{a^2} + \frac{y^2}{b^2} + \frac{z^2}{c^2}}{1 + 1 + 1} = \frac{1}{3}$$

Hence the extreme values of (x, y, z) are $\left(\frac{a}{\sqrt{3}}, \frac{b}{\sqrt{3}}, \frac{c}{\sqrt{3}}\right)$ avoiding $x = y = 0$, which gives $V = 0$,

So V_{in} maximum at $\left(\frac{a}{\sqrt{3}}, \frac{b}{\sqrt{3}}, \frac{c}{\sqrt{3}}\right)$ with $V_{max} = \frac{8abc}{3\sqrt{3}}$.

Example 1.123 A rectangular box is open at the top and has its volume 3z cubic ft. Find the dimensions of the box requiring least material for its construction.

Solution : If x, y, z be the edges

We have $V_0 = xyz = \text{constant}$ ...(1)

and surface area $S = xy + 2z(x + y)$ ...(2)

Eliminating z

$$S = xy + 2(x + y)\frac{V_0}{xy} = xy + \frac{2V_0}{x} + \frac{2V_0}{y} \qquad \text{...(3)}$$

For maxima or minima

$$\frac{\partial S}{\partial x} = 0, \frac{\partial S}{\partial y} = 0$$

So $$\frac{\partial S}{\partial x} = y - \frac{2V_0}{x^2} = 0,$$

$$\frac{\partial S}{\partial y} = x - \frac{2V_0}{y^2} = 0$$

Hence $x = y = (2V_0)^{1/3}$

$$r = \frac{\partial^2 S}{\partial x^2} = \frac{4V_0}{x^3}, \quad s = \frac{\partial^2 S}{\partial x\, \partial y} = 1, \quad t = \frac{\partial^2 S}{\partial y^2} = \frac{4V_0}{y^3}.$$

So $$rt - s^2 = \frac{16V_0^2}{(xy)^3} - 1 = 3 > 0$$

So S is minimum as $r > 0$.

When $V_0 = 32,\ x = y = (64)^{\frac{1}{3}} = 4$ ft and $Z = 2$ ft.

Example 1.124 **Find the minimum value of $r^2 = x^2 + y^2 + z^2$, subject to $ax + by + cz = p$.**

Solution : Let us define the Lagrangian function $L(x, y, z)$ as

$$L \equiv (x^2 + y^2 + z^2) + \lambda\,(ax + by + cz - p) = 0 \qquad ...(1)$$

For L to be stationary,

$$\frac{\partial L}{\partial x} = 0, \frac{\partial L}{\partial y} = 0, \frac{\partial L}{\partial z} = 0$$

So $$\frac{\partial L}{\partial x} = 2x + \lambda a = 0 \qquad ...(2)$$

$$\frac{\partial L}{\partial y} = 2y + \lambda b = 0 \qquad ...(3)$$

$$\frac{\partial L}{\partial z} = 2z + \lambda c = 0 \qquad ...(4)$$

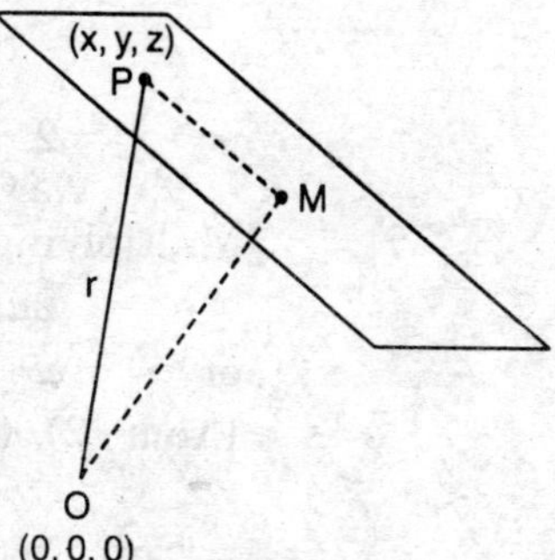

Fig. 1.12

Multiplying (2) by x, (3) by y (4) by z and adding

$$2r^2 + \lambda p = 0 \qquad ...(5)$$

Also from (2), (3), (4)

$$-\frac{\lambda}{2} = \frac{\frac{x}{a}}{1} = \frac{\frac{y}{b}}{1} = \frac{\frac{z}{c}}{1} = \frac{a^2 \frac{x}{a} + \frac{y}{b} b^2 + \frac{z}{c} c^2}{a^2 + b^2 + c^2} = \frac{p}{a^2 + b^2 + c^2}$$

The value is maximum or minimum where $(x, y, z) = \left(\frac{ap}{3}, \frac{bp}{3}, \frac{cp}{3}\right)$.

So $$r^2 = x^2 + y^2 + z^2$$

If OM be the length of perpendicular from O (0, 0, 0) upon the plane, then

$$OM = \left|\frac{0 + 0 + 0 - p}{\sqrt{a^2 + b^2 + c^2}}\right| = \frac{p}{\sqrt{a^2 + b^2 + c^2}}$$

Also the distance of any point on the plane from

$$(0, 0, 0) = (\sqrt{x^2 + y^2 + z^2})_{\min} = OM = \frac{p}{\sqrt{a^2 + b^2 + c^2}}$$

Hence $\quad r_{\min} = \dfrac{p}{\sqrt{a^2 + b^2 + c^2}}$

Example 1.125 **Prove that the stationary values of $u = \dfrac{x^2}{a^4} + \dfrac{y^2}{b^4} + \dfrac{z^2}{c^4}$ subject to the conditions :**

(i) $lx + my + nz = 0$,

(ii) $\dfrac{x^2}{a^2} + \dfrac{y^2}{b^2} + \dfrac{z^2}{c^2} = 1$ are the roots of $\dfrac{l^2a^4}{1 - a^2u} + \dfrac{m^2b^4}{1 - b^2u} + \dfrac{n^2c^4}{1 - c^2u} = 0$.

Solution : We define the Lagrangian function $L(x, y, z)$ as

$$L(x, y, z) = \frac{x^2}{a^4} + \frac{y^2}{b^4} + \frac{z^2}{c^4} + \lambda\,(lx + my + nz) + \mu\left(\frac{x^2}{a^2} + \frac{y^2}{b^2} + \frac{z^2}{c^2} - 1\right) \qquad ...(1)$$

L is stationary, where

$$\frac{\partial L}{\partial x} = 0, \frac{\partial L}{\partial y} = 0, \frac{\partial L}{\partial z} = 0$$

or

$$2\,\frac{x}{a^4} + \lambda l + \frac{2\mu x}{a^2} = 0 \qquad ...(2)$$

$$2\,\frac{y}{b^4} + \lambda m + \frac{2\mu y}{b^2} = 0 \qquad ...(3)$$

$$2\,\frac{z}{c^4} + \lambda n + \frac{2\mu z}{c^2} = 0 \qquad ...(4)$$

Multiplying (2) by x, (3) by y and (4) by z and adding we have

$$2u + \lambda(0) + 2\mu = 0$$

or

$$u + \mu = 0 \qquad ...(5)$$

From (2), (3), (4)

$$x = -\frac{\lambda l}{2\left(\dfrac{1}{a^4} + \dfrac{\mu}{a^2}\right)}$$

$$y = \frac{-\lambda m}{2\left(\dfrac{1}{b^4} + \dfrac{\mu}{b^2}\right)}$$

$$z = \frac{-\lambda n}{2\left(\dfrac{1}{c^4} + \dfrac{\mu}{c^2}\right)}.$$

Putting these values of x, y, z in $lx + my + nz = 0$ and writing $\mu = -u$.

$$-\frac{\lambda}{2}\left[\frac{l^2a^4}{1 - ua^2} + \frac{m^2b^4}{1 - ub^2} + \frac{n^2c^4}{1 - uc^2}\right] = 0$$

Hence u is given as the roots of the above equation.

PROBLEM SET 1.14

1. Find the stationary point of the function,

$$Z = f(x, y) = x^3 + 3xy^2 - 15x^2 - 15y^2 + 72x$$

Minima at (6, 0) $z_{min} = 108$,

Maxima at (4, 0) $Z_{max} = 112$.

2. Find the extreme values of the function $u = \sin x \sin y \sin (x + y)$.

Ans. maxima at $\left(\frac{\pi}{3}, \frac{\pi}{3}\right)$, $u_{max} = 3\sqrt{3}/8$.

3. Discuss the maxima and minima of $f(x, y) = x^3y^2 (1 - x - y)$.

Maxima at $\left(\frac{1}{2}, \frac{1}{3}\right) f_{max} = 1/432$, $f(0, 0)$ is not an extreme value.

4. Discuss the maxima or minima of $Z = x^3y^2 (1 - x - y)$.

Maxima at $\left(\frac{1}{2}, \frac{1}{3}\right)$ with $Z_{max} = \frac{1}{432}$.

5. Examine for maximum and minimum values, the function

$$Z = \sin x + \sin y + \sin (x + y)$$

Ans. Max at $\left(\frac{\pi}{3}, \frac{\pi}{3}\right)$, $Z_{max} = \frac{3\sqrt{3}}{2}$.

6. Find the maximum and minimum distances of the point (3, 4, 12) from the sphere

$$x^2 + y^2 + z^2 = 1.$$

Ans. Max = 14, Minimum = 12.

7. Find the minimum values of $r^2 = x^2 + y^2 + z^2$, when

(i) $x + y + z = 3a$

(ii) $xyz = a^3$

Ans. (i) $3a^2$, (ii) $3a^2$.

8. Examine for extreme values, the functions

(i) $x^2 + y^2 + 6x + 12$

(ii) $3x^2 - y^2 + x^3$

(iii) $x^3 + y^3 + 3xy$

Ans. (i) Min = 3 at (– 3, 0), (ii) Max = 4 at (– 2, 0), (iii) Max = 1 at (– 1, – 1).

9. The temperature T at any point in space is $T = 400\, xyz^2$. Find the highest temperature on the surface of the unit sphere $x^2 + y^2 + z^2 = 1$.

Ans. 50°.

10. Find the maximum value of $u = x^m y^n z^p$, subject to $x + y + z = a$.

Ans. $\frac{m^m n^n p^p a^{m+n+p}}{(m + n + p)^{m+n+p}}$.

11. Find the dimensions of the rectangular box open at top whose surface area is 108 sq. cm., when its capacity is maximum.

Ans. length = breadth = 6 cm, height = 3 cm.

12. Examine for extreme values the functions

(i) $x^2y^2 - 5x^2 - 8xy - 5y^2$,

(ii) $x^3 + 3xy^2 - 15x^2 - 15y^2 + 72x + 100$

Ans. (i) Max. value 0 at (0, 0),

(ii) Max. value 212 at (4, 0), Min. value 208 at (6, 0)

13. Given $f(x, y, z) = \dfrac{5xyz}{x + 2y + 4z}$, find the values of x, y, z for which $f(x, y, z)$ is maximum, subject to the condition $xyz = 8$.

Ans. $(4 \times 4 \times 2)\ (\text{cm})^3$.

14. A tent on a square base of side x has its sides vertical of height y and the top is a regular pyramid of height h. Find x and y in terms of h, if the canvas required for its construction is to be minimum for the tent to have a given capacity.

Ans. $x = \sqrt{5}\, h,\ y = \dfrac{1}{2} h$.

[**Hint** : vol of cube on square base $= x^2y$, Vol of pyramid $= \dfrac{1}{3}x^2h$., Surface $= 4xy + x\sqrt{x^2 + 4h^2}$]

15. Find the maximum and minimum distances from the origin to the curve

$$5x^2 + 6xy + 5y^2 - 8 = 0.$$

Ans. 4, 1.

16. If $u = a^3u^2 + b^3u^2 + c^3z^2$, where $\dfrac{1}{x} + \dfrac{1}{y} + \dfrac{1}{z} = 1$, show that the stationary value of u is given by $x = \dfrac{\Sigma a}{a},\ y = \dfrac{\Sigma a}{b},\ z = \dfrac{\Sigma a}{c}$.

17. Find the minimum value of $u = x^2 + y^2 + z^2$, subject to the condition

$$ax + by + cz = a + b + c$$

Ans. $\dfrac{(a + b + c)^2}{a^2 + b^2 + c^2}$

18. Find the miniumum value of $u = x^2 + y^2 + z^2$, subject to $xy + yz + zx = 3a^2$.

Ans. $3a^2$ at $x = y = z = a$ or $x = y = z = -a$.

19. Find the points on the surface $z^2 - xy = 1$, which are nearest to the origin.

Ans. $(0, 0, \pm 1)$

20. Find the minimum value of $u = ax + by + cz$ subject to $\dfrac{a^2}{x} + \dfrac{b^2}{y} + \dfrac{c^2}{z} = 1$.

Ans. $(a^{3/2} + b^{3/2} + c^{3/2})^2$.

21. Find the minimum value of $u = x + y + z$, subject to the condition $\dfrac{a}{x} + \dfrac{b}{y} + \dfrac{c}{z} = 1$.

Ans. $(\sqrt{a} + \sqrt{b} + \sqrt{c})^2$.

1.45 CURVE TRACING

The equation $f(x, y) = 0$, or $f(r, \theta) = 0$ represent a curve in a two dimensional $X - Y$ plane. Drawing the geometrical figure represented by the equation $f(x, y) = 0$ or $f(r, \theta) = 0$ is called tracing of Cartesian or polar curves.

The following points are **helpful for tracing a curve**.

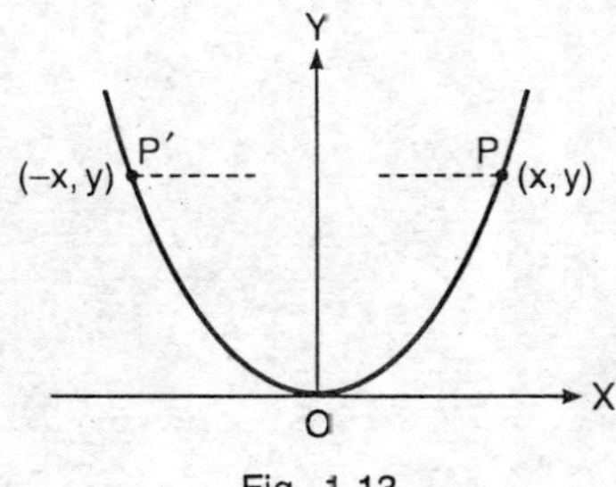

Fig. 1.13

1.45.1 Symmetry

Equations having only the even powers of x, as $x^2 = 4ay$, are $(\pm x)^2 = 4ay$ so if (x, y) satisfies then $(-x, y)$ also satisfies Hence the curve is symmetrical about the y axis. Equations as $y^2 = 4ax$ is $(\pm y)^2 = 4ax$. So when powers of y in the eqation are all even the curve will be of y are symmetrical about x axis. Equation as $\frac{x^2}{a^2} + \frac{y^2}{b^2} = 1$ has symmetry in all the four quadrants as they are symmetrical about both the axes.

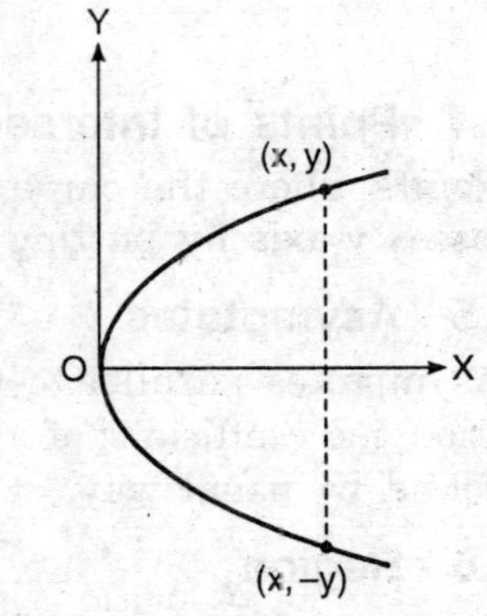

Fig. 1.14

Equation as $x^3 + y^3 = 3axy$, does not change on interchanging x with y, such curves are symmetrical about the line $y = x$. The polar curve $r = a(1 + \cos\theta)$, is symmetrical about the initial line $\theta = 0$ as $\cos(-\theta) = \cos\theta$. Here $f(\theta)$ is a periodic function of θ, so whole curve is known for $\theta = 0$ to $\theta = 2\pi$ be traced for values of θ from 0 to π only as $\cos\theta$ is an even function of θ.

1.45.2 Origin

If (0, 0) satisfy the equation $f(x, y) = 0$, then the curve passes through the origin. For a curve x may be supposed to lie for $-\infty < x < \infty$. From this those intervals are deleted from y becomes imaginary.

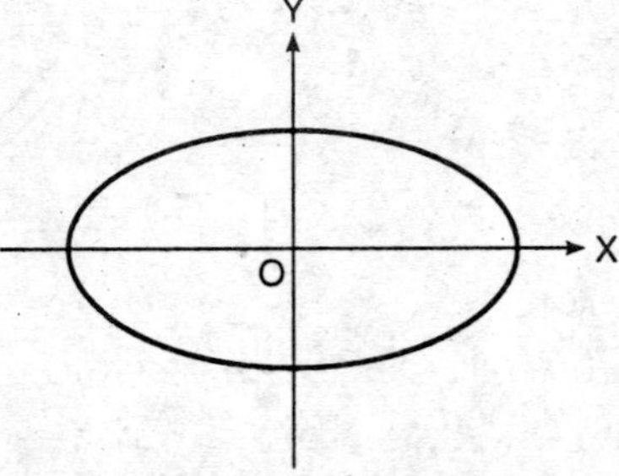

Fig. 1.15

$\frac{dy}{dx}$ the points where $\frac{dy}{dx} = 0$ and infinite, are the points on the curve where the tangents are parallel to x axis and y axis respectively.

1.45.3 TANGENTS AT ORIGIN

By equating the lowest degree terms to zero, we get the tangents at origin to the curve. Thus, the tangents at origin to the curve $x^3 + y^3 - 3axy = 0$ are $xy = 0$. If there be two tangents and they be distinct as here the tangents $x = 0$, $y = 0$ are distinct, the origin is a node.

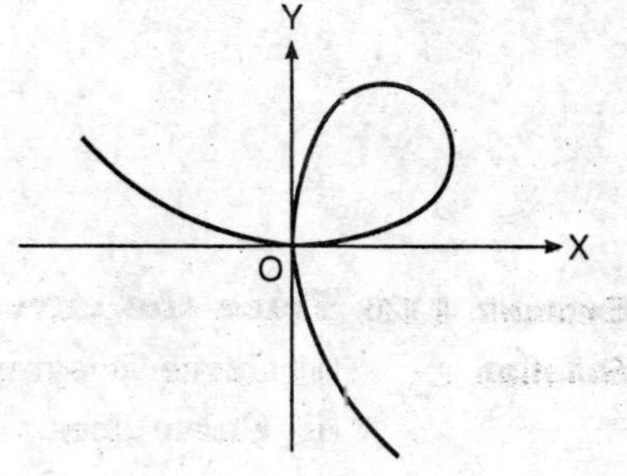

Fig. 1.16

But if there be two coincident tangents are for example in $x^3 + xy^2 - 2ay^2 = 0$

The tangents origin are $y^2 = 0$, which represent two coincident lines. Origin is then a cusp.

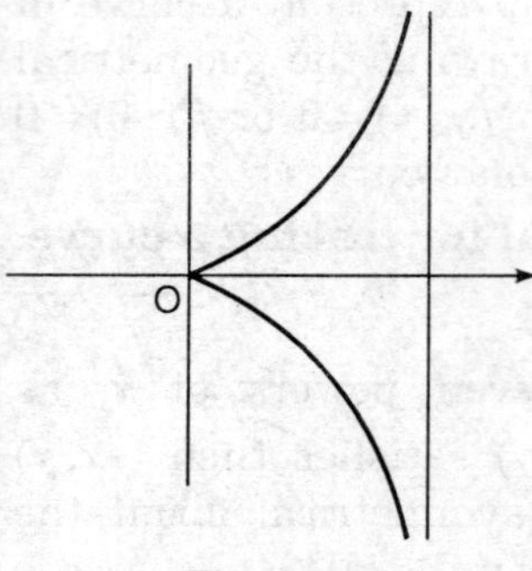

Fig. 1.17

1.45.4 Points of Intersection with Axis

Points where the curve crosses the x-axis are found by putting $y = 0$, and those where it crosses y-axis by putting $x = 0$, in the equation to the curve.

1.45.5 Asymptotes

Asymptotes parallel to the axes are obtained merely by inspection. They are given by equation the coefficient of x or y, provided they are not merely constants. Other asymptotes are found by usual way.

1.45.6 Region

Solving for y, the whole curve is the approximate shape of the curve as shown in the figure.

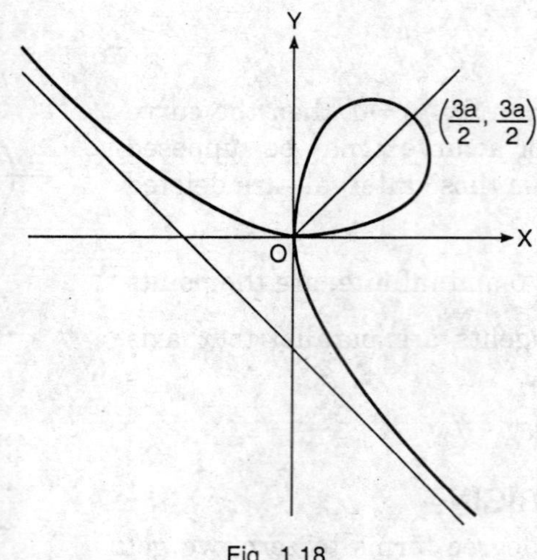

Fig. 1.18

Example 1.126 **Trace the curve $xy^2 = 4a^2(2a - x)$.**

Solution : (i) Curve is symmetrical about x-axis.

(ii) Curve does not pass through origin as (0, 0) does not satisfy the equation.

(iii) Solution for y,

$$y = 2a\sqrt{\frac{2a - x}{x}},$$

So y is real for $\frac{2a - x}{x} \geq 0$

So the Curve lies between $0 \leq x \leq 2a$ coeff. of highest power of y is x

(iv) As coeff. of highest power of y equated to zero gives $x = 0$.

$x = 0$ is the only real asymptote where $y \to \infty$

(v) at $x = 2a, y = 0$, for $x = 0$,

$$y = \pm 2a\sqrt{\frac{2a - x}{x}}$$

$$\frac{dy}{dx} = \frac{\pm 2a}{2x\sqrt{x}}\left\{\frac{-x - (2a - x)}{\sqrt{2a - x}}\right\} \to \infty \quad \text{as } x \to 2a$$

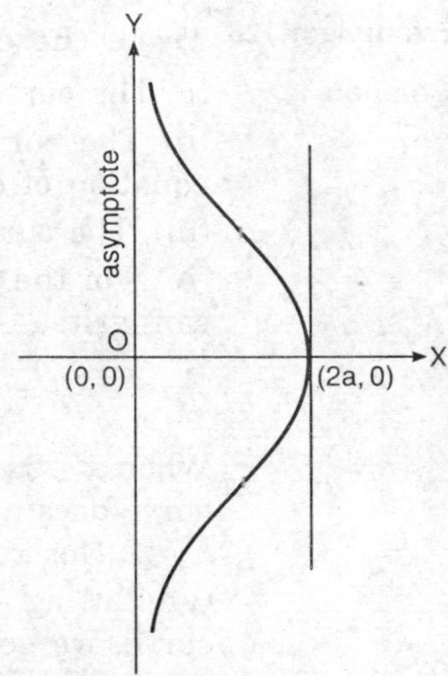

Fig. 1.19

So tangent at $(2a, 0)$ is parallel to y-axis. The approximate shape of the curve is as shown in the figure.

Example 1.127 **Trace the curve $y^2(2a - x) = x^3$.**

Solution : (i) Equation has only the even power of y so the curve is symmetrical about the x-axis.

(ii) The curve passes through (0, 0) as (0, 0) satisfies the equation.

(iii) Equating to zero the lowest degree terms of the equation the tangent at origin are given by $y^2 = 0$. It represents two coincident lines $y = 0, y = 0$ As there are two coincident tangents, so origin is a cusp.

(iv) By equating to zero the coeff. of highest power of y, the line $x = 2a$ is its asymptote which is parallel to y-axis.

Solving for y,

$$y = \pm x\sqrt{\frac{x}{2a - x}}$$

We note that y is real where $\frac{x}{2a - x} > 0$ or when $0 < x < 2a$.

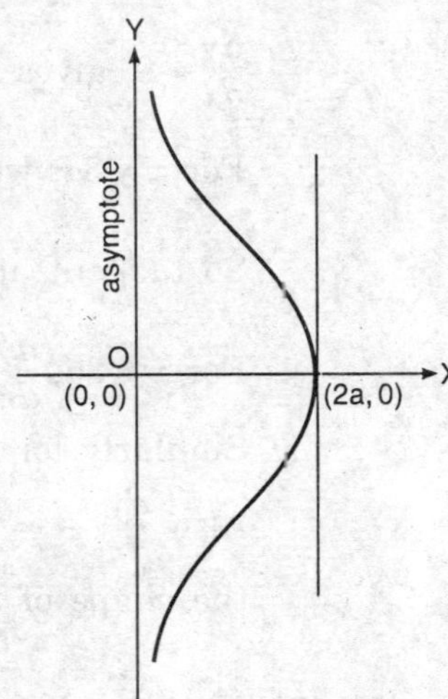

Fig. 1.20

The approximate shape of the curve is as shown in the Fig. 1.20. The curve is known as cissoid.

Example 1.128 **Trace the cruve $9ay^2 = (x - 2a)(x - 5a)^2$.**

Solution : (i) The curve is symmetrical about x-axis.

(ii) The curve does not pass through origin as (0, 0) does not satisfy the equation of curve.

(iii) The curve has no asymptote.

(iv) For the given equation of curve, we can write

$$y = \frac{1}{\sqrt{9a}}(x - 5a)\sqrt{x - 2a}$$

When $x < 2a$, y is imaginary, Thus, the curve does not lie to the left of the line $x = 2a$, for $x > 2a$, y is real.

(v) Putting $y = 0$, in the equation of curve, we get $x = 2a$, $5a$.

Thus, the curve meets x-axis in the points $(2a, 0)$ and $(5a, 0)$ putting $x = 0$, in the equation of curve, we get no real values of y, thus the curve does not intersect y-axis.

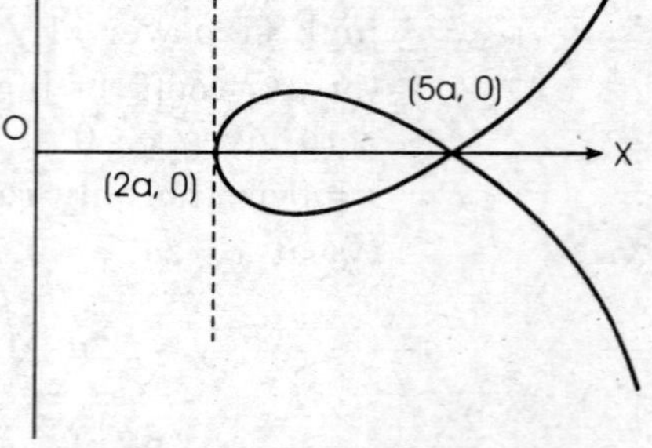

Fig. 1.21

(vi) For the given curve

$$\frac{dy}{dn} = \frac{1}{\sqrt{9a}}\left[(x - 5a)\frac{1}{2\sqrt{x - 2a}} + \sqrt{x - 2a}\right] = \frac{x - 3a}{2\sqrt{a}\sqrt{x - 2a}}$$

$\frac{dy}{dx} = 0$, gives $x = 3a$.

For $x = 3a$, we get $y = \pm\frac{2a}{3}$

So tangent at the points $\left(3a, \pm\frac{2a}{3}\right)$ are parallel to x-axis.

For $x > 3a$, $\frac{dy}{dn}$ is $+ve$ so the curve rises for all values of $x > 3a$.

Similarly for $x < 3a$, the curve is falling.

Also $\frac{dy}{dn} \to \infty$ as $x = 2a$, so the tangent at $(2a, 0)$ is parallel to y-axis. Thus, the shape of the curve is as in figure 1.21.

Example 1.129 **Trace the curve $x^2y^2 = x^2 - a^2$.**

Solution : The equation of the curve is

$$x^2y^2 = x^2 - a^2 \qquad ...(1)$$

(i) The curve is symmetrical about both the axis and in opposite quadrates

(ii) The curve does not pass through origin as (0, 0) does not satisfy the given equation.

(iii) $y = 1$ and $y = -1$ are asymptotes parallel to x-axis.

(iv) For $x = 0$, in equation (1), we get $x = \pm a$.

$\therefore$ The curve intersects x-axis at the points $(a, 0)$ and $(-a, 0)$ equating to zero the lowest degree form of the equation we get $x = a$ and $x = -a$. So $x = a$ and $x = -a$ are tangents to curve at $(a, 0)$ and $(-a, 0)$ respectively.

Obviously the curve does not intersect y-axis.

(v) From equation (1), we get

$$y = \frac{\sqrt{x^2 - a^2}}{x} \qquad ...(2)$$

For y be real, $x^2 - a^2$ must be positive,

i.e., $x^2 > a^2$ or $x < -a$ or $x > a$.

So curve does not line in between $x = -a$ and $x = a$.

Also from the same equation (1)

$$x = \frac{a}{\sqrt{1 - y^2}}$$

For real values of x, $1 - y^2 > 0$, *i.e.*, $y^2 < 1$. So $-1 < y < 1$.

Thus implus that curve lies between $y = -1$ and $y = 1$.

(iv) From equation (2), we get

$$\frac{dy}{dx} = \frac{a^2}{x^2\sqrt{x^2 - a^2}}$$

Obviously $\frac{dy}{dx} \neq 0$ and $\frac{dy}{dx}$ is always +ive.

Also $\frac{dy}{dx} \to \infty$ as $x^2 - a^2 = 0$, *i.e.*, $x = \pm a$, *i.e.*, $x = a$ and $x = -a$ are tangents parallel to y-axis.

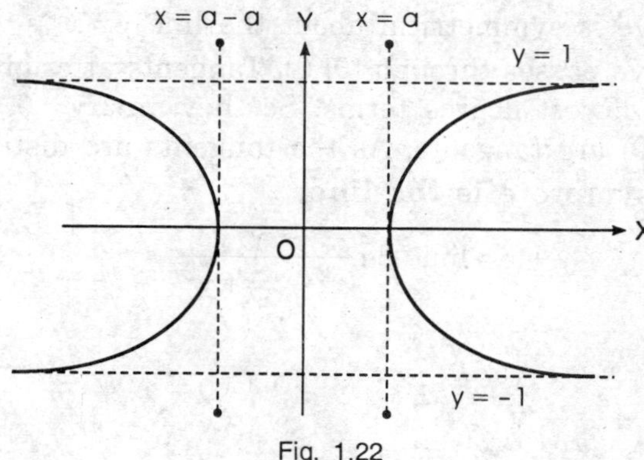

Fig. 1.22

Thus, the approximate shape of the curve is given in the figure.

Example 1.30 **Trace the curve $x^{1/2} + y^{1/2} = a^{1/2}$**

Solution : The given equation of curve is

$$x^{1/2} + y^{1/2} = a^{1/2} \qquad ...(1)$$

(i) The curve is symmetrical about the line $y = x$.

(ii) The curve does not pass through origin.

(iii) No asymptotes.

(iv) For $x = 0$ in equation (1), we get $y = a$ and for $y = 0, x = a$. Hence the curve meets the coordinate axes at $(0, a)$ and $(a, 0)$.

(v) Special points are as :

x :	a	$\frac{a}{4}$	$\frac{a}{2}$	$\frac{3a}{4}$	0
y :	0	$\frac{a}{4}$	$\left(\frac{\sqrt{2}-1}{\sqrt{2}}\right)^2 a$	$\left(\frac{2-\sqrt{3}}{2}\right)^2 a$	a

The approximate shape of the curve is given in the figure.

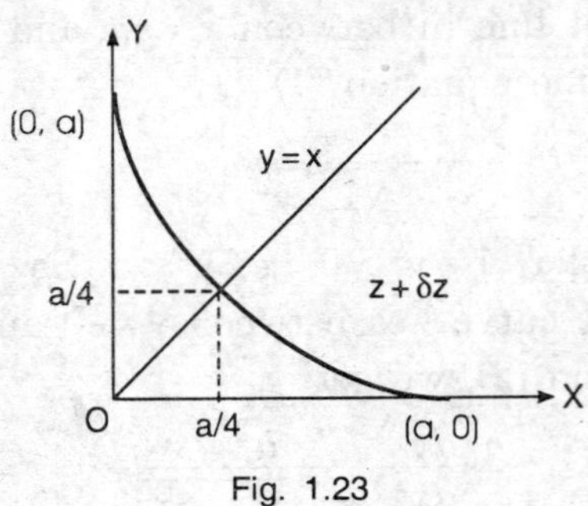

Fig. 1.23

Example 1.131 Trace the curve $x^3 + y^3 = 3axy$. *(Marathwada 1998)*

Solution : Putting $x = r\cos\theta$, $y = r\sin\theta$ the polar equation of the curve is

$$r = \frac{3a\sin\theta\cos\theta}{\sin^3\theta + \cos^3\theta}$$

(i) The curve is symmetrical about the line $y = x$.

(ii) The curve passes through (0, 0). Tangents at origin are given by equating to zero the lowest degree terms. So they are $xy = 0$. Thus x axis, $y = 0$ and y axis $(x = 0)$ are tangents, As the tangents are distinct so origin is a node.

(iii) **The asymptote is the line**

$$y + x = \lim_{x \to \infty} 3a\frac{xy}{x^2 - xy + y^2}$$

$$\frac{x}{y} \to -1$$

$$= 3a\frac{(-1)}{1+1+1}$$

or $\quad x + y + a = 0$ is the asymptote to the curve.

(iv) For $x = -$ ive $y = -$ ive, in third quadrant, the equation gives $-t^3 - t^3 = +3at^2$. But -ve = positive is true only when $x = 0, y = 0$. So no portion of the curve lies in the third quadrant.

When $\quad \theta \to \frac{3\pi}{4}$ or $\frac{5\pi}{4}$, $r \to \infty$.

Curve meets the line $y = x$ at $2x^3 = 3ax^2$, $x = 0$

or $$x = \frac{3}{2} a = y.$$

Curve passes through the origin.

The approximate shape of the curve is

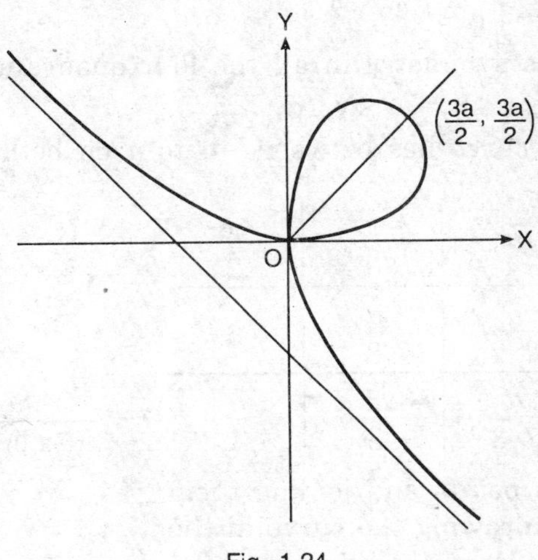

Fig. 1.24

Example 1.132 **Trace the curve represented by**

$$\boldsymbol{x = a \cos t + \frac{a}{2} \log \tan^2 \frac{t}{2}}$$

$$\boldsymbol{y = a \sin t.}$$

Solution : The equation giving x and y in terms of t are periodic in the parameter t. We have $x(-t) = x(t)$ and $y(-t) = -y(t)$

For $t \to 0$, $x \to \log 0 \to -\infty$, $y = 0$, so curve tends to the point $P_1(-\infty, 0)$ for $t = \pi/2$ $x = 0 + 0 + 0$, $y = a$. The point $P_2(0, a)$ is on the curve for $t = \pi$, $x \to \log \infty$, $y = 0$, so curve tends to $P_3(+\infty, 0)$ for $t = \frac{3\pi}{2}$, values of x remain same but $y < 0$, to curve tests from $(-\infty, 0)$ to $P_4(0, -a)$ remaining below the x-axis.

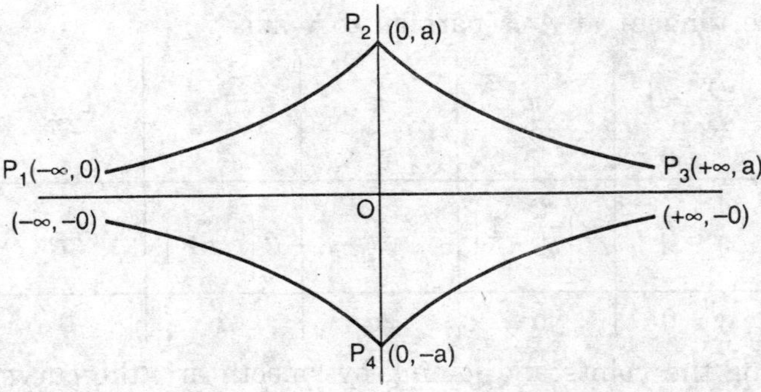

Fig. 1.25

For $t = 2\pi$ points tends to $(+\infty, 0)$ with y remaining – ive.

The curve has the above shape. It is known as tractrix.

Example 1.133 Trace the curve

$$\boldsymbol{x = a\cos^3 t, y = a\sin^3 t.}$$

Solution : The equation can also be expressed as

$$(x^2)^{1/3} + (y^2)^{1/3} = a^{2/3}$$

The curve has symmetry in all the four quadrants.

We have $-a \le x \le a$, $-a \le y \le a$.

So the whole curve lies between $-a$ to a on both the axes.

when $\theta = 0$	$\frac{\pi}{4}$	$\frac{\pi}{3}$	$\frac{\pi}{2}$
$x = a$	$\frac{a}{2\sqrt{2}}$	$\frac{a}{8}$	0
$y = 0$	$\frac{a}{2\sqrt{3}}$	$\frac{3\sqrt{3}\,a}{8}$	$\frac{a}{1}$

Plotting these points and joining them smoothly and drawing the curve in the remaining quadrants by symmetry the approximate shape is as shown in the Fig. 1.26.

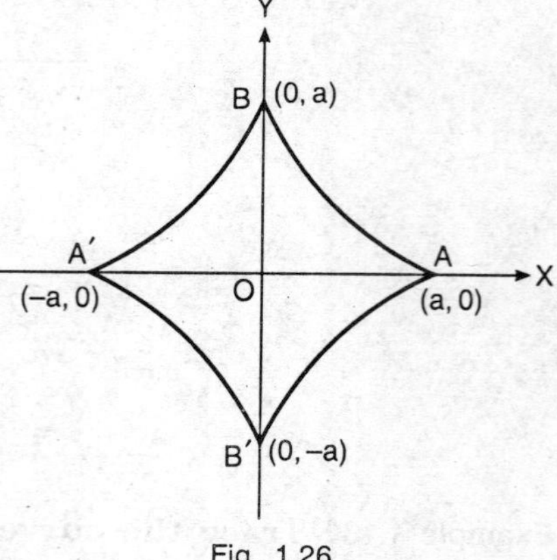

Fig. 1.26

The curve is known as astroid.

Example 1.134 Trace the curve

$$\boldsymbol{x = a\,(\theta - \sin\theta), y = a\,(1 - \cos\theta).}$$

Solution : $x = a\,\theta - a\sin\theta$. So as θ increases, x continuously increases because of the turn $a\,\theta$ in its expression. $y = a\,(1 - \cos\theta)$. So y lies between 0 to $2a$ and is always positive.

$$\frac{dy}{dx} = \frac{\frac{dy}{d\theta}}{\frac{dx}{d\theta}} = \frac{a\sin\theta}{a\,(1-\cos\theta)} = \frac{\cot\theta}{2} \text{ at } \theta = \pi,$$

$$\frac{dy}{dx} = 0.$$

So the tangent at A is parallel to x axis.

$\frac{dy}{dx} = 0$ $\theta = 0$	$\frac{\pi}{2}$	$\frac{\pi}{2}$	$\frac{3\pi}{2}$	2π
$x = 0$	$a\left(\frac{\pi}{2} - 1\right)$	$a\pi$	$a\left(\frac{\pi}{\cdot} + \pi\right)$	$2a\pi$
$y = 0$	a	$2a$	a	0

Plotting the points and joining by smooth arc, the curve is as shown above.

It is called a cycloid. The curve $x = a(\theta + \sin\theta), y = a(1 - \cos\theta)$, is known as inverted cycloid. $\frac{dy}{dx} = \tan\theta/2$. So at $\theta = \pi\ \frac{dy}{dx} = \infty$, and the tangent is parallel to y axis.

Considering as above the shape of the curve is

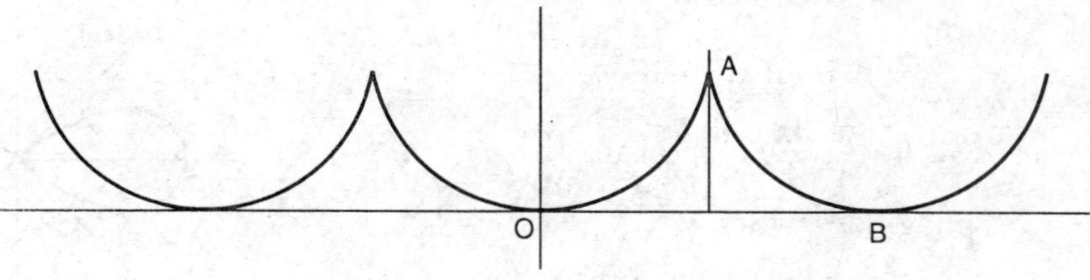

Fig. 1.27

Example 1.135 Trace the curve $r = a(1 + \cos\theta)$.

Solution : The curve is symmetrical about the initial line $\theta = 0$.

As $\cos(-\theta) = \cos\theta$.

The expression $f(\theta) = a(1 + \cos\theta)$ is a period function, so the whole curve is obtained by given values 0 to π too θ.

$0 =$	0	$\frac{\pi}{3}$	$\frac{\pi}{2}$	$\frac{2\pi}{3}$	π
$r =$	2α	$\frac{3\alpha}{2}$	α	$\frac{\alpha}{2}$	0

Plotting these points, the approximate shape of the curve is as shown in the Fig 1.28(i). The curve is known as cardioid.

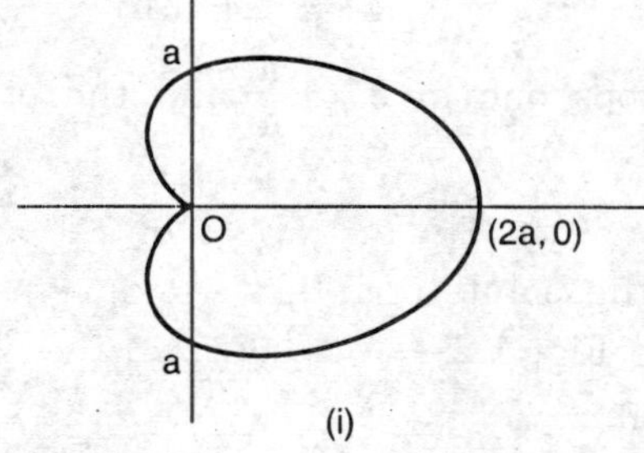

(i)

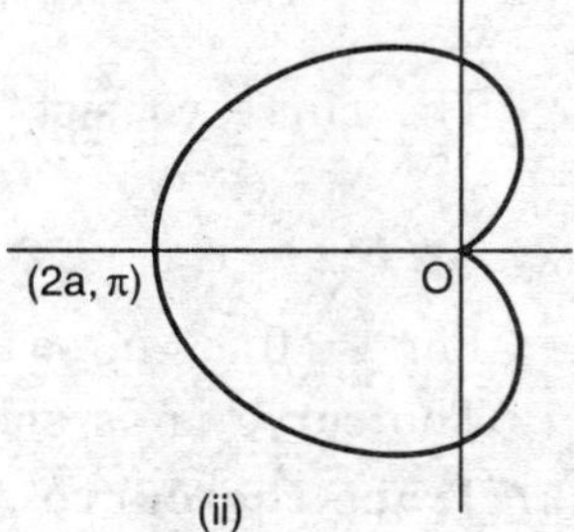

(ii)

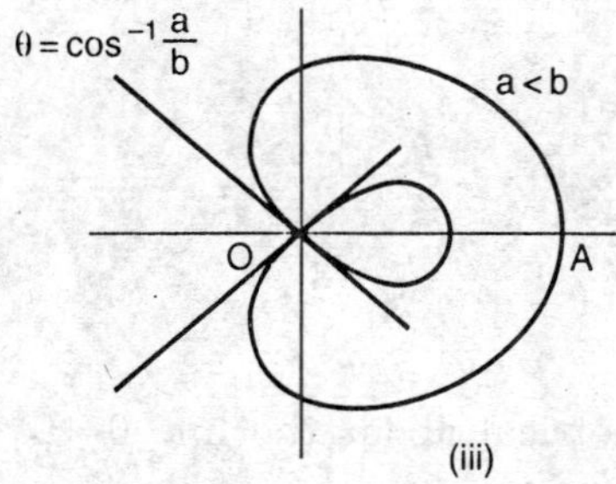

(iii)

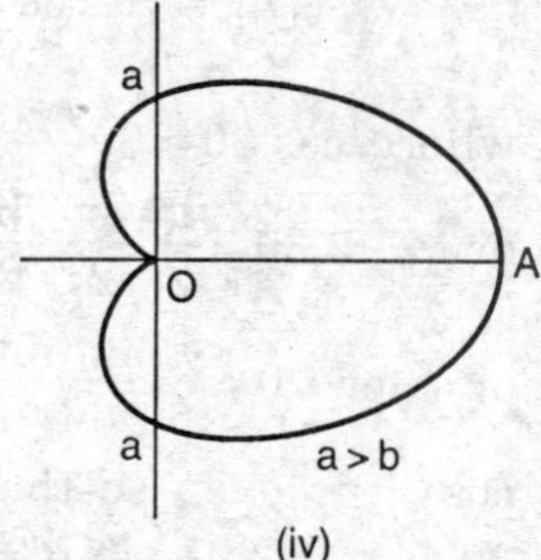

(iv)

Fig. 1.28

The curve $r = a(1 - \cos\theta)$ is also cardioid, its shape is given in Fig. 1.28(ii)

$r = a + b\cos\theta$, $a < b$ has the shape as shown in Fig. 1.28(iii).

$r = a + b\cos\theta$, $a > b$ has the shape given in Fig. 1.28(iv).

Example 1.136 **Trace the curve $(x^2 + y^2)^2 = a^2(x^2 - y^2)$ or $r^2 = a^2\cos 2\theta$.** *(Poona 1990)*

Solution : $r = 0$ when $\cos 2\theta = 0$

$$2\theta = \frac{\pi}{2}, \frac{3\pi}{2}, \frac{5\pi}{2}, \frac{7\pi}{2}$$

or $$\theta = \frac{\pi}{4}, \frac{3\pi}{4}, \frac{5\pi}{4}, \frac{7\pi}{4}.$$

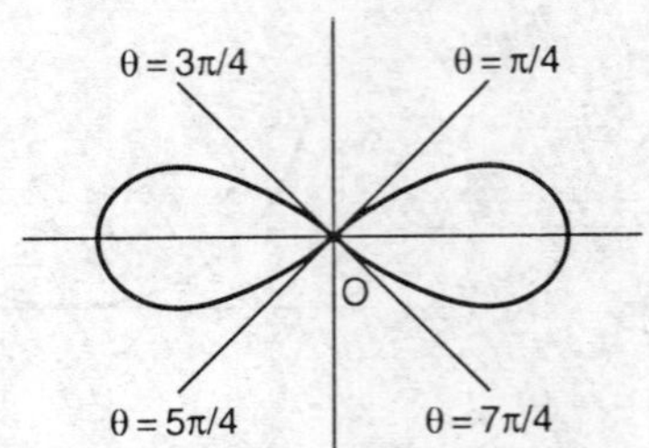

Fig. 1.29

r is imaginary when

$$r^2 = -\text{ive or } \cos 2\theta = -\text{ive}$$

So when $\frac{\pi}{2} < 2\theta < \frac{3\pi}{2}$ and when

$$\frac{5\pi}{2} < 2\theta < \frac{7\pi}{2}$$

or $$\frac{\pi}{4} < \theta \frac{3\pi}{4} \text{ and } \frac{5\pi}{4} < \theta < \frac{7\pi}{4}$$

The equation to the curve has only the even powers both in x and y. So the curve is symmetrical about both the axes. The curve lies only between $\theta = \pm\frac{\pi}{4}$ and $\theta = \frac{3\pi}{4}$ to $\theta = \frac{5\pi}{4}$.

$r = 0$ when $$\cos 2\theta = 0 \text{ or } 2\theta = \frac{\pi}{2}, \frac{3\pi}{2}, \frac{5\pi}{2}, \frac{7\pi}{2}$$

or $$\theta = \frac{\pi}{4}, \frac{3\pi}{4}, \frac{5\pi}{4}, \frac{7\pi}{4}.$$

The curve consists of two loops one in $\theta = \pm\frac{\pi}{4}$ and the other in $\theta = \frac{3\pi}{4}$ to $\frac{5\pi}{4}$.

For $y = 0$, $x = a$, so A, A' are the points $(\pm a, 0)$.

The curve is as shown in the Fig. 1.29.

Example 1.137 **Trace the curve $r = a\cos 3\theta$.** *(Kuvempu 1998)*

Solution : The curve lies in the region

$$|r| = |a\cos 3\theta| < a.$$

$$r = 0$$

where $\cos 3\theta = 0$

or $$3\theta = \frac{\pi}{2}, \frac{3\pi}{2}, \frac{5\pi}{2}, \frac{7\pi}{2}, \frac{9\pi}{2}, \frac{11\pi}{2}, \frac{13\pi}{2}$$

or for $$\theta = \frac{\pi}{6}, \frac{3\pi}{6}, \frac{5\pi}{6}, \frac{9\pi}{6}, \frac{11\pi}{6}.$$

$\cos 3\theta$ is even so the curve is symmetrical about the line $\theta = 0$

3θ in the following intervals.

$3\theta \to$	$\frac{\pi}{2} < 3\theta < \frac{3\pi}{2}$	$\frac{3\pi}{2} < 3\theta < \frac{5\pi}{2}$	$\frac{5\pi}{2} < 3\theta < \frac{7\pi}{2}$	$\frac{7\pi}{2} < 3\theta < \frac{9\pi}{2}$	$\frac{9\pi}{2} < 3\theta < \frac{11\pi}{2}$	$\frac{11\pi}{2} < 3\theta < \frac{13\pi}{2}$
$r = a \cos 3\theta$	−ve	+ve	−ve	+ve	−ve	+ve
	$\frac{\pi}{6} < \theta < \frac{3\pi}{6}$ r is −ve so loop lies in $\frac{7\pi}{6} < \theta < \frac{9\pi}{6}$	$\frac{3\pi}{6} < \theta < \frac{5\pi}{6}$ $r = +\text{ve}$	$\frac{5\pi}{6} < \theta < \frac{7\pi}{6}$ $r = -\text{ve}$ loop lies in $\frac{11\pi}{6} < \theta < \frac{13\pi}{6}$	$\frac{7\pi}{6} < \theta < \frac{9\pi}{6}$ $r = +\text{ve}$	$\frac{9\pi}{6} < \theta < \frac{11\pi}{6}$ $r = -\text{ve}$ $\frac{3\pi}{6} < \theta < \frac{7\pi}{6}$	$\frac{11\pi}{6} < \theta < \frac{13\pi}{6}$
	Ist loop	2nd loop	3rd loop	1st loop repeated	2nd repeated	3rd repeated

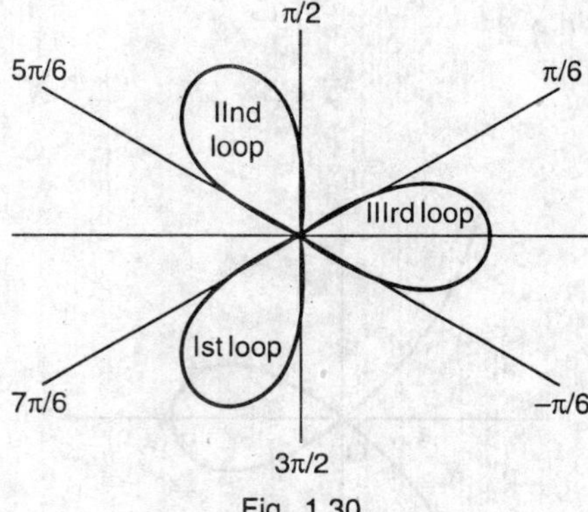

Fig. 1.30

Example 1.138 **Trace the curve $y^2 (a + x) = x^2 (a - x)$.** *(V.T.U. 2000)*

Solution : 1. Curve is symmetrical about x axis (only even powers in y)

2. Curves passes through (0, 0)

3. Tangent at origin are $y = \pm x$, origin in a node.

4. Asymptote $x + a = 0$

$$y = x\sqrt{\frac{a - x}{a + x}}$$

y is real only for $-a < x < a$.

So curve lies only in $-a$ to a

$y = 0$ at $x = 0, x = a$.

Shape is as shown in the Fig. 1.31.

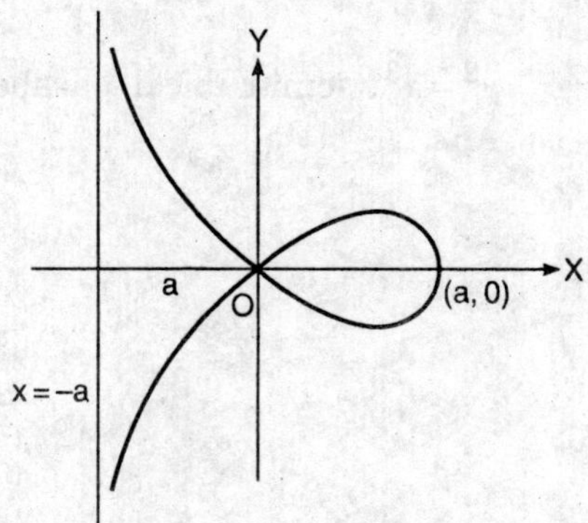

Fig. 1.31

Example 1.139 **Trace the curve**

$$y^2 (a^2 + x^2) = x^2 (a^2 - x^2).$$ *(Kuvempu'98)*

Solution : Equation has only even powers in both x and y. So there is symmetry in all the four quadrants.

Curve passes through origin.

Tangents at origin are $y = \pm x$.

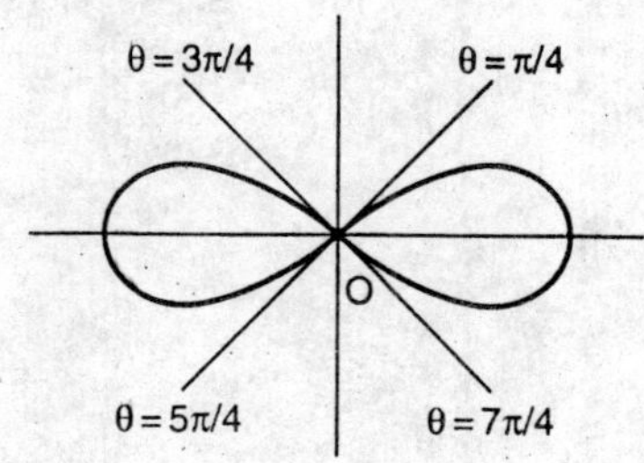

Fig. 1.32

So origin is a node.

y is real only for $a^2 - x^2 \geq 0$ or $-a < x < a$.

$$y = x\sqrt{\frac{(a-x)(a+x)}{a^2+x^2}}$$

curve meets x axis at $(\pm a, 0)$

Curve meets y axis only at (0, 0). There are no real asymtotes.

PROBLEM SET 1.15

Trace the following curves :

1. $ay^2 = x^2(a-x)$

Ans.

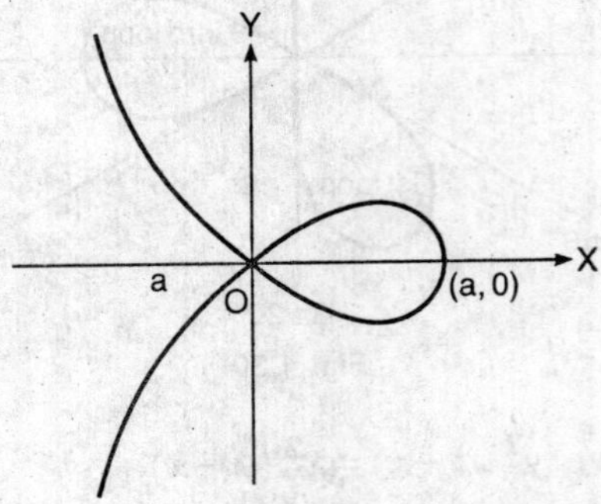

2. $y^2 = x^3$ (semi-cubical parabola).

Ans.

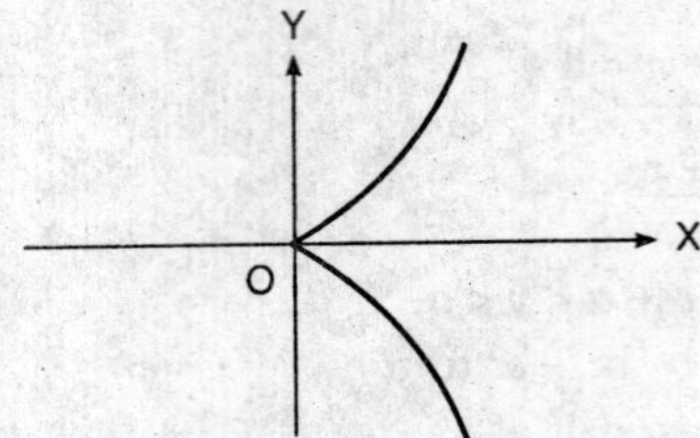

3. $y = \dfrac{x^2 + 1}{x^2 - 1}$

Ans.

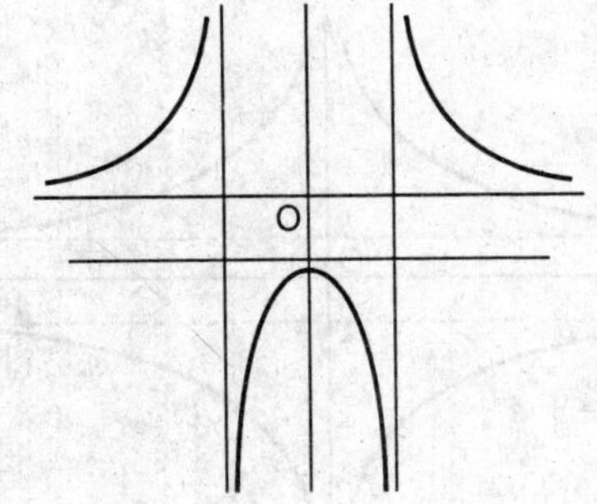

4. $a^4y^2 = a^2x^4 - x^6$

Ans.

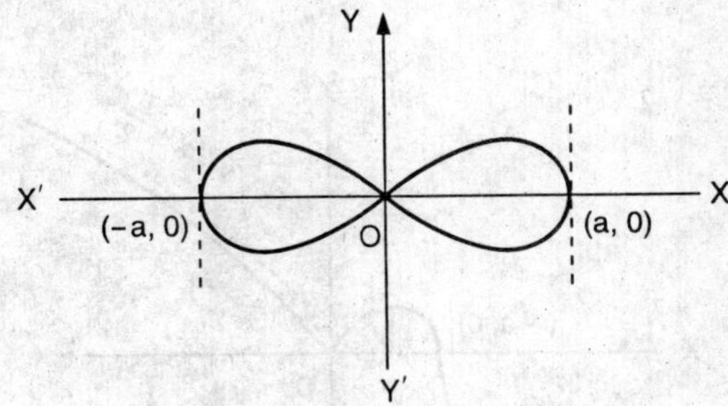

5. $xy^2 = 4a^2(2a - x)$.

Ans.

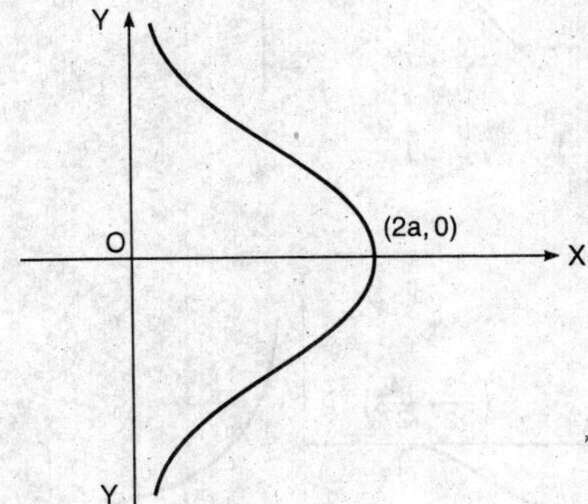

6. $x^2y^2 = x^2 + 1$

Ans.

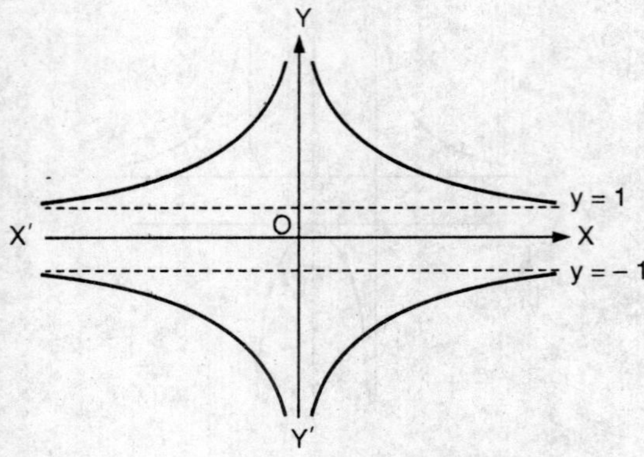

7. $y^3 = x^3 + ax^2$

Ans.

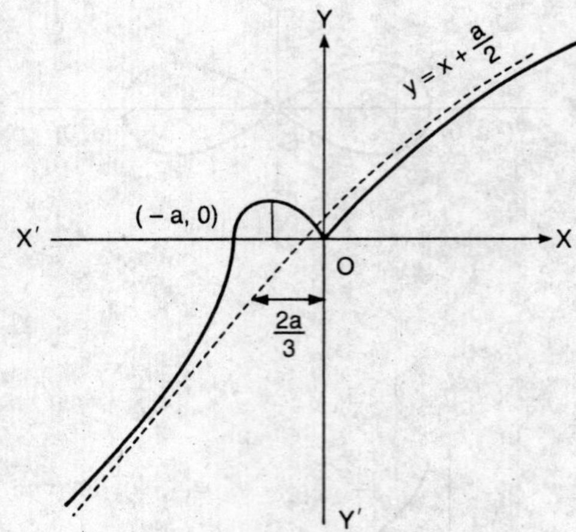

8. $x^3y - x - 1 = 0$

Ans.

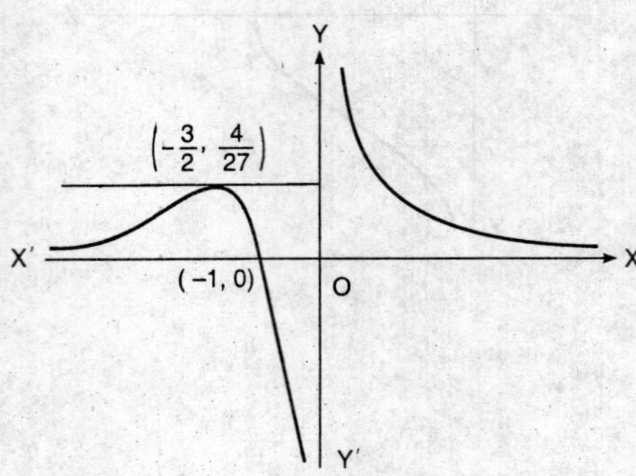

9. $x = \left(a \cos t + \log \tan \frac{1}{2}\right), \quad y = a \sin t$

Ans.

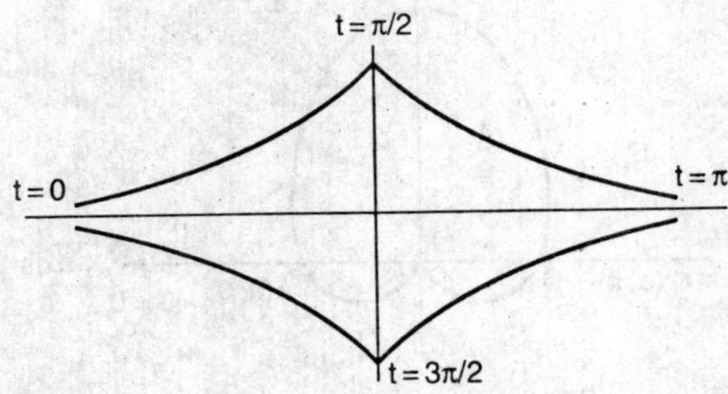

10. $x^2y^2 = a^2(y^2 - x^2)$

Ans.

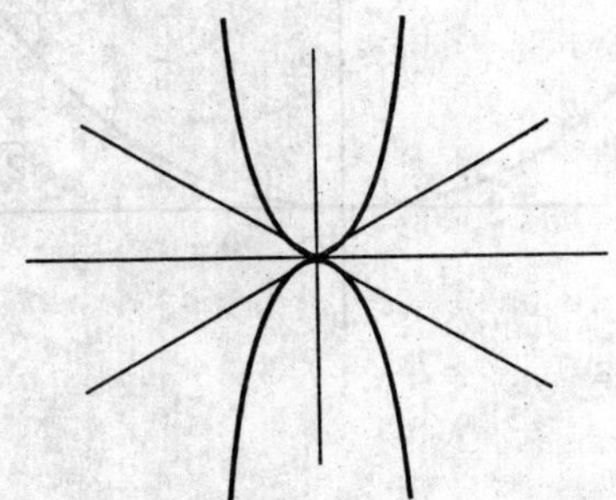

11. $\frac{2a}{r} = 1 + \cos\theta$

Ans.

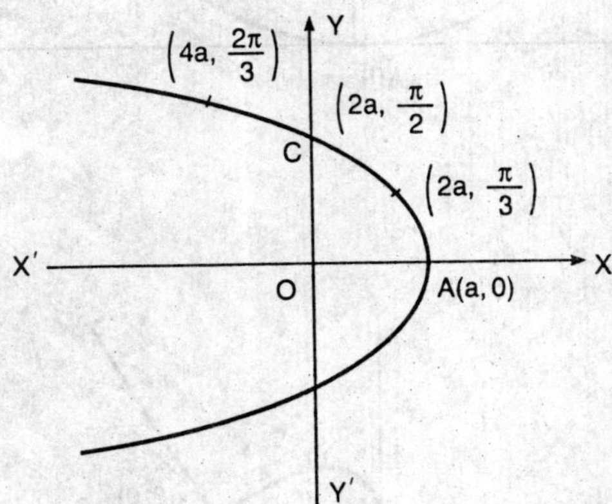

12. $\gamma = a(1 + \sin\theta)$

Ans.

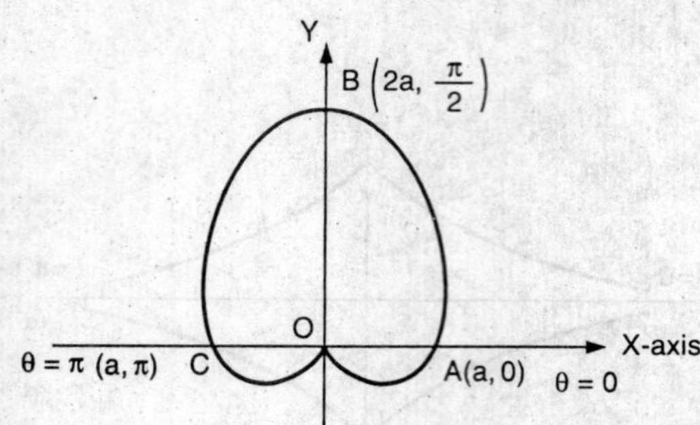

13. $x = a(\theta + \sin\theta),\ y = a(1 - \cos\theta)$

Ans.

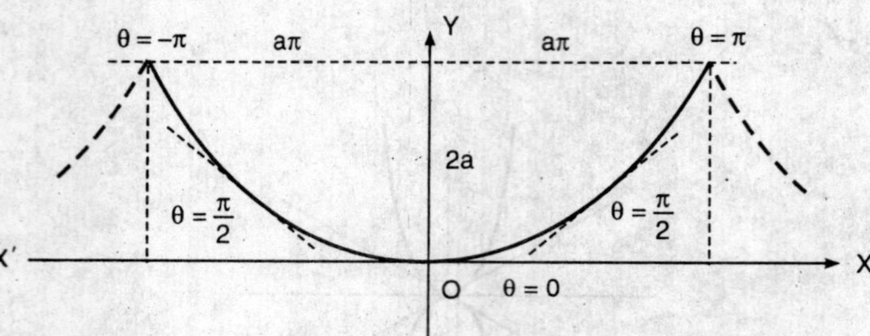

14. $x = a(\theta - \sin\theta),\ y = a(1 + \cos\theta)$

Ans.

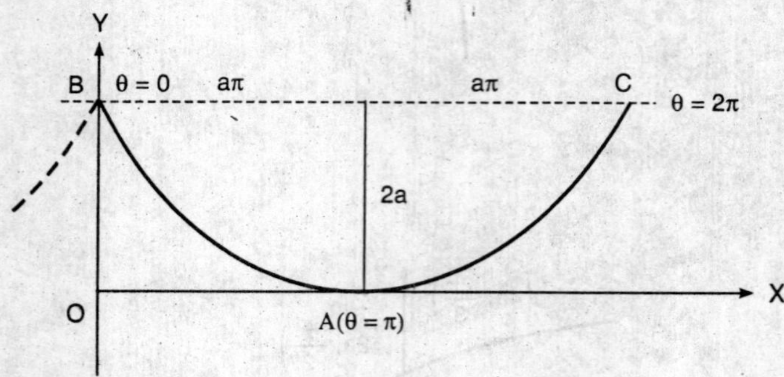

15. $x = t^2,\ y = t - \dfrac{t^3}{3}$

Ans.

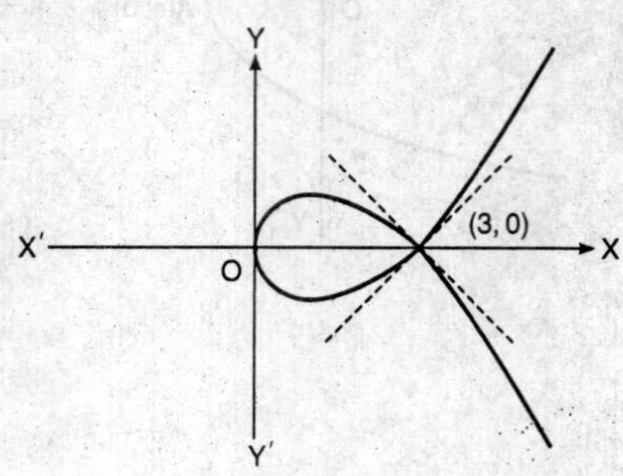

2

Integrals Calculus

2.1 INTEGRATION

Integration may be defined in two ways :

(i) **Integration as summation**; (ii) Integration as **the operation, reverse of differentiation**. Integration has wide applications. The length of **an arc of a curve, area under a curve, the surface and volume of solid formed by revolving a curve are all expressible as single variable definite integrals**. Volumes and surfaces of regular surfaces are found by use of **multiple integrals**.

2.2 INDEFINITE INTEGRATION (Reverse of Differentiation)

If $\frac{d}{dx} F(x) = f(x)$, then $F(x)$ is called the indefinite integral of $f(x)$, with respect to x and is expressed as

$$\int f(x)dx = F(x) + C \qquad \text{...(1)}$$

the arbitrary constant C is added because $\frac{d}{dx} \{F(x) + C\} = f(x) + 0 = f(x)$.

In (1), $F(x)$ is called the indefinite integral as C is arbitrary and does not have a definite value.

The readers are already familiar with the various methods of integration so the following results are given only as a summary.

(i) $\int x^n \, dx = \frac{x^{n+1}}{n+1} + C, \; n + 1 \neq 0$

(ii) $\int \frac{1}{x} \, dx = \log |x| + C$

(iii) $\int e^x\,dx = e^x + C$

(iv) $\int \log x\,dx = x\log_e x - x + C$ as $\frac{d}{dx}(x\log x - x) = 1\log x + \frac{x}{x} - 1 = \log x$

(v) $\int a^x\,dx = \frac{a^x}{\log_e a} + C$

(vi) $\int \sin x\,dx = -\cos x + C,$

(vii) $\int \cos x\,dx = \sin x + C$

(viii) $\int \tan x\,dx = \log\sec x + C,$

(ix) $\int \cot x\,dx = \log\sin x + C$

(x) $\int \operatorname{cosec} x\,dx = \log(\operatorname{cosec} x - \cot x) + C = \log\tan\frac{x}{2} + C,$

(xi) $\int \sec x\,dx = \log(\sec x + \tan x) + C = \log\tan\left(\frac{\pi}{4} + \frac{x}{2}\right)$

(xii) $\int \sec^2 x\,dx = \tan x + C$

(xiii) $\int \operatorname{cosec}^2 x\,dx = -\cot x + C$

(xiv) $\int \frac{dx}{\sqrt{a^2 - x^2}} = \sin^{-1}\frac{x}{a} + C$

(xv) $\int -\frac{dx}{\sqrt{a^2 - x^2}} = \cos^{-1}\frac{x}{a} + C$

(xvi) $\int \frac{dx}{\sqrt{a^2 + x^2}} = \log(x + \sqrt{x^2 + a^2}) + C = \sinh^{-1}\frac{x}{a} + C$

(xvii) $\int \frac{dx}{\sqrt{x^2 - a^2}} = \log(x + \sqrt{x^2 - a^2}) + C = \cosh^{-1}\frac{x}{a} + C$

(xviii) $\int \frac{dx}{x\sqrt{x^2 - a^2}} = \sec^{-1}\frac{x}{a} + C$

(xix) $\int \frac{-dx}{x\sqrt{x^2 - a^2}} = \operatorname{cosec}^{-1}\frac{x}{a} + C$

(xx) $\int \frac{1}{a^2 + x^2}\,dx = \frac{1}{a}\tan^{-1}\frac{x}{a}$

(xxi) $\int \frac{dx}{x^2 - a^2} = \frac{1}{2a} \log \left|\frac{x-a}{x+a}\right| + C$

(xxii) $\int \frac{dx}{a^2 - x^2} = \frac{1}{2a} \log \left|\frac{a+x}{a-x}\right| + C$

2.3 INTEGRATION BY PARTS

For deriving a formula {for integration of product $u \cdot v$ of two functions u and v} = $\int uv\, dx$, we see

$$\frac{d}{dx}\left\{u \int v\, dx\right\} = \frac{du}{dx} \int v\, dx + u \cdot v$$

as integration and differentiation are reverse operations.

On integrating both sides w.r.t. x, we have

$$\int \underset{\text{I}}{u}\, \underset{\text{II}}{v}\, dx = \underset{\text{I}}{u} \int \underset{\text{II}}{v}\, dx - \int \left[\frac{du}{dx} \cdot \int v\, dx\right] dx$$

Calling u as I and v as II, we have

$$\int \underset{\text{I}}{u} \cdot \underset{\text{II}}{v}\, dx = \text{First as } t \text{ is. (integral of II) } dx - \int \text{(Integral of product) of } \left(\frac{du}{dx} \cdot \int v\, dx\right) dx$$

The word 'ILATE' where I means inverse-trigonometric L = logarithmic A ≡ algebraic, T = trigonometric and E = exponential, helps in the selection of first function. The function occuring relatively on left in ILATE, w.r.t to other should be chosen as the first function u in evaluating $\int uv\, dx$.

2.4 SOME USEFUL INTEGRALS

1. Evaluate $I = \int \underset{\text{II}}{1} \underset{\text{I}}{\sqrt{a^2 - x^2}}\, dx$. Taking 1 as second function and integrating by parts,

$$I = \int 1 \cdot dx \cdot \sqrt{a^2 - x^2} - \int x \frac{(-x)}{\sqrt{a^2 - x^2}}\, dx$$

$$= x\sqrt{a^2 - x^2} - \int \left(\frac{a^2 - x^2}{\sqrt{a^2 - x^2}} - \frac{a^2}{\sqrt{a^2 - x^2}}\right) dx$$

$$= x\sqrt{a^2 - x^2} - I + a^2 \sin^{-1} x + 2C$$

Hence $$I = \frac{x\sqrt{a^2 - x^2}}{2} + \frac{a^2}{2} \sin^{-1} x + C$$

Similarly

2. $\int \underset{\text{II}}{1} \underset{\text{I}}{\sqrt{a^2+x^2}}\, dx = \dfrac{x\sqrt{a^2+x^2}}{2} + \dfrac{a^2}{2}\log\left(x+\sqrt{x^2+a^2}\right) + C$

3. $\int \underset{\text{II}}{1} \underset{\text{I}}{\sqrt{x^2-a^2}}\, dx = \dfrac{x\sqrt{x^2-a^2}}{2} - \dfrac{a^2}{2}\log\left(x+\sqrt{x^2-a^2}\right) + C.$

4. Evaluate $I = \underset{\text{II}}{e^{ax}} \underset{\text{I}}{\cos bx}\, dx$. Integrating by parts

$$I = \cos bx\, \frac{e^{ax}}{a} - \frac{(-b)}{a}\int \underset{\text{II}}{e^{ax}} \underset{\text{I}}{\sin bx}\, dx$$

Again integrating by parts, the integral occurring in R.H.S.

$$I = e^{ax}\left(\frac{\cos bx}{a} + \frac{b}{a^2}\sin bx\right) - \frac{b^2}{a^2}\int e^{ax}\cos bx\, dx$$

So

$$I = \frac{e^{ax}}{a^2}(a\cos bx + b\sin bx) - \frac{b^2}{a^2} I$$

or

$$\boldsymbol{I = \frac{e^{ax}}{a^2+b^2}(a\cos bx + b\sin bx) + C}$$

Similarly

$$\boldsymbol{\int e^{ax}\sin bx\, dx = \frac{e^{ax}}{a^2+b^2}(a\sin bx - b\cos bx) + C}$$

5. Evaluate $I = \int \underset{\text{II}}{1} \underset{\text{I}}{\log x}\, dx = x\log x - \int x \dfrac{d}{dx}(\log x)\, dx$

$= \boldsymbol{x\log x - x + C.}$

2.5 DEFINITE INTEGRALS

Definition. The integral $\int_a^b f(x)\, dx$ is called the definite integral of $f(x)$ between the limits a to b. If $\dfrac{d\,F(x)}{dx} = f(x)$ then the value of the definite integral is $F(b) - F(a)$.

Where indefinite integral of $f(x)$ does not exist, for the definite integral between given limits the function $f(x)$ should be expanded in a series and then integration be done term by term. Thus

$$\int_0^1 x^2\log(1-x^2)\, dx = \int_0^1 -x^2\left\{\frac{x^2}{1}+\frac{x^4}{2}+\frac{x^6}{3}+\ldots\right\} dx$$

$$= -\left[\frac{x^5}{1\cdot 5}+\frac{x^7}{2\cdot 7}+\frac{x^9}{3\cdot 9}+\ldots\right]_0^1 = -\left(\frac{1}{1\cdot 5}+\frac{1}{2\cdot 7}+\frac{1}{3\cdot 9}+\ldots\right).$$

2.6 PROPERTIES OF DEFINITE INTEGRALS

Property (i) $\int_a^b f(x)\,dx = \int_a^b f(t)\,dt$. If $\int f(x)\,dx = F(x) + C$.

Property (ii) $\int_a^b f(x)\,dx = -\int_b^a f(x)\,dx$.

The definite integral has the value $F(b) - F(a)$, it is immaterial whether the variable is denoted by x or t, as far as the function and the limits are same. Property (ii) is self evident as L.H.S. $= F(b) - F(a)$ and R.H.S. $= -\{F(a) - F(b)\}$.

Property (iii) $\int_a^b f(x)\,dx = \int_a^c f(x)\,dx + \int_c^b f(x)\,dx$, where $a < c < b$.

$$\text{R.H.S.} = [F(x)]_a^c + [F(x)]_c^b = F(c) - F(a) + F(b) - F(c)$$

$$= F(b) - F(a) = \int_a^b f(x)\,dx = \text{L.H.S.}$$

Property (iv) $\int_0^a f(x)\,dx = \int_0^a f(a-x)\,dx$

Let $x = a - t$ so $dx = -dt$, when $x = 0, t = a$, when $x = a, t = 0$.

So $$I = -\int_a^0 f(a-t)\,dt = \int_0^a f(a-t)\,dt = \int_0^a f(a-x)\,dx$$

on writing x, instead of t.

Example 2.1 **Evaluate :**

(i) $\displaystyle\int_0^{\pi/2} \frac{\sin^n x}{\sin^n x + \cos^n x}\,dx$

(ii) $\displaystyle\int_0^{\pi/2} \frac{1}{1+\tan x}\,dx$

(iii) $\displaystyle\int_0^{\pi/2} \frac{1}{1+\cot x}\,dx$

(Marathwada' 98S)

Solution : All these integrals are based on property (iv) proved above, so we prove only one of them.

Let $$I_n = \int_0^{\pi/2} \frac{\sin^n x}{\sin^n x + \cos^n x}\,dx.$$

Using $\int_0^a f(x)\,dx = \int_0^a f(a-x)\,dx$

$$I_n \text{ is also } = \int_0^{\pi/2} \frac{\sin^n\left(\frac{\pi}{2}-x\right)}{\sin^n\left(\frac{\pi}{2}-x\right)+\cos^n\left(\frac{\pi}{2}-x\right)}\,dx = \int_0^{\pi/2} \frac{\cos^n x}{\cos x^n + \sin x^n}$$

Adding these two values of I, we get

$$2\,I_n = \int_0^{\pi/2} 1\,dx = \frac{\pi}{2}$$

Hence $\quad I_n = \frac{\pi}{4}.$

Integrals (ii) and (iii) also have the value $\frac{\pi}{4}$.

Property (v). $I = \int_{-a}^{a} f(x)\,dx = 2\int_0^a f(x)\,dx$, if $f(x)$ is even or when $f(-x) = f(+x)$

$= 0$, if $f(x)$ is odd or when $f(-x) = -f(x)$

Proof : $\quad I = \int_{-a}^{0} f(x)\,dx + \int_0^a f(x)\,dx.$

Let $x = -t$ in the first integral, then

$$I = \int_a^0 f(-t)\,(-dt) + \int_0^a f(x)\,dx$$

$$= \int_0^a f(-t)\,dt + \int_0^a f(x)\,dx = \int_0^a \{f(-x) + f(x)\}\,dx$$

on writing x, instead of t.

Now $\quad f(-x) + f(x) = 2f(x)$ if $f(x)$ is even

$= -f(x) + f(x) = 0$, if $f(x)$ is odd.

So the property is proved.

Property (vi). $I = \int_0^{2a} f(x)\,dx = 2\int_0^a f(x)\,dx$, **if** $f(2a-x) = f(x)$

$= 0$, if $f(2a-x) = -f(x)$

PROOF : We have

$$I = \int_0^a f(x)\,dx + \int_a^{2a} f(x)\,dx$$

Making substitution $x = 2a - t$ in second integral when $x = a$, $t = a$ and when $x = 2a$, $t = 0$

So $\quad I = \int_0^a f(x)\,dx + \int_a^0 f(2a-t)\,(-dt)$

$$= \int_0^a f(x)\,dx + \int_0^a f(2a-t)\,dt. \text{ replacing } t \text{ by } x$$

$$= \int_0^a \{f(x) + f(2a-x)\}\,dx$$

$$= 2\int_0^a f(x)\,dx \text{ if } f(2a-x) = f(x)$$

$$= 0 \text{ if } f(2a-x) = -f(x).$$

Example 2.2 **Evaluate**

(i) $I_1 = \int_{-\pi/2}^{\pi/2} \sin^5 x\,dx$ (ii) $I_2 = \int_{-\pi/2}^{\pi/2} \cos^5 x\,dx$

(iii) $I_3 = \int_0^{\pi} \sin^5 x\,dx$ (iv) $I_4 = \int_0^{\pi} \cos^5 x\,dx.$

Solution : (i) $f(x) = \sin^5 x$,

$$f(-x) = (\sin(-x))^5 = -\sin^5 x = -f(x)$$

So the function is odd and $I_1 = 0$

(ii) In I_2, integrand $f(x) = \cos^5 x$, $f(-x) = (\cos(-x))^5 = \cos^5 x = f(x)$

So $$I_2 = 2\int_0^{\pi/2} \cos^5 x\,dx = 2 \times \frac{4}{5} \times \frac{2}{3} \times 1 = \frac{16}{15}.$$

(iii) In I_3, $f(x) = \sin^5 x$ and upper limit $2a = \pi$

$$f(2a-x) = \sin^5(\pi - x) = \sin^5 x = f(x)$$

So $$I_3 = 2\int_0^{\pi/2} \sin^5 x dx = 2 \times \frac{4}{5} \times \frac{2}{3} = \frac{16}{15}.$$

(iv) In I_4, $f(x) = \cos^5 x$, $2a = \pi$,

$$f(2a-x) = \{\cos(\pi - x)\}^5 = -\cos^5 x = -f(x)$$

So $$I_4 = 0$$

Example 2.3 **Prove** $I = \int_0^{\pi/2} \log \sin x = -\frac{\pi}{2} \log_e 2.$ **(Important Result)**

Solution : $$I = \int_0^{\pi/2} \log \sin x$$

Also $$= \int_0^{\pi/2} \log \sin\left(\frac{\pi}{2} - x\right) dx$$

So $$2I = \int_0^{\pi/2} (\log \sin x + \log \cos x)\, dx = \int_0^{\pi/2} \log \sin x \cos x\, dx$$

$$= \int_0^{\pi/2} \log \frac{\sin 2x}{2}\, dx = \int_0^{\pi/2} \log \sin 2x\, dx - \log_e 2 \int_0^{\pi/2} dx.$$

Let $2x = t$ in first integral, then

$$2I = \int_0^{\pi} \log \sin t \frac{dt}{2} - \frac{\pi}{2} \log_e 2.$$

Now $f(2a - t) = f(\pi - t) = \log \sin (\pi - t) = \log \sin t = f(t)$

So we multiply the value by 2 and reduce the limit to $\pi/2$.

So $$2I = \frac{2}{2} \int_0^{\pi/2} \log \sin t\, dt - \frac{\pi}{2} \log 2.$$

Similarly $\int_0^{\pi/2} \log \cos x\, dx$ is also $= -\frac{\pi}{2} \log 2$ and so $\int_0^{\pi/2} \log \tan x\, dx = 0.$

or $$2I = I - \frac{\pi}{2} \log_e 2$$

Hence $$I = -\frac{\pi}{2} \log_e 2.$$

Example 2.4 **Evaluate**

(i) $\boldsymbol{I_1 = \int_0^{\pi} \sin^3 x\, dx}$

(ii) $\boldsymbol{I_2 = \int_0^{\pi} \cos^6 x\, dx}$

Solution : (i) We have $I_1 = \int_0^{2a} f(x)\, dx = \int_0^{\pi} \sin^3 x\, dx$

So $\pi = 2a$, $f(x) = \sin^3 x$

$$f(2a - x) = \sin^3 (\pi - x) = \sin^3 x = f(x)$$

So $$I_1 = 2 \int_0^{\pi/2} \sin^3 x\, dx = 2 \frac{2}{3} = \frac{4}{3}.$$

(ii) $f(2a - x) = [\cos (\pi - x)]^6 = (-\cos x)^6 = \cos^6 x = f(x)$

So $$I_2 = 2 \int_0^{\pi/2} \cos^6 x\, dx$$

$$= 2 \frac{5}{6} \frac{3}{4} \frac{1}{2} \frac{\pi}{2} = \frac{5\pi}{16}$$

Example 2.5 **Show that**

$$I = \int_0^{\pi/2} \log \tan x \, dx = 0$$

Solution : $I = \int_0^{\pi/2} \log \tan x \, dx.$

Using $\int_0^{\pi/2} f(x)\, dx = \int_0^{\pi/2} f\left(\frac{\pi}{2} - x\right) dx$

$$I = \int_0^{\pi/2} \log \tan\left(\frac{\pi}{2} - x\right) dx$$

Adding $2I = \int_0^{\pi/2} \log(\tan x \cot x)\, dx = \int_0^{\pi/2} \log 1 \, dx = 0$ as $\log 1 = 0$

Hence $I = 0$.

Example 2.6 **Evaluate**

(i) $I_1 = \int_0^{\pi/4} \log(1 + \tan x)\, dx,$

(ii) $I_2 = \int_0^{\infty} \log\left(x + \frac{1}{x}\right) \frac{dx}{1 + x^2}$ ***(Madras 2000 PT)***

(iii) $I_3 = \int_0^{\pi} \log(1 + \cos x)\, dx.$

Solution : (i) $I_1 = \int_0^{\pi/4} \log(1 + \tan x)\, dx$, Applying $\int_0^{\pi/4} f(x)\, dx = \int_0^{\pi/4} f\left(\frac{\pi}{4} - x\right) dx$

$$= \int_0^{\pi/4} \log\left\{1 + \tan\left(\frac{\pi}{4} - x\right)\right\} dx$$

$$= \int_0^{\pi/4} \log\left\{1 + \frac{1 - \tan x}{1 + \tan x}\right\} dx$$

$$= \int_0^{\pi/4} \{\log 2 - \log(1 + \tan x)\}\, dx = \frac{\pi}{4} \log_e 2 - I_1$$

Hence $2I_1 = \frac{\pi}{4} \log_e 2$

or $I_2 = \frac{\pi}{8} \log_e 2.$

(ii) $$I_3 = \int_0^{\infty} \log\left(x + \frac{1}{x}\right) \frac{dx}{1+x^2},$$

Let $x = \tan\theta$

$$dx = \sec^2\theta \, d\theta \text{ when } x = 0,\ \theta = 0, \text{ when } x = \infty, \theta = \frac{\pi}{2}.$$

So $$I_3 = \int_0^{\pi/2} \log\left(\frac{\sec^2\theta}{\tan\theta}\right) \frac{\sec^2\theta\, d\theta}{\sec^2\theta}$$

$$= \int_0^{\pi/2} \log \frac{1}{\sin\theta\cos\theta}\, d\theta = -\int_0^{\pi/2} \log\sin\theta - \int_0^{\pi/2} \log\cos\theta\, d\theta$$

$$= -\left(-\frac{\pi}{2}\log 2 - \frac{\pi}{2}\log 2\right) = \pi \log_e 2.$$

(iii) $$I_4 = \int_0^{\pi} \log(1+\cos x)\, dx = \int_0^{\pi} \log 2\cos^2\frac{x}{2}\, dx$$

$$= \log 2 \int_0^{\pi} dx + 2\int_0^{\pi} \log\cos\frac{x}{2}\, dx$$

Let $\frac{x}{2} = t$

$$= \pi\log 2 + 2\int_0^{\pi/2} \log\cos t\, (2\, dt)$$

$$= \pi\log_e 2 + 4\left(-\frac{\pi}{2}\log 2\right) = -\pi\log_e 2.$$

Example 2.7 **Evaluate**

$$\boldsymbol{I = \int_0^1 \frac{\log(1+x)}{1+x^2}\, dx}$$

(Bangalore' 90)

Solution : Let $x = \tan\theta$

$$I = \int_0^{\pi/4} \log(1+\tan\theta)\, d\theta.$$

Using $$\int_0^{\pi/4} f\left(\frac{\pi}{4} - x\right) = \int_0^{\pi/4} f(x)\, dx$$

$$= \int_0^{\pi/4} \log\left\{1 + \tan\left(\frac{\pi}{4} - x\right)\right\} dx$$

$$= \int_0^{\pi/4} \log\left(1 + \frac{1-\tan x}{1+\tan x}\right) dx$$

$$= \log 2 \int_0^{\pi/4} dx - I$$

Hence $$2I = \frac{\pi}{4} \log_e 2$$

or $$I = \frac{\pi}{8} \log_e 2$$

Example 2.8 **Show that**

(i) $$I_1 = \int_0^{\pi} \frac{x \tan x \, dx}{\sec x + \tan x} = \pi\left(\frac{\pi}{2} - 1\right),$$

(ii) $$I_2 = \int_0^{\pi} \frac{x \tan x}{\sec x + \cos x} = \frac{\pi^2}{4}$$

(A.M.I.E.' 97 W, A.M.I.E.' 90)

Solution : (i) $I_1 = \int_0^{\pi} x \frac{\sin x}{1 + \sin x} dx$ Using the property $\int_0^{\pi} f(x)\, dx = \int_0^{\pi} f(\pi - x)\, dx$

$$I_1 = \int_0^{\pi} \frac{(\pi - x) \sin (\pi - x)}{1 + \sin (\pi - x)} dx$$

Adding the two values $2I_1 = \pi \int_0^{\pi} \frac{\sin x}{1 + \sin x} dx$

Using $\int_0^{2a} f(x)\, dx = 2 \int_0^{a} f(x)\, dx$

if $f(2a - x) = f(x)$

$$2I_1 = 2\pi \int_0^{\pi/2} \frac{\sin x}{1 + \sin x} dx$$

$$= 2\pi \int_0^{\pi/2} \frac{\sin x\,(1 - \sin x)}{\cos^2 x} dx$$

$$= 2\pi \int_0^{\pi/2} \{\sec x \tan x - (\sec^2 x - 1)\}\, dx$$

$$I_1 = \pi \log_e[\sec x - \tan x + x]_0^{\pi/2} = \pi \left[\lim_{x \to \pi/2} \left\{\frac{1 - \sin x}{\cos x}\right\}_{-1} + \pi/2 \right]$$

$$= \pi \left[\lim_{x \to \pi/2} \left\{\frac{- \cos x}{- \sin x}\right\} - 1 + \frac{\pi}{2} \right] = \pi\left(\frac{\pi}{2} - 1\right)$$

$$I_2 = \int_0^{\pi} \frac{x \tan x}{\sec x + \cos x} dx = \int_0^{\pi} \frac{x \sin x}{1 + \cos^2 x} dx$$

(Kurukshetra' 98)

Using $\int_0^{\pi} f(x)\,dx = \int_0^{\pi} f(\pi - x)\,dx$

$$I_2 = \int_0^{\pi} \frac{(\pi - x)\sin(\pi - x)}{1 + \cos^2(\pi - x)}\,dx$$

Adding there two values of I_2,

$$2I_2 = \pi \int_0^{\pi} \frac{\sin x}{1 + \cos^2 x}\,dx = 2\pi \int_0^{\pi/2} \frac{\sin x}{1 + \cos^2 x}\,dx$$

Let $\cos x = t - \sin x\,dx = dt$

So $I_2 = -\pi\left[\tan^{-1} t\right]_1^0 = -\pi\left[0 - \frac{\pi}{4}\right] = \frac{\pi^2}{4}.$

Example 2.9 **Prove that**

(i) $\boldsymbol{I_1 = \int_0^{\pi} \frac{x}{1 + \sin x}\,dx = \pi,}$

(Madras 2000)

(ii) $\boldsymbol{I_2 = \int_0^{\pi/2} \frac{x\,dx}{\sin x + \cos x} = \frac{\pi}{2\sqrt{2}} \log(\sqrt{2} + 1)}$

(Madras' 96)

Solution : (i) $I_1 = \int_0^{\pi} \frac{x}{1 + \sin x}\,dx.$

Using $\int_0^{\pi} f(x)\,dx = \int_0^{\pi} f(\pi - x)\,dx$

$$I_1 = \int_0^{\pi} \frac{(\pi - x)}{1 + \sin(\pi - x)}\,dx.$$

Adding the two values of I_1

$$2I_1 = \pi \int_0^{\pi} \frac{1}{1 + \sin x}\,dx = 2\pi \int_0^{\pi/2} \frac{1 - \sin x}{\cos^2 x}\,dx$$

$$I_1 = \pi\,[\tan x - \sec x]_0^{\pi/2}$$

$$= \pi\left\{\lim_{x \to \pi/2} \frac{\sin x - 1}{\cos x} - \frac{0 - 1}{1}\right\}$$

$$= \pi\left[\lim_{x \to \pi/2} \frac{\cos x}{-\sin x} + 1\right] = \pi(0 + 1) = \pi$$

(ii) $$I_2 = \int_0^{\pi/2} \frac{x\,dx}{\sin x + \cos x}$$

Using $\int_0^{\pi/2} f(x)\,dx = \int_0^{\pi/2} f\left(\frac{\pi}{2} - x\right) dx$

$$I_2 = \int_0^{\pi/2} \frac{\left(\frac{\pi}{2} - x\right) dx}{\cos x + \sin x}$$

Adding the two value of I_2,

$$2I_2 = \frac{\pi}{2}\int_0^{\pi/2} \frac{1}{\sin x + \cos x}\,dx$$

$$= \frac{\pi}{2}\int_0^{\pi/2} \frac{1}{\sqrt{2}\cos\left(\frac{\pi}{4} - x\right)}\,dx$$

$$= \frac{\pi}{2\sqrt{2}}\left[-\log\sec\left(\frac{\pi}{4} - x\right) + \tan\left(\frac{\pi}{4} - x\right)\right]_0^{\pi/2}$$

$$= \frac{\pi}{2\sqrt{2}}\left[-\log\left(\frac{\sec\frac{\pi}{4} - \tan\frac{\pi}{4}}{\sec\frac{\pi}{4} + \tan\frac{\pi}{4}}\right)\right]$$

$$= \frac{\pi}{2\sqrt{2}}\log\frac{\sqrt{2}+1}{\sqrt{2}-1}$$

$$2I_2 = \frac{\pi}{2\sqrt{2}}\log(\sqrt{2}+1)^2$$

So $$I_2 = \frac{\pi}{2\sqrt{2}}\log(\sqrt{2}+1).$$

PROBLEM SET 2.1

1. Evaluate

(i) $\int_0^{\pi/2} \frac{\sqrt{\sin x}\,dx}{\sqrt{\sin x} + \sqrt{\cos x}}$ (ii) $\int_0^{\pi/2} \frac{1}{1 + \sqrt{\cot x}}\,dx$

Ans. (i) $\frac{\pi}{4}$, (ii) $\frac{\pi}{4}$.

2. $\int_0^{\pi/2} \frac{\sin^2 x}{\sin x + \cos x}$

Ans. $\frac{1}{\sqrt{2}}\log(\sqrt{2}+1)$

3. $\int_0^{\pi} \frac{x\,dx}{a^2\cos^2 x + b^2\sin^2 x}$

Ans. $\frac{\pi^2}{2ab}$

4. $\int_0^{\pi} \theta \sin^3\theta\, d\theta$

Ans. $\frac{2}{3}\pi$

5. Prove that $\int_0^{\pi/2} \sin 2x \log \tan x\, dx = 0$

6. $\int_0^{\infty} \frac{x^7(1-x^{12})}{(1+x)^{28}}\,dx = 0$ $\left[\textbf{Hint : put } x = \frac{1}{t}\right]$

7. $\int_0^{a} \frac{dx}{x+\sqrt{a^2-x^2}} = \frac{\pi}{4}$

8. $\int_0^{a} \frac{\sin x}{\sin x + \cos x}\,dx = \frac{\pi}{4}$

9. $\int_0^{\pi/2} \frac{dx}{1+\sqrt{\tan x}} = \frac{\pi}{4}$

10. $\int_0^{\pi} \cos^6 x \cos^4 x\, dx$ ***(V.T.U. 2001)***

Ans. $\frac{3\pi}{256}$

11. $\int_0^{2\pi} \cos^6 x\, dx$

Ans. $\frac{5\pi}{8}$

12. $\int_0^{2\pi} \sin^4 x \cos^6 x\, dx$

Ans. $\frac{15\pi}{640}$

13. $\int_0^{\pi} \theta \sin^4 \theta \cos^6 \theta \, d\theta$ *(Delhi' 91)*

Ans. $\frac{3\pi^2}{512}$

Prove that

14. (i) $\int_0^{\pi} \frac{x\,dx}{1+\sin^2 x} = \frac{\pi^2}{2\sqrt{2}}$, (ii) $\int_0^{\pi} \frac{x\,dx}{a^2-\cos^2 x} = \frac{\pi^2}{2a\sqrt{a^2-1}}$ $(a > 1)$

15. (i) $\int_0^{\pi/2} \frac{x\,dx}{2\sin^2 x+\cos^2 x} = \frac{\pi^2}{2\sqrt{2}}$, (ii) $\int_0^{\pi} x \cot x \, dx = \frac{\pi}{2} \log_e 2$

16. (i) $\int_0^{\infty} \log(1+x^2) \frac{dx}{1+x^2} = \pi \log 2$ (ii) $\int_0^{\pi} \log(1+\cos x)\, dx = -\pi \log_e 2.$

2.7 INTEGRAL AS LIMIT OF A SUM

Let $y = f(x)$ be a curve. Then the area under the curve from $x = a$ to $x = b$ is $= \int_a^b f(x)\, dx.$

This area is also equal the to sum of elementary areas as seen below. Let the interval LR be divided in n equal sub intervals each width h, by introducing points $x_1, x_2 \dots x_{n-1}$ between $x_0 = a$ and $x_n = b_1$

so $$xh = b - a.$$

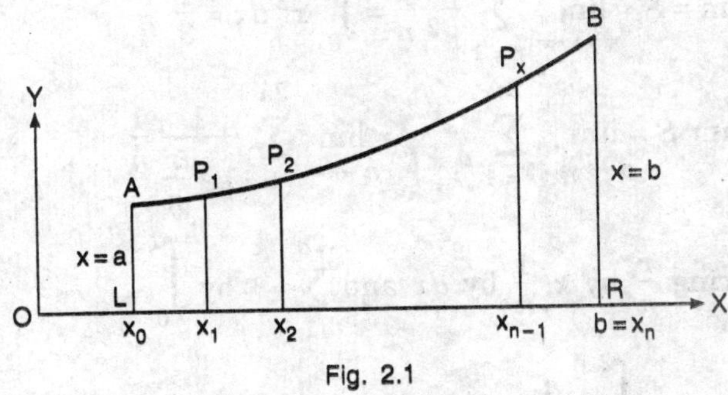

Fig. 2.1

Area of rectangle $A\, x_0\, x_1\, P_1 = hf(x_0)$

Area such angle $P_1\, x_1\, x_2\, P_2 = Qf(x_0 + h)$

Area of last rectangle $= hf(x_0 + \overline{n-1}\, h)$

Adding the total area $= h\, [f(x_0) + f(x_0 + h) + \dots n \text{ terms}]$

$$\lim_{h \to 0,\ \to \infty}$$

If we allow h to tend to zero and $n \to \infty$, such that $xh = b - a$, we have

$$\int_0^b f(x)\,dx = \text{Area under the curve} = h\left\{\sum_{r=0}^{n-1} f(a+rh)\right\}.$$

$$= \frac{b-a}{n}\sum_{r=0}^{n} f\left\{a + \frac{r}{n}(b-a)\right\},\ h \to 0, n \to \infty \text{ in such a way that } nh = (b-a)$$

Conversely the above sum can be converted into an integral by following substitutions.

If we replace $\frac{r}{n}$ by x, $\frac{1}{n}$ by dx and integrate from 0 to 1

$$\text{R.H.S.} = (b-a)\int_0^1 f\{a + x(b-a)\}\,dx$$

Let, $\quad (b-a)x + a = t = \dfrac{b-a}{b-a}\displaystyle\int_a^b f(t)\,dt = \text{L.H.S.}$

So, to sum of series, it can be changed in an integral by replacing $\frac{r}{n}$ by x, $\frac{1}{n}$ by dx and limits are 0 to 1 if r varies from 0 to n. If r varies from $r = 0$ to kn then limits will be from 0 to k.

Example 2.10 **Find the limit of the sum as $n \to \infty$, of the series**

(i) $\dfrac{1}{n^3}(1 + 4 + 9 + 16 + \ldots n^2)$

(ii) $\dfrac{1}{n} + \dfrac{1}{n+1} + \dfrac{1}{n+2} + \ldots + \dfrac{1}{2n}$.

Solution : (i) $\text{Sum} = S = \displaystyle\lim_{n\to\infty}\sum_{r=1}^{n}\frac{r^2}{n^2}\frac{1}{n} = \int_0^1 x^2\,dx = \frac{1}{3}.$

(ii) $\text{Sum } S = \displaystyle\lim_{n\to\infty}\sum_{r=1}^{2n}\frac{1}{n+r} = \lim_{n\to\infty}\sum_{1}^{2n}\frac{1}{1+\frac{r}{n}}\frac{1}{n}$

Replacing $\frac{r}{n}$ by x, $\frac{1}{n}$ by dx and $\displaystyle\sum_1^{2n}$ by $\displaystyle\int_0^2$

$$S = \int_0^2 \frac{1}{1+x}\,dx = [\log(1+x)]_0^2 = \log_e^3.$$

Example 2.11 **Find the limit as $n \to \infty$, of the sum of the series**

(i) $\dfrac{1}{n} + \dfrac{n^2}{(n+1)^3} + \dfrac{n^2}{(n+2)^3} + \ldots + \dfrac{1}{8n}$.

(ii) $\dfrac{1}{n} + \dfrac{1}{n+1} + \dfrac{1}{n+2} + \ldots + \dfrac{1}{3n}$.

Solution : (i) Sum $= S = \lim_{n \to \infty} \sum_{r=0}^{n} \frac{n^2}{(n+r)^3}$

$$= \lim_{n \to \infty} \sum_{r=0}^{n} \frac{1}{\left(1+\frac{r}{n}\right)^3} \frac{1}{n}$$

Replacing $\frac{r}{n}$ by x, $\frac{1}{n}$ by dx and limits for x as 0 to 1

$$S = \int_0^1 \frac{1}{(1+x)^3}\, dx = \left[-\frac{1}{2(1+x)^2}\right]_0^1 = \left[-\frac{1}{8}+\frac{1}{2}\right] = \frac{3}{8}$$

(ii) Sum $S = \lim_{n \to \infty} \sum_{r=0}^{2n} \frac{1}{n+r}$

$$= \lim_{n \to \infty} \Sigma \frac{1}{1+\frac{r}{n}} \cdot \frac{1}{n}$$

Replacing $\frac{1}{n}$ by dx, $\frac{r}{n} = x$, summation by integration, for limits when $r = 0$, $\frac{r}{n} = x = 0$, when

$$r = 2n,\ x = \frac{r}{n} = \frac{2n}{n} = 2$$

So $\quad S = \int_0^2 \frac{1}{1+x}\, dx = [\log(1+x)]_0^2 = \log_e 3.$

Example 2.12 **Find the limit as $n \to \infty$, of the sum of the series**

Solution : (i) $\frac{n^{\frac{1}{2}}}{n^{3/2}} + \frac{n^{1/2}}{(n+3)^{3/2}} + \frac{n^{1/2}}{(n+6)^{3/2}} + \ldots \frac{n^{1/2}}{\{n+3(n-1)\}^{3/2}}$

$$S = \sum_{r=0}^{n-1} \frac{n^{1/2}}{n^{3/2}\left(1+3\frac{r}{n}\right)^{3/2}} = \int_0^1 \frac{dx}{(1+3x)^{3/2}}$$

$$= \left[\frac{2(1+3x)^{-\frac{1}{2}}}{-3}\right]_0^1 = \left[-\frac{2}{3\sqrt{4}}+\frac{2}{3}\right] = \frac{1}{3}$$

(ii) $\lim_{n \to \infty} \left\{\frac{n}{n^2} + \frac{n}{n^2+1^2} + \ldots \frac{n}{n^2+(n+1)^2}\right\}$

$$S = \lim_{n \to \infty} \sum_{r=0}^{n+1} \frac{n}{n^2\left(1+\frac{r^2}{n^2}\right)} = \int_0^1 \frac{dx}{1+x^2}$$

$$= [\tan^{-1} x]_0^1 = \frac{\pi}{4}$$

(iii) $$S = \lim_{n \to \infty} \frac{1}{n}\left\{\sin^{2k}\frac{\pi}{2n} + \sin^{2k}\frac{2\pi}{2n} + \ldots \sin^{2k}\frac{\pi}{2}\right\}$$

$$= \lim_{n \to \infty} \sum_{r=1}^{n} \frac{1}{n} \sin^{2k}\frac{\pi}{2}\frac{r}{n} = \int_0^1 \left(\sin^{2k}\frac{\pi}{2}x\right) dx$$

Let $\frac{\pi x}{2} = t$

$$S = \frac{2}{\pi}\int_0^1 \sin^{2k} t\, dt = \frac{2}{\pi}\frac{\Gamma\frac{2k+1}{2}\Gamma\frac{1}{2}}{2\Gamma\frac{2k+2}{2}}$$

$$= \frac{1}{\pi\, !k}\left[\frac{(2k-1)}{2}\frac{(2k-3)}{2}\cdots\frac{5}{2}\cdot\frac{3}{2}\cdot\frac{1}{2}\sqrt{\pi}\sqrt{\pi}\right]$$

$$= \frac{1}{\pi\, !k}\left[\frac{\frac{2k}{k}\left(\frac{2k-1}{2}\right)\frac{2k-2}{2}\left(\frac{2k-3}{2}\right)\frac{2k-4}{2}\cdots\left(\frac{5}{2}\right)\frac{4}{2}\cdot\left(\frac{3}{2}\right)\left(\frac{2}{2}\right)\frac{1}{2}}{\frac{2k}{2}\cdot\frac{2k-2}{2}\cdots\frac{2k-4}{2}\cdots\frac{4}{2}\cdot\frac{2}{2}}\right]\pi$$

$$= \frac{1}{\pi\,\Gamma k}\frac{\Gamma 2k}{2^{2k}}\frac{1}{\Gamma k}\pi = \frac{\Gamma 2k}{(2^k\,\Gamma k)^2}$$

$$= \frac{\Gamma 2n}{\Gamma k\,\Gamma k}\cdot\frac{1}{2^{2k}}$$

Example 2.13 **Find the limit of the products**

(i) $$P_1 = \left\{\left(1+\frac{1}{n}\right)\left(1+\frac{2}{n}\right)\ldots\left(1+\frac{n}{n}\right)\right\}^{\frac{1}{n}} \text{ as } n \to \infty$$

(ii) $$P_2 = \left\{\left(1+\frac{1}{n^2}\right)\left(1+\frac{2^2}{n^2}\right)\ldots\left(1+\frac{n^2}{n^2}\right)\right\}^{\frac{1}{n}} \text{ as } n \to \infty$$

Solution : Taking log on both sides

(i) $$\log P_1 = \lim_{n \to \infty} \sum_{r=1}^{n} \frac{1}{n}\log\left(1+\frac{r}{n}\right) = \int_0^1 1 \log(1+x)\, dx$$

$$= [x \log(1+x)]_0^1 - \int_0^1 \frac{x}{1+x} dx$$

$$= \log_e 2 - \int_0^1 \left(1 - \frac{1}{1+x}\right) dx$$

$$= \log 2 - [x - \log(1+x)]_0^1 = \log 2 - 1 + \log 2$$

$$= \log (4\, e^{-1}) \text{ Hence } P_1 = 4e^{-1} = 4/e.$$

(ii) Taking logarithm of both sides, the product changes into a series, we get

$$\log P_1 = \lim_{n \to \infty} \sum_{r=0}^{n} \frac{1}{n} \log\left(1 + \frac{r^2}{n^2}\right) = \int_0^1 1 \log (1 + x^2)\, dx$$

Integrating by parts

$$\log P_1 = [x \log (1 + x^2)]_0^1 - 2 \int_0^1 \frac{x^2}{1 + x^2}\, dx$$

$$= \log_e 2 - 2 \int_0^1 \left(1 - \frac{1}{1 + x^2}\right) dx$$

$$= \log_e 2 - 2\, [x - \tan^{-1} x]_0^1$$

$$= \log_e 2 - 2 + \frac{2\pi}{4} = \log 2\, e^{\left(\frac{\pi}{2} - 2\right)}$$

So $\quad P_1 = 2\, e^{\left(\frac{\pi}{2} - 2\right)}$

Example 2.14 **Apply the definition of a definite integral as the limit of a sum to evaluate**

$$\lim_{n \to \infty} \left\{\frac{n!}{n^n}\right\}^{\frac{1}{n}}$$

Solution :

$$P = \lim_{n \to \infty} \left\{\frac{1}{n} \cdot \frac{2}{n} \cdot \ldots \frac{r}{n} \ldots \frac{n}{n}\right\}^{\frac{1}{n}}$$

So $\quad \log_e P = \sum_{r=1}^{n} \frac{1}{n} \log\left(\frac{r}{n}\right) = \int_0^1 \underset{\text{II}}{1} \underset{\text{I}}{\log x}\, dx$

$$= [x \log x]_0^1 - \int_0^1 \frac{x}{x}\, dx$$

$$= 0 - [x]_0^1 = -1$$

Hence $\quad P = e^{-1} = \frac{1}{e}$

PROBLEM SET 2.2

Find the limit as $n \to \infty$, of the series

1. $\dfrac{1}{1 + n^3} + \dfrac{4}{8 + n^3} + \dfrac{9}{n^3 + 27} + \ldots \dfrac{r^3}{n^3 + r^3} + \ldots \dfrac{1}{2n}$

Ans. $\frac{1}{3} \log_e 2$ *(Bangalore' 93)*

2. $\dfrac{\sqrt{n}}{(3+4\sqrt{n})^2}+\dfrac{\sqrt{n}}{\sqrt{2}\,(3\sqrt{2}+4\sqrt{n})^2}+\ldots\dfrac{1}{49n}.$

Ans. $\dfrac{1}{14}$ *(Bhopal 91)*

3. $\lim\limits_{n\to\infty}\ \sum\limits_{r=1}^{n}\dfrac{r}{n^2+r^2}$

Ans. $\dfrac{1}{2}\log_e 2.$

4. $\lim\limits_{n\to\infty}\left\{\dfrac{\sqrt{n}}{\sqrt{n^3}}+\dfrac{\sqrt{n}}{\sqrt{(n+3)^3}}+\ldots\dfrac{\sqrt{n}}{\sqrt{[n+3(n-1)]^3}}\right\}=\dfrac{1}{3}.$ *(Gorakhpur' 91)*

5. $\lim\limits_{n\to\infty}\left\{\dfrac{(n+1)\,(n+2)\,\ldots\,(n+n)_{n}^{\frac{1}{n}}}{n}\right\}\quad \lim\limits_{n\to\infty}\dfrac{\{(n+1)\,(n+2)\,\ldots\,(n+n)\}^{1/\pi}}{n}$

Ans. $\dfrac{1}{4}\log 2$

6. $\lim\limits_{n\to\infty}\left\{\left(1+\dfrac{1}{n^2}\right)\left(1+\dfrac{2^2}{n^2}\right)\ldots\left(1+\dfrac{n^2}{n^2}\right)\right\}^{\frac{1}{n}}$

Ans. $\dfrac{4}{e}$.

2.8 REDUCTION FORMULA

2.8.1 Reduction Formula for $\int \sin^n x\,dx$

$$I_n=\int \underset{\text{II}}{\sin x}\ \underset{\text{I}}{\sin^{n-1} x}\,dx, \text{ integrating by parts}$$

$$=(-\cos x)\sin^{n-1}x-(n-1)(-)\int \sin^{n-2}x\cos^2 x\,dx$$

$$=(-\cos x)\sin^{n-1}x-(n-1)(-1)\int \sin^{n-2}x\,(1-\sin^2 x)\,dx$$

$$=(-\cos x)\sin^{n-1}x-(n-1)(-1)(I_{n-2}-I_n)$$

$$=-\cos x\sin^{n-1}x+(n-1)I_{n-2}-(n-1)I_n$$

or $$(1+n-1)\,I_n=-\cos x\sin^{n-1}x+(n-1)\,I_{n-2}$$

$$I_n=\frac{-\cos x\sin^{n-1}x}{n}+\frac{n-1}{n}I_{n-2} \qquad \ldots(1)$$

2.8.2 Reduction Formula for $\int \cos^n x\,dx$

(ii) $$I_n=\int \underset{\text{II}}{\cos x}\ \underset{\text{I}}{\cos^{n-1} x}\,dx, \text{ Integrating by parts.}$$

$$=\sin x\cos x^{n-1}x-(-1)(n-1)\int \cos x^{n-2}x\sin^2 x\,dx$$

$$I_n = \sin x \cos^{n-1} x - (-1)(n-1) \int \cos^{n-2} x (1 - \cos^2 x)\, dx$$

$$= \sin x \cos^{n-1} x + (n-1)(I_{n-2} - I_n)$$

So $(1 + n - 1) I_n = \sin x \cos^{n-1} x + (n-1) I_{n-2}$

or $$I_n = \frac{\sin x \cos^{n-1} x}{n} + \frac{n-1}{n} I_{n-2} \quad ...(2)$$

Example 2.15 **Evaluate**

$$I = \int \sin^5 x\, dx$$

Solution :

$$I = \int \sin^5 x\, dx$$

$$= \frac{-\cos x \sin^4 x}{5} + \frac{4}{5} \int \sin^3 x\, dx$$

$$= \frac{-\cos \sin^4 x}{5} + \frac{4}{5}\left[-\cos \sin^2 x + \frac{2}{3} \int \sin x dx\right]$$

$$= \frac{-\cos x \sin^4 x}{5} - \frac{4}{5} \cos x \sin^2 x - \frac{8}{15} \cos x$$

Example 2.16 **Show that**

$$I_n = \int_0^{\pi/2} \sin^n x\, dx\; I_n = \int_0^{\pi/2} \cos^n x\, dx$$

$$= \frac{n-1}{n} \frac{n-3}{n-2} \ldots \times \frac{\pi}{2} \text{ (if } n \text{ be even)}$$

$$= \frac{n-1}{n} \frac{n-3}{n-2} \ldots \times I_1 \text{ (if } n \text{ is odd)}$$

Solution : We have the reduction formula

$$I_n = \int_0^{\pi/2} \sin^n x\, dx$$

$$= \left[-\frac{\sin^{n-1} x \cos x}{n}\right]_0^{\pi/2} + \frac{n-1}{n} \int_0^{\pi/2} \sin^{n-2} x\, dx$$

$$= 0 + \frac{n-1}{n} I_{n-2}$$

Let n be even, by repeated use of the above result, we get

$$I_n = \frac{n-1}{n} \cdot \frac{n-3}{n-2} \cdots \frac{3}{4} \cdot \frac{1}{2} \cdot I_0$$

and $$I_0 = \int_0^{\pi/2} \sin^0 x\, dx = [x]_0^{\pi/2} = \frac{\pi}{2}$$

$$= \frac{(n-1)(n-3)\ldots 5\cdot 3\cdot 1}{n(n-2)\ldots 4\cdot 2}\cdot\frac{\pi}{2}.$$

If n be an odd positive integer I_n ultimately reduces to I_1 by repeated use of reduction formula.

$$I_1 = \int_0^{\pi/2} \sin x\, dx = \int_0^{\pi/2} \cos x\, dx = 1$$

So when n is odd

$$I_n = \frac{n-1}{n}\cdot\frac{n-3}{n-2}\cdot\ldots\frac{4}{5}\cdot\frac{2}{3}\cdot I_1$$

Example 2.17 **Evaluate**

(i) $I_9 = \int_0^{\pi/2} \cos^9 x\, dx$

(ii) $I = \int_0^1 x^6 \sqrt{(1-x^2)}\, dx$

(Madras' 98S, AMIE' 90)

Solution : (i) Using reduction formula

$$I_9 = \frac{8}{9}\times\frac{6}{7}\times\frac{4}{5}\times\frac{2}{3}\times 1$$

$$= \frac{128}{315}.$$

(ii) $I = \int_0^1 x^6\, 1 - x^2\, dx.$

Let $x = \sin\theta$, $dx = \cos\theta\, d\theta$

$$= \int_0^{\pi/2} \sin^6\theta \cos\theta \cos\theta\, d\theta$$

$$= \int_0^{\pi/2} (\sin^6\theta - \sin^8\theta)\, d\theta$$

$$= \frac{5}{6}\cdot\frac{3}{4}\cdot\frac{1}{2}\frac{\pi}{2}\left(1-\frac{7}{8}\right)$$

$$= \frac{5\pi}{256}.$$

2.8.3 Reduction formula for $\int \sin x^m \cos x^n\, dx$

If one of m or n is odd, the integral can be found directly, through substitution and then by binomial expansion. Taking $n = 2p + 1 =$ odd,

$$I_{m,\,2p+1} = \int \sin^m x\, (\cos^2 x)^p \cos x\, dx$$

Let $\sin x = t$

$$I_{m,(2p+1)} = \int t^m (1-t^2)^p \, dt$$

$$= \int t^m \sum_{r=0}^{p} p_{c_r} (-1)^r t^{2r} \, dt$$

$$= \left[\sum_{r=0}^{p} p_{c_r} (-1)^r \frac{t^{m+2r+1}}{m+2r+1} + C \right].$$

Similar is the case if m be odd. Then first put $\cos x = t$.

If m and n be both even integers then the lesser of the two should be reduced. Suppose $n < m$. Then applying repeated reduction the integral will reduce to $\int \sin x^m \, dx$, which can be evaluated by reduction formula for it.

$$I_{m,n} = \int \sin^m x \cos^n x \, dx.$$

We shall reduce n.

$$= \int \underset{\text{II}}{(\sin^m x \cos x)} \underset{\text{I}}{\cos^{n-1} x} \, dx, \text{ by integrating by parts}$$

$$= \frac{\sin^{m+1} x}{m+1} \cdot \cos^{n-1} x - \frac{(-)(n-1)}{m+1} \int \sin^{m+1} x \cos^{n-2} x \sin x \, dx$$

$$= \frac{\sin^{m+1} x \cos^{n-1} x}{m+1} + \frac{n-1}{m+1} \int \sin^m x \cos^{n-2} x (1 - \cos x^2) \, dx$$

$$= T_1 + \frac{n-1}{m+1} (I_{m,n-2} - I_{m,n})$$

So $$\left(1 + \frac{n-1}{m+1}\right) I_{m,n} = T_1 + \frac{n-1}{m+1} I_{m,n-2}$$

or $$I_{m,n} = \frac{\sin^{m+1} x \cos^{n-1} x}{m+n} + \frac{n-1}{m+n} I_{m,n-2}$$

The formula is valid whether m, n be integers or not.

The term T_1 vanishes if limits 0 to $\pi/2$ be applied and we get

$$\int_0^{\pi/2} \sin^m x \cos^n x \, dx = 0 + \frac{n-1}{m+n} I_{m,n-2}.$$

Thus $$\int_0^{\pi/2} \sin^8 x \cos^2 x \, dx = \frac{2-1}{8+2} I_{8,2-2}$$

$$= \frac{1}{10} I_{(8,0)}$$

$$= \frac{1}{10} \times \frac{7}{8} \times \frac{5}{6} \times \frac{3}{4} \times \frac{1}{2} \times \frac{\pi}{2} = \frac{7\pi}{8}$$

Example 2.18 **Evaluate**

(i) $\int \sin^5 x \cos^3 x\, dx$

(ii) $I = \int_0^{\pi/2} \sin^{15} x \cos^3 x\, dx$

(Madras 96)

(iii) $I = \int_0^{\pi/6} \cos^4 3\theta \sin^3 6\theta\, d\theta$

Solution : (i) $\int \sin^5 x \cos^3 x dx = \frac{\sin^6 x \cos^2 x}{8} + \frac{2}{8}\int \sin^5 x \cos x dx$

$$= \frac{\sin^6 x \cos^2 x}{8} + \frac{1}{4}\frac{\cos^6 x}{6}$$

$$= \frac{\sin^6 x \cos^2 x}{8} + \frac{1}{24}\cos^6 x$$

$$= \frac{\cos^2 x}{8}\left[\sin^6 x + \frac{\cos^4 x}{3}\right].$$

(ii) We have $I = \int_0^{\pi/2} \sin^{15} x \cos^3 x dx$

$$\because \int_0^{\pi/2} \sin^{n-2} x \cos^n x\, dx = \frac{n-1}{m+n} I_{m,\, n-2}$$

$$\therefore \quad I = \frac{2}{18} \cdot I_{15,\,1} = \frac{2}{18}\int_0^{\pi/2} \sin^{15} x \cos x\, dx$$

$$= \frac{1}{9} \times \left[\frac{\sin^{16} x}{16}\right]_0^{\pi/2}$$

$$= \frac{1}{9} \cdot \frac{1}{16} = \frac{1}{144}$$

(iii) $I = \int_0^{\pi/6} \cos^4 3\theta\, (2 \sin 3\theta \cos 3\theta)^3\, d\theta$

Let $3\theta = t$

$$= \frac{8}{3}\int_0^{\pi/2} \cos^7 t \sin^3 t\, dt = \frac{8}{3}\left(\frac{2}{10}\right) \cdot I_{7,\,1}$$

$$= \frac{8}{15} \times \int_0^{\pi/2} \cos^2 x \sin x dx = \frac{8}{15} \times \frac{1}{8} = \frac{1}{15}$$

Example 2.19 **Evaluate**

(i) $I = \int_0^1 x^5 \sin^{-1} x \, dx$

(Bangalore' 90)

(ii) $I = \int_0^{2\alpha} \frac{x^3 \, dx}{\sqrt{2ax - x^2)}}$

(Madras 2000S)

Solution : (i) Let $x = \sin\theta,\ dx = \cos\theta \, d\theta,$

So $$I = \int_0^{\pi/2} (\underset{\text{II}}{\sin^5\theta \cos\theta})\, \underset{\text{I}}{\theta} \, d\theta$$

Integrating by parts

$$I = \left[\theta\left(\frac{\sin^6\theta}{6}\right)\right]_0^{\pi/2} - \int_0^{\pi/2} \frac{1 \cdot \sin^6\theta}{6} d\theta$$

$$= \frac{\pi}{12} - \frac{1}{6}\left\{\frac{5}{6} \cdot \frac{3}{4} \cdot \frac{1}{2} \times \frac{\pi}{2}\right\}$$

$$= \frac{\pi}{12} - \frac{5\pi}{12 \times 16} = \frac{(16-5)\,\pi}{192} = \frac{11\pi}{192}$$

(ii) $$I = \int_0^{2a} \frac{x^3}{\sqrt{x}\,\sqrt{2a - x}} dx$$

Let $x = 2a \sin^2\theta$

$dx = 4a \sin\theta \cos\theta \, d\theta$

When $x = 0, \theta = 0$ and when $x = 2a, \theta = \frac{\pi}{2}$

$$I = \int_0^{\pi/2} \frac{8a^3 \sin^6\theta\,(4a \sin\theta\cos\theta)\, d\theta}{\sqrt{2}\ a \sin\theta\,\sqrt{2}\ a\cos\theta}$$

$$= 16a^3 \int_0^{\pi/2} \sin^6\theta \, d\theta$$

$$= 16a^3\left(\frac{5}{6} \cdot \frac{3}{4} \cdot \frac{1}{2} \cdot \frac{\pi}{2}\right) = \frac{5\pi a^3}{2}$$

Example 2.20 **Evaluate**

(i) $I = \int_0^\infty \frac{dx}{(1 + x^2)^8}$

(Mangalore' 99)

(ii) $\int_0^{\pi/4} \sin^2\theta \cos^4\theta \, d\theta.$

Solution : (i) Let $x = \tan\theta$, then $dx = \sec^2\theta \, d\theta$, when $x = 0, \theta = 0, x \to \infty, \theta = \pi/2.$

So $$I = \int_0^{\pi/2} \frac{\sec^2\theta\, d\theta}{(\sec^2\theta)^8} = \int_0^{\pi/2} \cos^{14}\theta\, d\theta$$

$$= \frac{13\cdot 11\cdot 9\cdot 7\cdot 5\cdot 3\cdot 1}{14\cdot 12\cdot 10\cdot 8\cdot 6\cdot 4\cdot 2}\,\frac{\pi}{2} = \frac{429\pi}{512}$$

(ii) $$I = \int_0^{\pi/4} \sin^2\theta\cos^2\theta\cos^2\theta\, d\theta$$

$$= \int_0^{\pi/4} \frac{1}{4}\sin^2 2\theta\, \frac{1+\cos 2\theta}{2}\, d\theta \quad \text{Let } 2\theta = t$$

$$= \frac{1}{8}\int_0^{\pi/2} \sin^2 t\, \frac{dt}{2} + \frac{1}{16}\int_0^{\pi/2} \sin^2 t\cos t\, dt$$

$$= \frac{1}{16}\left[\frac{1}{2}\cdot\frac{\pi}{2}\right] + \frac{1}{16}\left[\frac{\sin^3 t}{3}\right]_0^{\pi/2}$$

$$= \frac{\pi}{64} + \frac{1}{16}\left[\frac{1}{3}\right] = \frac{\pi}{64} + \frac{1}{48}$$

Example 2.21 **Evaluate**

(i) $I_1 = \int \sin^{3/4} x \cos^3 x\, dx$

(ii) $I_2 = \int \sin^4 x \cos^2 x\, dx$

Solution : (i) $I_1 = \int \sin^{3/4} x \cos^2 x \cos x\, dx,$

Let $\sin x = t$ so $\cos x\, dx = dt$

$$= \int t^{3/4}(1 - t^2)\, dt = \frac{4}{7}t^{7/4} - \frac{4}{15}t^{15/4} + C$$

$$= \frac{4}{7}\sin^{7/4} x - \frac{4}{15}\sin^{15/4} x + C$$

(ii) $I_2 = \int \sin^4 x \cos^2 x\, dx$

Using reduction formula we have

$$I_2 = \frac{\sin^5 x\cos x}{6} + \frac{1}{6}\int \sin^4 x\, dx$$

$$= \frac{\sin^5 x\cos x}{6} + \frac{1}{6}\left[\frac{-\cos x\sin^3 x}{4} + \frac{3}{4}\int \sin^2 x\, dx\right]$$

$$= \frac{\sin^5 x\cos x}{6} - \frac{\cos x\sin^3 x}{24} + \frac{1}{8}\left[-\cos x\sin x + \frac{1}{2}\int dx\right]$$

$$= \frac{\sin^5 x\cos x}{6} - \frac{\cos x\sin^3 x}{24} - \frac{1}{8}\sin x\cos x + \frac{x}{16} + C$$

Example 2.22 **Integrate**

(i) $I = \int \frac{\sin^4 x}{\sec^3 x} dx$

(ii) $I = \int \frac{t^2}{\sqrt{1-t^2}} dt$

Solution : (i) $I = \int \sin^4 x \cos^3 x \sec^6 x \, dx$

$$= \frac{\sin^4 x \,.\, \cos^2 x}{7} + \frac{2}{7}\int \sin^4 x \cos x dx$$

$$= \frac{\sin^4 x \cos^2 x}{7} + \frac{2}{7} \cdot \frac{\sin^5 x}{5}$$

$$= \frac{\sin^4 x \cos^2 x}{7} + \frac{2}{35}\sin^5 x$$

$$= \frac{\sin^4 x}{7}\left[\cos^2 x + \frac{2}{5}\sin x\right]$$

(ii) We have $\int \frac{t^2}{\sqrt{1-t^2}} dt$

Let $t = \sin\theta \Rightarrow dt = \cos\theta \, d\theta$

$$\therefore \quad I = \int \frac{\sin^2\theta}{\sqrt{\cos^2\theta}} \cos\theta \, d\theta$$

$$= \int \sin^2\theta \, d\theta = \frac{-\cos x \sin x}{2} + \frac{1}{2}\int dx$$

$$= \frac{-\cos x \sin x}{2} + \frac{x}{2} + C$$

Example 2.23 **Integrate**

(i) $I = \int \sin^6 x \, dx$

(ii) $\int \cos^4 x \, dx$

(iii) $I = \int \frac{1}{\sqrt{\cos^3 x \sin^5 x}} dx$

Solution : (i) $I = \frac{-\cos x \sin^5 x}{6} + \frac{5}{6}\int \sin^4 x dx$

$$= \frac{-\cos x \sin^5 x}{6} + \frac{5}{8}\left[-\cos x \sin^3 x + \frac{3}{4}\int \sin^2 x \, dx\right]$$

$$= \frac{-\cos x \sin^5 x}{6} - \frac{5}{8}\cos x \sin^3 x + \frac{15}{32}\left[\frac{-\cos x \sin x}{2} + \int dx\right]$$

$$= \frac{-\cos \sin^5 x}{6} - \frac{5}{8}\cos x \sin^3 x - \frac{15}{32}\cos x \sin x + \frac{15}{32}x + C$$

(ii) $$\int \cos^4 x\, dx = \frac{\sin x \cos^3 x}{4} + \frac{3}{4}\int \cos^2 x\, dx$$

$$= \frac{\sin x \cos^3 x}{4} + \frac{3}{4}\left[\frac{\sin x \cos x}{2} + \frac{1}{2}\int dx\right]$$

$$= \frac{\sin x \cos^3 x}{4} + \frac{3}{8}\sin x \cos x + \frac{3}{8}x + C$$

PROBLEM SET 2.3

Evaluate the following integrals

1. $\int \sin^5 x\, dx$

   Ans. $-\cos x + \frac{2}{3}\cos^3 x - \frac{1}{5}\cos^5 x + C$

2. $\int \frac{\cos^3 x}{\sin^{2/3} x}\, dx$

   Ans. $3 \sin^{1/3} x \left(1 - \frac{1}{7}\sin^2 x\right) + C$

3. $\int \frac{1}{\sin x \cos^3 x}\, dx$

   Ans. $\log \tan x + \frac{1}{2}\tan^2 x + C$

4. $\int_0^{\pi/4} \sin^4 x\, dx$

   Ans. $\frac{3\pi - 8}{32}$

5. $\int_0^{\pi/3} \sin^6 x\, dx$

   Ans. $\frac{5}{32}\pi$

6. $\int_0^1 x^2 (1 - x^2)^{3/2}\, dx$

   Ans. $\frac{\pi}{32}$.

7. $\int_0^a \frac{x^4}{\sqrt{a^2 - x^2}}\,dx$

Ans. $\frac{3\pi\,a^4}{16}$.

8. $\int_0^1 x^{3/2}\,(1 - x)^{3/2}\,dx$

Ans. $\frac{3\pi}{128}$

9. $\int_0^{\pi/2} \sin^9 x\,dx$ *(Madras' 98S)*

Ans. $\frac{128}{315}$

10. $\int_0^1 \frac{x^9}{\sqrt{1 - x^2}}\,dx$

Ans. $\frac{128}{315}$

11. $\int_0^{\pi/6} \sin^5 3\theta\,d\theta$

Ans. $8/45$

12. $\int_0^2 x^{5/2}\sqrt{2 - x}\,dx$

Ans. $\frac{5\pi}{8}$

13. $\int_0^4 x^3\sqrt{4x - x^2}\,dx$

Ans. 28π

14. If $I_{m,n} = \int_0^{\pi/2} \sin^m x \cos^n x\,dx,\ m > 0, n > 0$. Show that $I_{m,n} = \frac{m-1}{m+n} I_{m-2,n}$. Hence evaluate $\int_0^{\pi/2} \sin^4 x \cos^8 x\,dx$.

Ans. $\frac{35\,\pi}{1024}$

2.9 OTHER REDUCTION FORMULAE

Find reduction formulae for

2.9.1 Reduction Formula for $\int \tan^n x\, dx$

$$\int \tan^n x = \int \tan^{n-2} x\,(\sec^2 x - 1)\, dx$$

$$= \frac{\tan^{n-1} x}{n-1} - I_{n-2}$$

2.9.2 Reduction Formula for $\int \sec^n x\, dx$

$$\int \sec^n x\, dx = \int \sec^{n-2} x \sec^2 x\, dx, \text{ integrating by parts}$$

$$= \sec^{n-2} x \tan x - (n-2)\int \sec^{n-3} x \sec x \tan x \tan x\, dx$$

$$= \sec^{n-2} x \tan x - (n-2)\int \sec^{n-2} x \tan^2 x\, dx$$

$$= \sec^{n-2} x \tan x - (n-2)\left[\int \sec^n x\, dx - \int \sec^{n-2} x\, dx\right]$$

$$= \sec^{n-2} x \tan x - (n-2)(I_n - I_{n-2})$$

So $\quad (1 + n - 2)\, I_n = \sec^{n-2} x \tan x + (n-2)\, I_{n-2}$

or $\quad I_n = \dfrac{\sec^{n-2} x \tan x}{n-1} + \dfrac{n-2}{n-1} I_{n-2}$

2.9.3 Reduction Formula for $\int x^n \sin mx\, dx$

$$I_n = \int \underset{\text{I}}{x^n}\, \underset{\text{II}}{\sin mx}\, dx, \text{ integrating by parts}$$

$$= x^n \frac{\cos mx}{-m} + \frac{n}{m}\int \underset{\text{I}}{x^{n-1}}\, \underset{\text{II}}{\cos mx}\, dx$$

$$= -\frac{x^n \cos mx}{m} + \frac{n}{m^2}\left[x^{n-1} \sin mx - (n-1)\int \sin mx\, x^{n-2}\, dx\right]$$

or $\quad I_n = -\dfrac{x^n \cos mx}{n} + \dfrac{n}{m^2} x^{n-1} \sin mx - \dfrac{n(n-1)}{m^2} I_{n-2}$

Similarly, we can derive reduction for $\int x^n \cos mx\, dx$

2.9.4 Reduction Formula for $\int \cos^m x \sin nx\, dx$

$$I_{m,n} = \int \underset{\text{I}}{\cos^m x}\, \underset{\text{II}}{\sin nx}\, dx, \text{ integrating by parts}$$

$$= \frac{-\cos^m x \cos nx}{n} - \frac{m}{n}\int \cos^{m-1} x \sin x \cos nx\, dx$$

$\because \quad \sin(n-1)x = \sin nx \cos x - \cos nx \sin x$

$\Rightarrow \quad \sin x \cos nx = \sin nx \cos x - \sin(n-1)x$

$$\therefore \quad I_{m,n} = \frac{-\cos^m x \cos nx}{n} - \frac{m}{n}\int \cos^{m-1} x\,[\sin nx \cos x - \sin (n -)x]\,dx$$

$$= \frac{-\cos^m x \cos nx}{n} - \frac{m}{n}\int \cos^m x \sin nx\,dx + \frac{m}{n}\int \cos^{m-1} x \sin (n-1)\,x\,dx$$

$$\Rightarrow \quad \left[1 + \frac{m}{n}\right] I_{m,n} = \frac{-\cos^m x \cos nx}{n} + \frac{m}{n}\int \cos^{m-1} x \sin (n-1)\,xdx$$

$$\Rightarrow \quad I_{m,n} = \frac{-\cos^m x \cos nx}{m+n} + \frac{m}{m+n}\int \cos^{m-1} x \sin (n-1)\,xdx$$

$$\Rightarrow \quad I_{m,n} = \frac{-\cos^m x \cos nx}{m+n} + \frac{m}{m+n} I_{m-1,\,n-1}$$

Example 2.23 **Show that**

$$\boldsymbol{I_{m,n} = \int_0^{\pi/2} \cos^m x \cos nx\,dx = \frac{m}{m+n}\int_0^{\pi/2} \cos^{m-1} x \cos (n-1)\,x\,dx}$$

Hence deduce that $\int_0^{\pi/2} \cos^n x \cos nx\,dx = \frac{\pi}{2^{n+1}}$.

Solution : $I_{m,n} = \int_0^{\pi/2} \underset{\text{I}}{\cos^m x}\,\underset{\text{II}}{\cos nx}\,dx$, Integrating by parts

$$= \left[\cos^m x \frac{\sin nx}{n}\right]_0^{\pi/2} + \frac{m}{n}\int_0^{\pi/2} \cos^{m-1} x \sin x \sin nx\,dx$$

$$= 0 + \frac{m}{n}\int_0^{\pi/2} \cos^{m-1} x \sin x \sin nx\,dx$$

Now $\quad \cos (n-1)\,x = \cos nx \cos x + \sin nx \sin x$

So $\quad \sin nx \sin x = \cos (n-1)\,x - \cos nx \cos x$

$$\int_0^{\pi/2} \cos^{m-1} x \sin nx \sin x\,dx = \int_0^{\pi/2} [\cos^{m-1} x \cos (n-1)\,x - \cos^m x \cos nx]\,dx$$

$$= I_{m-1,\,n-1} - I_{m,n}$$

Hence $\quad I_{m,n} = \frac{m}{n}(I_{m-1,\,n-1} - I_{m,n})$

or $\quad I_{m,n} = \frac{m}{m+n} I_{m-1,\,n-1}$

If we take $m = n$

then $\quad I_{n,n} = \frac{n}{2n} I_{n-1,\,n-1}$

$$= \left(\frac{1}{2}\right)^2 I_{n-2,\,n-2} = \ldots \frac{1}{2^n} I_{0,\,0}$$

$$I_{0,\,0} = \int_0^{\pi/2} dx = \frac{\pi}{2}$$

Hence $$I_{n,\,n} = \frac{1}{2^n}\,\frac{\pi}{2}.$$

Example 2.24 **(i) Evaluate $I_n = \int_0^{\infty} e^{-x}\, x^n\, dx$ where n is a positive integer,**

(ii) $I_n = \int_0^1 x^n (\log x)^2\, dx$

Solution : (i) $$I_n = \left[\frac{e^{-x}}{-1}\, x^n\right]_0^{\infty} + n\int_0^{\infty} e^{-x}\, x^{n-1}\, dx$$

$$= \left\{\lim_{x\to\infty} \frac{x^n}{e^x} - 0\right\} + n\, I_{n-1}$$

$$= n\, I_{n-1}$$

Thus, $I_n = n\,(n-1)\ldots 3\cdot 2\cdot 1\; I_0,$

where $I_0 = \int_0^{\infty} e^{-x}\, dx = 0 - (-1) = 1$

$\therefore$ $I_n = n!$

(ii) $$I_n = \left[\frac{x^{n+1}}{n+1}(\log x)^2\right]_0^1 - \frac{2}{n+1}\int_0^1 x^{n+1}\,\frac{\log x}{x}\, dx$$

$$\lim_{x\to 0} \frac{\log x}{\left(\dfrac{1}{x}\right)^{n+1}} = 0$$

$$I_n = 0 - \frac{2}{n+1}\int_0^1 x^n \log x\, dx$$

$$= -\frac{2}{n+1}\left\{\left[\frac{x^{n+1}}{n+1}\log x\right]_0^1 - \frac{1}{n+1}\int_0^1 x^{n+1}\,\frac{1}{x}\, dx\right\}$$

$$= \frac{2}{(n+1)^3}\left[x^{n+1}\right]_0^1$$

$$= \frac{2}{(n+1)^3}.$$

PROBLEM SET 2.4

1. If $I_n = \int \operatorname{cosec}^n x\, dx$ find the reduction formula and hence evaluate I_3 *(Madrurai' 90S)*

$$I_n = -\frac{\operatorname{cosec}^{n-2} x \cot x}{n-1} + \frac{n-2}{n-1} I_{n-2}$$

$$I_3 = -\frac{1}{2}\cot x \operatorname{cosec} x + \frac{1}{2}\log \tan x/2.$$

2. Evaluate : (i) $\int \operatorname{cosec}^4 x\, dx$, (ii) $\int \sec^4 x\, dx$

(Bangalore' 93)

Ans. (i) $-\frac{1}{3}(\operatorname{cosec}^2 x \cot x + 2\cot x) + C$, (ii) $\frac{1}{3}(\sec^2 x \tan x + 2\tan x) + C$

3. $\displaystyle\int_{\pi/4}^{\pi/3} \cot^4 x\, dx = \frac{3\pi - 8}{12}$

4. If $I_n = \displaystyle\int_0^{\pi/4} \tan^n x\, dx$, prove that $I_n = \frac{1}{n-1} - I_{n-2}$. Hence evaluate I_5

Ans. $-\frac{1}{4} + \frac{1}{2}\log 2.$

(Madras 2000 PT, Kuvempu '98S)

5. Prove that if $I_n = \displaystyle\int \frac{t^n}{1+t^2}\, dt$, then $I_{n+2} = \frac{t^{n+1}}{n+1} - I_n$. Hence evaluate I_6.

Ans. $\frac{t^5}{5} - \frac{t^3}{3} + t - \tan^{-1} t + C,$

6. Evaluate $\displaystyle\int_0^a x^5 (2a^2 - x^2)^{-3}\, dx$

Ans. $-\frac{1}{4} + \frac{1}{2}\log 2$

7. Evaluate : (i) $\displaystyle\int_{\pi/6}^{\pi/2} \operatorname{cosec}^5 x\, dx$, (ii) $\displaystyle\int_0^a (a^2 + x^2)^{5/2}\, dx$

Ans. (i) $\frac{11}{4}\sqrt{3} + \frac{3}{8}\log(2+\sqrt{3})$, (ii) $\left\{\frac{67\sqrt{2}}{48} + \frac{5}{16}\log(1+\sqrt{2})\right\} a^6$.

8. If $I_n = \int x^n e^x\, dx$ show that $I_n = x^n e^x - n I_{n-1}$. Hence find I_4. *(Madras 2000)*

Ans. $e^x(1 - x + x^2 - x^3 + x^4) + C$

9. If $I_n = \displaystyle\int_0^\infty e^{-x} \sin^m x\, dx$, where $m \ge 2$. Show $(1+m^2) I_m = m(m-1) I_{m-2}$. Hence evaluate I_4

(Gorakhpur' 99)

Ans. 24/85

10. If $I_{m,n} = \int_0^{\pi/2} \cos^m x \cos nx\, dx$, prove that $I_{m,n} = \frac{m(m-1)}{m^2 - n^2} I_{m-2,n}$.

(AMIE' 97W)

11. If $U_n = \int_0^{\pi/2} x \cos^n x\, dx\ (n > 1)$, prove $I_n = -\frac{1}{n^2} + \frac{n-1}{n} I_{n-2}$. Hence evaluate I_4.

Ans. $\frac{3\pi^2}{64} - \frac{1}{4}$.

12. If $I_n = \int_0^{\pi/2} x \sin^n x\, dx\ (n > 1)$, prove that $I_n = \frac{1}{n^2} + \frac{n-1}{n} I_{n-2}$. Hence evaluate $I_5 = \frac{149}{225}$.

(AMIE' 97)

13. Find a reduction formula for $\int e^{ax} \cos^n x\, dx$. Hence evaluate $\int_0^{\pi/2} e^{2x} \cos^3 x\, dx$

$$I_n (a^2 + n^2) + e^{ax} (a \cos x + n \sin x) \cos^{n-1} x + n(n-1) I_{n-2}.$$

Ans. $2/65\,(3\,e^{\pi} - 11)$.

14. Prove that $\int_0^{\pi/2} \cos^{n-2} x \sin nx\, dx = \frac{1}{n-1}, n > 1$.

15. If $I_n = \int_0^{\pi/2} x^n \sin x\, dx, n > 1$. Prove that $I_n = n\left(\frac{\pi}{2}\right)^{n-1} - n(n-1) I_{n-2}$.

Hence evaluate I_5. *(Marathwada' 98S, Madras 2000S)*

Ans. $\frac{5}{16}\pi^4 - 15\pi^2 + 120$.

16. If $U_n = \int_0^{a} x^n e^{-x}\, dx$. Prove that $U_n - (n + a) U_{n-1} + a(n-1) U_{n-2} = 0$.

(Pondicherry 98S)

17. Obtain a reduction formula for $\int x^m (\log x)^n\, dx, m > -1$

Show $\int_0^1 x^m (\log x)^n\, dx = \frac{(-1)^n\, n!}{(m+1)^{n+1}}$.

Find $I_{5,3} = -1/216$

$$I_{m,n} = \frac{x^{m+1}}{m+1} (\log x)^n - \frac{n}{m+1} \int x^m (\log x)^{n-1}\, dx.$$

2.10 INTEGRALS BASED ON REDUCTION FORMULA

A formula connecting an integral involving a positive integer n with another integral of the same form involving lower orders $n-1, n-2$ etc. is called a reduction formula.

Evaluate.

$$I_n = \int \underset{\text{II}}{1}\ \underset{\text{I}}{(x^2+a^2)^{n/2}}\,dx, \text{ Integrating by parts}$$

$$I_n = x\,(x^2+a^2)^{n/2} - \frac{n}{2} \times 2 \int x^2\,(x^2+a^2)^{n/2-1}\,dx$$

$$= x\,(x^2+a^2)^{n/2} - n \int (x^2+a^2-a^2)\,(x^2+a^2)^{n/2-1}\,dx$$

$$I_n = x\,(x^2+a^2)^{x/2} - n\,(I_n - a^2 I_{n-2})$$

Hence $\quad (1+n)\,I_n = x\,(x^2+a^2)^{n/2} + a^2 n\,I_{n-2}$ ...(1)

Example 2.25 **Evaluate $I_5 = \int (x^2+a^2)^{5/2}\,dx$.**

Solution : Using (1) above, with $n = 5$

$$I_5 = \frac{1}{6}\left\{x\,(x^2+a^2)^{5/2} + 5a^2 I_3\right\}, \text{ now taking } n = 3$$

$$I_3 = \frac{1}{4}\left\{x\,(x^2+a^2)^{3/2} + 3a^2 I_1\right\}$$

$$\text{So } I_5 = \frac{1}{6}\,x\,(x^2+a^2)^{5/2} + \frac{5}{6}\,a^2\left\{\frac{x(x^2+a^2)^{3/2}}{4} + \frac{3}{4}\,a^2 I_1\right\}$$

$$= \frac{1}{6}\,x\,(x^2+a^2)^{5/2} + \frac{5}{24}\,a^2 x\,(x^2+a^2)^{3/2}$$

$$+ \frac{15}{24}\,a^4\left\{\frac{x\sqrt{x^2+a^2}}{2} + \frac{a^2}{2}\log\left[x+\sqrt{x^2+a^2}\right]\right\}$$

Example 2.26 **If $I_n = \int x^n\,(a-x)^{\frac{1}{2}}\,dx$, prove that**

$$(2n+3)\,I_n = 2an\,I_{n-1} - 2x^n\,(a-x)^{3/2}$$

Solution : Taking x^n as first function, and integrating by parts,

$$I_n = \frac{x^n\,(a-x)^{3/2}}{-3/2} - n\left(-\frac{2}{3}\right)\int x^{n-1}\,(a-x)\,(a-x)^{\frac{1}{2}}\,dx$$

$$I_n = \frac{x^n(a-x)^{3/2}}{-3/2} - n\left(-\frac{2}{3}\right)(a\,I_{n-1} - I_n)$$

$$\text{So} \quad \left(1+\frac{2n}{3}\right)I_n = -\frac{2}{3}\,x^n\,(a-x)^{3/2} + \frac{2n\alpha}{3}\,I_{n-1}$$

Simplifying we get the result.

Example 2.27 **If $U_n = \int x^n (a^2 - x^2)^{1/2}\, dx$, prove that**

$$U_n = -\frac{x^{n-1}(a^2 - x^2)^{3/2}}{n+2} + \frac{n-1}{n+2} a^2 U_{n-2}.$$

Solution : Where n is to be decreased, choosing x^{n-1} as first function and integrating by parts, we have

$$U_n = \underset{\text{I}}{x^{n-1}} \left\{ \underset{\text{II}}{x (a^2 - x^2)^{1/2}} \right\} dx$$

$$= x^{n-1} \frac{(a^2 - x^2)^{3/2}}{-2(3/2)} - \frac{n-1}{-3} \int x^{n-2} (a^2 - x^2)(a^2 - x^2)^{1/2}\, dx$$

$$U_n = \frac{x^{n-1}(a^2 - x^2)^{3/2}}{-2(3/2)} - \frac{n-1}{-3} (a^2 U_{n-2} - U_n)$$

Hence

$$\left(1 + \frac{n-1}{3}\right) U_n = -\frac{x^{n-1}(a^2 - x^2)^{3/2}}{3} + \frac{n-1}{3} a^2 U_{n-2}$$

Simplifying we get the result.

Example 2.28 **If $I_n = \int 1 \frac{1}{(x^2 + a^2)^n} dx$, find a reduction formula connecting I_n with I_{n-1}.**

Solution : Integrating by parts, taking 1 as second function.

$$I_n = \frac{x}{(x^2 + a^2)^n} + 2n \int \frac{x^2}{(x^2 + a^2)^{n+1}} dx$$

$$= \frac{x}{(x^2 + a^2)^n} + 2n \int \frac{(x^2 + a^2) - a^2}{(x^2 + a^2)^{n+1}} dx$$

$$I_n = \frac{x}{(x^2 + a^2)^n} + 2n (I_n - a^2 I_{n+1})$$

So $(1 - 2n) I_n = \dfrac{x}{(x^2 + a^2)^n} - 2na^2 I_{n+1}$

So $\quad I_{n+1} = \dfrac{1}{2na^2} \dfrac{x}{(x^2 + a^2)^n} + \dfrac{2n-1}{2na^2} I_n$

Replacing n by $(n - 1)$

$$I_n = \frac{1}{2(n-1)a^2} \frac{x}{(x^2 + a^2)^{n-1}} + \frac{2n-3}{(2n-2)a^2} I_{n-1}$$

Exampe 2.29 **Evaluate $\int \frac{1}{(x^2 + 3)^3} dx$.**

Solution : Comparing with (1) above, $n = 3$, $a^2 = 3$.

So $\quad I_3 = \dfrac{1}{4 \times 3} \dfrac{x}{(x^2 + 3)^2} + \dfrac{3}{4 \times 3} I_2$

$$I_2 = \frac{1}{2 \times 1 \times 3} \frac{x}{(x^2+3)} + \frac{1}{2 \times 3} I_1$$

and $$I_1 = \int \frac{1}{x^2+3}\, dx$$

$$= \frac{1}{\sqrt{3}} \tan^{-1} \frac{x}{\sqrt{3}}$$

Hence $$I_3 = \frac{1}{12} \frac{x}{(x^2+3)^2} + \frac{1}{4}\left\{\frac{1}{6} \frac{x}{x^2+3} + \frac{1}{6\sqrt{3}} \tan^{-1} \frac{x}{\sqrt{3}}\right\} + C.$$

2.11 APPLICATIONS OF INTEGRATION

The length of an arc, area under a plane curve, the surface and volume of the surface formed by revolving a curve are all expressible as single integrals.

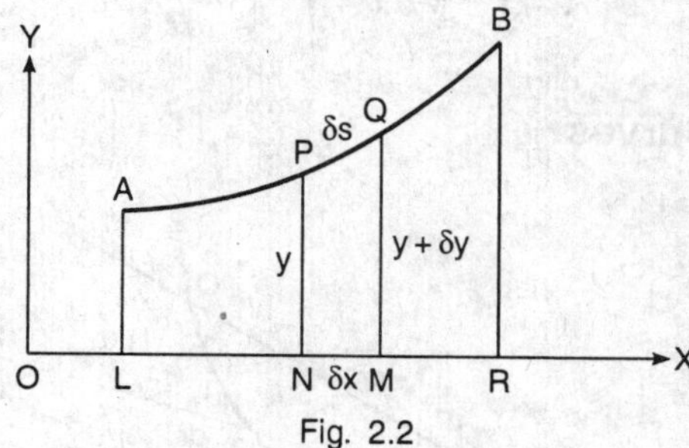

Fig. 2.2

2.11.1 Length of An Arc

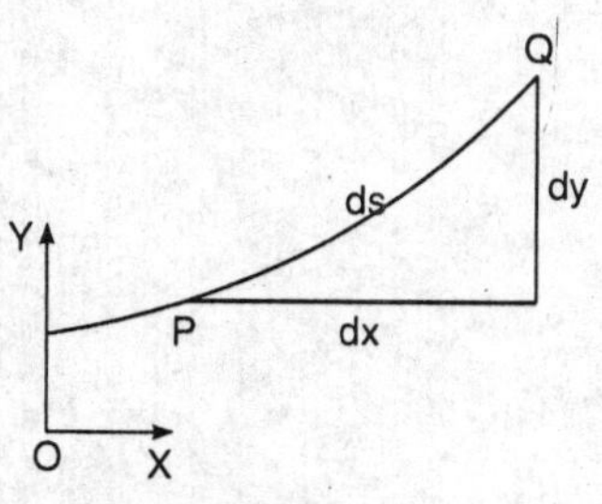

Fig. 2.3

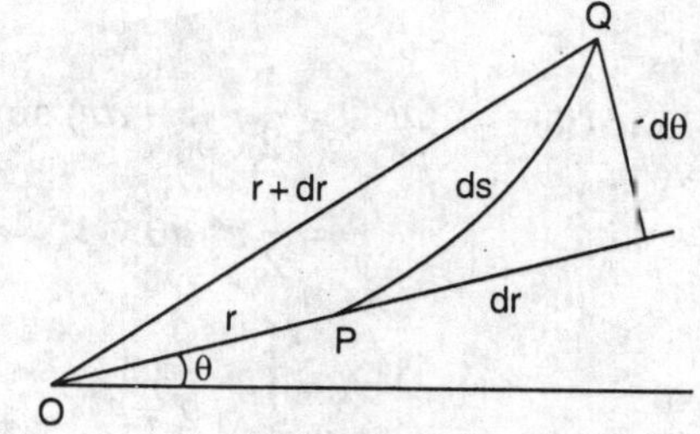

Fig. 2.4

$$(ds)^2 = (dx)^2 + (dy)^2$$

So $$ds = \sqrt{1 + \left(\frac{dy}{dx}\right)^2}\, dx = \sqrt{\left(\frac{dx}{dt}\right)^2 + \left(\frac{dy}{dt}\right)^2}\, dt$$

In terms of a parameter t

So length of an arc $s = \int \sqrt{1 + \left(\frac{dy}{dx}\right)^2}\, dx$ for equation $y = f(x)$ in Cartesian form.

$$s = \int \sqrt{\dot{x}^2 + \dot{y}^2}\, dt,$$

when the equation of curve is in parameteric form.

$$s = \int \sqrt{\left(\frac{dr}{d\theta}\right)^2 + r^2}\, d\theta$$

when the equation is in polar form.

The integrals are to be evaluated between the proper limits.

2.11.2 Area Under the Curve

The area of the elementary strip is $y\,dx$

So, area under the curve bounded by the x axis, and two bounding ordinates is given by

$$A = \int_{x_1}^{x_2} y\,dx = \int_{t_1}^{t_2} \left(y\,\frac{dx}{dt}\right) dt$$

2.11.3 Areas of Polar Curves

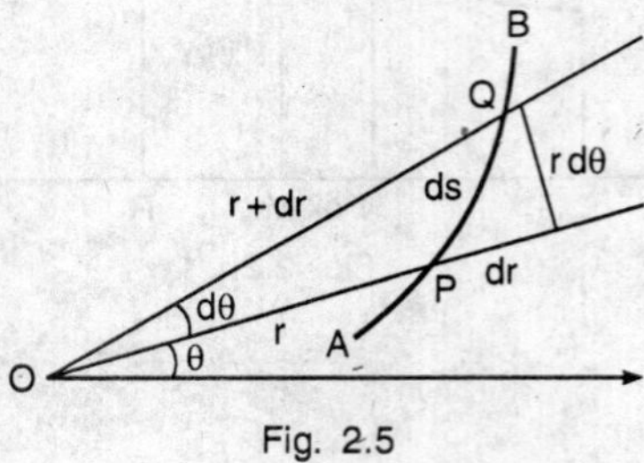

Fig. 2.5

Elementary Area $\quad \Delta\, OPQ = \frac{1}{2} r\,(r + dr) \sin d\theta$

$$= \frac{1}{2} r^2\, d\theta.$$

So area

$$A = \int_{\theta_1}^{\theta_2} \frac{1}{2} r^2\, d\theta$$

2.11.4 Volume of Revolution

The volume V of the solid formed by revolving the curve $y = f(x)$, about x axis is given by

$$V_x = \int_a^b \pi y^2\, dx.$$

On revolving about y-axis

$$V_y = \int_{y_A}^{y_B} \pi x^2\, dy.$$

2.11.5 Surface Of Revolution

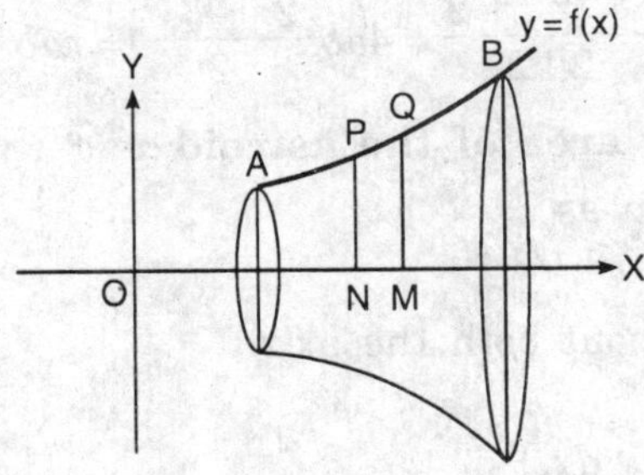

Fig. 2.7

If a curve $y = f(x)$ be revolved about x-axis.
The surface areas of the solid so formed is given by

$$\boldsymbol{S = \int_{s_A}^{s_B} 2\pi y\, ds = \int_{x_A = a}^{x_B = b} 2\pi y \sqrt{1 + \left(\frac{dy}{dx}\right)^2}\, dx}$$

On revolving about y-axis. Surface S is

$$S = \int_{s_A}^{s_B} 2\pi x\, ds$$

$$= 2\pi \int_{y_A}^{y_B} x \sqrt{\left(\frac{dx}{dy}\right)^2 + 1}\, dy$$

Area

Example 2.30 **Find the area of the ellipse $\boldsymbol{\frac{x^2}{a^2} + \frac{y^2}{b^2} = 1}$.**

Solution : Curve is symmetrical about both the axes.

So area $A = 4\int_0^a y\, dx$

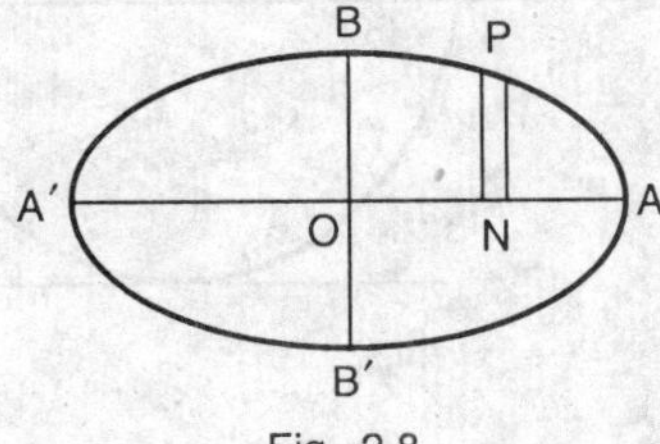

Fig. 2.8

expressing in parametric form

$x = a\cos t,\ y = b\sin t,$

$$|A| = 4\int_0^{\pi/2} y \left|\frac{dx}{dt}\right| dt$$

$$= 4ab \int_0^{\pi/2} \sin^2 t\, dt$$

$$= 4ab\,\frac{\Gamma 3/2\,\Gamma\frac{1}{2}}{2\,\Gamma 2} = 4ab\,\frac{\frac{1}{2}\sqrt{\pi}\,\sqrt{\pi}}{2} = \pi ab$$

Example 2.31 **Find the whole area of the astroid $x^{2/3} + y^{2/3} = a^{2/3}$.**

Solution : The curve written as

$(x^2)^{1/3} + (y^2)^{1/3} = (a^2)^{1/3}$

is symmetrical about both the axis.

So Area $A = 4\int_0^a y\,dx.$

Writing in parametric form

as $x = a\cos^3\theta,\ y = a\sin^3\theta,$

$$A = 4\int_0^{\pi/2} y\,\frac{dx}{d\theta}\,d\theta$$

$$A = 4\int_0^{\pi/2} a\sin^3\theta\,|-3a\cos^2\theta\sin\theta|\,d\theta$$

$$= 12a^2\,\frac{\Gamma 5/2\,\Gamma 3/2}{2\,\Gamma 4} = 12a^2\,\frac{\frac{3}{2}\frac{1}{2}\sqrt{\pi}\,\frac{1}{2}\sqrt{\pi}}{2\times 3\times 2} = \frac{3}{8}\pi a^2.$$

Fig. 2.8

If the curve be $\left(\frac{x}{a}\right)^{2/3} + \left(\frac{y}{b}\right)^{2/3} = 1,$

$$\text{Area} = 12ab\int_0^{\pi/2}\sin^4\theta\cos^2\theta\,d\theta = \frac{3}{8}\pi ab.$$

Example 2.32 **Find the area of the cycloid $x = a\,(\theta + \sin\theta)$, $y = a\,(1 - \cos\theta)$ and its base also. Also find the area between the curve and the x-axis.**

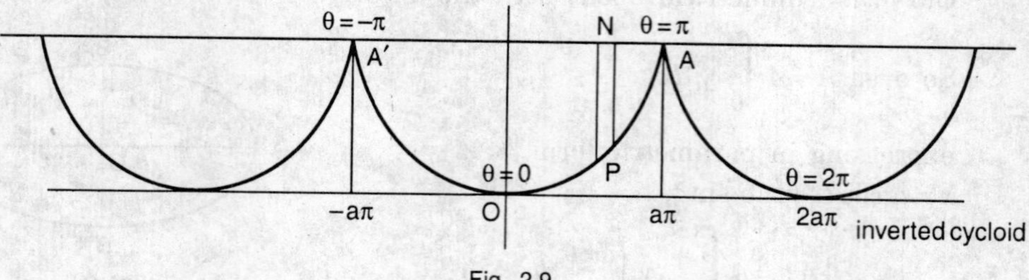

Fig. 2.9

Solution : We have for

$\theta = 0,$	$\frac{\pi}{2}$	π	$\frac{3\pi}{2}$	2π
$y = 0$	a	$2a$	a	0
$x = 0$	$a\frac{\pi}{2} + a$	$a\pi$	$\frac{3a\pi}{2} + a$	$2\pi a$

The curve has several branches, one branch may be taken from $\theta = -\pi$ to $\theta = +\pi$. Curve is symmetrical about y axis.

$$\text{Area between curve and its base} = \int_{-\pi}^{\pi} (2a - y) \frac{dx}{d\theta} d\theta$$

$$= 2 \int_0^{\pi} a (1 + \cos \theta)\, a (1 + \cos \theta)\, d\theta = 8a^2 \int_0^{\pi} \cos^4 \frac{\theta}{2} d\theta$$

Let $\theta/2 = t$

$$= 16a^2 \int_0^{\pi/2} \cos^4 t\, dt = 16a^2 \frac{\Gamma 5/2\, \Gamma 1/2}{2\, \Gamma 3} = 16a^2 \frac{\frac{3}{2} \cdot \frac{1}{2} \cdot \pi}{2 \cdot 2} = 3\pi a^2.$$

Area between curve and x axis

$$= 2 \int_0^{\pi} y \frac{dx}{d\theta} d\theta = 2a^2 \int_0^{\pi} (1 - \cos \theta)(1 + \cos \theta)\, d\theta.$$

$$= a^2 \int_0^{\pi} (1 - \cos 2\theta)\, d\theta = a^2 \left[\theta - \frac{\sin 2\theta}{2}\right]_{\pi} = \pi a^2.$$

Note. The above one is called the inverted cycloid. The equations $x = a (\theta - \sin \theta)$, $y = a (1 - \cos \theta)$ represent a cycloid

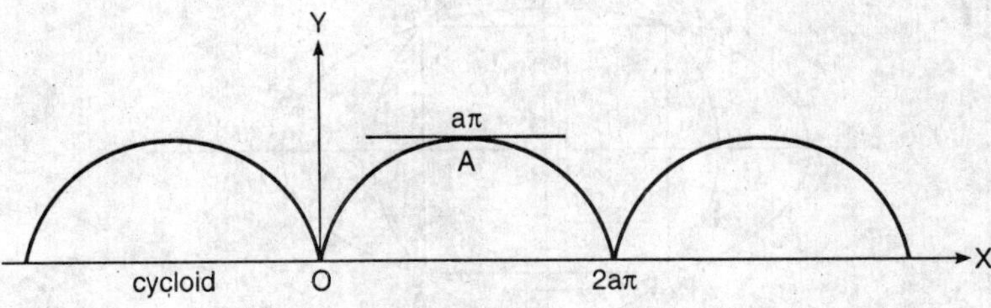

Fig. 2.10

In this case $\frac{dy}{dx} = \cot \frac{\pi}{2}$ at $A = 0$, and the tangent is parallel to x axis. For inverted cycloid $\frac{dy}{dx} = \tan \frac{\pi}{2}$ at A and the tangent is parallel to y axis.

Example 2.33 **Find the area included between the curve $xy^2 = 4a^2 (2a - x)$ and its asymptote.**

Solution : The curve is symmetrical about x axis. For $x > 2a$, y is imaginary. y axis $(x = 0)$ is its asymptote

Required area

$$= 2 \lim_{\varepsilon \to 0} \int_{\varepsilon}^{2a} y\, dx$$

$$= 4a \int_0^{2a} \sqrt{\frac{2a - x}{x}}\, dx$$

Let $x = 2a \sin^2 \theta$

$dx = 4a \sin \theta \cos \theta \, d\theta$

$$A = 16a^2 \int_0^{\pi/2} \frac{\cos \theta}{\sin \theta} \sin \theta \cos \theta \, d\theta$$

$$= 8a^2 \int_0^{\pi/2} (1 + \cos 2\theta)\, d\theta$$

$$= 8a^2 \left[\theta + \frac{\sin 2\theta}{2} \right]_0^{\pi/2}$$

$$= 4\pi a^2.$$

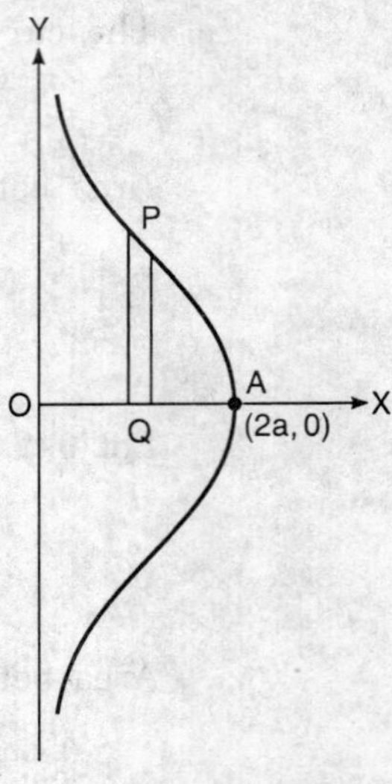

Fig. 2.11

Example 2.34 **Find the area common to the circle $x^2 + y^2 = 4$ and the ellipse $x^2 + 4y^2 = 9$.**

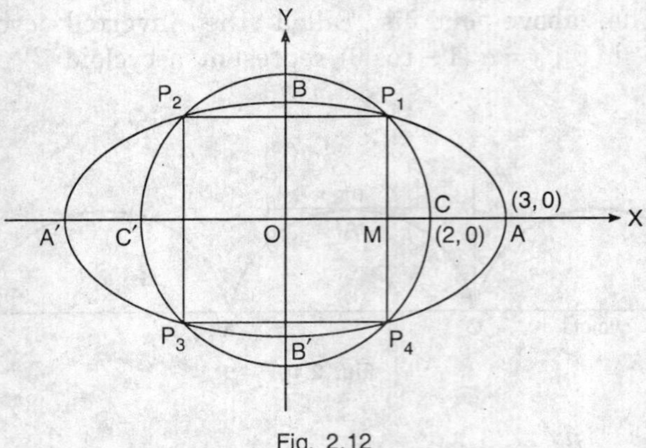

Fig. 2.12

Solution : The two curves cut at four points lying in the four quadrants.

By symmetry in all the four quadrants the required area

$= 4 \times \text{Area } OCP_1B$

$= 4$ [Area OMP_1B for ellipse + Area MCP_1M for circle]

at P_1 where the curves cut in first quadrant

$$x^2 + y^2 = 4, \; x^2 + 4y^2 = 9$$

So $\quad y = \sqrt{\frac{5}{3}},$

$$x = \sqrt{\frac{7}{3}}.$$

$$\text{Area } OMP_1B = \int_0^{\sqrt{7/3}} \frac{\sqrt{9-x^2}}{2}\,dx = \frac{1}{2}\left[\frac{x\sqrt{9-x^2}}{2} + \frac{9}{2}\sin^{-1}\frac{x}{3}\right]_0^{\sqrt{7/3}}$$

$$= \frac{1}{4}\left[\sqrt{\frac{7}{3}\times\frac{20}{3}} + 9\sin^{-1}\frac{\sqrt{7}}{3\sqrt{3}}\right] \quad ...(1)$$

$$\text{Area } MCP_1M = \int_{\sqrt{7/3}}^{2} \sqrt{4-x^2}\,dx = \frac{1}{2}\left[x\sqrt{4-x^2} + 4\sin^{-1}\frac{x}{2}\right]_{\sqrt{7/3}}^{2}$$

$$= \frac{1}{2}\left[4\,\frac{\pi}{2} - \sqrt{\frac{7}{3}\times\frac{5}{3}} - 4\sin^{-1}\sqrt{\frac{7}{12}}\right] \quad ...(2)$$

Hence required area

$$= \left[\frac{2}{3}\sqrt{35} + 9\sin^{-1}\frac{\sqrt{7}}{3\sqrt{3}} + 4\pi - \frac{2}{3}\sqrt{35} - 8\sin^{-1}\frac{\sqrt{7}}{2\sqrt{3}}\right]$$

$$= \left[4\pi + 9\sin^{-1}\frac{\sqrt{7}}{3\sqrt{3}} - 8\sin^{-1}\frac{\sqrt{7}}{2\sqrt{3}}\right]$$

Example 2.35 **Find the area common to the parabola $y^2 = ax$ and the circle $x^2 + y^2 = 4ax$.** *(Raipur' 98)*

Solution : The circle $(x-2a)^2 + y^2 = (2a)^2$, has centre at $(2a, 0)$ and radius $= 2a$. It cuts the parabola at $x^2 + ax = 4ax$ or $x = 0$ and $x = 3a$.

Both the curves are symmetrical about x axis

Required area $= 2$ [Area $OMPO$ of parabola + Area $MAPM$ of circle]

Fig. 2.13

$$\text{Area} \quad OMPO = \int_0^{3a} \sqrt{ax}\,dx$$

$$= \left[\sqrt{a}\,\frac{2}{3}x^{3/2}\right]_0^{3a} = 2a^2\sqrt{3} \quad ...(1)$$

$$\text{Area} \quad MAPM = \int_{3a}^{4a} \sqrt{x(4a-x)}\,dx.$$

Let $x = 4a\sin^2\theta$, $dx = 8a\sin\theta\cos\theta\,d\theta$

$$= \int_{\pi/3}^{\pi/2} (4a)(8a)\sin^2\theta\cos^2\theta\,d\theta = 8a^2\int_{\pi/3}^{\pi/2}\sin^2 2\theta\,d\theta$$

$$= 4a^2\int_{\pi/3}^{\pi/2}(1-\cos 4\theta)\,d\theta$$

$$= 4a^2 \left[\theta - \frac{\sin 4\theta}{4} \right]_{\pi/3}^{\pi/2} = 4a^2 \left[\left(\frac{\pi}{2} - \frac{\pi}{3} \right) - 0 + \frac{1}{4} \sin \frac{4\pi}{3} \right]$$

$$= 4a^2 \left(\frac{\pi}{6} - \frac{\sqrt{3}}{8} \right)$$

Hence $A = 2a^2 \left[2\sqrt{3} + \frac{2\pi}{3} - \frac{\sqrt{3}}{2} \right] = \left(3\sqrt{3} + \frac{4}{3}\pi \right) a^2.$

Example 2.36 **Find the area common to the circles $x^2 + y^2 = 4$ and $x^2 + y^2 = 4x$.**

Solution : The curves are symmetrical about x axis. They intersect at $(1, \pm\sqrt{3})$

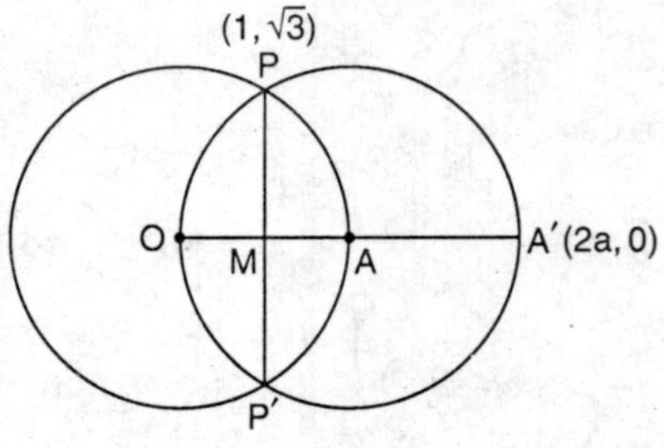

Fig. 2.14

Required area $= 2$ Area $OMPO$ of the circle $(x-2)^2 + y^2 = 2^2$ and the area $MAPM$ of the circle $x^2 + y^2 = 2^2$.

Area $POP'P = 2 \int_0^1 \sqrt{x(4-x)}\, dx.$

Let $x = 4\sin^2\theta$ $= 64 \int_0^{\pi/6} \sin\theta \cos\theta \sin\theta \cos\theta \, d\theta$

$$= 16 \int_0^{\pi/6} \sin^2 2\theta \, d\theta = 8 \int_0^{\pi/6} (1 - \cos 4\theta) \, d\theta$$

$$= 8 \left[\theta - \frac{\sin 4\theta}{4} \right]^{\pi/6} = 8 \left(\frac{\pi}{6} - \frac{\sqrt{3}}{8} \right) \qquad ...(1)$$

Area $PMP'AP = 2 \int_1^2 \sqrt{4 - x^2}\, dx$

$$= \left[x\sqrt{4-x^2} + 4\sin^{-1}\frac{x}{2} \right]_1^2$$

$$= -\sqrt{3} + \frac{4\pi}{3} = -\sqrt{3} + \frac{4\pi}{3}$$

Hence common area $= \frac{4\pi}{3} - \sqrt{3} - \sqrt{3} + \frac{4\pi}{3} = \left(\frac{8\pi}{3} - 2\sqrt{3} \right)$

Example 2.37 **Find the area included between the circle $x^2 + y^2 = 2ax$ and the parabola $y^2 = ax$.** *(Rewa' 90)*

Solution : Curves are symmetrical about x axis. They intersect at (0, 0) and (a, a)

Common Area = 2 [Area $OCPO$ of parabola + Area $CAPC$ of circle]

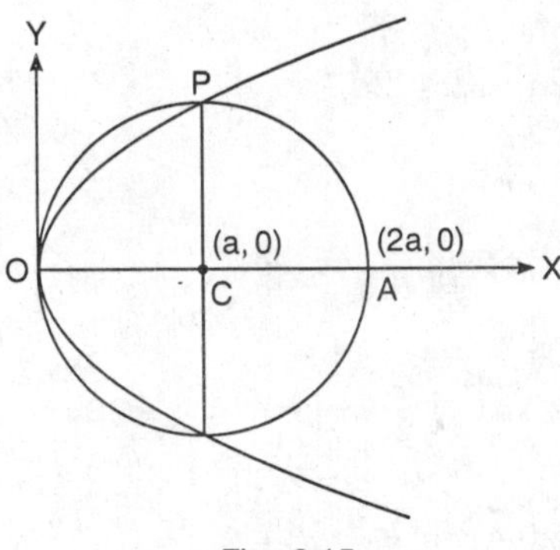

Fig. 2.15

$$\text{Area } OCPO \text{ of parabola} = \int_0^a \sqrt{ax}\, dx = \sqrt{a}\,\frac{2}{3}\left[x^{3/2}\right]_0^a = \frac{2}{3}a^2 \quad \text{...(1)}$$

$$\text{Area } CAPC \text{ of circle} = \int_a^{2a} \sqrt{x(2a-x)}\, dx \text{ Let } x = 2a\sin^2\theta$$

$$= \int_{\pi/4}^{\pi/2} 2a \sin\theta \cos\theta\, 4a \sin\theta\cos\theta\, d\theta$$

$$= a^2 \int_{\pi/4}^{\pi/2} 2\sin^2 2\theta\, d\theta = a^2 \int (1 - \cos 4\theta)\, d\theta$$

$$= a^2 \left[\theta - \frac{\sin 4\theta}{4}\right]_{\pi/4}^{\pi/2} = a^2\left(\frac{\pi}{4} - 0\right)$$

$$\text{Required area} = 2\left[\frac{2}{3}a^2 + \frac{\pi a^2}{4}\right] = a^2\left(\frac{4}{3} + \frac{\pi}{2}\right)$$

Example 2.38 **Prove that the area bounded by the parabola $y^2 = 4ax$ and the line $x + y = 3a$ is $\frac{64}{3}a^2$.** *(JNTU' 90)*

Solution : The parabola and the line intersect at

$L(a, 2a)$ and $P(9a, -6a)$

Required common area = Area $ASLA$ of parabola + area SML of Δ

+ Area $APNA$ of parabola − Area $\Delta\, MPN$.

$$\text{Area } ASLA = \int_0^a \sqrt{4ax}\, dx$$

$$= 2\sqrt{a} \times \frac{2}{3}\left[x^{3/2}\right]_0^a = \frac{4}{3}a^2 \qquad ...(1)$$

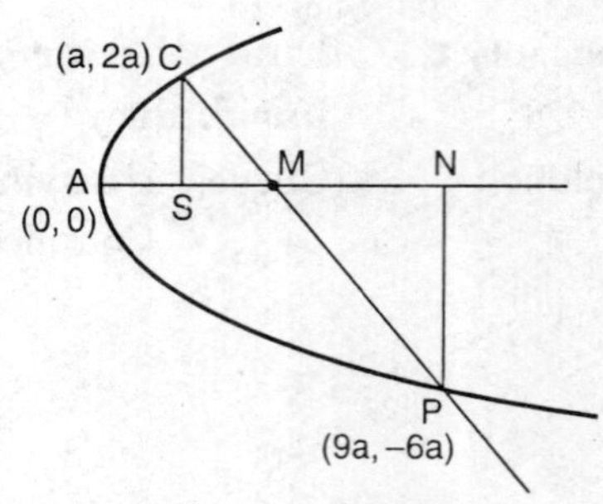

Fig. 2.16

at M $y = 0$ so from $x + y = 3a$, $x = 3a$

$SM = 3a - a = 2a$, $SL = 2a$

$$\text{Area } \Delta\, SML = \frac{2a \times 2a}{2} = 2a^2 \qquad ...(2)$$

$$\text{Area } \Delta\, MNP = \frac{6a}{2}(9a - 3a) = 18a^2 \qquad ...(3)$$

$$\text{Area } APNA = \int_0^{9a} 2\sqrt{ax} = \frac{4}{3}\sqrt{a}\left[x^{3/2}\right]_0^{9a} = 36a^2 \qquad ...(4)$$

$$\text{Required Area} = \frac{4}{3}a^2 + 2a^2 - 18a^2 + 36a^2 = \frac{64}{3}a^2.$$

Example 2.39 **Prove that the area common to the parabolas**

$$y^2 = 4ax \text{ and } x^2 = 4ay \text{ is } \frac{16a^2}{3}$$

(Kuvempu' 98)

Solution : The two curves intersect at

$x = 0$ and $x = 4a$

$$\text{Common area} = \int_0^{4a} (y_1 - y_2)\, dx$$

$$= \int_0^{4a} \left(\sqrt{4ax} - \frac{x^2}{4a}\right) dx$$

$$= \left[2\sqrt{a} \times \frac{2}{3}x^{3/2} - \frac{x^3}{12a}\right]_0^{4a}$$

$$= \frac{16}{3}a^2$$

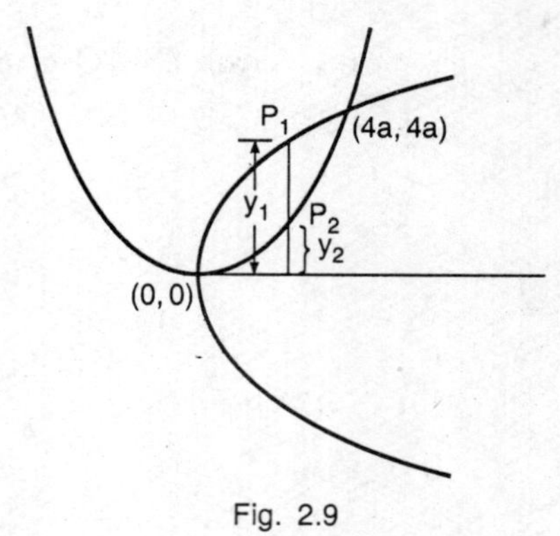

Fig. 2.9

PROBLEM SET 2.5

1. Find the area bounded by the x-axis and the following curves between given ordinates:

(i) $y = c\cosh\frac{x}{c}$; $x = 0$, $x = a$

Ans. $c^2 \sinh\frac{a}{c}$

(ii) $y = \log x$; $x = b$ $(b > a > 1)$

Ans. $b\log\frac{b}{e} - a\log\frac{a}{e}$.

(iii) $y = \sin^2 x$; $x = 0$ to $x = \frac{\pi}{2}$

Ans. $\frac{\pi}{4}$

2. Find the whole area of the curve $a^2y^2 = x^2 (a^2 - x^2)$.

Ans. $\frac{1}{3} a^2$.

3. Find the whole area of the curve $a^2x^2 = y^3 (2a - y)$.

Ans. πa^2

4. Find the area of the segment cu off from the parabola $y^2 = 2x$ and the line $y = 4x - 1$.

Ans. $\frac{9}{32}$

5. Find the area common to the curves $y^2 = ax$ and $x^2 + y^2 + 4ax$.

Ans. $\frac{1}{3} [9\sqrt{3} + 4\pi] a^2$

6. Find the area of the loop of the curve $ay^2 = x^2 (a - x)$. *(Osmania 2000S)*

Ans. $\frac{8}{15} a^2$.

7. Find the area enclosed by the curve $a^2x^2 = y^3 (2a - y)$.

Ans. πa^2

8. Find the area of the segment of the parabola cut off by the line $x - 2y + 8 = 0$.

Ans. 36

9. Find the area bounded by the parabola and its latus rectum $x = a$.

Ans. $\frac{8a^2}{3}$

10. Find the area of the loop of the curve $y^2 (a + x) = x^2 (a - x)$.

Ans. $(2 - \pi/2) a^2$ *(AMIE' 97)*

11. Find the area bounded by the curve $y = x(x - 3) (x - 5)$ and the x-axis. *(Ranchi' 90)*

Ans. $21\frac{1}{12}$.

12. Find the area of the loop of the curve $3ay^2 = x (x - a)^2$. *(Osmania' 95)*

Ans. $\frac{8}{15\sqrt{3}} a^2$.

13. Find the whole area of the following curves

(i) $y^2 = x^2 (1 - x^2)$ *(Nagpur' 97)*

Ans. $\frac{4}{3}$

(ii) $a^2y^2 = x^3(2a - x)$

Ans. πa^2

14. Find the area included between the curve and its asymptotes for the curves

(i) $xy^2 = a^2(a - x)$ *(AMIE 97W)*

Ans. πa^2

(ii) $x^2y^2 = a^2(y^2 - x^2)$

Ans. $4a^2$

15. Show that the area of the loop of the curve $y^2(a + x) = x^2(3a - x)$ is equal to the area between the curve and its asymptote.

16. Find the area bounded by the curve $\sqrt{x} + \sqrt{y} = \sqrt{a}$ and the coordinate axis.

Ans. $\frac{a^2}{6}$.

17. Find the area of the segment cut off from the parabola $y = 4x - x^2$ by the line $y = x$.

Ans. 9/2 *(AMIE' 90)*

[**Hint :** Area from 0 to 3 will be positive and that from 3 to 5 will be negative so whole area

$$= |A_1| + |A_2|$$

$$= \int_0^3 x(x-3)(x-5)\,dx + \int_5^3 -x(x-3)(x-5)\,dx$$

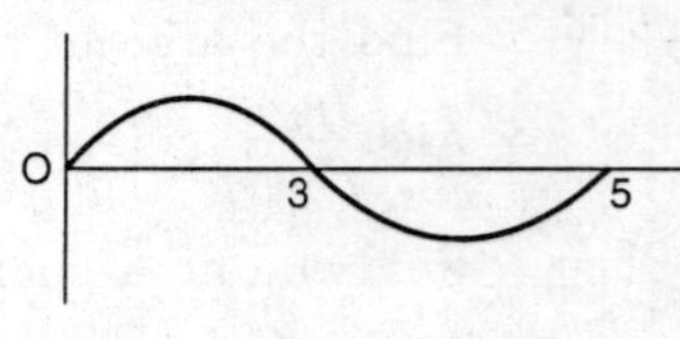

$$= \frac{63}{4} - \left(-\frac{16}{3}\right) = \frac{253}{12}$$]

2.11.6 Areas of Polar Curves

Area $$OAP = \frac{1}{2}\int_{\theta_A = \theta_1}^{\theta_B = \theta_2} r^2\,d\theta$$

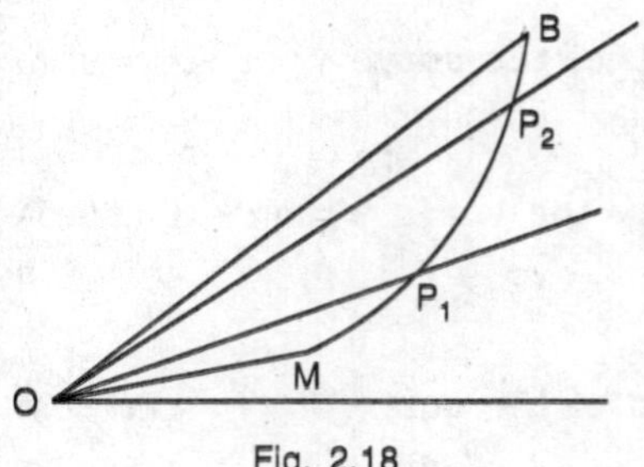

Fig. 2.18

Example 2.40 Find the area of the Cardioid :

(i) $r = a(1 + \cos\theta)$ and (ii) $r = a(1 - \cos\theta)$. *(Pondicherry' 98S)*

Solution : (i) The curve is symmetrical about the initial line. The function $f(\theta) = a(1 + \cos\theta)$ is periodic in θ. So whole curve is traced from $\theta = 0$ to $\theta = 2\pi$.

$$\text{Area} = \int_0^{2\pi} \frac{1}{2} r^2 \, d\theta$$

$$= a^2 \int_0^{\pi} (1 + \cos\theta)^2 \, d\theta$$

$$= 4a^2 \int_0^{\pi} \cos^4 \frac{\theta}{2} \, d\theta$$

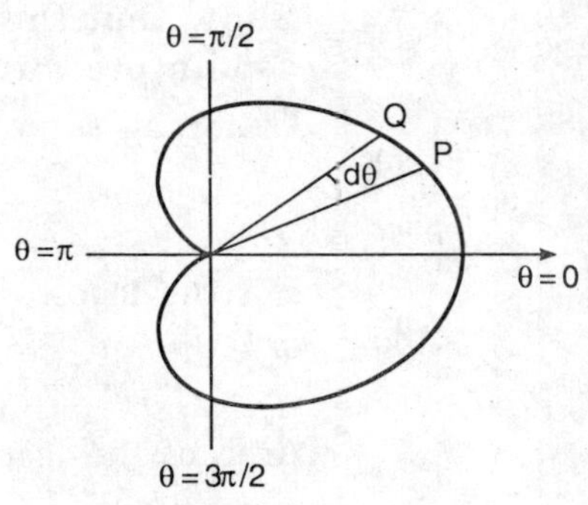

Fig. 2.19

Let $\frac{\theta}{2} = t$

$$A = 8a^2 \int_0^{\pi/2} \cos^4 t \, dt = \frac{8a^2 \, \Gamma 5/2 \, \Gamma 1/2}{2 \, \Gamma 3}$$

$$= 4a^2 \frac{\frac{3}{2}\frac{1}{2}\pi}{2 \times 1} = \frac{3}{2}\pi a^2.$$

(ii) For $r = a(1 - \cos\theta)$

$$A = 8a^2 \int_0^{\pi/2} \sin^4 \theta = \frac{3}{2}\pi a^2.$$

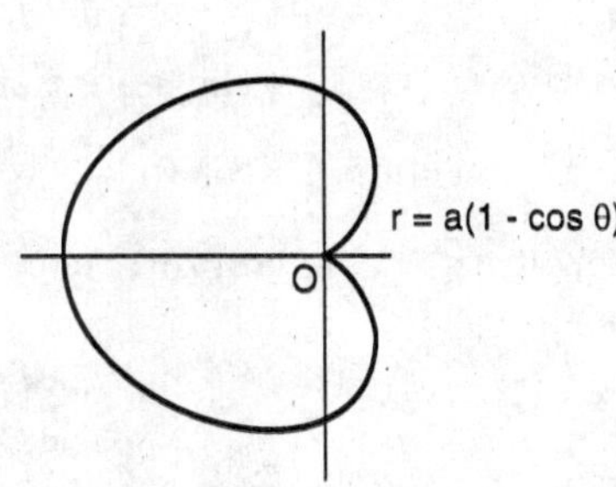

Fig. 2.20

Example 2.41 **Find the whole area of the lemniscate $r^2 = a^2 \cos 2\theta$.**

Solution : The curve consists of two loops of equal area.

First line is

$(r = 0, 2\theta = \pm \pi/2)$ from $-\frac{\pi}{4}$ to $\frac{\pi}{4}$.

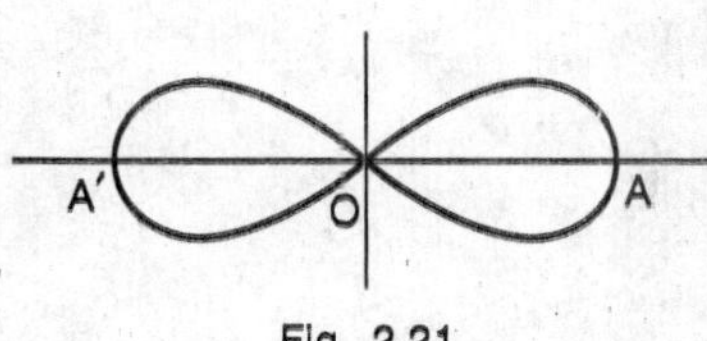

Fig 2.21

$$A = 4\frac{1}{2}\int_0^{\pi/4} r^2 \, d\theta$$

$$= 2a^2 \left[\frac{\sin 2\theta}{2}\right]_0^{\pi/4} = a^2.$$

Example 2.42 **Show that the area included between the folium $x^3 + y^3 = 3axy$ and its asymptote and the area of its loop.** *(Kurukshetra 98)*

Solution : Changing to polars,

$$r = 3a \frac{\sin\theta \cos\theta}{\cos^3\theta + \sin^3\theta}$$

For the loop $r = 0$ so $\sin\theta\cos\theta = 0$

or $\theta = 0$ to $\theta = \pi/2$

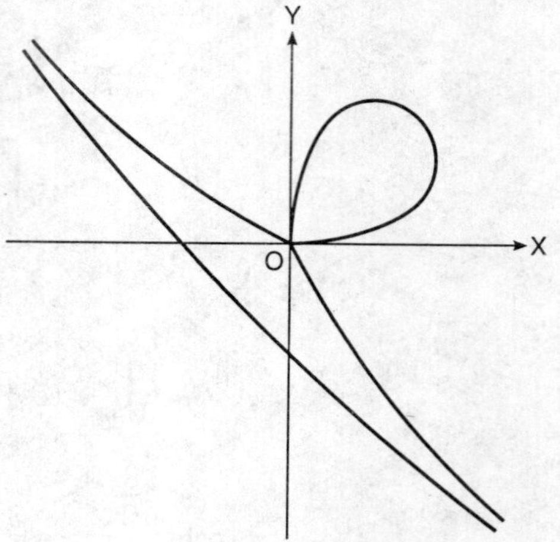

Fig. 2.22

$$\text{Area of the loop} = \frac{1}{2}\int_0^{\pi/2} r^2\, d\theta$$

$$= \frac{1}{2}\int_0^{\pi/2} 9a^2 \frac{\sin^2\theta\cos^2\theta}{(\cos^3\theta + \sin^3\theta)^2}\, d\theta$$

$$= \frac{9}{2}a^2 \int_0^{\pi/2} \frac{\tan^2\theta \sec^2\theta\, d\theta}{(1+\tan^3\theta)^2}$$

Let $1 + \tan^3\theta = t$

So $3\sec^2\theta\tan^2\theta\, d\theta = dt$

$$= \frac{9}{2}a^2 \int_1^{\infty} \frac{dt}{3t^2} = \frac{3a^2}{2}\left[\frac{1}{t}\right]_1^{\infty} = \frac{3}{2}a^2.$$

The asymptote to the curve has the equation $x + y = -a$

or $$\frac{a}{r} = -(\sin\theta + \cos\theta)$$

Area between curve any the asymptote

$$= 2\,\frac{1}{2}\int_{\theta=\pi/2}^{3\pi/4} (r_1^2 - r_2^2)\, d\theta$$

$$= 2\,\frac{a^2}{2}\int_{\pi/2}^{3\pi/4} \left\{\frac{1}{(\sin\theta+\cos\theta)^2} - \frac{9\sin^2\theta\cos^2\theta}{(\sin^3\theta + \cos^3\theta)^2}\right\} d\theta$$

$$= a^2 \int_{\pi/2}^{3\pi/4} \left\{\frac{1}{(1+\tan\theta)^2} - \frac{9\tan^2\theta}{(1+\tan^3\theta)^2}\right\} \sec^2\theta\, d\theta$$

Let $\tan\theta = t$

$$= a^2 \int_{\infty}^{-1} \left\{\frac{1}{(1+t)^2} - \frac{9t^2}{(1+t^3)^2}\right\} dt = a^2\left[\frac{-1}{1+t} + \frac{3}{1+t^3}\right]_{\infty}^{-1}$$

$$= a^2 \lim_{t\to -1}\left[\frac{3}{1+t^3} - \frac{1}{1+t}\right] = \left\{\frac{3-(1-t+t^2)}{1+t^3}\right\} a^2$$

$$= \lim_{t\to -1} a^2 \frac{2+t-t^2}{1+t^3} = \lim_{t\to -1} a^2 \frac{1-2t}{3t^2} = a^2.$$

Example 2.43 **Show that the area included between the cardioids $r = a(1 + \cos\theta)$ and $r = a(1 - \cos\theta)$ is $\frac{3\pi - 8}{2} a^2$.**

(Mangalore' 99)

Solution : The curves intersect on the line $\theta = \frac{\pi}{2}$

$$\text{Required area} = 2\,\frac{1}{2}\int_{\theta=0}^{\pi/2} r_1^2\, d\theta$$

$$r_1 = a(1 - \cos\theta) + 2\,\frac{1}{2}\int_{\pi/2}^{\pi} r_2^2\, d\theta$$

$$r_2 = a(1 + \cos\theta)$$

$\theta = \pi/2$

Fig. 2.23

$$= a^2 \int_0^{\pi/2} (1 - \cos\theta)^2\, d\theta + a^2 \int_{\pi/2}^{\pi} (1 + \cos\theta)^2\, d\theta$$

Let $\theta = \pi/2 - \phi$

$$= a^2 \int (1 - \cos\theta)^2\, d\theta + a^2 \int_0^{\pi/2} (1 - \cos\phi)^2\, d\theta$$

$$= 2a^2 \int_0^{\pi/2} (1 - 2\cos\phi + \cos^2\phi)\, d\phi$$

$$= 2a^2 \int_0^{\pi/2} \left(\frac{3}{2} - 2\cos\phi + \frac{\cos 2\phi}{2}\right) d\phi$$

$$= 2a^2 \left[\frac{3}{2}\phi - 2\sin\phi + \frac{\sin 2\phi}{4}\right]_0^{\pi/2}$$

$$= a^2 \left[\frac{3\pi}{2} - 4\right]$$

Example 2.44 **Find the area of a loop of $r = a \sin 3\theta$. $r = 0$ when $\sin 3\theta = 0$, $3\theta = 0$ or $3\theta = \pi$.**

Solution : One loop is from $\theta = 0$ to $\theta = \pi/3$.

$$\text{Area of loop} = \frac{1}{2}\int_0^{\pi/3} a^2 \sin^2 3\theta\, d\theta$$

$$= \frac{a^2}{4}\int_0^{\pi/3} (1 - \cos 6\theta)\, d\theta$$

$$= \frac{a^2}{4}\left[\theta - \frac{\sin 6\theta}{6}\right]_0^{\pi/3} = \frac{a^2\pi}{12}.$$

PROBLEM SET 2.6

Find the area between the following curves and radii vectors :

1. The spiral $r\,\theta^{1/2} = a,\ \theta = \alpha,\ \theta = \beta$

   Ans. $\frac{1}{2}a^2 \log \frac{\beta}{\alpha}$.

2. The parabola $\frac{l}{r} = 1 + \cos\theta,\ \theta = 0$ to $\theta = \alpha$

   Ans. $\frac{l^2}{4}\left[\tan\frac{\alpha}{2} + \frac{1}{3}\tan^3\frac{\alpha}{2}\right]$

3. The equiangular spiral $r = a\,e^{m\theta};\ \theta = \alpha,\ \theta = \beta$.

   Ans. $\frac{a^2}{4m}(e^{2m\beta} - e^{2m\alpha})$.

4. Find the area of a loop of $r = a\cos 4\theta$

   Ans. $\frac{\pi a^2}{16}$

5. Find the area out side the circle $r = 2a\cos\theta$ and inside the cardioid $r = a(1 + \cos\theta)$.

   Ans. $\frac{\pi a^2}{2}$.

6. Prove that the area of the loop of the curve $x^3 + y^3 = 3axy$ is three times the area of a loop of $r^2 = a^2\cos 2\theta$.

7. Find the area of one loop of the curves :
   (i) $r = a\sin 2\theta$, (ii) $r = a\cos 3\theta$.

   Ans. (i) $\frac{\pi a^2}{8}$, (ii) $\frac{\pi a^2}{12}$

2.11.7 Length of Arc

$$\text{Arc length } AB = \int_a^b \sqrt{1 + \left(\frac{dy}{dx}\right)^2}\, dx$$

for $y = f(x)$

$$= \int_{y_A}^{y_B} \sqrt{1 + \left(\frac{dx}{dy}\right)^2}\, dy$$

for the curve $x = \phi(y)$

$$= \int_{\theta_1}^{\theta_2} \sqrt{r^2 + \left(\frac{dr}{d\theta}\right)^2}\, d\theta$$

for polar curve $r = f(\theta)$.

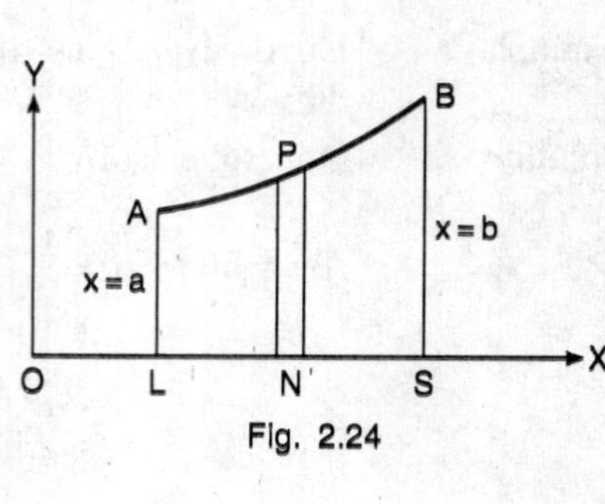

Fig. 2.24

Example 2.45 **Find the entire length of the Cardioid $r = a\ (1 + \cos\theta)$.**

(Mysore 99, AMIE 97W, Mangalore 99)

Solution : The curve is symmetrical about the line $\theta = 0$ and is periodic. Sc whole length is included between $\theta = 0$ to $\theta = 2\pi$ or twice of that between $\theta = 0$ to $\theta = \pi$

$$r = a\ (1 + \cos\theta),\ \frac{dr}{d\theta} = -\ 2a \sin\theta,$$

$$r^2 + \left(\frac{dr}{d\theta}\right)^2 = 4a^2 \cos^2\frac{\theta}{2}\left(\cos^2\frac{\theta}{2} + \sin^2\frac{\theta}{2}\right)$$

$$s = 2\int_0^{\pi} 2a \cos\frac{\theta}{2}\, d\theta = 4a\left[2\sin\frac{\theta}{2}\right]_0^{\pi} = 8a.$$

Fig. 2.25

Also show that the upper half bisected by $\theta = \frac{\pi}{3}$.

$$\text{Length of upper half} = \frac{8a}{2} = 4a$$

$$\text{Length (from } \theta = 0 \text{ to } \theta = \pi/3) = \int_0^{\pi/3} 2a \cos\frac{\theta}{2} = 4a\left[\sin\frac{\theta}{2}\right]_0^{\pi/3} = 2a.$$

Example 2.46 **Find the length an arc of the cycloid** ($\theta = 0$ **to** $\theta = 2\pi$)

$$x = a\ (\theta - \sin\theta),\ y = a\ (1 - \cos\theta).$$

(Kuvempu' 98, Hamirpur 96S)

Solution : Required length

$$s = \int_0^{2\pi} \sqrt{\dot{x}^2 + \dot{y}^2}\, d\theta$$

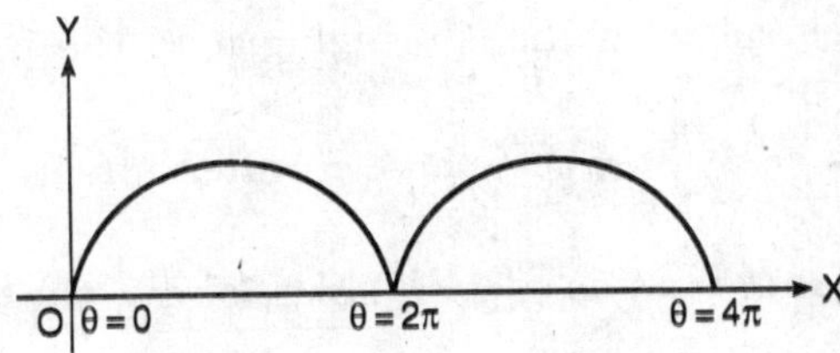

Fig. 2.26

$$\dot{x}^2 = a\ (1 - \cos\theta),$$

$$\dot{y} = a \sin\theta = 2a \sin\theta/2 \cos\theta/2$$

$$= 2a \sin^2\theta/2$$

$$\sqrt{\dot{x}^2 + \dot{y}^2} = 2a \sin\theta/2$$

$$s = \left|\int_0^{2\pi} 2a \sin\theta/2\, d\theta\right| = \left|-\ 4a\ [\cos\theta/2]_0^{2\pi}\right| = 8a.$$

Example 2.47 **Find the perimeter of the loop of the curve $3ay^2 = x\ (x - a)^2$.**

(Ranchi' 98)

Solution :

$$y = \pm\frac{x - a}{3a}\sqrt{x}.$$

Curve is symmetrical about x-axis. Curve has real y only for $x > 0$. When $x \to \infty$, $y \to \infty$. For loop $y = 0$

So $x(x-a)^2 = 0$, $x = 0$ or $x = a$.

Length of arc of loop $= 2\int_0^a \sqrt{1+\left(\frac{dy}{dx}\right)^2}\, dx$

Fig. 2.27

Let $x = a\cos^2\theta$.

$$y^2 = \frac{a\cos^2\theta}{3a}\cdot a^2\sin^4\theta.$$

So $$y = \frac{a}{\sqrt{3}}\sin^2\theta\cos\theta.$$

$$\dot{x} = -2a\sin\theta\cos\theta,\quad \dot{y} = \frac{a}{\sqrt{3}}(2\sin\theta\cos^2\theta - \sin^3\theta)$$

$$L = 2\int_0^{\pi/2} a\sqrt{4\sin^2\theta\cos^2\theta + \frac{1}{3}\sin^2\theta\,(2\cos^2\theta - \sin^2\theta)^2}\; d\theta.$$

$$= \frac{2a}{\sqrt{3}}\int_0^{\pi/2} \sin\theta\sqrt{12\cos^2\theta + (3\cos^2\theta - 1)}\; d\theta$$

$$= \frac{2a}{\sqrt{3}}\int_0^{\pi/2} (3\cos^2\theta + 1)\sin\theta\, d\theta$$

$$= \frac{2a}{\sqrt{3}}\Big[-(\cos^3\theta + \cos\theta)\Big]_0^{\pi/2} = \frac{4a}{\sqrt{3}}.$$

It can also be solved without bringing it to the parameter form.

It reduces to $\frac{1}{\sqrt{3}\,a}\int_0^a \frac{3x+a}{\sqrt{x}}\,dx = \frac{4a}{\sqrt{3}}$, after integration.

Example 2.48 **In the ellipse $x = a\cos\phi$, $y = b\sin\phi$, show that**

$$ds = a\sqrt{1 - e^2\cos^2\phi}\; d\phi$$

and hence show that the whole length of the ellipse is

$$2\pi a\left[1 - \left(\frac{1}{2}\right)^2\frac{e^2}{1} - \left(\frac{1\cdot 3}{2\cdot 4}\right)^2\frac{e^4}{3} - \left(\frac{1\cdot 3\cdot 5}{2\cdot 4\cdot 6}\right)^2\frac{e^6}{5} - \ldots\right]$$

Solution : For an ellipse we know

$$b^2 = a^2(1 - e^2)$$

$$\frac{ds}{d\phi} = \sqrt{\dot{x}^2 + \dot{y}^2}$$

$$= \sqrt{a^2\sin^2\phi + b^2\cos^2\phi}$$

$$= a\sqrt{\sin^2\phi + (1 - e^2)\cos^2\phi}$$

$$= a\sqrt{1 - e^2\cos^2\phi}$$

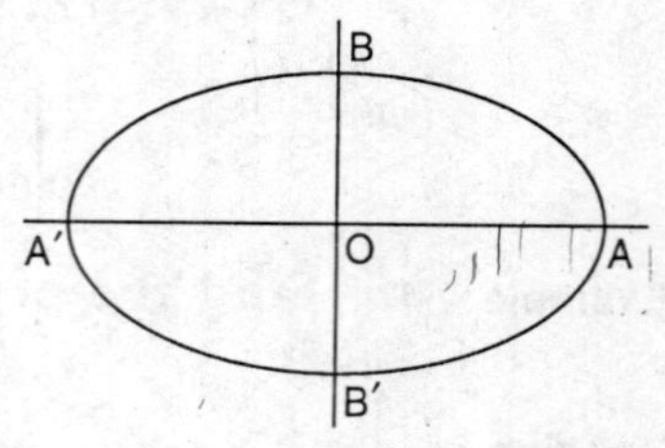

Fig. 2.28

Whole length by symmetry in all 4 quadrants

$$= 4\int_0^{\pi/2} \frac{ds}{d\phi}\, d\phi$$

$$= 4a\int_0^{\pi/2} (1 - e^2 \cos^2 \phi)^{1/2}\, d\phi$$

$$= 4a\int_0^{\pi/2} \left\{1 - \frac{e^2}{2}\cos^2\phi + \frac{\frac{1}{2}\left(\frac{1}{2}-1\right)}{1\cdot 2} e^4 \cos^4\phi - \frac{\frac{1}{2}\left(\frac{1}{2}-1\right)\left(\frac{1}{2}-2\right)}{6} e^6 \cos^6\phi\right\} d\phi$$

$$= 4a\left[\frac{\pi}{2} - \frac{e^2}{2}\frac{\Gamma 3/2\,\Gamma 1/2}{2\,\Gamma\tau 2} - \frac{e^4}{8}\frac{\Gamma 5/2\,\Gamma 1/2}{2\,\Gamma 3} - \frac{\frac{1}{2}\left(-\frac{1}{2}\right)\left(-\frac{3}{2}\right)}{1\cdot 2\cdot 3} e^6 \frac{\Gamma 7/2\,\Gamma 1/2}{2\,\Gamma 4} + \ldots\right]$$

$$= 2\pi a\left[1 - \frac{1}{4}e^2 - \frac{1\cdot 3}{2\cdot 4}\frac{e^4}{8} - \frac{1\cdot 3}{2\cdot 4\cdot 6} \ldots e^6 \left(\frac{1\cdot 3\cdot 5}{2\cdot 4\cdot 6}\right) + \ldots\right]$$

$$= 2\pi a\left[1 - \left(\frac{1}{2}\right)^2 e^2 - \left(\frac{1\cdot 3}{2\cdot 4}\right)^2 \frac{e^4}{3} - \left(\frac{1\cdot 3\cdot 5}{2\cdot 4\cdot 6}\right)^2 \frac{e^6}{5} + \ldots\right]$$

Example 2.49 **Show that the lengths of an arc of the curve $x \sin\theta + y\cos\theta = f'(\theta)$, $x\cos\theta - y\sin\theta = f''(\theta)$, is given by $s = f(\theta) + f''(\theta) + C$.**

Solution : Solving the equations

$$\frac{x}{f''\cos\theta + f'\sin\theta} = \frac{y}{f'\cos\theta - f''\sin\theta} = \frac{1}{\sin^2\theta + \cos^2\theta}$$

So $\quad x = f''\cos\theta + f'\sin\theta$

So $\quad \dot{x} = f'''\cos\theta - f''\sin\theta + f''\sin\theta + f'\cos\theta$

$$= (f''' + f')\cos\theta$$

$$y = -f''\sin\theta + f'\cos\theta$$

$$\dot{y} = -f'''\sin\theta - f''\cos\theta + f''\cos\theta - f'\sin\theta$$

$$= -\sin\theta\,(f''' + f')$$

Hence $\quad \sqrt{\dot{x}^2 + \dot{y}^2} = (f' + f''')\sqrt{\sin^2\theta + \cos^2\theta}$

So $$s = \int \sqrt{\dot{x}^2 + \dot{y}^2}\, d\theta = \int (f' + f''')\, d\theta$$

$$= f(\theta)\, t + f''(\theta) + C.$$

Example 2.50 **Find the length of an arc of the parabola $y^2 = 4ax$ from vertex to one end of latus reaction.**

Solution : $$2y\frac{dy}{dx} = 4a$$

So $$s = \int_0^a \sqrt{1 + \left(\frac{2a}{y}\right)^2}\, dx$$

$$= \int_0^a \sqrt{1 + \frac{a}{x}}\, dx$$

Let $x = a \cot^2 \theta$

$$dx = -\, 2a \cot \theta \operatorname{cosec}^2 \theta \, d\theta$$

$$= \int_{\pi/2}^{\pi/4} \sec \theta \, (-\, 2a) \frac{\cos \theta}{\sin \theta} \operatorname{cosec}^2 \theta \, d\theta$$

$$= 2a \int_{\pi/4}^{\pi/2} \operatorname{cosec}^3 \theta \, d\theta$$

$$= \int_{\pi/4}^{\pi/2} \underset{\text{II}}{\operatorname{cosec} \theta} \, \underset{\text{II}}{\operatorname{cosec}^2 \theta}$$

$$= 2a \left[-\frac{\operatorname{cosec} \theta \cot \theta}{2} - \frac{1}{2} \log (\operatorname{cosec} \theta + \cot \theta) \right]_{\pi/4}^{\pi/2}$$

$$= a \, [0 + \sqrt{2} + \log (\sqrt{2} + 1)]$$
$$= a \, [\sqrt{2} + \log (\sqrt{2} + 1)]$$

L

x = a

(0, 0) A

(a, 2a)

Fig. 2.29

PROBLEM SET 2.7

1. If s be the length of the arc of the catenary $y = c \cosh \frac{x}{c}$ from the vertex $(0, c)$ to the point (x, y), show that $s^2 = y^2 - c^2$.

2. Find the length of the arc of the semi cubical parabola $ay^2 = x^3$ from the vertex to the ordinate $x = 5a$.

Ans. $12 \frac{11}{27} a$.

3. Find the perimeter of the loop of the curves :

(i) $9y^2 = (x - 2)(x - 5)^2$ ***(Osmania 95)***

(ii) $ay^2 = x^2 (a - x)$ ***(S. Gujarat)***

Ans. (i) $4\sqrt{3}$ (ii) $4a\sqrt{3}$.

4. Find the length of the curve $y = \log \frac{e^x - 1}{e^x + 1}$ from $x = 1$ to $x = 2$. ***(Bhopal 91)***

Ans. $\log_e (e + e^{-1})$.

5. Find the length of the arc of the cycloid $x = a(\theta + \sin \theta)$, $y = a(1 - \cos \theta)$.

Ans. $8a$ ***(Kuvempu' 96)***

6. Find the length of the loop of the curve $x = t^2$, $y = t - \frac{t^3}{3}$. ***(Marathwada' 98S)***

Ans. $4\sqrt{3}$

7. Show that the whole length of the curve $x^2(a^2 - x^2) = 8a^2y^2$ is $\pi a\sqrt{2}$. *(Kurukshetra' 98)*

8. Find the length of the arc of the parabola $y^2 = 4ax$ from
(i) the vertex to one end of the latus rectum,
(ii) cut off by the line $3y = 8x$ *(Nagpur' 98S)*

Ans. (i) $a[\sqrt{2} + \log(1 + \sqrt{2})]$; (ii) $\left(\frac{15}{16} + \log 2\right) a$.

9. Show that the length of arc of the cardioid $r = a(1 - \cos\theta)$ from $\theta = 0$ to θ is given by $s = 8a\sin^2\frac{\theta}{4}$. Hence find the whole length.

Ans. $s = 8a$.

2.11.8 Volume of Revolution

If the curve $y = f(x)$ is revolved about the x axis, the volume generated is

$$V = \int_{x=a}^{b} \pi y^2\, dx.$$

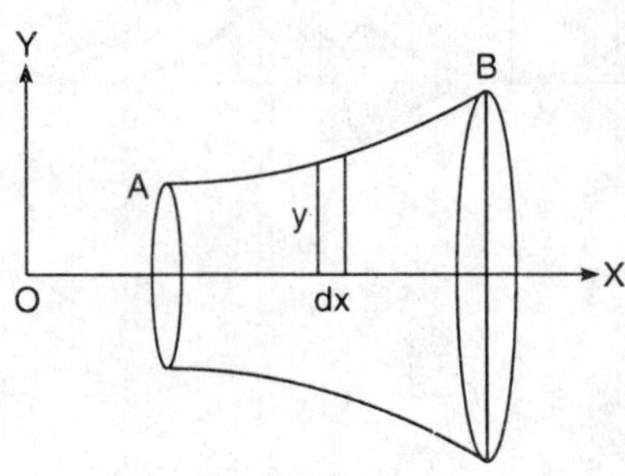

Fig. 2.30

Example 2.51 **Find the volume of the ellipsoid formed by revolving the ellipse $\frac{x^2}{a^2} + \frac{y^2}{b^2} = 1$**

(i) about x axis, (ii) about the y axis

Solution : (i) On revolving about x-axis

$$V_x = \int_{-a}^{a} \pi y^2\, dx$$

$$= 2\pi \int_0^a b^2\left(1 - \frac{x^2}{a^2}\right) dx$$

$$= 2\pi b^2\left[x - \frac{x^3}{3a^2}\right]_0^a$$

$$= 2\pi b^2 \frac{2a}{3} = \frac{4\pi ab^3}{3}$$

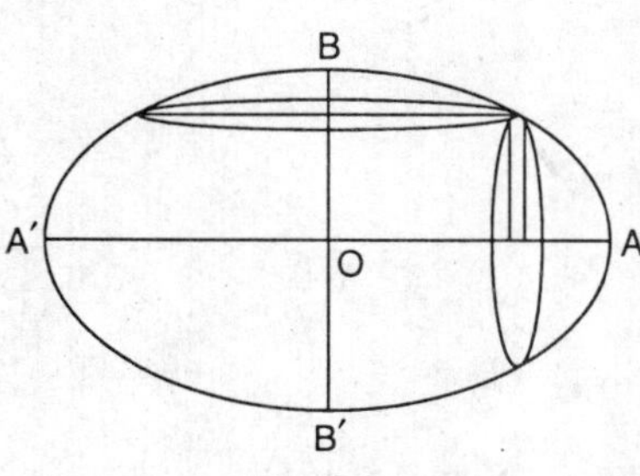

Fig. 2.31

(ii) On revolving about y axis.

$$V_y = \int_{-b}^{b} \pi x^2 \, dy = 2\pi \int_0^b a^2 \left(1 - \frac{y^2}{b^2}\right) dy$$

$$= 2\pi a^2 \left[y - \frac{y^3}{3b^2} \right]_0^b = \frac{4\pi a^2 b}{3}.$$

In the case of a circle $a = b$

$$V_x = V_y = \frac{4}{3} \pi a^2 = \frac{4}{3} \pi a^3$$

= Volume of a sphere.

Example 2.52 **Prove that the volume of the real formed by revolving the cycloid $x = a\,(\theta + \sin \theta, y = a\,(1 - \cos \theta)$, about the tangent at the vertex is $\pi^2 a^3$.**

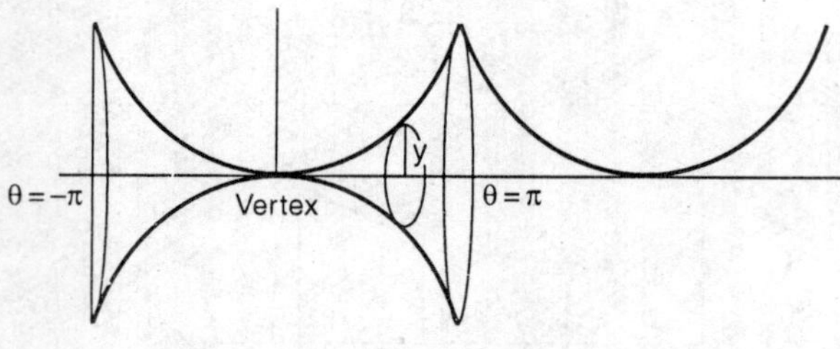

Fig. 2.32

Solution : The required volume

$$V = 2 \int_0^{a\pi} \pi y^2 \, dx = 2 \int_0^{\pi} \pi y^2 \frac{dx}{d\theta} d\theta$$

$$= 2\pi \int_<^{\pi} a^2 (1 - \cos \theta)^2 \, a\,(1 + \cos \theta) \, d\theta$$

$$= 2\pi a^3 \int_0^{\pi} (2 \sin^2 \theta/2)^2 \, 2 \cos^2 \theta/2 \, d\theta \text{ Let } \theta/2 = t$$

$$= 16\pi a^3 \cdot 2 = \int_0^{\pi/2} \sin^4 t \cos^2 t \, dt$$

$$= 32\, \pi^3 \frac{\Gamma\frac{5}{2} \Gamma\frac{3}{2}}{2\,\Gamma 4}$$

$$= 32\pi a^3 \frac{\frac{3}{2}\frac{1}{2}\sqrt{\pi}\,\frac{1}{2}\sqrt{\pi}}{2 \times 3 \times 2 \times 1} = \pi^2 a^3.$$

Example 2.53 **The curve $y^2 (a + x) = x^2 (3a - x)$ revolves about the x-axis. Find the volume of the solid formed by the loop.**

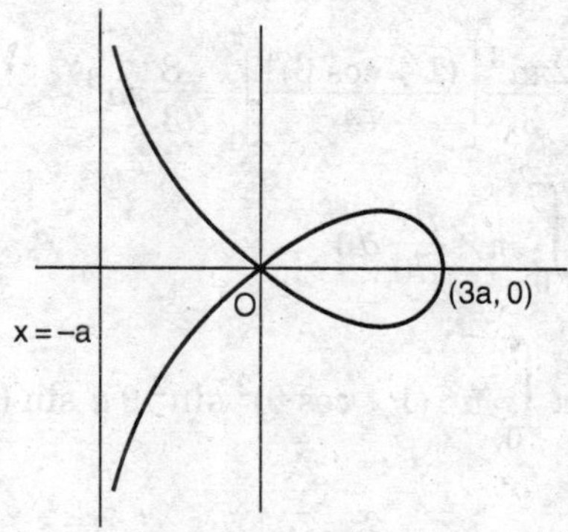

Fig. 2.33

Solution : Required Volume $V = \pi \int_0^{3a} y^2\, dx$

$$= \pi \int_0^{3a} \frac{x^2 (3a - x)}{a + x}\, dx$$

$$= \pi \int_0^{3a} \left(- x^2 + 4ax - 4a^2 + 4a^3 \frac{1}{x + a} \right) dx$$

$$= \pi \left[- \frac{x^3}{3} + \frac{4ax^2}{2} - 4a^2 x + 4a^3 \log (x + a) \right]_0^{3a}$$

$$= \pi \left[- 9a^3 + 18a^3 - 12a^3 + 4a^3 \log \frac{4}{a} a \right]$$

$$= \pi a^3 (8 \log 2 - 3)$$

Example 2.54 **Find the volume of the solid formed by revolving the cardioid $r = a (1 + \cos \theta)$, about the initial line.**

Solution : Required Volume $= \pi \int y^2\, dx$

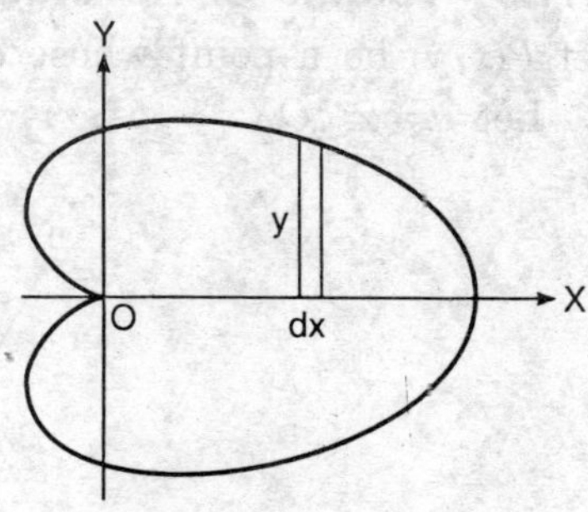

Fig. 2.34

$$y = r \sin \theta$$

$$= a (1 + \cos \theta) \sin \theta$$

$$r = a (1 + \cos \theta)$$

$$\frac{dr}{d\theta} = - a \sin \theta$$

$$x = r \cos \theta$$

$$dx = dr \cos \theta - r \sin \theta\, d\theta$$

$$= \{- a \sin \theta \cos \theta - a (1 + \cos \theta) \sin \theta\}\, d\theta$$

$$= - a \sin \theta (1 + 2 \cos \theta)\, d\theta$$

$$= \frac{2\pi a^3}{3} \int_0^{\pi} (1 + \cos\theta)^3 \sin\theta \, d\theta$$

$$= \frac{2\pi a^3}{3} \left[\frac{(1 + \cos\theta)^4}{-4} \right]_0^{\pi} = \frac{8}{3} \pi a^3.$$

$$V = \int_0^{\pi} \pi y^2 \frac{dx}{d\theta} d\theta$$

$$= \pi \int_0^{\pi} a^2 (1 + \cos\theta)^2 \sin^2\theta \, a \sin\theta \, (1 + 2\cos\theta) \, d\theta$$

Let $\cos\theta = t$

$$= \left| \pi a^3 \int_1^{-1} (1 + t)^2 (1 + t)(1 - t)(1 + 2t) \, dt \right|$$

$$= \pi a^3 \int_{-1}^{1} (1 + 3t + 3t^2 + t^3)(1 + t - 2t^2) \, dt$$

$$= \pi a^3 \int_{-1}^{1} (1 + 4t + 4t^2 - 2t^3 - 5t^4 - t^5) \, dt$$

the integral of odd powers is zero.

$$= \pi a^3 \, 2 \int_0^{1} (1 + 4t^2 - 5t^4) \, dt$$

$$2\pi a^3 \left[t + \frac{4t^3}{3} - \frac{5t^5}{5} \right]_0^1 = \frac{8}{3} \pi a^3.$$

It can more easily be solved using $V = 2/3 \, \pi \int_0^{\pi} r^3 \sin\theta \, d\theta$.

2.11.9 Volume of Revolution About Any Axis

Let $P(x, y)$ be a point whose coordinates relation to axes $O\xi$ and $O\eta$ through O are (ξ, η).

Let $\alpha = \angle \xi OX$

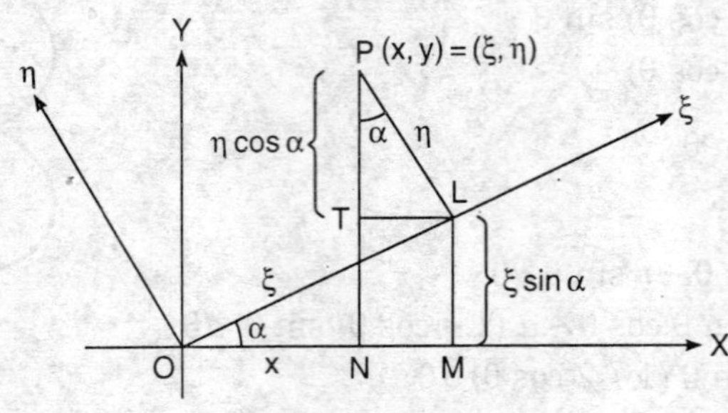

Fig. 2.35

Then $\quad x = ON = OM - MN$

$= \xi \cos\alpha - \eta \sin\alpha$

$y = NP = NT + TP$

$= \xi \sin\alpha + \eta \cos\alpha$

also $\quad \xi = x\cos\alpha + y\sin\alpha$

$\eta = -x\sin\alpha + y\cos\alpha$

Volume of Revolution (Polar Curves)

Volume of revolution about initial line ($\theta = 0$)

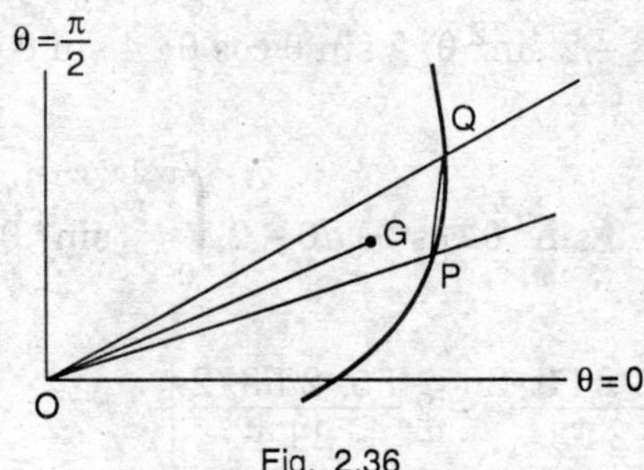

Fig. 2.36

$$= \int_{\alpha}^{\beta} \text{(Area of element } OPQ) \times \text{path traced by the centroid in one revolution}$$

$$= \int_{\alpha}^{\beta} \left(\frac{1}{2} r^2 \, d\theta\right) \left\{\left(\frac{0 + r + r}{3}\right) \sin\theta \; 2\pi\right\} = \frac{2}{3} \int_{\alpha}^{\beta} r^3 \sin\theta \, d\theta$$

Volume of revolution about the line ($\theta = \pi/2$)

$$= \int_{\alpha}^{\beta} \frac{1}{2} r^2 \, d\theta \left(\frac{2r}{3} \cos\theta \cdot 2\pi\right) = \frac{2}{3} \int_{\alpha}^{\beta} r^3 \cos\theta \, d\theta.$$

If curve the revolved about axis $O\xi$ then

$$V = \int \pi n^2 \, d\xi$$

$$= \pi \int (-x\sin\alpha + y\cos\alpha)^2 \, d\,(x\cos\alpha + y\sin\alpha)$$

$$V = \pi \int (-x\sin\alpha + y\cos\alpha)^2 \left\{\cos\alpha + \sin\alpha \frac{dy}{dx}\right\} dx.$$

Example 2.55 **Find the volume of the solid obtained by revolving the ellepsoid $y^2(2a - x) = x^3$ about its asymptote.**

Solution : Asymptote to the curve is $x = 2a$

So $\quad V = 2\pi \int_{x=0}^{2a} PN^2 \frac{dy}{dx} \, dx$

$$= 2\pi \int_0^{2a} (2a - x)^2 \frac{[3x^2(2a - x) + x^3]}{(2a - x)^2 \, 2y} \, dx$$

$$= 2\pi \int_0^{2a} \frac{2x^2 (3a - x)}{2} \sqrt{\frac{2a - x}{x^3}} \, dx$$

$$= 2\pi \int_0^{2a} (3a - x) \sqrt{x (2a - x)} \, dx$$

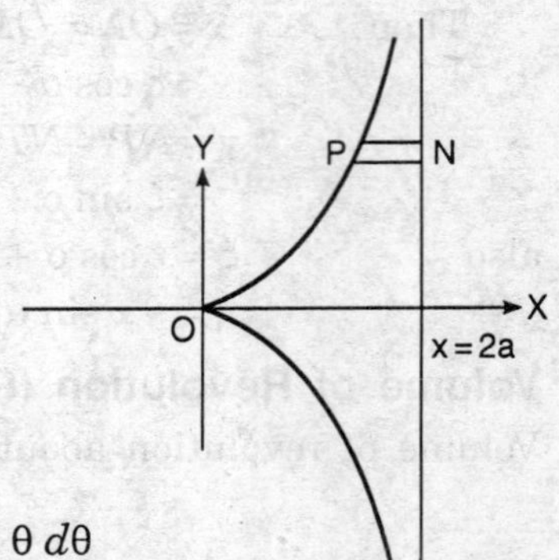

Fig. 2.37

Let $x = 2a \sin^2 \theta$

$$= 2\pi a^3 \int_0^{\pi/2} (3 - 2 \sin^2 \theta) \, 2 \sin \theta \cos \theta \cdot 4 \sin \theta \cos \theta \, d\theta$$

$$= 16\pi a^3 \left[\int_0^{\pi/2} 3 \sin^2 \theta \cos^2 \theta \, d\theta - 2 \int_0^{\pi/2} \sin^4 \theta \cos^2 \theta \, d\theta \right]$$

$$= 16\pi a^3 \left[3 \frac{\Gamma 3/2 \, \Gamma 3/2}{2 \, \Gamma 3} - 2 \frac{\Gamma 5/2 \, \Gamma 3/2}{2 \, \Gamma 4} \right]$$

$$= 16\pi a^3 \left[\frac{3\pi}{16} - \frac{\pi}{16} \right] = 2\pi^2 a^3$$

Example 2.56 **Find the volume of a spherical segment of height h cut off from a sphere of radius a.** *(Tirupati 98S; AMIE 97W)*

Solution : Let the circle be $x^2 + y^2 = a^2$ from which a cap of height h is cut.

Volume of the cap

$$= \pi \int_{a-h}^{a} y^2 \, dx = \pi \int_{a-h}^{a} (a^2 - x^2) \, dx$$

$$V = \pi \, a^2 \, [x]_{a-h}^{a} - \pi \left[\frac{x^3}{3} \right]_{a-h}^{a}$$

Fig. 2.38

$$= \pi a^2 (a - a + h) - \frac{1}{3} \pi \{a^3 - (a^3 - 3a^2 h + 3ah^2 - h^3\}$$

$$= \frac{\pi}{3} [3a^2 h - 3a^2 h + 3ah^2 - h^3] = \pi h^2 \left(a - \frac{h}{3} \right)$$

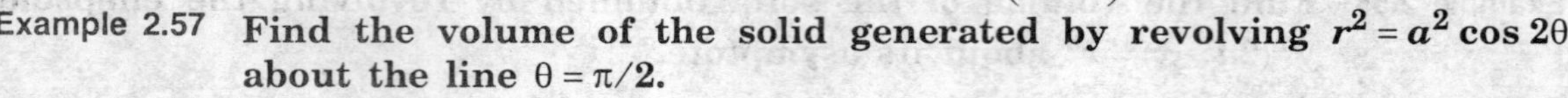

Example 2.57 **Find the volume of the solid generated by revolving $r^2 = a^2 \cos 2\theta$ about the line $\theta = \pi/2$.**

Solution :

$$V = 2 \int_0^{\pi/4} \frac{2}{3} \pi r^3 \cos \theta \, d\theta$$

$$= 4\pi \frac{a^3}{3} \int_0^{\pi/4} \cos^{3/2} 2\theta \cos \theta \, d\theta$$

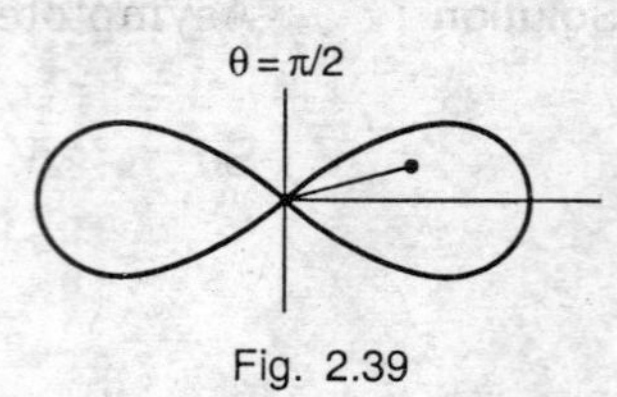

Fig. 2.39

$$= \frac{4\pi a^3}{3}\int_0^{\pi/4} (1-2\sin^2\theta)^{3/2}\cos\theta\, d\theta$$

Let $\sqrt{2}\sin\theta = \sin t$

So $\sqrt{2}\cos\theta\, d\,\theta = \cos dt$

$$= \frac{4\pi a^3}{3}\frac{1}{\sqrt{2}}\int_0^{\pi/2} \cos^3 t\cos t\, dt$$

$$= \frac{4\pi a^3}{3\sqrt{2}}\frac{\Gamma 5/2\,\Gamma 1/2}{2\,\Gamma 3} = \frac{\pi a^3}{4\sqrt{2}}$$

Example 2.58 **Find the volume of the spindle shaped solid generated by revolving the astroid $x^{2/3}+y^{2/3}=a^{2/3}$ about the x-axis.** *(JNTU' 90)*

Solution : Required Volume $= 2\int_{x=0}^{a} \pi y^2\, dx$

Changing to parametric form

Let $x = a\cos^3\theta,\ y = a\sin^3\theta$

$$V = 2\pi a^3\int_0^{\pi/2} \sin^6\theta\,(-3\cos^2\theta\sin\theta)\, d\theta$$

$$= 6\pi a^3\frac{\Gamma\frac{8}{2}\Gamma\frac{3}{2}}{2\Gamma\frac{11}{2}}$$

$$= 6\pi a^3\frac{3\cdot 2}{2\times\frac{9}{2}\times\frac{7}{2}\times\frac{5}{2}\times\frac{3}{2}}$$

$$= \frac{32}{105}\pi a^3.$$

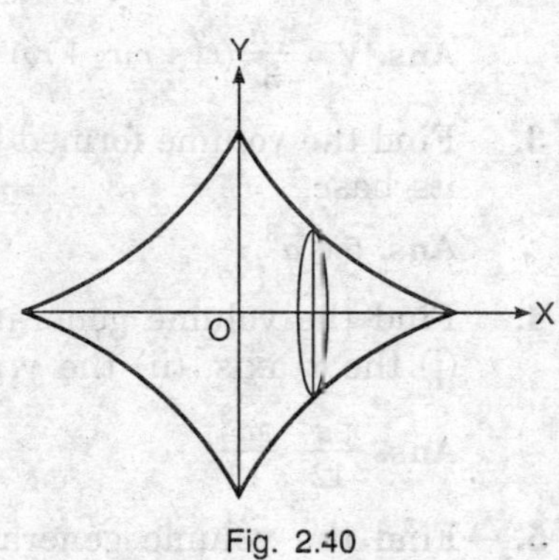

Fig. 2.40

Example 2.59 **Find the volume obtained by revolving $x = a(\theta+\sin\theta)$, $y = a(1+\cos\theta)$ about the x-axis.** *(Mangalore 99)*

Solution : $V = \int_{-\pi}^{\pi} \pi y^2\frac{dx}{d\theta}\, d\theta$

$$= 2\pi a^3\int_0^{\pi} (1+\cos\theta)^2(1+\cos\theta)\, d\theta$$

$$= 2\pi a^3\int_0^{\pi} (2\cos^2\theta/2)^3\, d\theta$$

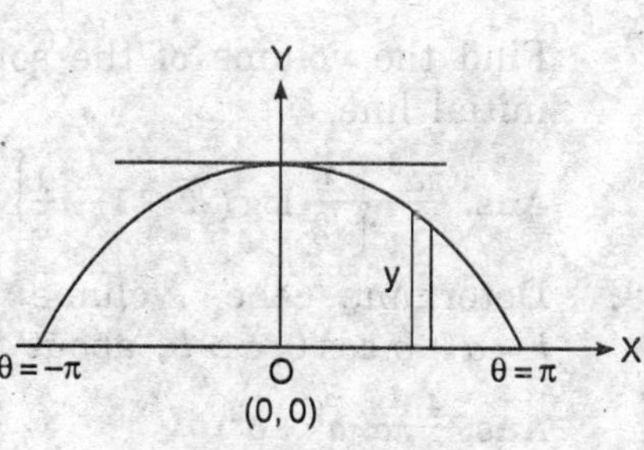

Fig. 2.41

Let $\frac{\theta}{2} = t$

$$16\pi a^3\, 2\int_0^{\pi/2} \cos^6 t\, dt = 32\,\pi a^3 \frac{\Gamma 7/2\, \Gamma\frac{1}{2}}{2\,\Gamma 4}$$

$$= 16\pi a^3 \frac{5}{2}\times\frac{3}{2}\times\frac{1}{2}\frac{\pi}{3\times 2\times 1}$$

$$= 5\pi^2 a^3$$

PROBLEM SET 2.8

1. Find the volume generated by revolving the portion of the parabola $y^2 = 4ax$, cut off by the ratus-rectum about its axis.

   Ans. $V = 2\pi a^3$

2. Find the volume of a frustum of a cone of height h and end radii r_1 and r_2.

   Ans. $V = \frac{\pi h}{3}(r_1^2 + r_1 r_2 + r_2^2)$

3. Find the volume formed by revolving the cycloid $x = a(\theta - \sin\theta)$, $y = a(1 - \cos\theta)$, about its base.

   Ans. $5\pi^2 a^3$. *(Osmania 95)*

4. Find the volume generated by revolving the area bounded by $\sqrt{x} + \sqrt{y} = \sqrt{a}$, about (i) the x axis, (ii) the y axis.

   Ans. $\frac{\pi a^3}{12}, \frac{\pi a^3}{12}$.

5. Find the volume generated by the revolution of the area bounded by the catenary $y = a \cosh \frac{x}{c}$ and the ordinates $x = \pm c$, about the x-axis.

   Ans. $\pi c^3 \left(1 + \frac{e^2 - e^{-2}}{4}\right)$

6. Find the volume of the solid formed by the revolution of the area enclosed between the curve $xy^2 = a^2(a - x)$ and its asymptote about the y axis. *(Ranchi' 90)*

   Ans. $\frac{1}{2}\pi^2 a^3$

7. Find the volume of the solid formed by revolving a loop of $r^2 = a^2 \cos 2\theta$ about the initial line.

   Ans. $\frac{\pi a^3}{4}\left\{\frac{1}{\sqrt{2}}\log(\sqrt{2} + 1) - \frac{1}{3}\right\}$

8. Determine the volume of the solid formed by revolving the lemicon $r = a + b\cos\theta$, $a > b$, about the initial line. *(Gorakhpur' 99)*

   Ans. $\frac{4}{3}\pi a(a^2 + b^2)$

9. Find the volume generated by revolving $r = 2a \cos\theta$ about the initial line.

Ans. $\frac{4}{3}\pi a^3$

10. Prove that the volume of the solid formed by revolving the curve $y(a^2 + x^2) = a^3$ about its a symptote is $\frac{1}{2}\pi^2 a^3$.

2.11.10 Surface of Revolution

Surface S of the solid formed by revolving the curve $y = f(x)$ about the x-axis is

$$S = \int_{s_A}^{s_B} 2\pi y \, ds = \int_a^b 2\pi y \frac{ds}{dx} dx$$

$$= \int_{y_A}^{y_B} 2\pi y \frac{ds}{dy} dy$$

$$= 2\pi \int_a^b y \sqrt{1 + \left(\frac{dy}{dx}\right)^2} \, dx$$

For polar curve $S = \int_{\theta_1}^{\theta_2} 2\pi y \frac{ds}{d\theta} d\theta$

$$= \int_{\theta_1}^{\theta_2} 2\pi y \sqrt{r^2 + \left(\frac{dr}{d\theta}\right)^2} \, d\theta.$$

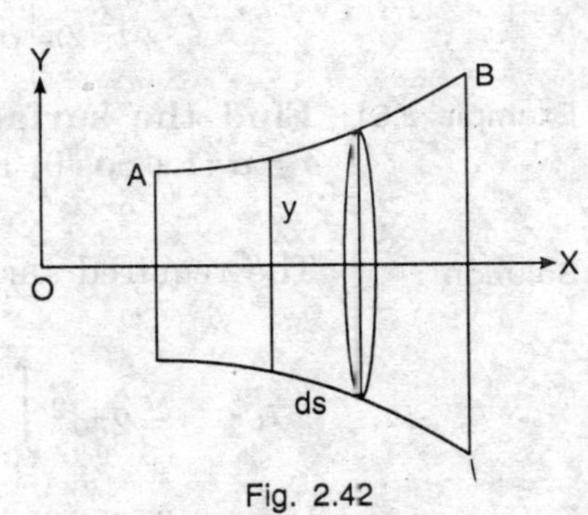

Fig. 2.42

For Revolution About y-axis

Surface Area generated $= \int_{s_A}^{s_B} 2\pi x \, ds$. If the equation to the curve be $x = f(y)$,

$$= \int_{y_A}^{y_B} 2\pi x \sqrt{1 + \left(\frac{dx}{dy}\right)^2} \, dy$$

Example 2.60 **Find the surface of the solid generated by revolving the ellipse $\frac{x^2}{a^2} + \frac{y^2}{b^2} = 1$, about x axis.**

$$S = 2\int_0^a 2\pi y \frac{ds}{dx} dx$$

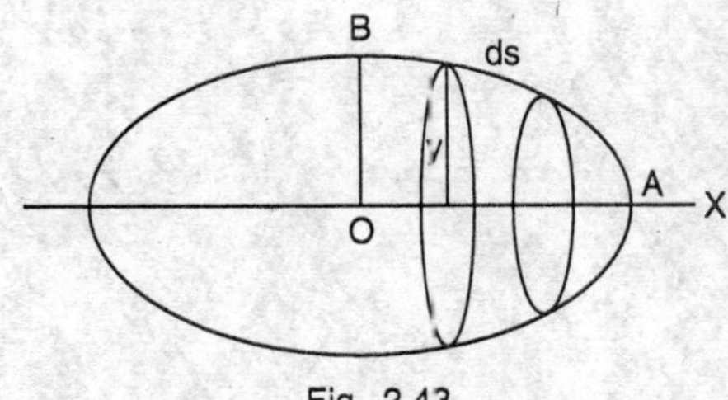

Fig. 2.43

Solution : Changing to parametric form.

$$x = a\cos\phi, \; y = b\sin\phi$$

$$\frac{ds}{d\theta} = \sqrt{\dot{x}^2 + \dot{y}^2} = \sqrt{a^2 \sin^2\phi + b^2 \cos^2\phi}$$

$$= \sqrt{a^2 \sin^2 \phi + a^2 (1-e^2) \cos^2 \phi} = a\sqrt{1-e^2\cos^2\phi}$$

$$S = 4\pi ab \int_0^{\pi/2} \sin\phi \sqrt{1-e^2\cos^2\phi}\, d\phi$$

Let $e \cos \phi = t$

$$- e \sin \phi \, d\phi = dt$$

$$S = 4\pi ab \int_e^0 -\frac{1}{e}\sqrt{1-t^2}\, dt$$

$$= \frac{4\pi ab}{e}\left[\frac{t\sqrt{1-t^2}}{2} + \frac{1}{2}\sin^{-1} t\right]_0^e$$

$$= \frac{2\pi ab}{e}\left[e\sqrt{1-e^2} + \sin^{-1} e\right]$$

$$= 2\pi\left[b^2 + \frac{ab}{e}\sin^{-1} e\right]$$

Example 2.61 **Find the surface of the solid generated by revolving the cardioid $r = a(1+\cos\theta)$ about the initial line.** *(Nagpur 98S Marathwara' 98)*

Solution : The required surface $= \displaystyle\int_0^{\pi} 2\pi y \frac{ds}{d\theta}\, d\theta = 2\pi \int_0^{\pi} r \sin\theta \sqrt{r^2 + \left(\frac{dr}{d\theta}\right)^2}\, d\theta$

$$= 2\pi a^2 \int_0^{\pi} (1+\cos\theta)\sin\theta \sqrt{(1+\cos\theta)^2 + \sin^2\theta}\, d\theta$$

Let $\cos\theta = t$

$$= 2\pi a^2 \int_{-1}^{1} (1+t)\sqrt{2(1+t)}\, dt$$

$$= 2\pi a^2 \frac{2}{5}\left[(1+t)^{5/2}\right]_{-1}^{1} = \frac{4\pi a^2\sqrt{2}}{5}\cdot 4\sqrt{2} = \frac{32\pi a^2}{5}$$

Example 2.62 **A circular arc subtanding an angle 2α at the centre is revolved about its axis. Find the surface area of the solid thus generated at P $x = a\cos\theta$, $y = a\sin\theta$, $s = a\theta$.**

Solution : Surface denerated $S = 2\pi \displaystyle\int_0^{\alpha} y \frac{ds}{d\theta}\, d\theta$

$$= 2\pi \int_0^{\alpha} a\sin\theta \, a\, d\theta$$

$$= 2\pi a^2 [-\cos\theta]_0^{\alpha}$$

$$= 2\pi a^2 (1 - \cos\alpha)$$

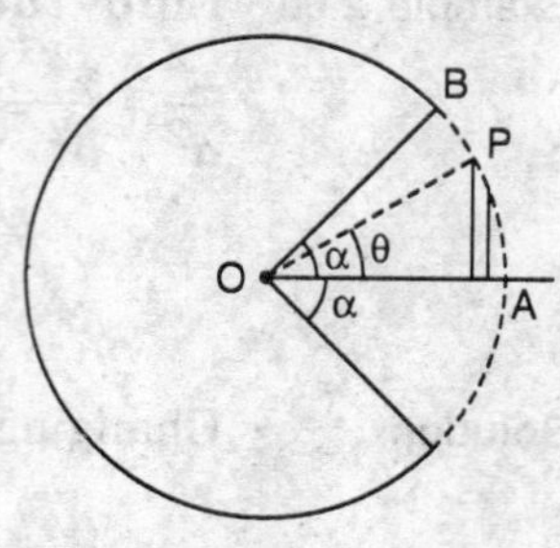

Fig. 2.44

Example 2.63 **A circular arc subtanding an angle 2α at the centre is revolved about the bounding chord. Find the surface area of the solid thus generated.**

Solution :

$$S = \int_{\theta=0}^{\alpha} 2\pi\, PN \frac{ds}{d\theta}\, d\theta$$

$$= 2\pi a^2 \int_0^{\alpha} (\cos\theta - \cos\alpha)\, d\theta$$

$$PN = a\cos\theta - a\cos\alpha$$

$$\frac{ds}{d\theta} = \alpha$$

$$= 2\pi a^2 [\sin\theta - \theta\cos\alpha]_0^{\alpha}$$

$$= 2\pi a^2 [\sin\alpha - \alpha\cos\alpha]$$

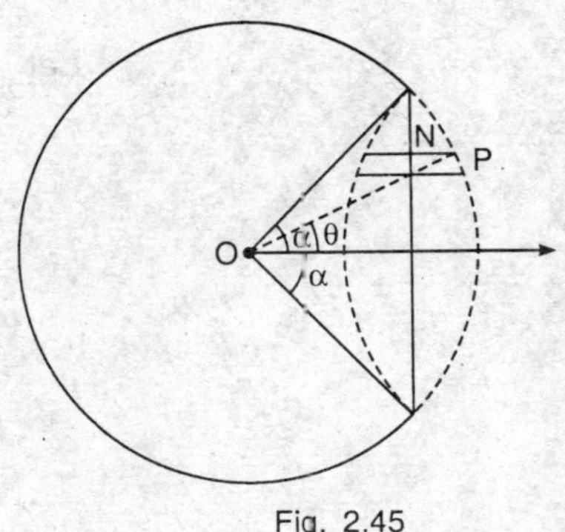

Fig. 2.45

Example 2.64 **Find the surface of the solid generated by revolving the astroid $x = a\cos^3 t, y = a\sin^3 t$, about the x-axis.**

Solution : Surface generated

$$S = 2\int_0^{\pi/2} 2\pi y \frac{ds}{dt}\, dt$$

$$= 4\pi a^2 \int_0^{\pi/2} \sin^3 t \cdot 3\sqrt{(\cos^2 t\sin t)^2 + (\sin^2 t\cos t)^2}\, dt$$

$$= 12\pi a^2 \int_0^{\pi/2} \sin^4 t\cos t\, dt$$

$$= 12\pi a^2 \left[\frac{\sin^5 t}{5}\right]_0^{\pi/2}$$

$$= \frac{12\pi a^2}{5}$$

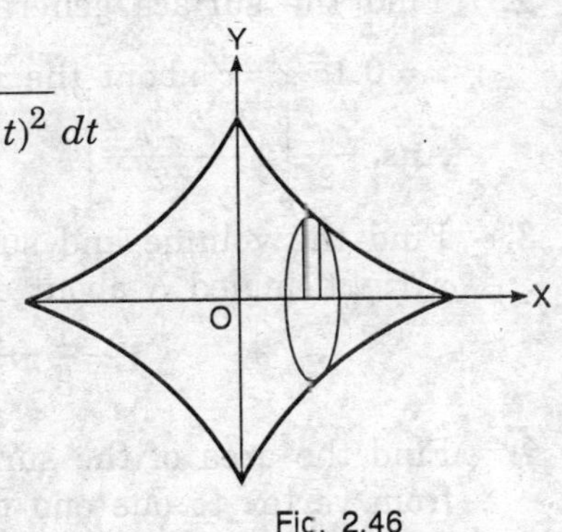

Fig. 2.46

Example 2.65 **Find the surface of the solid formed by revolving the cycloid $x = a(\theta - \sin\theta), y = a(1 - \cos\theta)$, about the x-axis.**

Solution :

$$\text{Surface generated} = \int_0^{2\pi} 2\pi y \frac{ds}{d\theta}\, d\theta$$

$$\left(\frac{ds}{d\theta}\right)^2 = \dot{x}^2 + \dot{y}^2$$

$$= a^2\{(1 - \cos\theta)^2 + \sin^2\theta\}$$

$$= 2a^2(1 - \cos\theta)$$

$$= 4a^2\sin^2\theta/2$$

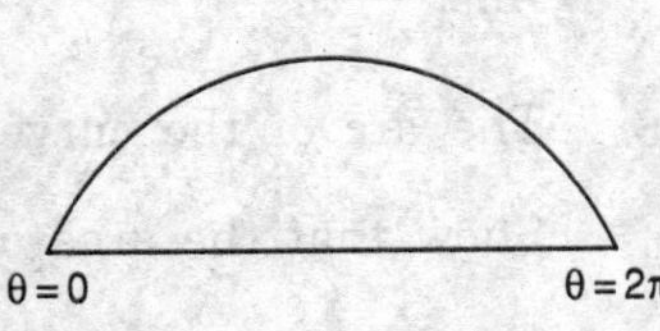

Fig. 2.47

$$S = 2\pi \int_0^{2\pi} a\ 2 \sin^2 \frac{\theta}{2}\ 2a \sin \theta/2\ d\theta$$

Let $\theta/2 = t$

$$= 16\pi a^2 \int_0^{\pi} \sin^3 t\ dt = 32\pi a^2 \int_0^{\pi/2} \sin^3 t\ dt$$

$$= 32\pi a^2 \left[\frac{\Gamma 4/2\ \Gamma 1/2}{2\ \Gamma 5/2}\right]$$

$$= 32\ \pi a^2 \times \frac{2}{3} = \frac{64\pi a^2}{3}$$

PROBLEM SET 2.9

1. Find the surface area of a sphere of radius a.

   Ans. $S = 4\pi a^2$ *(Andhra' 98)*

2. Find the surface generated by revolving the arc of the catenary $y = c \cosh \frac{x}{c}$ from $x = 0$ to $x = c$ about the x axis.

   Ans. $\frac{\pi a^2}{2}\left[2 + \frac{e^2 - e^{-2}}{2}\right]$

3. Find the volume and surface of the right circular cone formed by the revolution of a right angled Δ about a side containing the right angle.

   $V = \frac{1}{3}\pi r^2 h,\ S = \pi r\sqrt{r^2 + h^2}$

4. Find the area of the surface formed by revolving $y^2 = 4ax$ about its axis, by the arc from vertex to one end of the latus rectum.

   $\frac{8\pi a^2}{3}(2\sqrt{2} - 1)$

5. Find the surface of the real formed by the revolution round the tangent at the vertex of an arc of the cycloid $x = a(\theta + \sin\theta)$, $y = a(1 - \cos\theta)$. *(V.T.U. 2000S)*

   Ans. $\frac{32}{3}\pi a^2$.

6. The arc of the curve $x^{2/3} + y^{2/3} = a^{2/3}$ in the first quadrant revolves about x-axis. Show that the area of the surface generated is $\frac{6\pi a^2}{5}$. *(Marathwada 98S)*

7. Find the surface of the solid generated by the revolution of $x = a\cos\phi, y = b\sin\phi$ about the x axis.

   Ans. $2\pi ab\left\{\frac{b}{a} + \frac{a}{\sqrt{a^2 - b^2}}\sin^{-1}\frac{\sqrt{a^2 - b^2}}{a}\right\}$

8. Prove that the surface and volume of the solid generated by the revolution of the loop of $x = t^2, y = t - t^3/3$ about x axis is 3π. *(Delhi 91)*

9. Prove that the surface of the solid generated by the revolution of a loop of $9y^2 = x(x-3)^2$ about x-axis is 3π. *(Delhi 91)*

10. Prove that the surface of the solid generated by the revolution of the cardioid $r = a(1 - \cos\theta)$ about the initial line is $\frac{32}{5}\pi a^2$. *(INTU 90)*

11. Prove that the surface of the solid generated by the revolution of the tractrix. $x = a\cos t + \frac{a}{2}\log\tan^2 t/2,\ y = a\sin t$, about x-axis is $4\pi a^2$.

12. Find the surface of the solid generated by the revolution of $r^2 = a^2\cos 2\theta$ about the initial line.

Ans. $4\pi a^2(1 - 1/\sqrt{2})$.

2.12 BETA AND GAMMA FUNCTIONS

Eulerian Integrals

Beta l, m is defined as

$$\beta(l, m) = \int_0^1 x^{l-1}(1-x)^{m-1}\,dx = \frac{\Gamma l\ \Gamma m}{\Gamma(l+m)}$$

and gamma n as ,
$$\Gamma n = \int_0^\infty e^{-x}x^{n-1}\,dx = \Gamma n$$

are called Beta and Gamma Functions respectively, where $l, m, n \geq 0$

Value of gamma $n \equiv \Gamma n$

It is defined by properties

$$\Gamma n = (n-1)\,\Gamma(n-1)$$
$$\Gamma 1 = 1$$
$$\Gamma(1/2) = \Gamma\pi$$

The value of $\beta(l, m)$ is expressed in terms of gamma functions as $\beta(l, m) = \frac{\Gamma l\ \Gamma m}{\Gamma(l+m)}$.

By symmetry in the value $\beta(l, m) = \beta(m, l)$ making the substitution $x = \sin^2\theta$,

$$\beta(m, n) = 2\int_0^{\pi/2} \sin^{2m-1}\theta\cos^{2n-1}\theta\,d\theta.$$

$$= \frac{2\,\Gamma\frac{2m-1+1}{2}\,\Gamma\frac{2n-1+1}{2}}{2\,\Gamma\frac{2m+2n}{2}} = \frac{\Gamma m\ \Gamma l}{\Gamma(m+l)},$$

By putting $\qquad y = \frac{1}{1+x}$

$$\beta = (l, m) = \int_0^\infty \frac{x^{m-1}}{(1+x)^{l+m}} dx.$$

Example 2.66 **Show that**

(i) $\Gamma n = \int_0^1 \left(\log \frac{1}{y}\right)^{n-1} dy \quad n > 0$

(Mangalore 97)

(ii) $\beta(p, q) = \int_0^\infty \frac{y^{q-1}}{(1+y)^{p+q}} dy$

(V.T.U. 2001, Gauhati 99)

(iii) $= \int_0^1 \frac{x^{p-1} + x^{q-1}}{(1+x)^{p+q}} dx$

(Mangalore 99, Marathwada 98 S)

Solution : (i) $\Gamma n = \int_0^\infty e^{-x} x^{n-1} dx$

Let

$$e^{-x} = y$$

$$e^x = \frac{1}{y}$$

$$x = \left(\log \frac{1}{y}\right)$$

$$-e^{-x} dx = dy$$

when $x = 0$, $y = -1$

when $x = \infty$, $y = 0$

$$\Gamma n = \int_1^0 (-dy)\left(\log \frac{1}{y}\right)^{n-1} = \int_0^1 \left(\log \frac{1}{y}\right)^{n-1} dy$$

(ii)

$$\beta(p, q) = \int_0^1 x^{p-1}(1-x)^{q-1} dx$$

Let

$$x = \frac{1}{1+y}$$

$$dx = -\frac{1}{(1+y)^2} dy$$

$$1 - x = \frac{y}{1+y}$$

when $x = 0$, $y = \infty$, when $x = 1$, $y = 0$

$$\beta(p, q) = \int_\infty^0 \frac{1}{(1+y)^{p-1}} \left(\frac{y}{1+y}\right)^{q-1} \left(\frac{-1}{(1+y)^2} dy\right)$$

$$= \int_0^\infty \frac{y^{q-1}}{(1+y)^{p+q}} dy$$

$$= \int_0^1 \frac{y^{q-1}}{(1+y)^{p+q}}\,dy + \int_1^{\infty} \frac{y^{q-1}}{(1+y)^{p+q}}\,dy$$

Let $y = \frac{1}{t}$ in second integral

$$= \int_0^1 \frac{y^{q-1}}{(1+y)^{p+q}}\,dy + \int_1^0 \frac{t^{-(q-1)} \cdot t^{(p+q)}}{(1+t)^{(p+q)}} \left(-\frac{1}{t^2}\right) dt$$

$$= I_1 + \int_0^1 \frac{t^{p-1}}{(1+t)^{p+q}}\,dt = \int_0^1 \frac{x^{q-1} + x^{p-1}}{(1+x)^{p+q}}\,dx$$

Example 2.67 **Show $\Gamma(1/2) = \sqrt{\pi}$.**

Solution : We know $\Gamma(1/2) = \int_0^{\infty} e^{-x} x^{\frac{1}{2}-1}\,dx.$

Multiplying $\Gamma\left(\frac{1}{2}\right)$ with gamma $\Gamma\left(\frac{1}{2}\right)$ again,

So $\Gamma(1/2)\,\Gamma(1/2) = \int_0^{\infty} e^{-x} x^{-\frac{1}{2}}\,dx \int_0^{\infty} e^{-y} y^{-\frac{1}{2}}\,dy$

Let $x = \xi^2$ $y = \eta^2$

$$= \int_0^{\infty} e^{-\xi^2} \xi^{-1}\, 2\xi\, d\xi \int_0^{\infty} e^{-\eta^2} \eta^{-1}\, 2\eta\, d\eta$$

$$= 4 \int_0^{\infty} \int_0^{\infty} e^{-(\xi^2+\eta^2)}\, d\xi\, d\eta$$

Changing to polars

$\xi^2 + \eta^2 = r^2$ and $d\xi\, d\eta = r\, d\theta\, dr$

Limits for Ist quadrant extending upto ∞ are

$$0 \le \theta \le \frac{\pi}{2} \quad r \to \infty$$

$$= 4 \int_0^{\pi/2} \int_0^{\infty} e^{-r^2}\, r\, d\theta\, dr$$

$$\Gamma(1/2)^2 = 4\, [\theta]_0^{\pi/2} \left[\frac{e^{-r^2}}{-2}\right]_0^{\infty}$$

$$= 4\, \frac{\pi}{2} \times \frac{1}{2} \times 1 = \pi$$

Hence $\Gamma(1/2) = \sqrt{\pi}$.

Example 2.68 **Evaluate in terms of gamma functions, the integral**

$$\boldsymbol{I = \int_0^\infty e^{-x^4}\, dx}$$

Solution : We have

$$I = \int_0^\infty e^{-x^4}\, dx$$

Let $x^4 = t$ or $x = t^{1/4}$

$$dx = \frac{1}{4} t^{\frac{1}{4}-1}\, dt$$

$$I = \frac{1}{4}\int_0^\infty e^{-t}\, t^{\frac{1}{4}-1}\, dt = \frac{1}{4}\Gamma(1/4)$$

$$= \Gamma\left(\frac{1}{4}+1\right) = \Gamma(5/4)$$

Example 2.69 **Evalaute** $\boldsymbol{\int_0^1 \sqrt{1-x^4}\, dx}$

Solution : We have

$$\int_0^1 \sqrt{1-x^4}\, dx$$

Let
$$x^2 = \sin\theta \text{ or } x = \sin^{1/2}\theta$$

$$dx = \frac{1}{2}\sin^{1/2-1}\theta\cos\theta\, d\theta$$

$$I = \int_0^{\pi/2} \cos\theta\, \frac{1}{2}\sin^{1/2-1}\theta\cos\theta\, d\theta$$

$$= \frac{1}{2}\int_0^{\pi/2} \sin^{-1/2}\theta\cos^2\theta\, d\theta$$

$$= \frac{1}{2}\,\frac{\Gamma\left(\frac{2+1}{2}\right)\Gamma(1/4)}{2\,\Gamma(7/4)}$$

$$= \frac{1}{2}\cdot\frac{\frac{1}{2}\sqrt{\pi}}{2}\,\frac{\Gamma(1/4)}{\frac{3}{4}\Gamma(3/4)}$$

$$= \frac{\sqrt{\pi}\,\Gamma(1/4)}{6\,\Gamma\, 3/4}$$

$$= \frac{\sqrt{\pi}\,.\,\Gamma(1/4)}{6\,\Gamma(3/4)}$$

Example 2.70 **Express the following integrals in terms of gamma functions :**

(i) $I_1 = \int_0^{\pi.2} \sqrt{\tan x}\, x$

(Mysore 95, Andhra 94)

(ii) $I_2 = \int_0^{\infty} e^{-a^2x^2} x^n \, dx.$

Solution : (i) $I_1 = \int_0^{\pi/2} \sin^{1/2}\theta \, d\theta$

$$= \int_0^{\pi/2} \sin^{3/2-1} x \cos^{1/2-1} dx$$

$$= \frac{\Gamma(3/4)\,\Gamma(1/4)}{2\,\Gamma 1} = \frac{\Gamma(3/4)\,\Gamma(1/4)}{2}$$

Example 2.71 **Prove that**

$$I = \int_0^1 \frac{x^2\,dx}{\sqrt{1-x^4}} \times \int_0^1 \frac{dx}{\sqrt{1-x^4}} = \frac{\pi}{4}$$

Solution : Let $x = \sqrt{\sin\theta}$

$$I = \int_0^{\pi/2} \frac{\sin\theta}{\cos\theta} \frac{1}{2} \sin^{-1/2}\theta \cos\theta \, d\,\theta \int_0^{\pi/2} \frac{1}{2} \sin^{-1/2}\theta \frac{\cos\theta}{\cos\theta} d\theta$$

$$= \frac{1}{2}\int_0^{\pi/2} \sin^{1/2}\theta \, d\theta \,.\, \frac{1}{2}\int_0^{\pi/2} \sin^{-1/2}\theta \, d\,\theta$$

$$= \frac{1}{2} \frac{\Gamma\left(\dfrac{\frac{1}{2}+1}{2}\right)\Gamma(1/2)}{2\,\Gamma(5/4)} \times \frac{1}{2} \frac{\Gamma(1/4)\,\Gamma(1/2)}{2\,\Gamma(3/4)}$$

$$= \frac{\pi}{16 \times \frac{1}{4}} = \frac{\pi}{4}$$

Example 2.72 **Prove** $\int_0^1 \frac{x^2\,dx}{\sqrt{1-x^4}} \times \int_0^1 \frac{dx}{\sqrt{1+x^4}} = \frac{\pi}{4\sqrt{2}}$

Solution : L.H.S. $= I_1 \times I_2$

Let $x^2 = \sin\theta$ or $x = \sqrt{\sin\theta}$

$$I_1 = \int_0^1 \frac{x^2\,dx}{\sqrt{1-x^4}} = \int_0^{\pi/2} \frac{\sin\theta}{\cos\theta} \frac{1}{2\sqrt{\sin\theta}} \cos\theta \, d\theta$$

$$= \frac{1}{2}\int_0^1 \sin^{1/2}\theta \, d\theta$$

$$= \frac{1}{2} \frac{\Gamma\left(\dfrac{\frac{1}{2}+1}{2}\right)\Gamma(\frac{1}{2})}{2\,\Gamma(5/4)} = \frac{1}{4}\sqrt{\pi}\,\frac{\Gamma(3/4)}{\Gamma(5/4)} \quad \ldots(1)$$

$$I_2 = \int_0^1 \frac{dx}{\sqrt{1+x^4}}$$

Let $x^2 = \tan\theta$ or $x = \sqrt{\tan\theta}$, $dx = \dfrac{\sec^2\theta\, d\theta}{2\sqrt{\tan\theta}}$

$$= \int_0^{\pi/4} \frac{1}{2} \frac{\sec^2\theta\, d\theta}{\sec\theta\sqrt{\tan\theta}} = \frac{1}{2}\int_0^{\pi/4} \frac{1}{\sqrt{\sin\theta\cos\theta}}\, d\theta$$

Let $2\theta = t$

$$\frac{1}{2}\int_0^{\pi/2} \frac{1}{\sqrt{\frac{1}{2}\sin t}} \frac{dt}{2} = \frac{\sqrt{2}}{4}\int_0^{\pi/2} \sin^{-1/2} t\, dt$$

$$= \frac{\Gamma 2}{4} \frac{\Gamma(1/4)\,\Gamma(1/2)}{2\,\Gamma(3/4)} = \frac{\sqrt{2}\,\sqrt{\pi}}{8} \frac{\Gamma(1/4)}{(3/4)}$$

$$I_1 I_2 = \frac{\sqrt{\pi}}{4} \frac{\Gamma(3/4)}{\frac{1}{4}\,\Gamma(1/4)} \times \frac{\sqrt{2}\,\sqrt{\pi}}{8} \frac{\Gamma(1/4)}{\Gamma(3/4)}$$

$$= \frac{\sqrt{2}\pi}{8} = \frac{\pi}{4\sqrt{2}}.$$

Example 2.73 **Prove that (i)** $\beta\left(m, \frac{1}{2}\right) = 2^{2m-1}\beta(m, n)$

(ii) $\Gamma m\, \Gamma\left(m + \frac{1}{2}\right) = \dfrac{\sqrt{\pi}}{2^{2m-1}}\Gamma 2m$ *(V.T.U. 2001, Andhra 94)*

Solution : (i) LH.S. $= \beta\left(m, \frac{1}{2}\right) = \int_0^1 x^{m-1}(1-x)^{-1+\frac{1}{2}}\, dx$

Let $x = \sin^2\theta$

$$= \int_0^1 \sin^{2m-2}\theta\cos^{-1}\theta\, 2\sin\theta\cos\theta\, d\theta$$

$$= 2\int_0^1 \sin^{2m-1}\theta\, d\theta$$

$$= 2\frac{\Gamma\left(\dfrac{2m-1+1}{2}\right)\Gamma(1/2)}{2\,\Gamma\left(m+\frac{1}{2}\right)} = \frac{\Gamma m\,\sqrt{\pi}}{\Gamma\left(m+\frac{1}{2}\right)}.$$

$$\text{R.H.S.} = 2^{2m-1}\int_0^1 x^{m-1}(1-x)^{m-1}\,dx$$

Let $x = \sin^2\theta$

$$= 2^{2m-1}\int_0^{\pi/2} \sin^{2m-2}\theta\cos^{2m-2}\theta\, 2\sin\theta\cos\theta\, d\theta$$

$$= \int_0^{\pi/2} (2\sin\theta\cos\theta)^{2m-1}\, 2\, d\theta$$

Let $2\theta = t$

$$= \int_0^{\pi} \sin^{2m-1} t\, 2\,\frac{dt}{2}$$

$$= 2\int_0^{\pi/2} \sin^{2m-1} t\, dt \qquad \text{(By property of definite integral.)}$$

$$= \frac{2\ \Gamma(2m/2)\,\Gamma(1/2)}{2\,\Gamma\left(m+\frac{1}{2}\right)} = \frac{\Gamma m\,\sqrt{\pi}}{\Gamma\left(m+\frac{1}{2}\right)} = \text{L.H.S.}$$

(ii) From (i)

$$\text{L.H.S.} = \beta\left(m, \frac{1}{2}\right) = \frac{\Gamma m\ \Gamma(1/2)}{\Gamma\left(m+\frac{1}{2}\right)}$$

$$\text{R.H.S.} = 2^{2m-1}\,\beta(m, n)$$

$$= 2^{2m-1}\,\frac{\Gamma m\ \Gamma m}{\Gamma 2m}$$

Hence $$\frac{\Gamma m\,\sqrt{\pi}}{\Gamma\left(m+\frac{1}{2}\right)} = 2^{2m-1}\,\frac{\Gamma m\,\Gamma m}{\Gamma 2m}$$

or $$\Gamma\left(m+\frac{1}{2}\right)\Gamma m = \frac{\sqrt{\pi}\ \Gamma 2m}{2^{2m-1}}. \text{ (Proved)}$$

This is important formula and is known as duplication formula for gamma function.

Example 2.74 $\Gamma(1/4)\,\Gamma(3/4)$. ***(Bhopal 91)***

Solution :

$$\Gamma m\,\Gamma\left(m+\frac{1}{2}\right) = \frac{\sqrt{\pi}\ \Gamma 2m}{2^{2m-1}}$$

for $m = \frac{1}{4}$

$$\Gamma(1/4)\,\Gamma(3/4) = \frac{\sqrt{\pi}\ \Gamma(1/2)}{2^{1/4-1}}$$

$$= \sqrt{\pi}\ \sqrt{\pi}\ \sqrt{2} = \pi\sqrt{2}$$

2.13 DIRICHLET'S INTEGRAL

$$I = \iint_D x^{l-1} y^{m-1}\, dx\, dy = \frac{\Gamma l\, \Gamma m}{\Gamma(l+m+1)}$$

where D is the area of the $\Delta\, OAB$

$$I = \int_{x=0}^{h} \int_{y=0}^{h-x} x^{l-1} y^{m-1}\, dx\, dy$$

Fig. 2.48

Let
$$x = hX$$
$$y = hY$$

$$I = \int_0^1 \int_0^{1-X} h^{l+m} X^{l-1} Y^{m-1}\, dX\, dY,\ 0 \le X + Y \le 1$$

$$= \int_0^1 h^{l+m} X^{l-1} \frac{\left[y^m\right]_0^{1-X}}{m}\, dX$$

$$= \frac{h^{l+m}}{m} \int_0^1 X^{l-1} (1-X)^m\, dX$$

$$= \frac{h^{l+m}}{m} = \frac{\Gamma l\, \Gamma(m+1)}{\Gamma(l+m+1)} + \frac{\Gamma l\, \Gamma m\, h^{l+m}}{\Gamma(l+m+1)}.$$

Similarly,

(ii) $$\iiint_V x^{l-1} y^{m-1} z^{n-1}\, dx\, dy\, dz = \frac{\Gamma l\, \Gamma m\, \Gamma n}{\Gamma(1+l+m+n)}$$

where V is the volume $x \ge 0,\ y \ge 0,\ z \ge 0$ and $0 \le x + y + z \le 1$.

Example 2.75 **Find the mass of an octant of the sphere $x^2 + y^2 + z^2 = 1$, if the density varies as xyz.**

Solution : $V = \iiint xyz\, dx\, dy\, dz$ subject to $x \ge 0,\ y \ge 0,\ z \ge 0$

$$x^2 + y^2 + z^2 \le 1.$$

Let $x = \sqrt{X},\ y = \sqrt{Y},\ z = \sqrt{Z}$ then

$$dx = \frac{1}{2} X^{1/2-1}\, dX,\ dy = \frac{1}{2} Y^{1/2-1}\, dY,\ dz = \frac{1}{2} Z^{1/2-1}\, dZ \text{ and } X + Y + Z \le 1$$

So $$V = \iiint \frac{1}{8} X^{\frac{1}{2}+\frac{1}{2}-1} Y^{1-1} Z^{1-1}\, dx\, dy\, dz = \frac{1}{8} \frac{\Gamma 1\, \Gamma 1\, \Gamma 1}{\Gamma(1+1+1)+1}$$

$$= \frac{1}{8\, \Gamma 4} = \frac{1}{48}.$$

as $\Gamma 4 = 3.2 \times 1 = 6$

Example 2.76 **Find the mass of the ellipsoid $\frac{x^2}{a^2}+\frac{y^2}{b^2}+\frac{z^2}{c^2}=1$, if the density varies as $x^2y^2z^2$.**

Solution : Mass $M = \iiint_V kx^2y^2z^2\,dx\,dy\,dz$, where volume

$V = 8 \times$ Volume of the positive octant.

Let $x = aX^{1/2},\ y = bY^{1/2},\ z = cZ^{1/2}$ so

$$dx = \frac{a}{2}X^{-1/2}\,dX,\ dy = \frac{b}{2}Y^{-1/2}\,dY,\ dz = \frac{c}{2}Z^{-1/2}\,dZ$$

Subject to $X \geq 0,\ Y \geq 0,\ Z \geq 0,\ X + Y + Z \leq 1$.

So $$M = 8k \iiint a^2b^2c^2\,XYZ\,\frac{abc}{8}\,X^{-1/2}\,Y^{-1/2}\,Z^{-1/2}\,dX\,dY\,dZ$$

$$= ka^3b^3c^3 \iiint X^{3/2-1}\,Y^{3/2-1}\,Z^{3/2-1}\,dX\,dY\,dZ$$

$$= ka^3b^3c^3\,\frac{\Gamma(3/2\,\Gamma(3/2)\,\Gamma(3/2)}{\Gamma\left(1+\frac{9}{2}\right)}$$

$$= \frac{4\pi}{945}\,ka^3b^3c^3.$$

Example 2.77 **Given $I = \int_0^\infty \frac{x^{n-1}}{1+x}\,dx = \frac{\pi}{\sin n\pi}$, show that**

$$\Gamma n\,\Gamma(1-n) = \frac{\pi}{\sin n\pi}$$ ***(Assam 98)***

Solution : We have

$$I = \int_0^\infty \frac{x^{n-1}}{1+x}\,dx$$

Let $x = \frac{1}{y} - 1,\ dx = -\frac{1}{y^2}\,dy$

$$\therefore \quad I = \int_1^0 \frac{(1-y)^{n-1}}{y^{n-1}}\,y\left(-\frac{1}{y^2}\right)dy$$

$$= \int_0^1 (1-y)^{n-1}\,y^{1-n-1}\,dy$$

$$= \frac{\Gamma n\,\Gamma(1-n)}{\Gamma 1}$$

Hence $\Gamma n\,\Gamma 1 - n = \frac{\pi}{\sin n\pi}$

Using the result, we can evaluate

$$I = \int_0^\infty \frac{dy}{1+y^4}$$

(Assam 98, Rewa 90)

Let $y^4 = t$

$$dy = \frac{1}{4} t^{-3/4}\, dt$$

$$I = \frac{1}{4}\int_0^1 \frac{t^{1/4-1}}{1+t}\, dt$$

$$= \frac{1}{4} \frac{\pi}{\sin\frac{\pi}{4}} = \frac{\sqrt{2}\pi}{4}.$$

Example 2.78 **Prove** $I \int_0^{\pi/2} \sqrt{\sin\theta}\, d\theta \int_0^{\pi/2} \frac{1}{\sqrt{\sin\theta}}\, d\theta = \pi.$

(V.T.U. 2000 S)

Solution : We have

$$I = \int_0^{\pi/2} \sin^{1/2}\theta\, d\theta \int_0^{\pi/2} \sin^{-1/2}\theta\, d\theta$$

$$= \int_0^{\pi/2} \sin^{3/2-1}\theta \int_0^{\pi/2} \sin^{1/2-1}\theta\, d\theta$$

$$= \frac{\Gamma\left(\frac{3/2}{2}\right)\Gamma(1/2)}{2\Gamma(5/4)} \times \frac{\Gamma(1/4)\,\Gamma(1/2)}{2\Gamma(3/4)}$$

$$= \frac{\sqrt{\pi}\,\sqrt{\pi}}{2\times 2\times \frac{1}{4}} = \pi$$

PROBLEM SET 2.10

Express in terms of gamma functions, the values of the following integrals

1. (i) $\int_0^1 \frac{dx}{\sqrt{1-x^4}}$,

(ii) $\int_0^\infty e^{-bx^2}\, dx$,

(iii) $\int_0^1 x^4 \left(\log\frac{1}{x}\right)^3 dx$

Ans. (i) $\frac{\sqrt{\pi}}{4}\frac{\Gamma(1/4)}{\Gamma(3/4)}$ (ii) $\frac{1}{2\sqrt{b}}\sqrt{\pi}$ (iii) $\frac{6}{625}$

(Madras 2000)

2. (i) $\int_0^\infty x^{p-1} e^{-kx} \; k > 0$

Ans. $\frac{\Gamma p}{k^p}$.

(ii) $\int_0^\infty \frac{x^c}{c^x} dx = \frac{\Gamma c + 1}{(\log c)^{c+1}}$.

Hint : $[c^x = e^{\log e^x} = e^{x \log c}]$

3. Compute :

(i) $\frac{\Gamma 7}{2}$ *(Assam 98)*

(ii) $\frac{\Gamma 9}{2}$ *(Calicut)*

(iii) $\beta\left(\frac{5}{2}, \frac{3}{2}\right)$

(iv) $\beta\left(\frac{9}{2}, \frac{7}{2}\right)$ *(Andhra 2000)*

Ans. (i) $\frac{5}{2} \cdot \frac{3}{2} \cdot \frac{1}{2} \cdot \sqrt{\pi} = 3.323$, (ii) $\frac{7}{2} \cdot \frac{5}{2} \cdot \frac{3}{2} \cdot \frac{1}{2} \sqrt{\pi} = 11.629$, (iii) $\frac{\Gamma(5/2)\,\Gamma(3/2)}{\Gamma 4} = \frac{3\pi}{48} = .1964$.

(iv) .1227.

4. Prove the following :

(i) $\int_0^\infty e^{-a^2x^2} x^n \, dx = \frac{1}{2a^{n+1}} \Gamma\left(\frac{n+1}{2}\right)$

(ii) $\int_1^\infty \frac{dx}{x^{p+1} + (x-1)^q} \; -p < q < 1$

$= 2^{p+q-1} \beta(p, q)$ *(Poona 89)*

(iii) $\int_0^{\pi/2} \sqrt{\cot \theta} \, d\theta = \frac{1}{2} \Gamma(1/4)\,\Gamma(3/4)$

(iv) $\int_0^\infty \frac{e^{-\sqrt{x}}}{x^{7/4}} dx = \frac{8}{3} \sqrt{\pi}$

(v) $\int_0^{\pi/2} \sqrt{\sin \theta} \, d\theta \int_0^{\pi/2} \frac{1}{\sqrt{\sin \theta}} d\theta = \pi$

(vi) $\int_0^{\pi/2} \{\sqrt{\tan \theta} + \sqrt{\sec \theta}\} d\theta = \left[\frac{1}{2} \frac{1}{2}! \frac{3}{4}! + \frac{\sqrt{\pi}}{\frac{3}{4}!}\right]$ *(J.N.T.U. 2000)*

5. Prove :

(i) $\int_a^b (x-a)^m (b-a)^n \, dx + (b-a)^{m+n+1} \beta(m+1, n-1)$ *(Marathdawa 98 S)*

(ii) $\int_0^1 \frac{x\,dx}{\sqrt{1-x^5}} = \frac{1}{5}\beta\left(\frac{2}{5}, \frac{1}{2}\right)$

(Banalore 90)

6. Show that the area in the first quadrant enclosed by the curve $\left(\frac{x}{a}\right)^n + \left(\frac{y}{b}\right)^m$, $n > 0$, $m > 0$, is given by $\frac{ab}{m+n} \frac{\frac{1}{n}!\,\frac{1}{n}!}{\left(\frac{1}{n}+\frac{1}{n}\right)!}$.

7. Prove that $\int_0^1 x^m (\log x)^n\,dx + \frac{(-1)^n\,n!}{(m+1)^{n+1}}$, where n is positive integer and $m > -1$,

(Assam, 99, Delhi 92)

8. Show that $\iint x^{m-1} y^{n-1}\,dx\,dy$ over the positive quadrant of the ellipse $\frac{x^2}{a^2} + \frac{y^2}{b^2} = 1$ is $\frac{a^m b^n}{2n}\beta\left\{\frac{m}{2}, \frac{n}{2}+1\right\}$.

9. Show that $\int_0^1 y^{n-1}\left(\log\frac{1}{y}\right)^{p-1} dy = \frac{p!}{n^p}$ where n, $p > 0$.

10. The plane $\frac{x}{a} + \frac{y}{b} + \frac{z}{c} = 1$ meets the axis in ABC. Apply Dirichlet's integrals to find the volume of the tetrahedron $OABC$. Also find its mass if the density varies as xyz.

Ans. $\frac{abc}{6}$; $k\,\frac{a^2bc^2}{720}$,

11. Express $\int_0^1 x^m (1-x^m)^p\,dx$ in terms of gamma functions and evaluate $\int_0^1 x^5(1-x^3)^{10}\,dx$

(Kottayam 97)

Ans. $\frac{\left(\frac{m+1}{m}\right)!\,(p+1)!}{n\left(\frac{m+1}{n}+p+1\right)!}$, $\frac{!}{396}$.

12. Prove that

(i) $\beta(n, n) = \frac{\sqrt{\pi}\,n!}{2^{2n-1}\left(n+\frac{1}{2}\right)!}$

(Nagpur 97, Calicut 91)

(ii) $\left(n+\frac{1}{2}\right)!\,(n+1)! = \frac{\sqrt{\pi}\,(2n+1)!}{2^{2n}}$

(Bangalore)

(iii) $\frac{\beta(m+1, m)}{m} = \frac{\beta(m, n+1)}{n} = \frac{\beta(m, n)}{m+n}$

(Kuvempur 96; Bhopal 91)

2.14 MULTIPLE INTEGRALS

Introduction. We have seen that integral of single variable function is an analytic substitute for area. If a differentiable function $y = f(x)$ be represented graphically, it represents a continuous curve in the two dimensional X-Y plane.

So the integral $\int_a^b f(x)\,dx =$ the area $ARSBPA$,

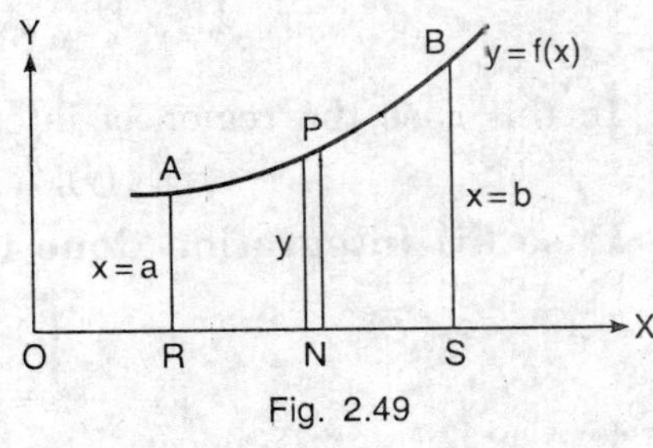

Fig. 2.49

bounded by the curve, x-axis and the ordinates $x = a$ and $x = b$. Three of the boundaries of this area are parallel to the coordinate axes. If such simple conditions are not present then the area is found by double integration, defined as follows.

2.15 DOUBLE INTEGRALS

Let $z = f(x, y)$ be a function of two independent variable x and y. If x and y be the coordinates (x, y) of a point P in the $X - Y$ plane and P be allowed to lie anywhere in a certain given area A, then dividing the area A into small elementary area $\delta A_1, \delta A_2, \ldots \delta A_n$, each tending to zero, the limit of the sum

$$f(x_1, y_1)\,\delta A_1 + f(x_2, y_2)\,\delta A_2 + \ldots + f(x_n, y_n)\,\delta A_n$$

$$= \sum_{r=1}^{n} f(x_r, y_r)\,\delta A_r,$$

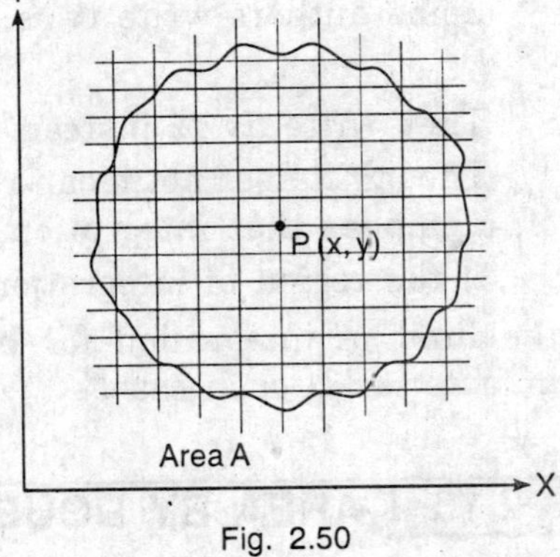

Fig. 2.50

As $n \to \infty$, (x_r, y_r) being a points in the area δA_r, is called the double integral of $f(x, y)$ over the area A and is denoted by $\iint_A f(x, y)\,dx\,dy$. As area is two dimensional, two integrations will be required in evaluating the sum. So dividng area A by drawing lines parallel to x and y axes, we get $\delta A_r = \delta x_r \cdot \delta y_r$. The summation will involve once summation w.r.t. x and then summation w.r.t y or vice versa. But limit of the sum will be independent of the order in which the summation is done. In other words, we have

$$\iint_A f(x, y)\,dx\,dy = \iint_A f(x, y)\,dy\,dx.$$

In our notations evaluating $\iint_A f(x, y)\,dx\,dy$, integration is done first with respect to y, the limits of integration for it will be functions of x and lastly w.r.t x, whose limits will be simply constants. While for $\iint f(x, y)\,dy\,dx$, integration is done first w.r.t. x, whose limits will be functions of y and lastly w.r.t y, whose limits will then be constant.

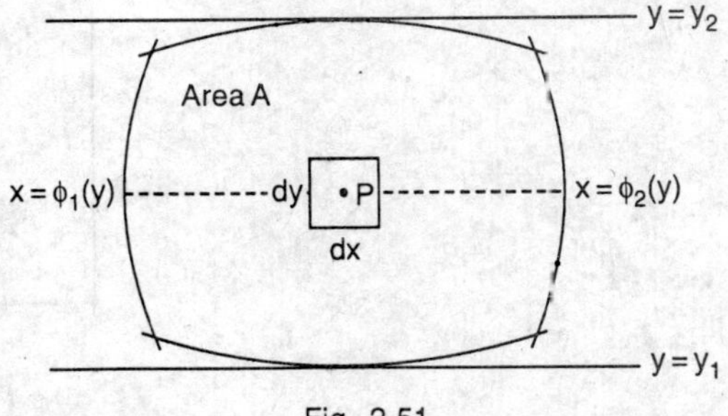

Fig. 2.51

Case (i) Integration done first w.r.t. *x*. The integral is expressed as

$$\int_{y=y_1}^{y_2}\int_{x=\phi_1(y)}^{x=\phi_2(y)} f(x, y)\, dy\, dx$$

In this case the region of integration is expressed in the form

$$x = \phi_1(y),\ x = \phi_2(y),\ y = y_1,\ y = y_2$$

Case (ii) Integration done first w.r.t. *y*. The integral for this case is written as

$$\int_{x=x_1}^{x=x_2}\int_{y=\psi_1(x)}^{y=\psi_2(x)} f(x, y)\, dx\, dy.$$

For this case the region of integration is defined in the form

$$y = \psi_1(x) \text{ to } y = \psi_2(x) \text{ and } x = x_1 \text{ to } x = x_2.$$

Note : We follow the notations $\int_{x_1}^{x_2}\int_{y=\psi_1(x)}^{\psi_2(x)} f(x, y)\, dx\, dy$ for doing integration first with respect to y and then w.r.t. x.

Some authors write it as $\int_{x_1}^{x_2}\int_{y=\psi_1(x)}^{\psi_2(x)} f(x, y)\, dy\, dx.$

They write $dy\, dx$ instead of $dx\, dy$.

Though it does not mean reversal in the order of integration. If limits for y are functions of, it implies that integration is to be done first with respect to y.

If the region of integration be a rectangle bounded by $x = x_1$, $x = x_2$, $y = y_1$, $y = y_2$, then the limit of integration for both x and y are constants and the order of integration is immaterial in such case.

2.16 AREA BY DOUBLE INTEGRALS

If $f(x, y) = 1$, then the double integral $\iint_A dx\, dy$ represents the area of the region of integration.

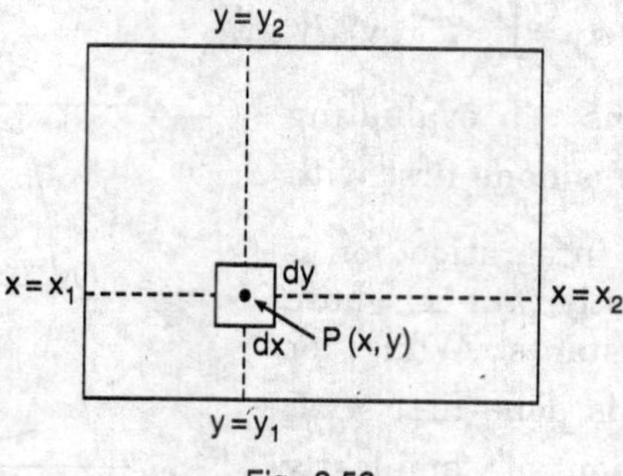

Fig. 2.52

Example 2.79 **Evaluate** $I = \int_{x=0}^{3} \int_{y=1}^{2} xy(x+y)\, dx\, dy.$

Solution :

$$I = \int_{x=0}^{3} \left[\frac{x^2 y^2}{2} + \frac{xy^3}{3} \right]_{y=1}^{2} dx$$

$$= \int_{0}^{3} \left[x^2 \frac{(2^2 - 1)}{2} + \frac{x}{3}(2^3 - 1^3) \right] dx$$

$$= \int_{0}^{3} \left(\frac{3}{2} x^2 + \frac{7}{3} x \right) dx$$

$$= \left[\frac{x^3}{2} + \frac{7}{3} \frac{x^2}{2} \right]_0^3 = \frac{27}{2} + \frac{7}{2} \times 3 = 24$$

Example 2.80 **Find by double integration, the areas enclosed by the following curves: (i) $x^2 + y^2 = a^2$, (ii) $x^4 = a^2(x^2 - y^2)$, (iii) $a^2y^2 = x^3(2a - x)$**

Solution : (i) By symmetry of circle w.r.t. the axes, the requied area is 4 times the area lying in the first quadrant. The area in the first quadrant is bounded by $OA(y = 0)$ the point $A(x = a)$, the arc

$$AB\ (y = \sqrt{a^2 - x^2}) \text{ and } BO(x = 0)$$

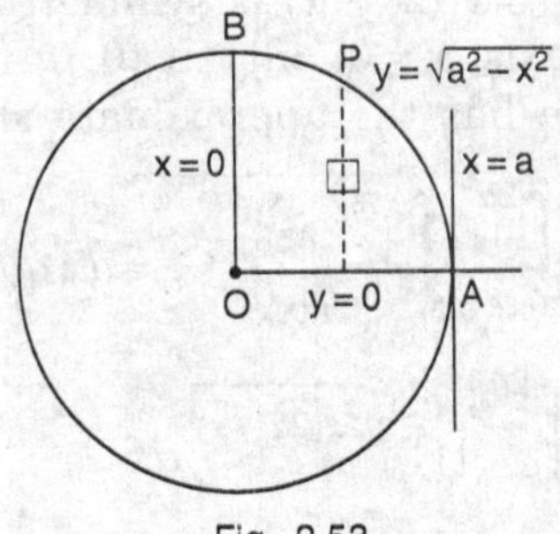

Fig. 2.53

Integrating first w.r.t. y, here $f(x, y) = 1$ and

$$\text{Area} = 4 \int_{x=0}^{a} \int_{y=0}^{y=\sqrt{a^2 - x^2}} dx\, dy$$

$$= 4 \int_{x=0}^{a} [y]_0^{\sqrt{a^2 - x^2}} dx = \int_0^a \sqrt{a^2 - x^2}\, dx$$

(It has now become the integral for area, in single integration)

$$A = 4 \left[\frac{x\sqrt{a^2 - x^2}}{2} + \frac{a^2}{2} \sin^{-1} \frac{x}{a} \right]_0^a = 4 \left(0 + \frac{a^2}{2} \frac{\pi}{2} \right) = \pi a^2.$$

(ii) The curve $x^4 = a^2(a^2 - x^2)$ is symmetrical about both the axes. The equation can be written as $ay = \pm x\sqrt{a^2 - x^2}$ so curve is real when $-a \le x \le a$.

Writing the equations as $x^4 - a^2x^2 + a^2y^2 = 0$

$$x^2 = \frac{a^2}{2}\left(1 \pm \sqrt{1 - \frac{y^2}{a^2}}\right). \text{ So } x \text{ is real for } -a \le y \le a$$

Curve passes through the origin.

Tangent at origin are given by $x = \pm y$. So origin is a node. Curve meets y-axis $x = 0$ only at $(0, 0)$. It meets x axis at $x = 0$, and $x = \pm a$.

$$\text{Area } A = 4\int_{x=0}^{a}\int_{y=0}^{\frac{x}{a}\sqrt{a^2 - x^2}} dx\, dy$$

$$= 4\int_0^a [y]_0^{\frac{x}{a}\sqrt{a^2 - x^2}} dx$$

$$= \frac{4}{a}\int_0^a x\sqrt{a^2 - x^2}\, dx$$

$$= \frac{4}{a}\left[\frac{(a^2 - x^2)^{3/2}}{-3}\right]_{x=0}^{a} = \frac{4}{3}a^2.$$

Fig. 2.54

(iii) The curve $a^2y^2 = x^3(2a - x)$ is symmetrical about x-axis $y = 0$ at $x = 0$ and $x = 2a$, it meets y-axis $(x = 0)$ only at $(0, 0)$ for $x > 2a$, y is imaginary and also for $x < 0$, the curve has the approximate shape as in fig. 2.55

$$\text{Required area} = 2\int_{x=0}^{2a}\int_{y=0 \text{ at } N}^{y=\frac{x}{a}\sqrt{x(2a-x)}} dx\, dy$$

$$= 2\int_0^{2a} [y]_0^{\frac{x}{2}\sqrt{x(2a-x)}} dx$$

$$= \frac{2}{a}\int_0^{2a} x\sqrt{x(2a - x)}\, dx$$

Fig. 2.55

Let $\quad x = 2a\sin^2\theta.$

$$\text{Area } = \frac{2}{a}\int_0^{\pi/2} 2a\sin^2\theta\, 2a\sin\theta\cos\theta\,(4a\sin\theta\cos\theta)\, d\theta$$

$$= 32a^2\int_0^{\pi/2}\sin^4\theta\cos^2\theta\, d\theta = 32a^2\,\frac{\Gamma\frac{5}{2}\,\Gamma\frac{3}{2}}{2\,\Gamma 4} = 32a^2\,\frac{\frac{3}{2}\frac{1}{2}\cdot\frac{1}{2}\pi}{2 \times 3 \times 2}$$

$$= \pi a^2.$$

2.17 EVALUATION OF DOUBLE INTEGRALS IN POLAR CO-ORDINATIES

In polar co-ordinates, element of area $\delta A = dr\, r\, d\theta$. If the region of integration is bounded by $r = f_1(\theta)$, $r = f_2(\theta)$, $\theta = \alpha$, $\theta = \beta$. The limits for θ are constant so integration w.r.t. θ is to be done in the last. Integral of the function $F(r, \theta)$ over the region is the value of the double integral

$$\int_{\theta=\alpha}^{\theta=B} \int_{r=f_1(\theta)}^{f_2(\theta)} = F(r, \theta)\, r\, d\theta\, dr$$

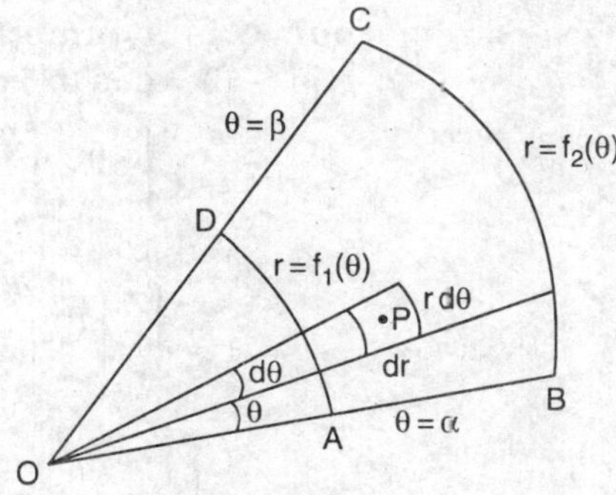

Fig. 2.56

Change from carterian to polar form.

Let $f(x, y)$ be the function whose double integral $\iint_A f(x, y)\, dx\, dy$ is to be evaluated over the given region A. Some times the integration becomes easier, when expressed in polar form. For this, replace x by $r \cos \theta$, y by $r \sin \theta$, the elementary area $dA = dx\, dy$ be replaced by another element $r\, d\theta\, dr$. The limits for r and θ are to be taken such that the whole region A of integration is covered.

Example 2.81 **Evaluate $\iint \sqrt{a^2 - r^2}\, r\, d\theta\, dr$, over the semicircle $r = a \cos \theta$, $0 \le \theta \le \dfrac{\pi}{2}$.**

Solution : Let $P(r, \theta)$ be point in the element dr, $r\, d\theta$ limits for integration first w.r.t. are $r_{\text{at } 0} = 0$ to r at $Q = a \cos \theta$ limits for θ are from 0 to $\dfrac{\pi}{2}$. So

$$I = \int_0^{\pi/2} \int_0^{a \cos \theta} \sqrt{a^2 - r^2}\, r\, d\theta\, dr$$

$$I = \int_0^{\pi/2} \frac{\left[(a^2 - r^2)^{3/2}\right]_0^{a \cos \theta}}{-\dfrac{3}{2} \cdot 2}\, d\theta$$

$$= \frac{1}{3} \int_0^{\pi/2} a^3 (1 - \sin^3 \theta)\, d\theta$$

Fig. 2.57

$$= \frac{a^3}{3} [\theta]_0^{\pi/2} - \frac{a^3}{3} \frac{\dfrac{4}{2}!\, \dfrac{1}{2}!}{2 \cdot \dfrac{5}{2}!} = \frac{a^3 \pi}{6} - \frac{a^3}{3} \times \frac{1 \,.\, \sqrt{\pi}}{2 \,.\, \dfrac{3}{2} \,.\, \dfrac{1}{2} \sqrt{\pi}} = \left(\frac{\pi a^3}{6} - \frac{2}{9} a^3\right).$$

Example 2.82 **Evaluate $I = \int_0^{\infty} \int_0^{\infty} e^{-(x^2 + y^2)}\, dx\, dy.$**

(MREC 2002)

Solution : The region of integration is the whole positive quadrant. Which is also a quadrant of circle r, where $r \to \infty$, So $I = \lim\limits_{r \to \infty} \int_{\theta=0}^{\pi/2} \int_0^{r} e^{-r^2}\, r\, d\theta\, dr.$

$$I = \int_0^{\pi/2} \lim_{r \to \infty} \left[\frac{e^{-r^2}}{2} \right]_0^r d\theta = \frac{1}{2} \int_0^{\pi/2} d\theta = \frac{1}{2} [\theta]_0^{\pi/2} = \frac{\pi}{4}.$$

Example 2.83 **Find the area of the cardiod $r = a(1 + \cos \theta)$, by double integration curve is symmetrical about the initial line $\theta = 0$ and is periodic. As $\cos(-\theta) = \cos \theta$, required area**

$$A = 2 \int_{\theta=0}^{\pi} \int_{r=0}^{a(1+\cos\theta)} r \, d\theta \, dr$$

$$= 2 \int_0^{\pi} \left[\frac{r^2}{2} \right]_0^{a(1+\cos\theta)} d\theta$$

$$= \int_0^{\pi} a^2 \left(2 \cos^2 \frac{\theta}{2} \right)^2 d\theta$$

Fig. 2.58

Solution : Let $\frac{\theta}{2} = t$

then $$= 8 \int_0^{\pi/2} a^2 \cos^4 t \, dt = 8a^2 \frac{\Gamma \frac{5}{2} \Gamma \frac{1}{2}}{2 \, \Gamma 3}$$

$$= \frac{a^2}{2} \frac{3}{2} \cdot \frac{1}{2} \pi = \frac{3\pi a^2}{8}.$$

Example 2.84 **Find by double integration the area of the region enclosed by the curves $y = \frac{1}{2}x$, $y = \frac{8a^3}{x^2 + 4a^2}$, the y-axis ($x = 0$) and the point ($2a, a$). The region of integration is OAB_2B_1O.**

Solution : If integration is to be done first w.r.t. x, the region has to be divided in two parts by the line $B_1A(y = a)$.

For the region OAB_1O

$$\text{Area} = A_1 = \int_{y=0}^{a} \int_{x=0}^{2y} dy \, dx$$

For region $B_1AB_2B_1$

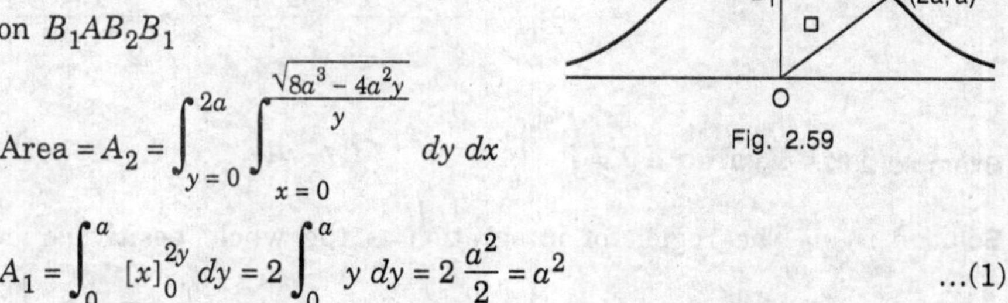

Fig. 2.59

$$\text{Area} = A_2 = \int_{y=0}^{2a} \int_{x=0}^{\sqrt{\frac{8a^3 - 4a^2 y}{y}}} dy \, dx$$

$$A_1 = \int_0^a [x]_0^{2y} \, dy = 2 \int_0^a y \, dy = 2 \frac{a^2}{2} = a^2 \qquad \text{...(1)}$$

$$A_2 = \int_{a}^{2a} 2a \sqrt{\frac{2a-y}{y}}\, dy$$

Let $y = 2a \sin^2 \theta$,

$$A_2 = \int_{\pi/4}^{\pi/2} 2a \frac{\cos\theta}{\sin\theta} 4a \sin\theta \cos\theta \, d\theta$$

$$= 8a^2 \int_{\pi/4}^{\pi/2} \frac{1+\cos 2\theta}{2} d\theta = 4a^2 \left[\theta + \frac{\sin 2\theta}{2}\right]_{\pi/4}^{\pi/2}$$

$$= 4a^2 \left[\left(\frac{\pi}{2} - \frac{\pi}{4}\right) + 0 - \frac{1}{2}\right] = \pi a^2 - 2a^2$$

Hence $\quad A = A_1 + A_2 = a^2(\pi - 1)$

Example 2.85 **Find by double integration the area lying inside the circle $r = a\sin\theta$ and outside the cardiod $r = a(1 - \cos\theta)$.**

Solution : The curve intersect where $r = a\sin\theta = a(1-\cos\theta)$

or $$\sin\theta + \cos\theta = 1$$

where $\theta = 0$ or $\theta = \frac{\pi}{2}$ at O and A.

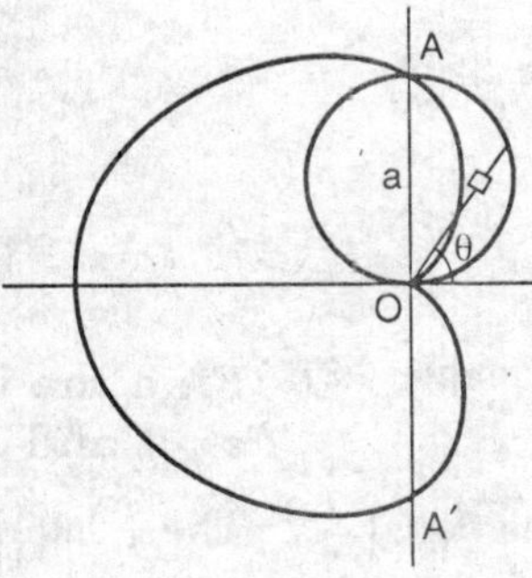

Fig. 2.60

Required area

$$A = \int_{\theta=0}^{\pi/2} \int_{r=a(1-\cos\theta)}^{r=a\sin\theta} r \, d\theta \, dr$$

$$= \int_0^{\pi/2} \left[\frac{r^2}{2}\right]_{a(1-\cos\theta)}^{a\sin\theta} d\theta$$

$$= \frac{1}{2} a^2 \int_0^{\pi/2} \{\sin^2\theta - (1-\cos\theta)^2\}\, d\theta$$

$$= \frac{a^2}{2} \int_0^{\pi/2} (2\cos\theta + \sin^2\theta - 1 - \cos^2\theta)\, d\theta$$

$$= a^2 \int_0^{\pi/2} (\cos\theta - \cos^2\theta)\, d\theta$$

$$= a^2 \int_0^{\pi/2} \left(\cos\theta - \frac{1+\cos 2\theta}{2}\right) d\theta$$

$$= a^2 \left[\sin\theta - \frac{\theta}{2} - \frac{\sin 2\theta}{4}\right]_0^{\pi/2} = a^2\left[1 - \frac{\pi}{4}\right].$$

Example 2.86 **Find the area included between the curve $r = a(\sin\theta + \cos\theta)$ and its asymptote.**

Solution : Changing to cartesian

$$x^2 + y^2 = a\left(\frac{x^2 + y^2}{x} + x\right),$$

($r = 0$ is not true so curve does not pass though (0, 0))

or $\quad x(x^2 + y^2) - a(2x^2 + y^2) = 0,\ x \neq 0,\ y \neq 0$

The line $x - a = 0$ is its asymptote.

$$2\int_0^{\pi/2}\int_{a\sec\theta}^{a(\sec\theta + \cos\theta)} r\,d\theta\,dr$$

$$= \frac{2a^2}{2}\int_0^{\pi/2} \{(\sec\theta + \cos\theta)^2 - \sec^2\theta\}\,d\theta$$

$$= a^2\int_0^{\pi/2} (\cos^2\theta + 2)\,d\theta$$

$$= a^2\int_0^{\pi/2} \left\{\frac{1 + \cos 2\theta}{2} + 2\right\} d\theta$$

$$= a^2\left[\frac{5\theta}{2} + \frac{\sin 2\theta}{4}\right]_0^{\pi/2} = \frac{5a^2\pi}{4}.$$

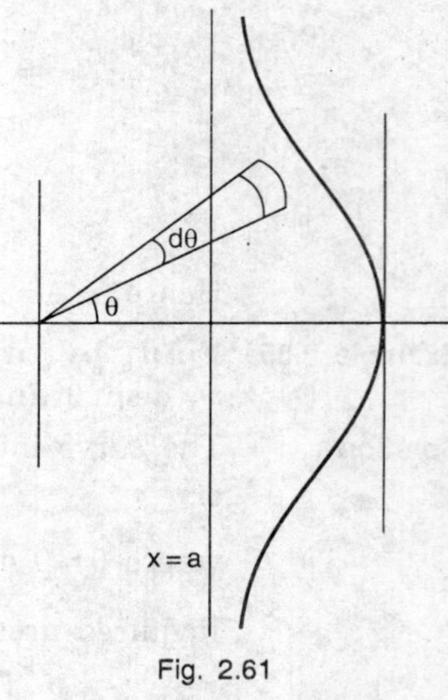

Fig. 2.61

Example 2.87 **Find the volume of the cylinder $x^2 + y^2 - ay = 0$, bounded by the planes $z = 0$ and $z = y$.**

Solution : The required volume $V = \iint [z]_0^y\,dx\,dy$ over the area of the circle $x^2 + y^2 - ay = 0$

$$= \int_{-a/2}^{a/2}\int_{y=\frac{a}{2}-\frac{\sqrt{a^2-4x^2}}{2}}^{\frac{a}{2}+\frac{a^2-4x^2}{2}} y\,dx\,dy$$

$$= 2\int_0^a\int_0^{\sqrt{ay-y^2}} y\,dy\,dx$$

$$= 2\iint_0^a y\sqrt{y(a-y)}\,dy$$

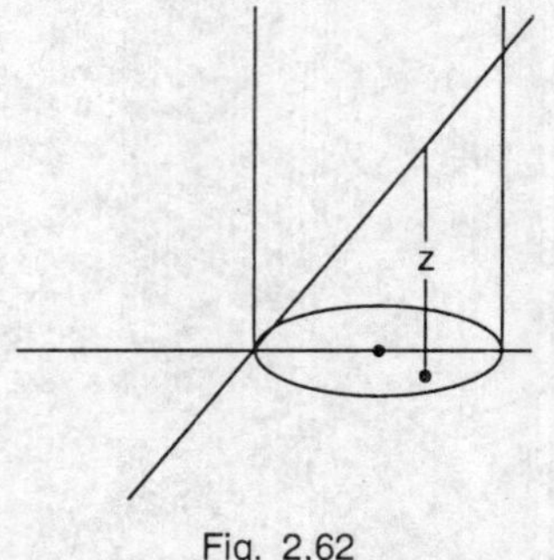

Fig. 2.62

Let $\quad y = a\sin^2\theta$

$$= 2a^3\int_0^{\pi/2} 2\sin^4\theta\cos^2\theta\,d\theta = \frac{\pi a^3}{8}$$

Example 2.88 **Evaluate the integrals**

(i) $\int_0^1 \int_x^{\sqrt{2x-x^2}} (x^2 + y^2)\, dx\, dy$,

(ii) $\int_0^9 \int_0^{\sqrt{a^2-x^2}} y^2 \sqrt{x^2+y^2}\, dx\, dy$ **by changing them in polar forms.**

Solution : (i) The region of integration is bounded by $y = x$, $y^2 = 2x - x^2$, $x = 0, x = 1$. It is the area bounded by $OABO$.

In polar co-ordinates it is defined by $r = 0$ to $r = 2 \cos \theta$; $\theta = \frac{\pi}{4}$ to $\theta = \frac{\pi}{2}$.

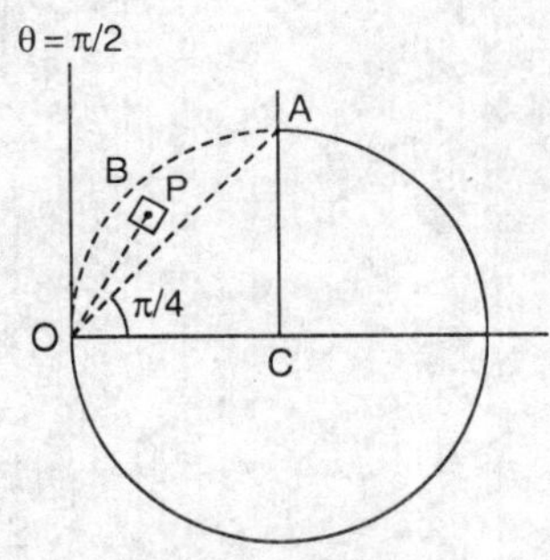

Fig. 2.62

So $$I = \int_{\pi/4}^{\pi/2} \int_0^{2\cos\theta} r^2\, r\, d\theta\, dr$$

$$= \int_{\pi/4}^{\pi/2} \left[\frac{r^4}{4}\right]_0^{2\cos\theta} d\theta$$

$$= \frac{16}{4} \int_{\pi/4}^{\pi/2} \cos^4 \theta\, d\theta$$

$$= 4 \int_{\pi/4}^{\pi/2} \left(\frac{1 + \cos 2\theta}{2}\right)^2 d\theta$$

$$= \int_{\pi/4}^{\pi/2} \left(1 + 2\cos 2\theta + \frac{1 + \cos 4\theta}{2}\right) d\theta$$

$$= \left[\frac{3}{2}\theta + \sin 2\theta + \frac{\sin 4\theta}{8}\right]_{\pi/4}^{\pi/2}$$

$$= \left[\frac{3}{2}\left(\frac{\pi}{2} - \frac{\pi}{4}\right) + (0 - 1) + \frac{1}{8}(0 - 0)\right]$$

$$= \left(\frac{3\pi}{8} - 1\right)$$

(ii) The region of integration is bounded by $y = 0$, $y^2 = a^2 - x^2$, $x = 0$, $x = a$.
It is the first quadrant of the circle defined by $0 \le r \le a$, $0 \le \theta \le \frac{\pi}{2}$.

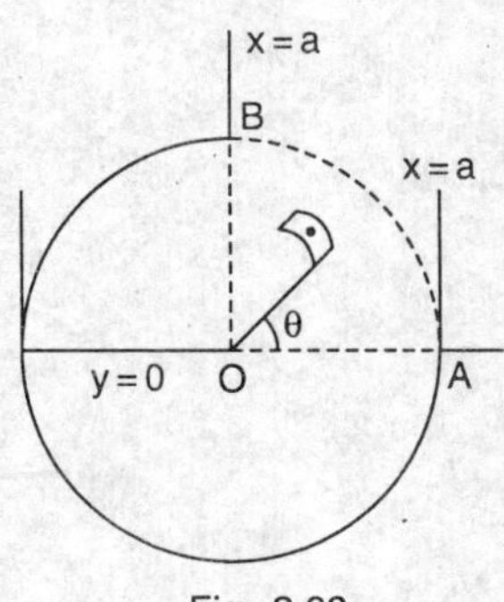

Fig. 2.63

So $$I = \int_0^{\pi/2} \int_0^a r^3 \sin^3 \theta\, r\, d\theta\, dr$$

$$= \int_0^{\pi/2} \sin^2\theta \left[\frac{r^5}{5}\right]_0^a d\theta$$

$$= \frac{a^5}{5}\int_0^{\pi/2} \frac{1-\cos 2\theta}{2}\, d\theta = \frac{a^5}{10}\left[\theta - \frac{\sin 2\theta}{2}\right]_0^{\pi/2} = \frac{\pi a^5}{20} - 0$$

Example 2.89 **Evaluate $I = \iint_R y\,dx\,dy$ over the region R bounded by the parabolas $y^2 = 4ax$ and $x^2 = 4ay$.**

Solution : Region R is the area $OLMNO$.

The two curves intersects at $O(0, 0)$ and $M(4a, 4a)$

$$I = \int_{x=0}^{4a}\int_{y_1 = \frac{x^2}{4a}}^{y_2 = \sqrt{4ax}} y\,dx\,dy$$

$$= \int_0^{4a} \left[\frac{y^2}{2}\right]_{\frac{x^2}{4a}}^{\sqrt{4ax}} dx$$

$$= \frac{1}{2}\int_0^{4a} \left(4ax - \frac{x^4}{16a^2}\right) dx$$

$$= \frac{1}{2}\left[2ax^2 - \frac{x^5}{80a^2}\right]_0^{4a}$$

$$= \frac{1}{2}\left[32a^3 - \frac{2^{10}}{80}a^3\right] = \frac{48}{5}a^3.$$

Fig. 2.64

Example 2.90 **Evaluate $I = \iint \frac{r\,d\theta\,dr}{\sqrt{a^2 + r^2}}$ over one loop of the lemniscate $r^2 = a^2\cos 2\theta$.**

(Kuvempu 96; Madurai M.E. 90)

Solution : Taking the loop $ABACO$ over which integration is done the limits for r are $r = 0$ to $r = a\sqrt{\cos 2\theta}$ and for θ, they are from $-\frac{\pi}{4}$ to $+\frac{\pi}{4}$.

So $$I = \int_{-\pi/4}^{\pi/4}\int_0^{a\sqrt{\cos 2\theta}} \frac{r}{\sqrt{a^2 + r^2}}\, d\theta\, dr$$

Fig. 2.65

$$= \int_{-\pi/4}^{\pi/4} \left[\sqrt{a^2 + r^2} \right]_0^{a\sqrt{\cos 2\theta}} d\theta$$

$$= a \int_{-\pi/4}^{\pi/4} \{\sqrt{1 + \cos 2\theta} - 1\}\, d\theta$$

$$= \sqrt{2}a \int_{-\pi/4}^{\pi/4} \cos\theta\, d\theta - a\, [\theta]_{-\pi/4}^{\pi/4}$$

$$= \sqrt{2}a \,.\, 2\, [\sin\theta]_0^{\pi/4} - \frac{2a\pi}{4} = 2a\left(1 - \frac{\pi}{4}\right).$$

Example 2.91 **Evaluate** $I = \int_0^1 \int_0^{\sqrt{1+x^2}} \frac{dx\, dy}{1 + x^2 + y^2}$

(Andhra 94; Kerala 90)

Solution :

$$I = \int_0^1 \frac{1}{\sqrt{1+x^2}} \left[\tan^{-1} \frac{y}{\sqrt{1+x^2}} \right]_{y=0}^{\sqrt{1+x^2}} dx$$

$$= \int_0^1 \frac{1}{\sqrt{1+x^2}} [\tan^{-1} 1 - 0]\, dx$$

$$= \frac{\pi}{4} \left[\frac{x\sqrt{1+x^2}}{2} + \log x + \sqrt{1+x^2} \right]_0^1 = \frac{\pi}{4} \log(\sqrt{2} + 1).$$

Example 2.92 **Evaluate** $I = \iint (x+y)^2\, dx\, dy$ **over the area bounded by the ellipse** $\frac{x^2}{a^2} + \frac{y^2}{b^2} = 1$

(Andhra 1994; Madras 93)

Solution : We write $I = I_1 + I_2$

where $I_1 = \iint (x^2 + y^2)\, dx\, dy$ and $I_2 = 2 \iint xy\, dx\, dy$.

The function $f_1(x, y) = x^2 + y^2$ is an even function. So its integral = 4 × integral over first qaudrant. The function $f_2(x, y) = xy$, satisfy $f_2(x, y) = xy$, satisfy

$$f_2(-x, y) = -xy, \quad f_2(x, -y) = -xy, \quad f_2(-x, -y) = xy.$$

So the value of the integral will be + for first and 3rd and – (minus) for second and 4th quadrant and the whole value will be zero so

$$I = 4 \iint (x^2 + y^2)\, dx\, dy$$

for $y = 0$ to $y = b\sqrt{1 - \frac{x^2}{a^2}}$ and $x = 0$ to a.

$$= 4 \int_0^a \int_0^{b\sqrt{1 - \frac{x^2}{a^2}}} (x^2 + y^2)\, dx\, dy.$$

$$= x \int_0^1 \left[x^2 y + \frac{y^3}{3} \right]_0^{b\sqrt{1-\frac{x^2}{a^2}}} dx$$

$$= 4 \int_0^1 b\sqrt{1 - \frac{x^2}{a^2}} \left[x^2 + \frac{b^2}{2}\left(1 - \frac{x^2}{a^3}\right) \right] dx$$

Let $x = a \sin \theta$

$$I = 4b \int_0^{\pi/2} \cos\theta \left[\frac{b^2}{3} + \left(1 - \frac{b^2}{3a^2}\right) a^2 \sin^2\theta \right] a \cos\theta \, d\theta$$

$$= 4ab \left[\frac{\frac{3}{2}!\,\frac{1}{2}!}{2\,2!} + \frac{b^2}{3} + \frac{3a^2 - b^2}{3} \frac{\frac{3}{2}!\,\frac{3}{2}!}{2\,3!} \right]$$

$$= 4ab \left[\frac{1}{2}\frac{\pi}{2}\frac{b^2}{3} + \frac{3a^2 - b^2}{3} \cdot \frac{1}{2} \cdot \frac{1}{2} \frac{\pi}{2 \cdot 2} \right]$$

$$= \frac{\pi ab}{3} \left[b^2 + \frac{3a^2 - b^2}{4} \right] = \frac{\pi ab}{4} (a^2 + b^2).$$

Note : If the integrand is positive in one part and negative in the other, if the value comes out to be $U - V$, on integration, the value of changing the order will come out to be $V - U$.

Example 2.93 **Show that the two integrals**

$$I_1 = \int_{x=0}^{1} \int_{y=0}^{1} \frac{x - y}{(x + y)^3} \, dx \, dy$$

and $I_2 = \int_{y=0}^{1} \int_{x=0}^{1} \frac{x - y}{(x + y)^3} \, dy \, dx$ **obtained by changing their order of integration differ in sign.**

Solution :

$$I_1 = \int_0^1 \int_0^1 2x\,(x + y)^{-3} \, dx \, dy + \int_0^1 \int_0^1 \frac{-(x + y)}{(x + y)^3} \, dx \, dy$$

$$= \int_0^1 2x \left[\frac{(x + y)^{-2}}{-2} \right]_{y=0}^{1} dx - \int_0^1 \left[\frac{(x + y)^{-1}}{-1} \right]_{y=0}^{1} dx$$

$$= \int_0^1 \left\{ -x\left(\frac{1}{(x + 1)^2} - \frac{1}{x^2} \right) + \left(\frac{1}{x + 1} - \frac{1}{x} \right) \right\} dx$$

$$= \int_0^1 + \frac{1}{(1 + x)^2} \, dx = \left[-\frac{1}{1 + x} \right]_0^1 = \left(\frac{1}{2} - 1 \right) = + \frac{1}{2} \qquad \text{...(1)}$$

$$I_2 = \int_{y=0}^{1} \int_{x=0}^{1} \frac{x+y-2y}{(x+y)^3}\, dy\, dx$$

$$= \int_{y=0}^{1} \left[\frac{(x+y)^{-1}}{-1} - \frac{2y(x+y)^{-2}}{-2} \right]_{x=0}^{1} dy$$

$$= \int_0^1 \left\{ -\left(\frac{1}{1+y} - \frac{1}{y} \right) + y \left(\frac{1}{(1+y)^2} - \frac{1}{y^2} \right) \right\} dy \qquad \ldots(2)$$

Thus, the two values as given by (1) and (2) differ in signs

PROBLEM SET 2.11

1. $\int_{x=0}^{3} \int_{y=1}^{2} xy(x+y+1)\, dx\, dy$

Ans. $\frac{123}{4}$

2. $\int_{x=0}^{5} \int_{y=0}^{x^2} x(x^2+y^2)\, dx\, dy.$

Ans. $5^6 \times \frac{29}{24} \approx 18880.2.$

3. $\int_{x=3}^{4} \int_{y=1}^{2} (e^x + xy)\, dx\, dy.$

Ans. $e^4 - e^3 + \frac{21}{4}$

4. (i) $\int_0^1 \int_0^1 \frac{dx\, dy}{\sqrt{1-x^2}\,\sqrt{1-y^2}}$

Ans. $\frac{\pi^2}{4}$

(ii) $\int_0^3 \int_0^1 (x^2 + 3y^2) dx\, dy$

Ans. 12

(iii) $\int_1^a \int_1^b \frac{dx\, dy}{xy}$

Ans. $\log a \log b$

(iv) $\int_0^1 dx \int_0^x e^{y/x}\, dy.$

Ans. $\frac{e-1}{2}$

(v) $\int_0^{2a} \int_0^{\sqrt{2ax - x^2}} (x^2 + y^2)\, dx\, dy.$

Ans. $\frac{3\pi a^4}{4}$.

5. Using double integration, find the area of the ellipse $\frac{x^2}{a^2} + \frac{y^2}{b^2} = 1.$

Ans. πab.

6. $r^2 = a^2 \cos^2 \theta + b^2 \sin^2 \theta.$

Ans. $\frac{\pi}{2}(a^2 + b^2)$

7. Evaluate $\iint xy\, dx\, dy$ over the region of the positive quadrant bounded by $x + y = 1$.

Ans. $\frac{1}{24}$.

8. Evaluate $\iint xy\, dx\, dy$ over the positive quadrant of the circle $x^2 + y^2 = a^2$.

Ans. $\frac{a^4}{8}$.

(Madras 96; AMIE 97)

9. Evaluate $\iint xy(x + y)\, dx\, dy$ over the area bounded by the curves $y^2 = x$ and $y = x$.

Ans. $\frac{3}{56}$.

(Hamirpur 94S, Karnataka 88, Poona 87 S)

10. Evaluate $\iint (x + y)^2\, dx\, dy$ over the area of the ellipse $\frac{x^2}{a^2} + \frac{y^2}{b^2} = 1.$

Ans. $\frac{\pi ab}{4}(a^2 + b^2)$.

(Raj. 87; Andhra 94; Madras 93)

11. Evaluate $\iint r^2\, r\, d\theta\, dr$ over the region between $r = 2 \cos \theta$ and $r = 4 \cos \theta$

Ans. $\frac{45\pi}{2}$

12. Evaluate $\iint_A xy\, dx\, dy$, where A is the region bounded by x-axis, the ordinate $x = 2a$ and the curve $x^2 = 4ay$.

(Gulbarga 99 S; Marathwada 93)

Ans. $\frac{a^4}{3}$.

13. Evaluate $\int_0^a \int_{x=0}^{\sqrt{a^2 - y^2}} \sqrt{a^2 - x^2 - y^2}\, dy\, dx$

Ans. $\frac{\pi a^3}{6}$

14. Evaluate $\iint_R y\,dx\,dy$ over the region R, bounded by the parabolas $y^2 = 4x$, $x^2 = 4y$.

Ans. $\frac{48}{5}$.

15. Evaluate $\int e^{2x+3y}\,dx\,dy$ over the region bounded by $x = 0$, $y = 0$, $x + y = 1$.

Ans. $\frac{1}{6}(e-1)^2(2e+1)$

16. Evaluate the following integrals

(i) $\int_0^{\pi}\int_0^{a\sin\theta} r\,d\theta\,dr$

Ans. $\frac{\pi a^2}{4}$.

(ii) $\int_0^{\pi/2}\int_{a(1+\cos 0)}^{a} r\,d\theta\,dr$

Ans. $a^2\left(1+\frac{\pi}{g}\right)$.

(iii) $\int_0^{\pi}\int_0^{a(1+\cos\theta)} r^2\cos\theta\,d\theta\,dr$

Ans. $\frac{5}{8}\pi a^3$.

2.18 CHANGE OF ORDER OF INTEGRATION

In evaluation of the double integral $I = \iint_A f(x, y)\,dx\,dy$ over a region A, the order of integration is immaterial if limits for x and y be simply constants and the region A is a rectangle, bounded by $x = a$, $x = b$ and $y = c$, $y = d$. But if the limits are not simply constants, on changing the order of integration in a double integral, care has to be taken to cover the whole region A. The region A may be covered fully in one order expressing $\iint_A f(x, y)\,dA$ by a single integral I, but on changing the order, in the changed order, it may have to be sometimes broken in regions $A_1, A_2, \ldots$ over which the integrals of the given function $f(x, y)$ may be I_1, I_2, etc. Thus, we would then have $I = I_1 + I_2 + \ldots$, on changing the order.

For example, region bounded by $y = x^2$, $y = 2 - x$, $x = 0$, $x = 1$. The region is covered by only one integral if integration be done first with respect to y and then w.r.t x.

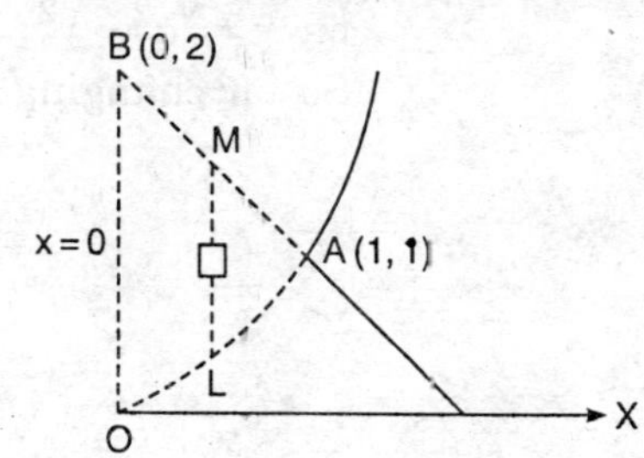

Fig. 2.66

$$I = \int_{x=0}^{1} \int_{y=y_L=x^2}^{y_M=2-x} f(x, y)\, dx\, dy$$

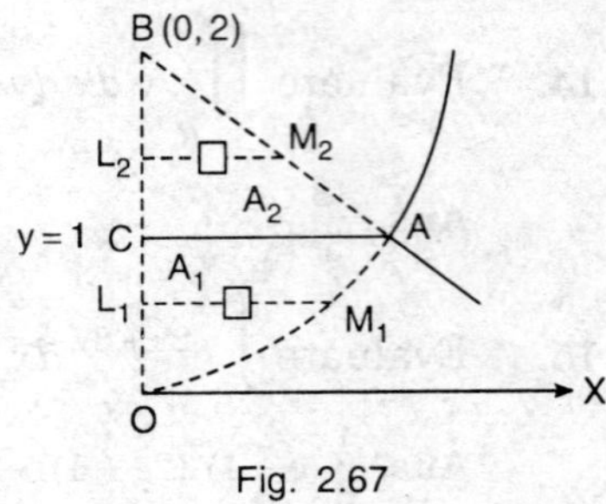

Fig. 2.67

But on doing integration first w.r.t. x. The regions $OACO$ and $CABC$ are of different nature.

In region A_1 and A_2 both L_1 and L_2 lie on $x = 0$. But M_1 in region A_1 lies on the parabola, while it lies on the line AB in A_2. So the integral I will be the sum of the two integrals I_1 on $OABO$ and I_2 on $CABC$ given by

$$I_1 = \int_{y=0}^{1} \int_{x=9}^{\sqrt{y}} f(x, y)\, dy\, dx$$

and $$I_2 = \int_{y=1}^{2} \int_{x=0}^{2-y} f(x, y)\, dy\, dx \text{ and } I = I_1 + I_2.$$

Example 2.94 **Change the order of integration in**

$$I = \int_{x=0}^{a} \int_{y=0}^{\sqrt{a^2 - x^2}} f(x, y)\, dx\, dy.$$

Solution : The boundary of the region of integration are given by $y = 0$, $y^2 = a^2 - x^2$, $x = 0$, $x = a$ region is as shown in figure.

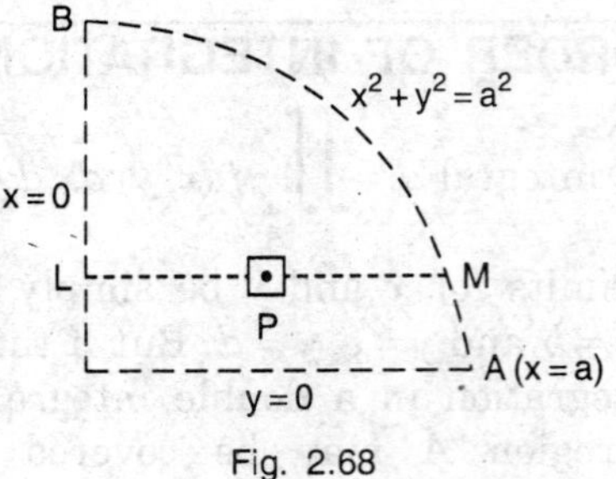

Fig. 2.68

On changing order of integration, integration has to be done first w.r.t. x. The limits for x will be from x_L to x_M. L always lies on $x = 0$, and M always on $x = \sqrt{a^2 - y^2}$.

So on changing the order there will again be only one integral given as

$$I = \int_{y=0}^{a} \int_{x=0}^{\sqrt{a^2 - y^2}} f(x, y)\, dx\, dy.$$

Example 2.95 **Change the order of integration in**

$$I = \int_0^{2a} \int_{y=x^2/4a}^{3a-x} f(x, y)\, dx\, dy.$$

Solution : The boundaries of the region of integration are given by

$x = 0,\ x = 2a,\ y = \dfrac{x^2}{4a^2},\ y = 3a - x$. The region is $OABO$ as shown in the figure.

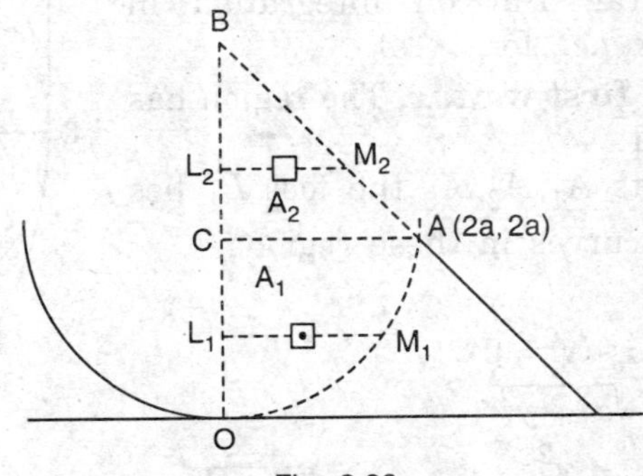

Fig. 2.69

On chainging the order, integration has first to be done with respet to x, for which the limit will be from L_1 to M_1 in region $OACO = A_1$ and L_2 to M_2 in $CABC = A_2$

$$A = A_1 + A_2$$

At both L_1 and L_2, $x = 0$, But at M_1 $x = \sqrt{4ay}$, while $M_2 = 3a - y$.

Hence $I = I_1 + I_2$

$$I_1 = \int_{y=0}^{a} \int_{x=0}^{\sqrt{4ay}} f(x, y)\, dy\, dx$$

$$I_2 = \int_{y=a}^{2a} \int_{x=0}^{3a-y} f(x, y)\, dy\, dx$$

Example 2.96 **Change the order of integration in** $\displaystyle\int_{x=0}^{2a} \int_{y=\sqrt{2ax-x^2}}^{\sqrt{2ax}} f(x, y)\, dx\, dy$ **integration done first inhibit** y.

Solution : The region of integration is bounded by $x = 0,\ x = 2a,\ y^2 + x^2 - 2ax = 0,\ y^2 = 2ax$.
It consists of the region shown graphically by $OEABCDO$. O is the points $(0, 0)$, AB the line $x = 2a$, OEA the upper half of the circle $(x - a)^{2^{2'}} + y^2 = a^2$, and ODC the arc of the parabola $y^2 = 2ax$.

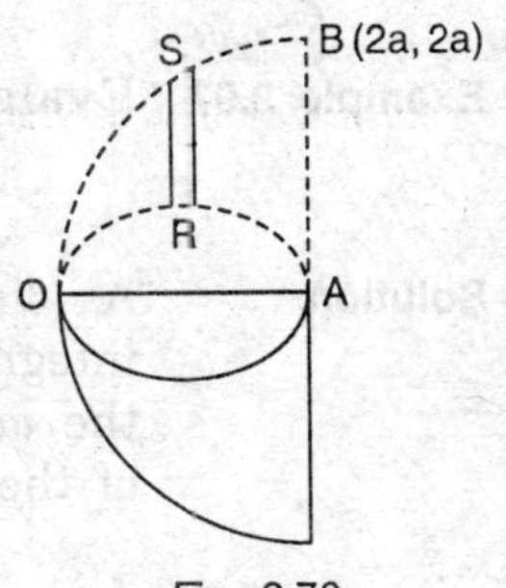

Fig. 2.70

Integration first w.r.t *y*.

$$\int_{x=0}^{x_A=2a}\int_{y_R=\sqrt{2ax-x^2}}^{\sqrt{2ax}} f(x, y)\, dx\, dy.$$

The foot y_R always lines on circle and top on the parabola so it is represented by a single integral. But on integration in reverse order *i.e.*, for,

Integration first w.r.t. *x*. The region has to be divided in three parts A_1, A_2, A_3 the foot L_1 lies on different curves in these regions.

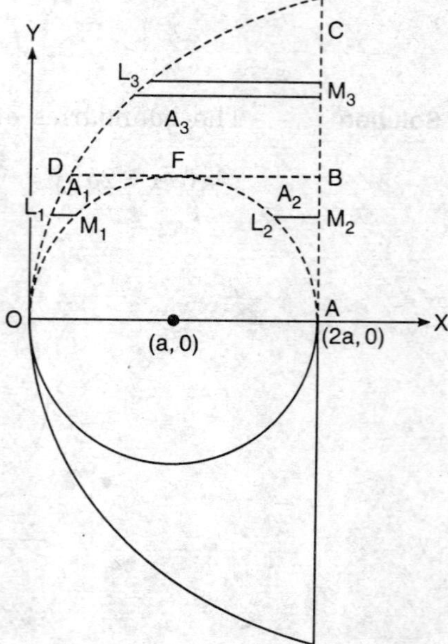

Fig. 2.71

From

$$x^2 - 2ax + y^2 = 0$$
$$x = a \pm \sqrt{a^2 - y^2}$$

(i) on A_1. $x_{L_1} = \dfrac{y^2}{2a}$, $x_{M_1} = a - \sqrt{a^2 - y^2}$.

y varies from 0 to *a*. So I_1 is given by

$$I_1 = \int_{y=0}^{a}\int_{x=y^2/2a}^{a-\sqrt{a^2-y^2}} f(x, y)\, dy\, dx.$$

(ii) In A_2. $x_{L_2} = a + \sqrt{a^2 - y^2}$, $x_{M_2} = 2a$,

$y_1 = 0,\ y_2 = a$

So $$I_2 = \int_0^a \int_{a+\sqrt{a^2-y^2}}^{2a} f(x, y)\, dy\, dx$$

(iii) In A_3. $x_{L_3} = \dfrac{y^2}{2a}$, $x_{M_3} = 2a, y_1 = a,\ y_2 = 2a$ etc

So $$I_3 = \int_a^{2a}\int_{y^2/2a}^{2a} f(x, y)\, dx\, dy$$

and $$I = I_1 + I_2 + I_3.$$

Example 2.97 **Evalaute** $I = \displaystyle\int_0^1 \int_{e^x}^{e} \frac{1}{\log y}\, dx\, dy.$

(MREC 2002)

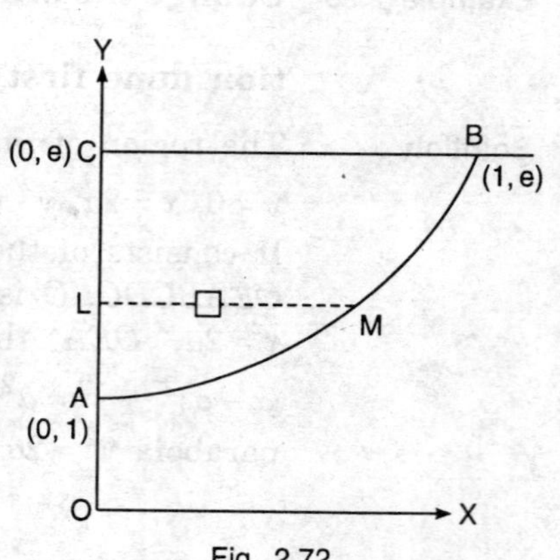

Fig. 2.72

Solution : As integration is not possible when integrated first w.r.t. *y*, so we first change the order of integration. The boundaries of the region of integration are

$$x = 0,\ x = 1,\ y = e^x,\ y = e$$

We have $x_L = 0,\ x_M = \log y$

$y_A = 1,\ y_{BC} = e$

$$I = \int_1^e \int_0^{\log y} \frac{1}{\log y}\, dy\, dx$$

$$= \int_1^e \frac{1}{\log y} [x]_0^{\log y}\, dy$$

$$= \int_0^e dy = [y]_1^e = e - 1$$

Example 2.98 **Evaluate**

$$I = \int_0^\infty \int_x^\infty \frac{1}{y} e^{-y}\, dx\, dy.$$

(Raj. 93; Nagpur 97, Mysore 97 S)

Solution : The region of integration is defined by

$$x = 0,\ x = \infty,\ y = x,\ y = \infty$$

Integration of $\frac{e^{-y}}{y}$ w.r.t. y is not possible so order should be changed.

On changing the order

$$I = \int_{y=0}^\infty \int_{x=0}^{y} \frac{1}{y} e^{-y}\, dy\, dx$$

$$= \int_0^\infty \frac{1}{y} e^{-y} [x]_0^y\, dy = \int_0^\infty e^{-y}$$

$$= \left[\frac{e^{-y}}{1}\right]_0^\infty = [-e^{-\infty} + 1] = 1.$$

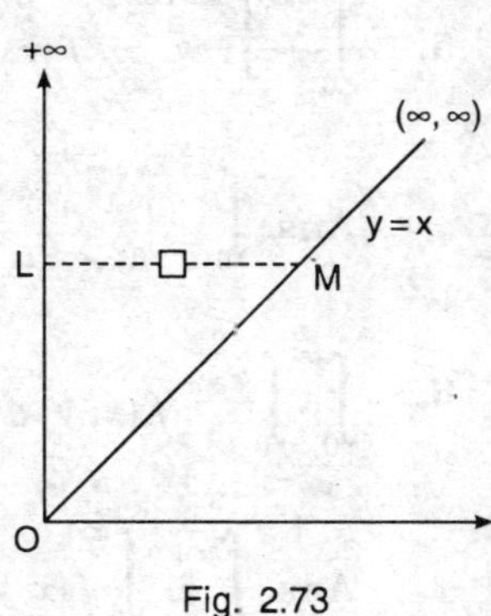

Fig. 2.73

PROBLEM SET 2.12

Change the order of integration in the following integrals :

1. $\int_0^2 \int_0^y f(x, y)\, dy\, dx$

Ans. $\int_0^a \int_x^a f(x, y)\, dx\, dy$

2. $\int_0^4 \int_x^{2\sqrt{x}} f(x, y)\, dx\, dy$

Ans. $\int_0^4 \int_{y^2/4}^{y} f(x, y)\, dy\, dx.$

3. $\int_0^a \int_x^{2x} f(x, y)\, dx\, dy$

Ans. $\int_0^a \int_{y/2}^{y} f(x, y)\, dy\, dx + \int_a^{2a} \int_{y/2}^{a} f(x, y)\, dy\, dx$

4. $\int_\varepsilon^a \int_x^{a^2/x} f(x, y)\, dx\, dy,\ \varepsilon \to 0$

Ans. $\int_0^a \int_0^{y} f(x, y)\, dy\, dx + \int_a^{\infty} \int_0^{a^2/y} f(x, y)\, dy\, dx$

5. $\int_0^{\pi/2} \int_0^{2a\cos\theta} f(r, \theta)\, r\, d\theta\, dr$

Ans. $\int_0^{2a} \int_0^{\cos^{-1}\frac{r}{2a}} f(r, \theta)\, r\, dr\, d\theta$

6. $\int_0^a \int_0^{\frac{b}{b+x}} f(x, y)\, dx\, dy$

Ans. $\int_0^{\frac{b}{a+b}} \int_0^{a} f(x, y)\, dy\, dx + \int_{\frac{b}{a+b}}^{1} \int_0^{\frac{b(1-y)}{y}} f(x, y)\, dy\, dx$

Evaluate the following integrals by changing their order :

7. $\int_0^1 \int_x^{\sqrt{2-x^2}} \frac{x}{\sqrt{x^2+y^2}}\, dx\, dy = \int_0^1 \int_0^{y} \frac{x}{\sqrt{x^2+y^2}}\, dy\, dx + \int_1^{\sqrt{2}} \int_0^{\sqrt{2-y^2}} \frac{x}{\sqrt{x^2+y^2}}\, dy\, dx$

$$= \left(\frac{\sqrt{2}}{2} - \frac{1}{2}\right) + \left(\frac{3}{2} - \sqrt{2}\right) = \left(1 - \frac{1}{\sqrt{2}}\right)$$

8. Evaluate $\int_0^1 \int_0^{\sqrt{2-x^2}} \frac{x\, dx\, dy}{\sqrt{x^2+y^2}}$ ***(MREC 2001)***

Ans. $\frac{3}{2} - \frac{1}{\sqrt{2}} + \log\sqrt{2} + 1$

9. Change the order of integration and evaluate

$$\int_0^1 \int_{x^2}^{2-x} xy\, dx\, dy = \int_0^1 \int_0^{\sqrt{y}} xy\, dy\, dx + \int_1^2 \int_0^{2-y} xy\, dy\, dx$$

(Madras 93, Andhra 91, Kerla 90, Ranchi 90, Mysore 94)

Ans. 3/8 .

10. Change the order in $I = \int_0^{4a} \int_{x^2/4a}^{2\sqrt{ax}} dy\, dx$ and hence evalute. *(Mangalore 97)*

Ans. $\frac{16}{3} a^2$.

11. Evaluate $\int_0^1 \int_x^{2-x} \frac{x}{y}\, dx\, dy$, by changing its order.

Ans. 2 log 2 – 1

12. Evaluate $\int_0^a \int_y^a \frac{x\, dx\, dy}{x^2 + y^2}$ by changing its order evalaute.

Ans. $\frac{\pi a}{4}$

13. Evaluate $\int_0^a \int_{\sqrt{ax}}^a \frac{y^2\, dx\, dy}{\sqrt{y^4 - a^2 x^2}}$. *(Delhi 97)*

Ans. $\frac{\pi a^2}{6}$

14. Evaluate $\int_0^\infty \int_0^x x e^{-x^2/y}\, dx\, dy$. *(Bhopal 91, Gorakhpur 91)*

Ans. 1/2

15. Evaluate by changing order $\int_0^3 \int_1^{\sqrt{4-y}} (x + y)\, dy\, dx$. *(Kerala 95, Mysore 94 S, Andhra 92)*

Ans. $\frac{241}{60}$.

16. Evaluate $\int_0^{a/\sqrt{2}} \int_y^{\sqrt{a^2 - y^2}} \log (x^2 + y^2)\, dy\, dx$.

Ans. $\frac{\pi a^2}{4}\left(\log a - \frac{1}{2}\right)$

17. $\int_0^a \int_{x/a}^{\sqrt{x/a}} (x^2 + y^2)\, dx\, dy$.

Ans. $\left(\frac{a^3}{28} + \frac{a}{20}\right)$

2.19 TRIPLE INTEGRALS

Let $f(x, y, z)$ be a continuous functions of x, y, z, defined over a volume V in a three dimensional space. Let the volume V be divided into elementary volumes $\delta V_1, \delta V_2 \dots \delta V_n$ each tending to zero.

If (x_r, y_r, z_r) be an arbitrary point in δV_r, then the limit of the sum $\sum_{r=1}^{n} f(x_y, y_r, z_r)\, \delta V_r$, as $n \to \infty$, if it exists is called the triple integral of f over V and is denoted by $\iint_V \int f(x, y, z)\, dV$.

Dividing V by drawing planes parallel to the co-ordinate planes, $dV = dx\, dy\, dz$ and the triple integral becomes $\iint_V \int f(x, y, z)\, dx\, dy\, dz$.

The order of integration is taken t be in such a way that the whole volume V is covered. If V be a rectangular parallelopiped bounded by $x = a_1$, $x = a_2$; $y = b_1$, $y = b_2$; $z = c_1$, $z = c_2$, the triple integral would then be

$\int_{a_1}^{a_2} \int_{b_1}^{b_2} \int_{c_1}^{c_2} f(x, y, z)\, dx\, dy\, dz$, where the order of integration would be immaterial.

2.20 VOLUME BY TRIPLE INTEGRALS

If $f(x, y, z) = 1$, then $\iint_V \int dx\, dy\, dz$ gives the volume V.

Example 2.101 **Find the volume of the region bounded by $y = x^2$, $x = y^2$, and the plane $z = 0$, $z = 3$.**

Solution :

$$V = \int_0^1 \int_{y_1 = x^2}^{\sqrt{x}} \int_{z=0}^{3} dx\, dy\, dz$$

$$= \int_0^1 \int_{x^2}^{\sqrt{x}} [dz]_0^3\, dx\, dy = 3 \int_0^1 [y]_{x^2}^{\sqrt{x}}\, dx$$

$$= 3 \int_0^1 (\sqrt{x} - x^2)\, dx = 3 \left[\frac{2}{3} x^{3/2} - \frac{x^3}{3} \right]_0^1$$

$$= 3 \left(\frac{2-1}{3} \right) = 1$$

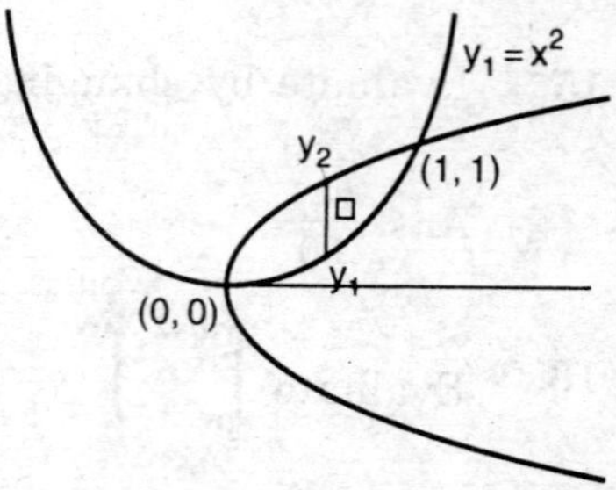

Fig. 2.74

Example 2.100 **Find the volume bounded by the co-ordinate planes $x = 0$, $y = 0$, $z = 0$ and the plane $\frac{x}{a} + \frac{y}{b} + \frac{z}{c} = 1$.** *(Raj 98)*

Solution :

$$V = \int_{x=0}^{a} \int_{y=0}^{b(1-\frac{x}{a})} \int_{z=0}^{c(1-\frac{x}{a}-\frac{y}{b})} dx\, dy\, dz$$

$$= \int_0^a \int_0^{b(1-\frac{x}{a})} [z]_0^{c(1-\frac{x}{a}-\frac{y}{b})} dx\, dy$$

$$= \int_0^a \int_0^{b(1-\frac{x}{a})} c\left(1-\frac{x}{a}-\frac{y}{b}\right) dx\, dy$$

$$= \int_0^a c\left[y\left(1-\frac{x}{a}\right)-\frac{y^2}{2b}\right]_0^{b(1-\frac{x}{a})} dx$$

$$= \int_0^a c\left[b\left(1-\frac{x}{a}\right)^2 - \frac{b^2}{2b}\left(1-\frac{x}{a}\right)^2\right] dy$$

$$= c\,\frac{b}{2}\left[\frac{a\left(1-\frac{x}{a}\right)^3}{-3}\right]_0^a$$

$$= \frac{abc}{6}[0+1] = \frac{abc}{6}.$$

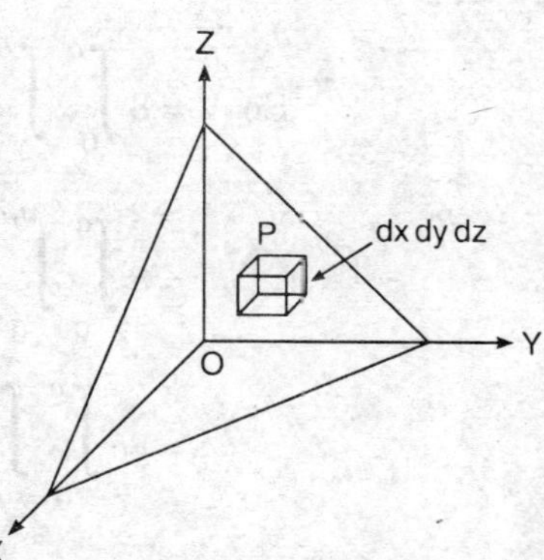

Fig. 2.75

Aliter : Let $x = au,\ y = bv,\ z = cw$ so $dx\, dy\, dz = abc\ du\, dv\, dw$ subject to $u = 0,\ v = 0,\ w = 0,\ u + v + w + \leq 1$

(by Dirichlet's intergral $\int u^{l-1}\, u^{m-1}\, w^{n-1}\, du\, dv\, dw = \frac{\Gamma l\, \Gamma m\, \Gamma n}{\Gamma 1 + l + m + n}$).

Here $l = m = n = 1$.

So $$I = abc\,\frac{\Gamma 1\, \Gamma 1\, \Gamma 1}{\Gamma_4} = \frac{abc}{3.2.1} = \frac{abc}{6}$$

Example 2.101 Find the volume of the ellipsoid

$$\frac{x^2}{a^2} + \frac{y^2}{b^2} + \frac{z^2}{c^2} = 1.$$

Solution : The volume is symmetrical in all the 8 octants.

So $$V = 8 \iiint_V dx\, dy\, dz$$

V is the volume in the positive octant, bound by $x = 0,\ y = 0,\ z = 0$ and the surface

$$\frac{x^2}{a^2} + \frac{y^2}{b^2} + \frac{z^2}{c^2} = 1$$

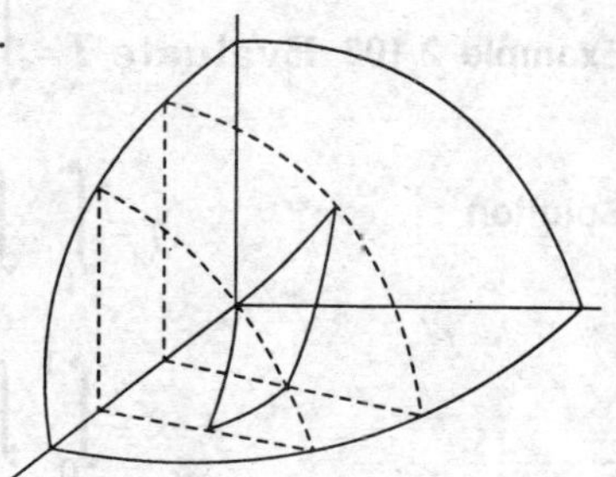
Fig. 2.76

$$\text{So } V = 8\int_0^a \int_{y=0}^{b\sqrt{1-x^2/a^2}} \int_{z=0}^{c\sqrt{1-x^2/a^2-y^2/b^2}} dx\,dy\,dz$$

$$= 8\int_0^a \int_0^{b\sqrt{1-x^2/a^2}} [z]_0^{c\sqrt{1-\frac{x^2}{a^2}-\frac{y^2}{b^2}}} dx\,dy$$

$$= 8c\int_0^a \int_0^{b\sqrt{1-x^2/a^2}} \frac{1}{b}\sqrt{b^2\left(1-\frac{x^2}{a^2}\right)-y^2}\,dx\,dy$$

$$= \frac{8c}{b}\int_0^a \left[\frac{y\sqrt{b^2\left(1-\frac{x^2}{a^2}\right)-y^2}}{2} + \frac{b^2\left(1-\frac{x^2}{a^2}\right)}{2}\sin^{-1}\frac{y}{b\sqrt{1-\frac{x^2}{a^2}}}\right]_0^{b\sqrt{1-\frac{x^2}{a^2}}}$$

$$= \frac{8c}{b}\int_0^a \left\{0 + \frac{b^2\left(1-\frac{x^2}{a^2}\right)}{2}\times\frac{\pi}{2}\right\} dx$$

$$= \frac{2\pi c}{b}\,b^2\left\{x - \frac{x^3}{3a^2}\right\}_0^a = 2\pi bc \times \frac{2}{3}a$$

$$= \frac{4\pi abc}{3}.$$

or let $x^2 = a^2u,\ y^2 = b^2v,\ z^2 = c^2w$, then

$$v = abc\,.\,8\int \frac{abc}{8}\,u^{-1/2}\,v^{-1/2}\,w^{-1/2}\,du\,dv\,dw$$

$$u \ge 0, v \ge 0, w \ge 0, u + v + w \le 1\,\frac{\Gamma\frac{1}{2}\Gamma\frac{1}{2}\Gamma\frac{1}{2}}{\Gamma 1 + 3/2} = abc\,\frac{4}{3}\pi$$

Example 2.102 **Evaluate** $\boldsymbol{I = \int_0^1 \int_0^{\sqrt{1-x^2}} \int_0^{\sqrt{1-x^2-y^2}} \frac{1}{\sqrt{1-x-y^2-z^2}}\,dx\,dy\,dz.}$

Solution :

$$I = \int_0^1 \int_0^{\sqrt{1-x^2}} \left[\sin^{-1}\frac{z}{\sqrt{1-x^2-y^2}}\right]_{z=0}^{\sqrt{1-x^2-y^2}} dx\,dy$$

$$= \int_0^1 \int_0^{\sqrt{1-x^2}} \frac{\pi}{2}\,dx\,dy = \frac{\pi}{2}\int_0^1 [y]_0^{\sqrt{1-x^2}}\,dx$$

$$= \frac{\pi}{2}\int_0^1 \sqrt{1-x^2} = \frac{\pi}{4}\left[x\sqrt{1-x^2} + \sin^{-1} x\right]_0^1 = \frac{\pi}{4}\left(0+\frac{\pi}{2}\right) = \frac{\pi^2}{8}$$

Example 2.103 **Evaluate** $I = \int_{y=0}^{1}\int_{x=y^2}^{1}\int_{z=0}^{1-x} x\,dy\,dx\,dz$

(Karnataka 88)

Solution :

$$\int_0^1\int_{y^2}^1 x\,[z]_0^{1-x}\,dy\,dx = \int_0^1\int_{y^2}^1 (x-x^2)\,dy\,dx$$

$$= \int_0^1 \left[\frac{x^2}{2}-\frac{x^3}{3}\right]_{y^2}^1 dy = \int_0^1 \left\{\left(\frac{1}{2}-\frac{1}{3}\right)-\frac{y^4}{2}+\frac{y^6}{3}\right\} dy$$

$$= \left[\frac{1}{6}y - \frac{y^5}{10}+\frac{y^7}{21}\right]_0^1 = \frac{1}{6}-\frac{1}{10}+\frac{1}{21} = \frac{4}{35}$$

Example 2.104 **Evaluate** $I = \int_{\theta=0}^{\pi/2}\int_{r=0}^{a\sin\theta}\int_{z=0}^{\frac{a^2-r^2}{a}} r\,d\theta\,dr\,dz.$

(Mysore 97, Gulbarga 99)

Solution :

$$I = \int_0^{\pi/2}\int_0^{a\sin\theta} r\,[z]_0^{\frac{a^2-r^2}{a}}\,d\theta\,dr$$

$$= \int_0^{\pi/2}\int_0^{a\sin\theta} \frac{a^2r-r^3}{a}\,d\theta\,dr$$

$$= \frac{1}{a}\int_0^{\pi/2}\left[\frac{a^2r^2}{2}-\frac{r^4}{4}\right]_0^{a\sin\theta} d\theta = a^5\int_0^{\pi/2}\left(\frac{\sin^2\theta}{2}-\frac{1}{4}\sin^4\theta\right)d\theta$$

$$= a^5\left[\frac{\frac{3}{2}!\,\frac{1}{2}!}{4} - \frac{1}{4}\,\frac{\frac{5}{2}!\,\frac{1}{2}!}{2\,3!}\right]$$

$$= a^5\left[\frac{\pi}{8}-\frac{3\pi}{64}\right] = \frac{5\pi a^5}{64}$$

Example 2.105 **Find the volume of the cylinder $x^2+y^2=4$ cut off by the plane $y+z=4$ and $z=0$.**

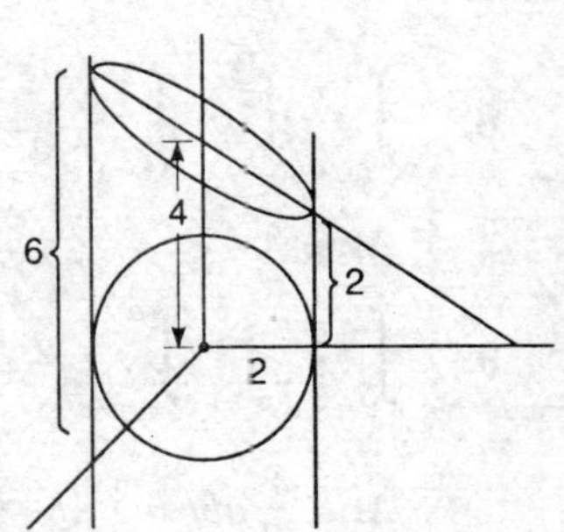

Fig. 2.77

Solution : Required Volume

$$V = \iint_A\int_{z=0}^{4-y} dx\,dy\,dz = \iint_A [z]_0^{4-y}\,dx\,dy$$

$$= 4\iint_A dx\,dy - \iint y\,dx\,dy$$

$4\times$ Area of circle $-\ 0$ as y is odd $= 4\times 4\pi = 16\pi$ **Ans.**

PROBLEM SET 2.13

Evaluate the following integrals :

1. $\int_{y=1}^{c} \int_{x=1}^{\log y} \int_{z=1}^{e^x} \log z \, dy \, dx \, dz$

Ans. $\frac{1}{4}(e^2 - 8e + 13)$

2. $\int_{x=0}^{a} \int_{y=0}^{x} \int_{z=0}^{x+y} e^{x+y+z} \, dx \, dy \, dz$

Ans. $\frac{1}{8}(e^{4a} - 6e^{2a} + 8e^a - 3)$

[Andhra 94, Ranchi 90]

3. $\int_0^1 \int_0^1 \int_0^1 e^{x+y+z} \, dx \, dy \, dz,$

Ans. $(e-1)^3$.

4. $\int_0^a \int_0^a \int_0^a (xy + yz + zx) \, dx \, dy \, dz.$

Ans. $\frac{3}{4} a^5$.

5. $\int_{z=-1}^{1} \int_{x=0}^{z} \int_{y=x-z}^{x+y} (x + y + z) \, dz \, dx \, dy.$

Ans. 0

6. $\int_{x=0}^{1} \int_{y=0}^{\sqrt{1-x^2}} \int_0^{\sqrt{1-x^2-y^2}} xyz \, dx \, dy \, dz$

Ans. $\frac{1}{48}$

7. $\int_{z=0}^{4} \int_{x=0}^{2\sqrt{z}} \int_{y=0}^{\sqrt{4z-x^2}} dz \, dx \, dy$

(Gauhati 99)

8. $\int_{z=0}^{1} \int_{y=0}^{2} \int_{x=0}^{2} x^3 yz \, dz \, dy \, dx$

Ans. 4

9. $\int_{z=-c}^{c} \int_{-b}^{b} \int_{x=-a}^{a} (x^2 + y^2 + z^2) \, dz \, dy \, dx$

Ans. $\frac{8}{3} abc (a^2 + b^2 + c^2)$

2.21 VOLUME OF SOLIDS OF REVOLUTIONS

(a) For Polar Curves

Volume of revolution of an area A about
(i) the axis of x is

$$V_x = \iint_A 2\pi y \, dx \, dy$$

and (ii) about the y-axis is

$$V_y = \iint_A 2\pi x \, dx \, dy.$$

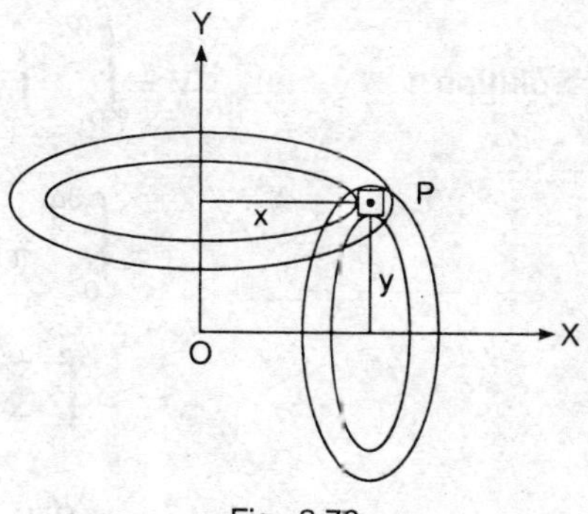

Fig. 2.78

(b) For Polar Curve

$$V_x = \iint_A 2\pi r \sin\theta \, r \, d\theta \, dr$$

and

$$V_y = \iint_A 2\pi r \cos\theta \, r \, d\theta \, dr.$$

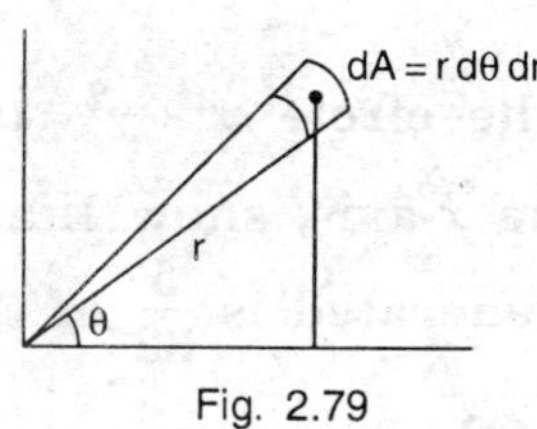

Fig. 2.79

Example 2.106 **Calculate by double integrations the volume generated by revolving the cardiod $r = a(1 - \cos\theta)$ about its axis.**

Solution :

$$V = 2\pi \int_{\theta=0}^{\pi} \int_{r=0}^{a(1-\cos\theta)} r \sin\theta \, r \, d\theta \, dr$$

$$= 2\pi \int_0^{\pi} \sin\theta \, \frac{a^3 (1-\cos\theta)^3}{3} \, d\theta$$

$$= 2\pi \frac{a^3}{3} 16 \int_0^{\pi} \sin\frac{\theta}{2} \sin^6\frac{\theta}{6} \cos\frac{\theta}{2} \, d\theta$$

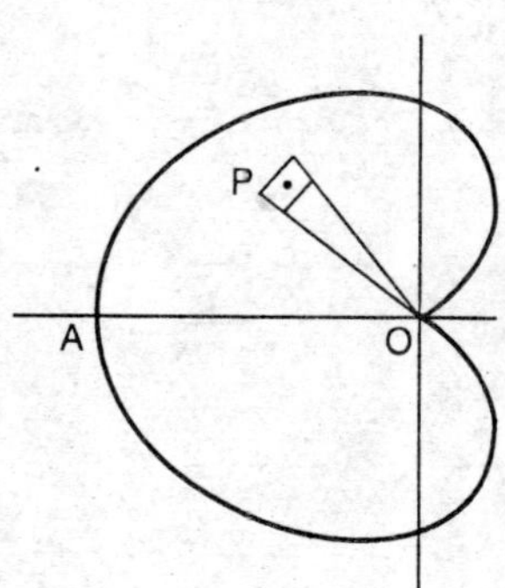

Fig. 2.80

Let $\frac{\theta}{2} = t$

then $V = \frac{64\,\pi a^3}{3} \left[\frac{t^8}{8}\right]_0^1$

$$= \frac{8\pi a^3}{3}$$

Example 2.107 The curve $y^2(a+x) = x^2(3a-x)$ revolve about x-axis. Find the volume generated by its loop. *(Osmania 95)*

Solution :

$$V = \int_0^{3a} \int_{y=0}^{\sqrt{x^2\left(\frac{3a-x}{a+x}\right)}} 2\pi y \, dx \, dy$$

$$= \int_0^{3a} \pi \left[y^2\right]_0^{\sqrt{\frac{x^2(3a-x)}{a+x}}} dx$$

$$= \pi \int_0^{3a} \frac{3ax^2 - x^2}{a+x} dx$$

$$= \pi \int_0^{3a} \left(-x^2 + 4ax - 4a^2 + \frac{4a^3}{a+x}\right) dx$$

$$= \pi \left[-\frac{x^3}{3} + \frac{4ax^2}{2} - 4a^2x + 4a^3 \log(x+a)\right]_0^{3a}$$

$$= \pi \left[-9a^3 + 18a^3 - 12a^3 + 4a^3 \log \frac{4a}{a}\right] = \pi (8 \log 2 - 3)a^3$$

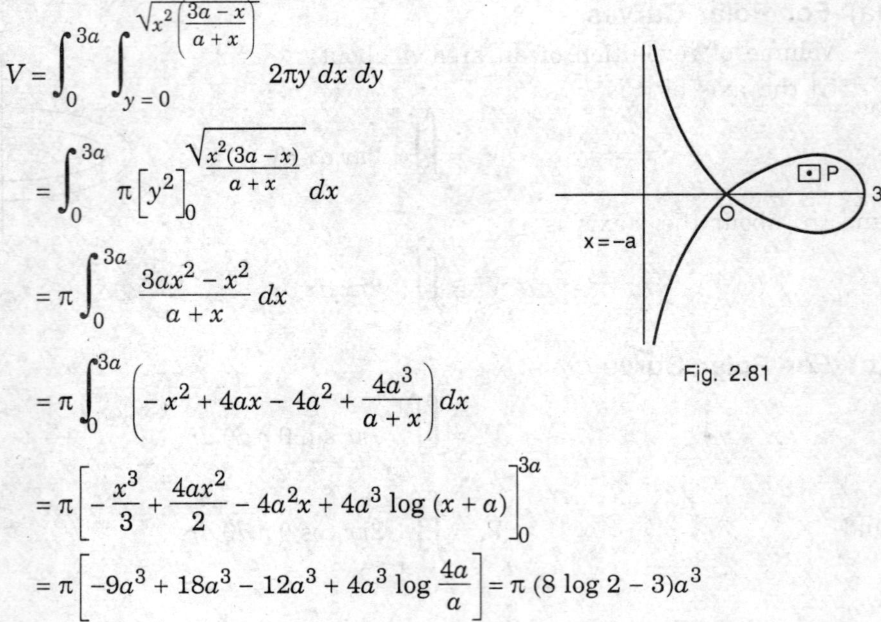

Fig. 2.81

Example 2.108 The segment of the circle $x^2 + y^2 = a^2$, cut off by the line $x = \frac{a}{2}$, is revolved about the x-axis, show that the volume of the segment of the sphere thus generated is $\frac{5}{32}$ of the volume of the whole sphere.

Solution :

$$V = \int_{a/2}^{a} \int_0^{\sqrt{a^2 - x^2}} 2\pi y \, dx \, dy$$

$$= 2\pi \int_{a/2}^{a} \frac{(a^2 - x^2)}{2} dx$$

$x = a \sin \theta$

$$= \pi \left[a^2 x - \frac{x^3}{3}\right]_{a/2}^{a}$$

$$= \pi a^3 \left[\left(1 - \frac{1}{2}\right) - \frac{1}{3}\left(1 - \frac{1}{8}\right)\right]$$

$$= \frac{5}{24} \pi a^3 = \frac{5}{32}\left(\frac{4}{3} \pi a^3\right) = \frac{5}{32} \text{ (Vol. of a sphere).}$$

O

x = a/2

Fig. 2.82

Change of Variables. If the variables x, y in a double integral be changed to u and v given by $x = \phi(u, v)$, $y = \psi_1(u, v)$ than $dx\, dy = \frac{\partial(\phi, \psi)}{\partial(u, v)} du\, dv$. By this change of variables, the evaluation of $\iint_A f(x, y)\, dx\, dy$, some times becomes easier when it is changed to $\iint_A f\{x(u), y(v)\} \frac{\partial(x, y)}{\partial(u, v)} du\, dv.$

Similarly for a triple integral, if x, y, z be changed to u, v, w, then $dx\, dy\, dz = \frac{\partial(x, y, z)}{\partial(u, v, w)} du\, dv\, dw$ and we have

$$\iiint_V f(x, y, z)\, dx\, dy\, dz = \iiint_V f[x(y), y(v), z(w)] \times \frac{\partial(x, y, z)}{\partial(u, v, w)} du\, dv\, dw$$

Provided the Jacobians of the transformations are not zero $\frac{\partial(x, y)}{\partial(u, v)} \neq 0$ and for three variables $\frac{\partial(x, y, z)}{\partial(u, v, w)} \neq 0$.

Example 2.109 **Evaluate $I = \int_0^\infty \int_0^\infty e^{-(x^2+y^2)} dx\, dy$, by changing to polar co-ordinates.**

Hence show that $\int_0^\infty e^{-x^2} dx = \frac{\sqrt{\pi}}{2}$.

(J.N.T.U. 2000, Mysore 94 S, 97 S, Rahchi 89)

Solution : Let $x = r\cos\theta,\ y = r\sin\theta$

$$\frac{\partial(x, y)}{\partial(r, \theta)} = \begin{vmatrix} \frac{\partial x}{\partial r} & \frac{\partial x}{\partial \theta} \\ \frac{\partial y}{\partial r} & \frac{\partial y}{\partial \theta} \end{vmatrix} = \begin{vmatrix} \cos\theta & -r\sin\theta \\ \sin\theta & r\cos\theta \end{vmatrix} = r$$

Hence $$I = \int_{\theta=0}^{\pi/2} \int_{r=0}^{\infty} e^{-r^2} r\, d\theta\, dr$$

$$= \int_0^{\pi/2} \left[\frac{e^{-r^2}}{-2}\right]_0^\infty d\theta$$

$$= -\frac{1}{2}\int_0^{\pi/2} d\theta = \frac{\pi}{4}$$

$$\frac{\pi}{4} = I = \left[\int_0^\infty e^{-x^2} dx\right]\left[\int_0^\infty e^{-y^2} dy\right]$$

$$= \left\{\int_0^\infty e^{-x^2} dx\right\}^2$$

Hence $$\int_0^\infty e^{-x^2} dx = \frac{\sqrt{\pi}}{2}.$$

Example 2.110 **Evaluate $\iint_R (x+y)^2\,dx\,dy$ where R is the interior of the parallelogram having vertices $A(1, 0)$, $B(3, 1)$, $C(2, 2)$, $D(0, 1)$.**

Solution : On changing the variables by the substitutes

$$x + y = u \text{ and } x - 2y = v$$

or $$x = \frac{2u+v}{3},\ y = \frac{u-v}{3}$$

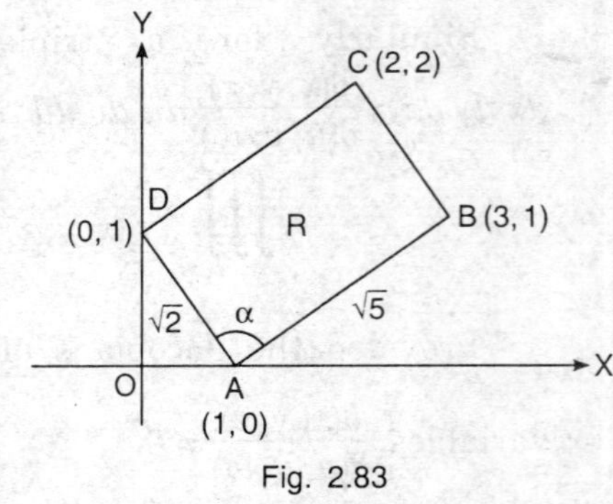

Fig. 2.83

A, B, C, D change to

$$A'(1+0, 1-2\times 0) = (1, 1)$$
$$B'(3+1, 3-2) = (4, 1)$$
$$C' = (4, 2-4) = (4, -2)$$
$$D' = (1, -2) = (1, -2)$$

$$\frac{\partial(x,y)}{\partial(u,v)} = \frac{1}{9}\begin{vmatrix} 2 & 1 \\ 1 & -1 \end{vmatrix} = -\frac{1}{3}$$

So the terms for u are 1 to 4 for $v = -2$ to 1.

$$I = \int_1^4 \int_{-2}^1 u^2 \frac{1}{3}\,du\,dv = \frac{1}{3}\left[\frac{u^3}{3}\right]_1^4 [v]_{-2}^{+1} = \frac{1}{3}\times\frac{63}{3}\times 3 = 21$$

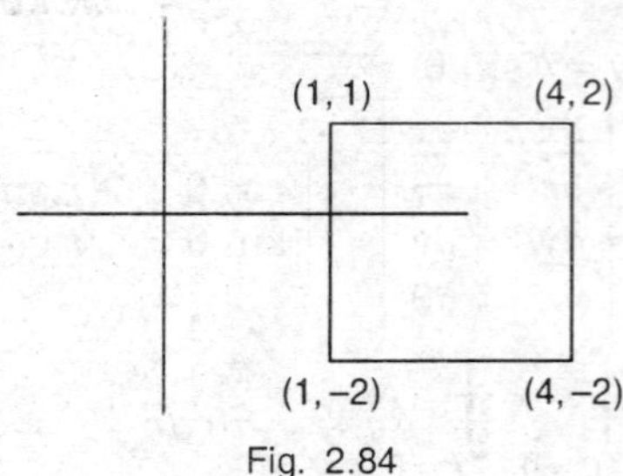

Fig. 2.84

Example 2.111 **Evaluate $I = \iint \sqrt{\frac{1-x^2-y^2}{1+x^2+y^2}}\,dx\,dy$, over the positive quadrant of the circle $x^2 + y^2 = 1$.**

Solution : Let $x = r\cos\theta$, $y = r\sin\theta$

$$\frac{\partial(x,y)}{\partial(r,\theta)} = \begin{vmatrix} \frac{\partial x}{\partial r} & \frac{\partial x}{\partial \theta} \\ \frac{\partial y}{\partial r} & \frac{\partial y}{\partial \theta} \end{vmatrix} = \begin{vmatrix} \cos\theta & -r\sin\theta \\ \sin\theta & r\cos\theta \end{vmatrix} = r$$

$$I = \int_{\theta=0}^{\pi/2}\int_{r=0}^{1}\sqrt{\frac{1-r^2}{1+r^2}}\,r\,d\theta\,dr = [\theta]_0^{\pi/2}\int_0^1 \frac{r\,dr\,(1-r^2)}{\sqrt{1-r^4}}$$

$$= \frac{\pi}{2}\left\{\int_0^1 \frac{r\,dr}{\sqrt{1-(r^2)^2}} - \int_0^1 \frac{r^3\,dr}{(1-r^4)^{1/2}}\right\}$$

$$= \frac{\pi}{2}\left[\frac{1}{2}\sin^{-1}(r^2)\right]_0^1 - \left[\frac{\pi}{8}\frac{(1-r^4)^{1/2}}{-\frac{1}{2}}\right]_0^1$$

$$= \frac{\pi}{4}\frac{\pi}{2} + \frac{\pi}{4}(0-1) = \left(\frac{\pi^2}{8} - \frac{\pi}{4}\right)$$

Example 2.112 **Evaluate $I = \int_0^2 \int_0^{\sqrt{2x-x^2}} \frac{x\,dx\,dy}{\sqrt{x^2+y^2}}$ by changing to polar cordinates.**

Solution : Let $x = r\cos\theta$, $y = r\sin\theta$

$$\frac{\partial(x, y)}{\partial(r, \theta)} = \begin{vmatrix} \cos\theta & -r\sin\theta \\ \sin\theta & r\cos\theta \end{vmatrix} = r$$

Circle $y^2 = 2x - x^2$ is $r^2 = 2r\cos\theta$

$$r = 2\cos\theta$$

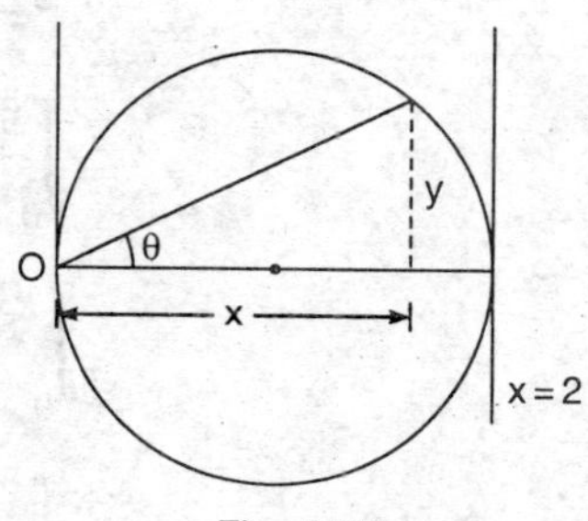

Fig. 2.85

$$I = \int_{\theta=0}^{\pi/2} \int_{r=0}^{2\cos\theta} \frac{r\cos\theta\, r}{r}\, d\theta\, dr$$

$$= \int_0^{\pi/2} \cos\theta \left[\frac{r^2}{2}\right]_0^{2\cos\theta} d\theta$$

$$= 2\int_0^{\pi/2} \cos^3\theta\, d\theta = \frac{2\,\Gamma\frac{3+1}{2}\,\Gamma\frac{1}{2}}{2\,\Gamma\frac{5}{2}} = \frac{1}{\frac{3}{2}\times\frac{1}{2}} = \frac{4}{3}.$$

Example 2.113 **Evaluate $I = \int_0^1 \int_0^{\sqrt{1-x^2}} \int_0^{\sqrt{1-x^2-y^2}} \frac{dx\,dy\,dz}{\sqrt{1-x^2-y^2-z^2}}$ by changing to polar co-ordinates.** *(Punjab 97, Gorakhpur 91, Madras 94, 96, Marathwada 93)*

Solution : Let $x = r\sin\theta\cos\phi$, $y = r\sin\theta\sin\phi$, $z = r\cos\theta$

$$\frac{\partial(x, y, z)}{\partial(r, \theta, \phi)} = \begin{vmatrix} \sin\theta\cos\phi & r\cos\theta\cos\phi & -r\sin\theta\sin\phi \\ \sin\theta\sin\phi & r\cos\theta\sin\phi & r\sin\theta\cos\phi \\ \cos\theta & -r\sin\theta & 0 \end{vmatrix}$$

$$= \cos\theta\, r^2 \sin\theta\cos\theta \begin{vmatrix} \cos\phi & -\sin\phi \\ \sin\phi & \cos\phi \end{vmatrix} + r\sin\theta\, r\sin^2\theta \begin{vmatrix} \cos\phi & -\sin\phi \\ \sin\phi & \cos\phi \end{vmatrix}$$

$$= r^2\cos^2\theta\sin\theta + r^2\sin^3\theta = r^2\sin\theta$$

Hence $dx\,dy\,dz = I\, d\phi\, d\theta\, dr$

$$= r^2\sin\theta\, d\phi\, d\theta\, dr$$

So $$I = \int_0^{\pi/2}\int_0^{\pi/2}\int_{r=0}^{1} \frac{1}{\sqrt{1-r^2}}\, r^2\sin\theta\, d\phi\, d\theta\, dr.$$

$$= [\phi]_0^{\pi/2} [-\cos\theta]_0^{\pi/2} \int_0^{\pi/2} \frac{\sin^2 t \cos t}{\cos t} dt = \left(\frac{\frac{3}{2}!\frac{1}{2}!}{2\,2!} = \frac{\pi}{4} \right)$$

$$= \frac{\pi}{2} [0+1] \frac{\pi}{4} = \frac{\pi^2}{8}.$$

Example 2.114 **Find the volume of the parabolic cylinder bounded by the cylinder $y = 4x - x^2$ and $y = x$ and $z = 0$ and $z = 1$.**

Solution :

$$r = \int_{x=0}^{3} \int_{y=x}^{y=4x-x^2} \int_{z=0}^{1} dx\, dy\, dz$$

$$= \int_0^3 \int_x^{4x-x^2} [z]_1^1 \, dx\, dy$$

$$= \int_0^3 [dy]_x^{4x-x^2} dx$$

$$= \int_0^3 (4x - x^2 - x)\, dx = \left[\frac{3x^2}{2} - \frac{x^3}{3} \right]_0^3 = \frac{9}{2}.$$

Example 2.115 **Find the volume common to the cylinders $x^2 + y^2 = a^2$ and $x^2 + z^2 = a^2$.** *(Andhra 1990, Kerala 1990)*

Solution :

$$\iiint_{z=-a}^{a} dx\, dy\, dz$$

$$\iint_A 2\sqrt{a^2 - x^2}\, dx\, dy$$

A represents the circle $x^2 + y^2 = a^2$

$$= 8 \int_{x=0}^{a} \int_{y=0}^{\sqrt{a^2-x^2}} \sqrt{a^2 - x^2}\, dx\, dy$$

$$= 8 \int_0^a \sqrt{a^2 - x^2}\, [y]_0^{\sqrt{a^2-x^2}} dx$$

$$= 8 \int_0^a (a^2 - x^2)\, dx$$

$$= 8 \left[a^2 x - \frac{x^3}{3} \right]_0^a = \frac{16}{3} a^3.$$

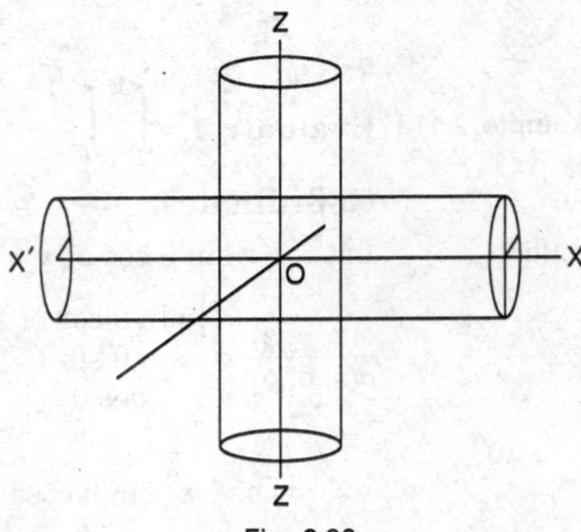

Fig. 2.86

Example 2.116 Find the volume common to the sphere $x^2 + y^2 + z^2 = a^2$ and the cylinder $x^2 + y^2 - ay = 0$. *(Kerala 90)*

Solution :

$$V = \iiint_{A - \sqrt{a^2 - x^2 - y^2}}^{+\sqrt{a^2 - x^2 - y^2}} dx\, dy\, dz$$

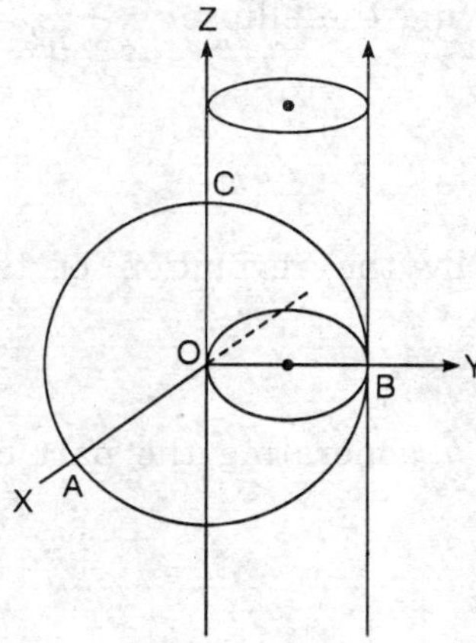

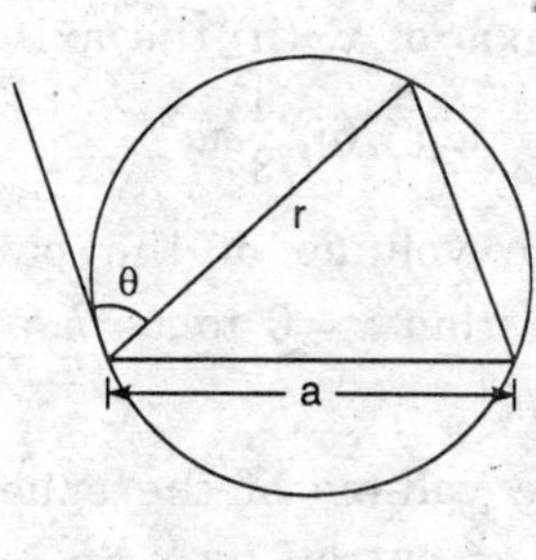

Fig. 2.87

where Z varies from $-\sqrt{a^2 - x^2 - y^2}$ to $+\sqrt{a^2 - x^2 - y^2}$

$$= 2 \iint_A \sqrt{a^2 - x^2 - y^2}\, dx\, dy$$

where A is the area of the circle.

$$x^2 + y^2 - ay = 0$$

or $$r = a \sin \theta$$

Changing to polars

$$V = 2 \times 2 \int_0^{\pi/2} \int_{r=0}^{a \sin \theta} \sqrt{a^2 - r^2}\, r\, d\theta\, dr$$

$$= 4 \int_0^{\pi/2} \left[\frac{(a^2 - x^2)^{3/2}}{-\frac{3}{2} \times 2} \right]_0^{a \sin \theta} d\theta$$

$$= \frac{4}{3} a^3 \int_0^{\pi/3} (1 - \cos^3 \theta)\, d\theta$$

$$= \frac{4}{3} a^3 \left[\frac{\pi}{2} - \frac{\frac{4}{2}!\, \frac{1}{2}!}{2\, \frac{5}{2}!} \right] = \frac{4}{3} a^3 \left(\frac{\pi}{2} - \frac{2}{3} \right).$$

PROBLEM SET 2.14

1. Show that the volume generated by revolving the circle $x^2 + y^2 = a^2$ about the x-axis is $\frac{4}{3}\pi a^3$.

2. Find the volume of the solid generated by revolving the ellipse $\frac{x^2}{a^2} + \frac{y^2}{b^2} = 1$ about (i) the axis of x, (ii) the axis of y.

   Ans. (i) $\frac{4}{3}\pi ab^2$, (ii) $\frac{4}{3}\pi a^2 b$.

3. Find the volume of the paraboloid generated by the revolution of the parabola $y^2 = 4ax$ from $x = 0$ to $x = h$ about the x-axis.

   Ans. $2\pi a h^2$

4. Find the volume of the spherical cap of height h generating the part of the circle $x^2 + y^2 = a^2$ cut off by $x = a - h$, about the x-axis.

   Ans. $\pi h^2 \left(a - \frac{1}{3}h\right)$

5. Find the volume generated by revolving the curve $xy^2 = 4(2 - x)$ about the y-axis.

   Ans. $4\pi^2$.

6. Find the volume of the spindle shaped solid generated by revolving the astroid $x^{2/3} + y^{2/3} = a^{2/3}$ about the x-axis.

   Ans. $\frac{32}{105}\pi a^3$.

7. Find the volume of the solid formed by revolving the cirroid $y^2(2a - x) = x^3$ about $x = 2a$.

   Ans. $2\pi^2 a^3$. ***(V.T.U. 2000)***

8. Find the volume of the solid generated by revolving the lemniscate $r^2 = a^2 \cos 2\theta$ about the line $\theta = \frac{\pi}{2}$.

   Ans. $\frac{\pi a^3}{4\sqrt{2}}$.

9. Evaluate $\iint \sqrt{a^2 - x^2 - y^2}\, dx\, dy$ over the semicircle $x^2 + y^2 = ax$, about the x-axis.

   Ans. $\frac{a^3}{3}\left(\frac{\pi}{2} - \frac{2}{3}\right)$.

10. Evaluate $\iiint z(x^2 + y^2 + z^2)\, dx\, dy\, dz$ over the volume of the cylinder $x^2 + y^2 = a^2$ intercepted between $z = 0$ to $z = h$.

    Ans. $\frac{\pi a^2 h^2}{4}(a^2 + h^2)$

11. Evaluate the following changing into polar co-ordinates.

(i) $\int_0^a \int_0^{\sqrt{a^2-y^2}} (x^2+y^2)\, dx\, dy$

(J.N.T.U 89)

Ans. $\pi \frac{a^4}{8}$

(ii) $\int_0^a \int_y^a \frac{x^2\, dx\, dy}{\sqrt{x^2+y^2}}$

(Delhi 97)

Ans. $\frac{a^3}{3} \log(\sqrt{2}+1)$

(iii) $\int_0^{4a} \int_{y^2/4a}^{y} \frac{x^2-y^2}{x^2+y^2}\, dx\, dy$

Ans. $8\left(\frac{\pi}{2}-\frac{5}{3}\right)a^2$

(iv) $\int_0^a \int_y^a \frac{x\, dx\, dy}{x^2+y^2}$

Ans. $\frac{\pi a}{4}$.

12. Transform the following to cartesian form and evaluate $\int_0^{\pi} \int_0^a r^3 \sin\theta \cos\theta\, dr\, d\theta$.

Ans. 0

13. Evaluate $\iint \sqrt{\frac{1-x^2/a^2-y^2/b^2}{1+x^2 ? a^{2+} y^2/b^2}}\, dx\, dy$ over positive quadrant of the ellipse $\frac{x^2}{a^2}+\frac{y^2}{b^2}=1.$

Ans. $8\left(\frac{\pi}{2}-\frac{5}{3}\right)a^2$

14. Evaluate the following integrals by changing to spherical polar coordinaes over $x^2+y^2+z^2=1$:

(i) $\iiint (x^2+y^2+z^2)^m\, dx\, dy\, dz,$

Ans. $\frac{4\pi}{2m+3}$

(ii) $\iiint z^2\, dx\, dy\, dz$

Ans. $\frac{4\pi}{15}$.

15. Evaluate $\iint (a^2 - x^2 - y^2)\, dx\, dy$ over the semi circle $x^2 + y^2 = ax$ above the x-axis by changing to polars.

Ans. $\frac{5\pi a^4}{64}$.

16. Evaluate $\iint_D e^{-(x^2+y^2)}\, dx\, dy$ over the region D of the circle $x^2 + y^2 = a^2$.

Ans. $\pi(1 - e^{-a^2})$.

17. Evaluate $\iint xy\,(x^2 + y^2)^{n/2}\, dx\, dy$ over the positive quadrants of $x^2 + y^2 = 4$, for $x + 3 > 0$.

Ans. $\frac{2^{n+3}}{n+4}$.

18. Evaluate $\iiint \frac{dx\, dy\, dz}{\sqrt{1-(x^2+y^2+z^2)}}$ over the positive octant of the sphere $x^2 + y^2 + z^2 = 1$.

Ans. $\frac{\pi^2}{8}$.

19. Find the ratio of the parts in which the ellipic cylinder $4x^2 + 9y^2 = 36$ is divided by the plane $2x + 3y = 6$.

Ans. $\frac{\pi - 2}{3\pi + 2}$.

20. Find the double integration, the area lying inside the circle $r = a \sin\theta$ and outside the cardiod $r = a(1 - \cos\theta)$

Ans. $a^2\left(1 - \frac{\pi}{4}\right)$

21. Show by double integration that the volume of the cylinder of base area as $y^2 = 4ax$ and $x^2 = 4ay$ and unit height is $\frac{16}{3}a^2$ ***(Kerala 1987, Mysore 94, Kernataka 93)***

22. Find the volume bounded by the xy plane, the paraboloid $2z = x^2 + y^2$ and the cylinder $x^2 + y^2 = 4$. ***(Hamirpur 95)***

Ans. 4π

23. Find the volume bounded by the paraboloid $x^2 + y^2 = az$ and the cylinder $x^2 + y^2 = 2ay$ and the plane $z = 0$

Ans. $\frac{3\pi a^3}{2}$ ***(Poona 87, Mysore 95)***

24. Find the volume bounded by the cylinder $x^2 + y^2 = 4$ and the hyperboloid $x^2 + y^2 - z^2 = 1$.

Ans. $4\sqrt{3}\,\pi$

25. Find the volume inclosed between the cylinder $x^2 + y^2 = 2ax$ and $z^2 = 2ax$

Ans. $\frac{128a^3}{15}$ *(Delhi 97)*

26. Find the volume bounded by the *xy*-plane, the cylinder $x^2 + y^2 = 1$ and the plane $x + y + z = 3$.

Ans. $3\pi - \frac{4}{3}$.

2.22 MASS AND VOLUME AS TRIPLE INTEGRAL

Volume V in cartlesian co-ordinates is

$$V = \iiint_V dx\, dy\, dz$$

where the integral is to be evaluated so as to include the whole volume V. If density $\rho = P(x, y, z)$ then the mass M is given by

$$M = \iiint \rho(x, y, z)\, dx\, dy\, dz.$$

In cylindrical polar co-ordinates $x = r\cos\theta$, $y = r\sin\theta$, $z = z$

$$dx\, dy\, dz = \frac{\partial(x, y, z)}{\partial(r, \theta, z)}\, dr\, d\theta\, dz = r\, dr\, d\theta\, dz$$

as the value of the Jacobian

$$\frac{\partial(x, y, z)}{\partial(r, \theta, z)} = \begin{vmatrix} \frac{\partial x}{\partial r} & \frac{\partial x}{\partial \theta} & \frac{\partial x}{\partial z} \\ \frac{\partial y}{\partial r} & \frac{\partial y}{\partial \theta} & \frac{\partial y}{\partial z} \\ \frac{\partial z}{\partial r} & \frac{\partial z}{\partial \theta} & \frac{\partial z}{\partial z} \end{vmatrix} = r.$$

So $$M = \iiint \rho(r, \theta, z)\, r\, dr\, d\theta\, dz$$

In spherical polar co-ordiantes $x = r\sin\theta\cos\phi$, $y = r\sin\theta\sin\phi$, $z = r\cos\theta$

$$dx\, dy\, dz = \frac{\partial(x, y, z)}{\partial(r, \theta, \phi)} = r^2 \sin\theta\, dr\, d\theta\, d\phi$$

So $$\text{Mass} = \iiint \rho(r, \theta, \phi)\, r^2 \sin\theta\, dr\, d\theta\, d\phi$$

Example 2.117 Find the mass of the tetrahedron bounded by the ordinate plane and the plane $\frac{x}{a} + \frac{y}{b} + \frac{z}{c} = 1$, if the density varies and is given by $\rho = kxyz$.

Solution : $$\text{Mass} = \iiint_V kxyz\, dx\, dy\, dz$$

Let $\quad \frac{x}{a} = X, \frac{y}{b} = Y, \frac{z}{c} = z$

$$V = K\int_0^1 \int_0^{1-x} a^2b^2c^2 \int_0^{1-x-y} xyz\, dx\, dy\, dz$$

$$= Ka^2b^2c^2 \int_0^1 \int_0^{1-x} \underset{\text{I}}{xy} \underset{\text{II}}{\frac{(1-x-y)^2}{2}}\, dx\, dy$$

It will make calculations simple if we integrate by parts instead of expanding and multiplying all the terms

$$= \frac{ka^2b^2c^2}{2}\int_0^1 x\left\{\frac{y(1-x-y)^3}{-3} - \int \frac{(1-x-y)^3}{-3}\, dy\right\} dx$$

$$= k\frac{a^2b^2c^2}{2}\int_0^1 \frac{x}{3}\left\{-(1-x-y)^3 - \frac{(1-x-y)^4}{4}\right\}_0^{1-x} dx$$

$$= k\frac{a^2b^2c^2}{6}\int_0^1 x\left\{0 + \frac{1}{4}(1-x)^4\right\} dx$$

$$= k\frac{a^2b^2c^2}{24}\int_0^1 \underset{\text{I}}{x}\, \underset{\text{II}}{(1-x)^4}\, dx$$

$$= k\frac{a^2b^2c^2}{24}\left[\frac{x(1-x)^5}{-5} - \int \frac{(1-x)^5}{-5}\, dx\right]$$

$$= k\frac{a^2b^2c^2}{24}\left[-\frac{1}{5}\times(1-x)^5 - \frac{(1-x)^6}{30}\right]_0^1$$

$$= k\frac{a^2b^2c^2}{24}\left[0 + 0 + \frac{1}{30}\right] = k\frac{a^2b^2v^2}{720}.$$

Example 2.118 **Find the mass of the plane in the shape of $\left(\frac{x}{a}\right)^{2/3} + \left(\frac{y}{b}\right)^{2/3} = 1$, if the density varies as $\rho = \mu xy$.**

Solution : $\quad M = \iint \mu xy\, dx\, dy$

x and y are odd functions so the product xy = +ive in Ist quadrant xy = – in second, xy = + in this and xy = – in 4th. But if $\rho\ |x|\ |y|$ then it is always positive and mass = 4 × mass is Ist quadrant. Hence

$$M = 4\mu \int_0^a \int_{y=0}^{b(1-(x/a)^{2/3})^{3/2}} xy\,dx\,dy$$

$$= 4\mu \int_0^a \frac{xb^2 \left\{1-\left(\frac{x}{a}\right)^{2/3}\right\}^{\frac{3}{2}\times 2}}{2}\,dx$$

$$= \frac{4\mu b^2}{2} \int_0^a x \left\{1-\left(\frac{x}{a}\right)^{2/3}\right\}^{3} dx$$

Fig. 2.83

Let $\quad x = a \sin^3 \theta$

Then $\; M = 2\mu b^2 \int_0^{\pi/2} a \sin^3\theta \cos^6\theta\,(3a \sin^2\theta \cos\theta)\,d\theta$

$$= 6\mu a^2 b^2 \frac{\Gamma(6/2)\,\Gamma(8/2)}{2\,\Gamma(7/2)}$$

$$= 6\mu a^2 b^2 \frac{2\,.\,1 \ldots 3.\,2\,.\,1}{2\,.\ldots 6\,.\,5\,.\,4\,.\,3\,.\,2\,.\,1} = \frac{\mu a^2 b^2}{20}.$$

2.23 AREA OF A CURVED SURFACE

Let $F(x, y, z) = z - f(x, y) = 0$ be a curved surface. The direction ratios of the normal at $P(x, y, z)$ on the surfac are $-\frac{\partial f}{\partial x}, -\frac{\partial f}{\partial y}, 1.$

So the direction cosines are

$$\cos\alpha = -\frac{-f_x}{\sqrt{1+f_x^{\,2}+f_{y2}}},$$

$$\cos\beta = \frac{-f_y}{\sqrt{1+f_x^{\,2}+f_y^{\,2}}},$$

$$\cos\gamma = \frac{1}{\sqrt{1+f_x^{\,2}+f_y^{\,2}}}.$$

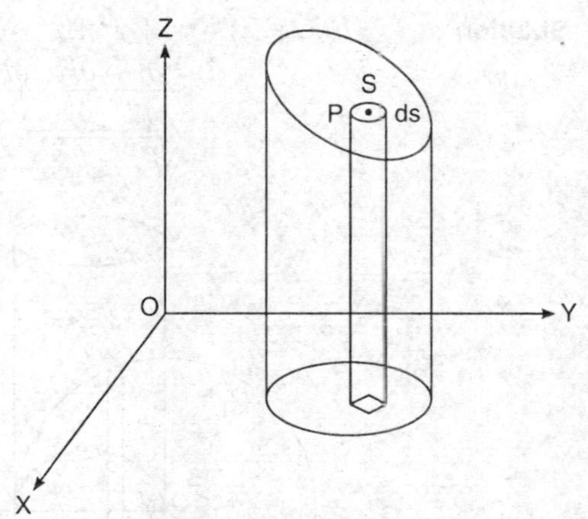

Fig. 2.89

If the projection of dS, an element of S at P, on the plane $z = 0$ be $dx\,dy$ then as normal to the plane $z = 0$ is the z-axis, with which the normal at P makes an angle γ.

So $\qquad dS \cos\gamma = dx\,dy$

or $\qquad dS = \frac{dx\,dy}{\cos\gamma} = \sqrt{1+f_x^{\,2}+f_y^{\,2}}\;dx\,dy$

So $\qquad S = \int dS = \iint \sqrt{1+f_x^{\,2}+f_y^{\,2}}\;dx\,dy.$

Example 2.119 **Find the area of the portion of the curved surface of the cylinder $x^2 + z^2 = 4$, lying inside the cylinder $x^2 + y^2 = 4$.**

Solution : The curved surface $F(x, y, z) = x^2 + z^2 - 4 = 0$ has direction ratios $2x$, 0, $2z$, so the direction cosiens are :

$$\cos \alpha \frac{x}{\sqrt{x^2 + z^2}}, \cos \beta = 0, \cos \gamma = \frac{z}{\sqrt{x^2 + y^2}}$$

$$= \frac{z}{2} = \frac{\sqrt{4 - x^2}}{2}$$

The upper surface will have projections in all the 4 quadrants and similarly the lower. So the projection of $\frac{1}{8}$th of the surface is in the positive quadrant of x, y plane

$$\text{area} \quad \frac{S}{8} = \int_{x=0}^{2} \int_{y=0}^{\sqrt{4-x^2}} \frac{dx\,dy}{\cos r}$$

$$= \int_0^2 \int_0^{\sqrt{4-x^2}} \frac{2\,dx\,dy}{\sqrt{4 - x^2}}$$

$$= 2 \int_0^2 \frac{\sqrt{4 - x^2}}{\sqrt{4 - y^2}} dx = 2\,[x]_0^2 = 4$$

So total surface area $= 8 \times 4 = 32$.

Example 2.120 **Find the area of the portion of the sphere $x^2 + y^2 + z^2 = 9$, lying inside the cylinder $x^2 + y^2 - 3y = 0$.** *(Kerala 90)*

Solution : $F(x, y, z) = x^2 + y^2 + z^2 - 9 = 0$

$$\frac{\frac{\partial F}{\partial x}, \frac{\partial F}{\partial y}, \frac{\partial F}{\partial z}}{\sqrt{f_x^2 + f_y^2 + f_z^2}} = \text{Direction cosines of the normal} = \frac{x}{3}, \frac{y}{3}, \frac{z}{3}.$$

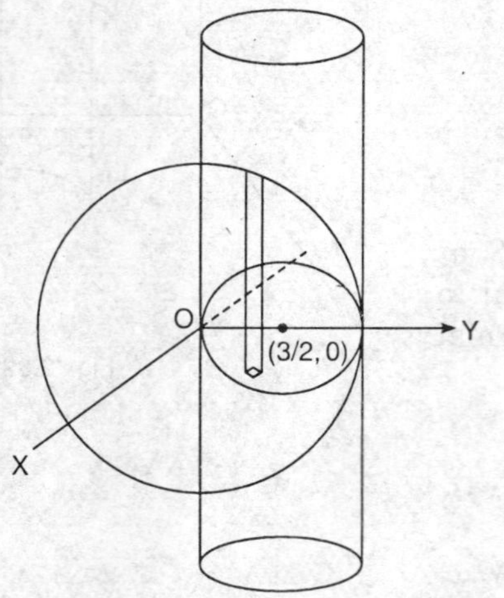

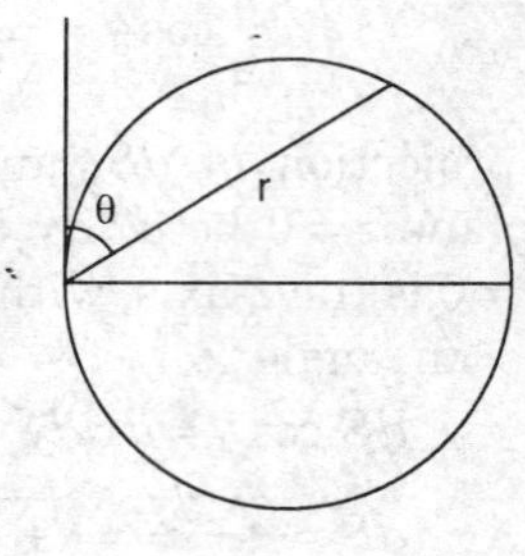

Fig. 2.90

$$\cos\alpha = \frac{1}{3}\sqrt{9 - y^2 - z^2},\ \cos\beta = \frac{1}{3}\sqrt{9 - z^2 - x^2},\ \cos\gamma = \frac{1}{3}\sqrt{9 - x^2 - y^2}.$$

Let $\frac{S}{2}$ be the upper part of the surface then

$$\frac{S}{2} = \iint_A \frac{dx\,dy}{\cos\gamma}$$

where A is the area of the circle $x^2 + y^2 - 3y = 0$

$$S = 2 \times 3 \iint_A \frac{dx\,dy}{\sqrt{9 - x^2 - y^2}}$$

changing to polars

$$r = 3\sin\theta$$

$$= 6 \times 2 \int_0^{\pi/2} \int_0^{3\sin\theta} \frac{r\,d\theta\,dr}{\sqrt{9 - r^2}}$$

$$12 \int_0^{\pi/2} \left[\frac{\sqrt{9 - r^2}}{-1}\right]_0^{3\sin\theta} d\theta = 12 \int_0^{\pi/2} 3(1 - \cos\theta)\,d\theta$$

$$= 36\,[\theta - \sin\theta]_0^{\pi/2}$$

$$= 36\left(\frac{\pi}{2} - 1\right).$$

PROBLEM SET 2.15

1. Find the mass of the lamina in the form of the cardiod $r = a(1 + \cos\theta)$ whose density varies as sqaures of its distance from the initial line.

   Ans. $\frac{21}{32}\mu\pi a^4$.

2. A lamina is bounded by the curve $y = x^2 - 3x$ and $y = 2x$. If the density at any points is given by λxy, find by double integration the mass of the lamina.

   Ans. $182\frac{\lambda}{24}\lambda$

3. Find the mass of the solid in the form of the positive octant of the sphere $x^2 + y^2 + z^3 = 9$, if the density at any pointis is $2xyz$. *(Gorakhpur 91)*

   Ans. $30\frac{3}{8}$

4. Find the mass of the area bounded by the curves $y^2 = x$ and $y = x^3$, if $\rho = \mu(x^2 + y^2)$.

   Ans. $\frac{23\mu}{105}$

5. Find the mass of the solid bounded by the planes $z = 0$, $z = 0$, $z = h$, and the cylinder $x^2 + y^2 = a^2$, $x > 0$, if $\rho = \mu yz$.

Ans. $\dfrac{\mu a^3 h^2}{3}$

6. Find the mass of the plate $\dfrac{x^2}{a^2} + \dfrac{y^2}{b^2} = 1$, if the density at (x, y) is μxy.

Ans. $\mu \dfrac{a^2 b^2}{2}$.

7. Find the mass of a plate in the form of one loop of $r^2 = a^2 \cos 2\theta$, if the density at a point varies as sqaure of the distance of the points from the pole.

Ans. $\dfrac{\mu \pi a^4}{16}$.

8. Find the mass of a rectangular parallelopiped bounded by the planes $x = 0$, $x = a$; $y = 0$, $y = b$ and $z = 0$, $z = c$, if the density ρ at $P(x, y, z)$ be $k(x + y + z)$.

Ans. $M = \dfrac{kabc}{2}(a + b + c)$

3

Ordinary Differential Equations

3.1 DIFFERENTIAL EQUATIONS OF FIRST ORDER AND FIRST DEGREE

Differential equations of the form $f(x, y) + g(x, y)\dfrac{dy}{dx} = 0$ or $f(x, y)\, dx + g(x, y)\, dy = 0$ or $f\left(x, y, \dfrac{dy}{dx}\right) = 0$ are known as differential equations of first order and first degree. Here $f(x, y)$ and $g(x, y)$ are functions of x and y or constants.

It is not always possible to solve all differential equations of first order. The solution of these differential equations is possible only if they belong to any one of the following standard forms:

(i) Equations of the form $\dfrac{dy}{dx} = f(x)$.

(ii) Equations of the form $\dfrac{dy}{dx} = f(y)$.

(iii) Equations in variable separable form.

(iv) Equations reducible to variable separable form.

(v) Homogeneous differential equations.

(vi) Equations reducible to homogeneous form.

(vii) Linear differential equations.

(viii) Equations reducible to linear form.

(ix) Exact differential equations.

(x) Equations reducible to exact form.

3.2 DIFFERENTIAL EQUATIONS OF THE TYPE $\frac{dy}{dx} = f(x)$

To solve such type of equations we integrate both sides and add an arbitrary constant of integration to any one of the two sides.

We have $\frac{dy}{dx} = f(x) \Rightarrow dy = f(x)\,dx$

One integrating both sides, we get

$$\int dy = \int f(x)\,dx + c \quad \text{or} \quad y = \int f(x)\,dx + c$$

Example 3.1 **Solve :**

(i) $\frac{dy}{dx} = \frac{x}{x^2+1}$ (ii) $(e^x + e^{-x})\frac{dy}{dx} = (e^x - e^{-x})$

Solution : (i) We have,

$$\frac{dy}{dx} = \frac{x}{x^2+1} \Rightarrow dy = \frac{x}{x^2+1}\,dx$$

Integrating both sides w.r. to x, we get

$$\int dy = \int \frac{x}{x^2+1}\,dx + c \Rightarrow y = \frac{1}{2}\log(x^2+1) + c$$

(ii) We have,

$$\frac{dy}{dx} = \frac{e^x - e^{-x}}{e^x + e^{-x}} \Rightarrow dy = \frac{e^x - e^{-x}}{e^x + e^{-x}}\,dx$$

Integrating both sides w.r. to x, we get

$$\int dy = \int \frac{e^x - e^{-x}}{e^x + e^{-x}}\,dx + c$$

Let $e^x + e^{-x} = t \Rightarrow (e^x - e^{-x})\,dx = dt$

$$\therefore \quad \int dy = \int \frac{dt}{t} + c \Rightarrow y = \log t + c = \log(e^x + e^{-x}) + c$$

Example 3.2 **Solve :**

(i) $\frac{dy}{dx} = \frac{1}{\sin^4 x + \cos^4 x}$ (ii) $\frac{dy}{dx} = \frac{1 - \cos x}{1 + \cos x}$

Solution : (i) We have,

$$\frac{dy}{dx} = \frac{1}{\sin^4 x + \cos^4 x} \Rightarrow dy = \frac{1}{\sin^4 x + \cos^4 x}\,dx$$

Integrating both sides w.r. to x, we get

$$\int dy = \int \frac{dx}{\sin^4 x + \cos^4 x} + c = \int \frac{\sec^4 x\,dx}{\tan^4 x + 1} + c$$

$$= \int \frac{\sec^2 x \cdot \sec^2 x}{\tan^4 x + 1} dx + c = \int \frac{(1 + \tan^2 x) \cdot \sec^2 x}{\tan^4 x + 1} dx + c$$

Let $\tan x = t \Rightarrow \sec^2 x \, dx = dt$

$$\therefore \quad \int dy = \int \frac{1 + t^2}{1 + t^4} dt + c$$

Divide numerator and denominator by t^2

$$y = \int \frac{1 + \frac{1}{t^2}}{t^2 + \frac{1}{t^2}} dt + c = \int \frac{1 + \frac{1}{t^2}}{\left(t - \frac{1}{t}\right)^2 + 2} dt + c$$

Again, let $z = t - \frac{1}{t} \Rightarrow dz = \left(1 + \frac{1}{t^2}\right) dt$

$$\therefore \quad y = \int \frac{dz}{z^2 + 2} + c$$

$$\Rightarrow \quad y = \int \frac{dz}{z^2 + (\sqrt{2})^2} + c = \frac{1}{\sqrt{2}} \tan^{-1}\left(\frac{z}{\sqrt{2}}\right) + c$$

$$\Rightarrow \quad y = \frac{1}{\sqrt{2}} \tan^{-1}\left(\frac{t - \frac{1}{t}}{\sqrt{2}}\right) + c \Rightarrow y = \frac{1}{\sqrt{2}} \tan^{-1}\left(\frac{\tan x - \cot x}{\sqrt{2}}\right) + c$$

This is the required solution.

(ii) We have,

$$\frac{dy}{dx} = \frac{1 - \cos x}{1 + \cos x} \Rightarrow dy = \frac{1 - \cos x}{1 + \cos x} dx$$

Integraeting both sides, we get

$$\int dy = \int \frac{1 - \cos x}{1 + \cos x} dx + c \Rightarrow y = \int \frac{1 - \cos x}{1 + \cos x} dx + c$$

$$= \int \frac{1 - (1 - 2\sin^2 \frac{x}{2})}{1 + 2\cos^2 \frac{x}{2} - 1} dx + c = \int \frac{\sin^2 \frac{x}{2}}{\cos^2 \frac{x}{2}} dx + c$$

$$= \int \tan^2 \frac{x}{2} dx + c$$

$$= \int \left(\sec^2 \frac{x}{2} - 1\right) dx + c \qquad \left[\because \tan^2 \frac{x}{2} = \sec^2 \frac{x}{2} - 1\right]$$

$$= 2 \tan \frac{x}{2} - x + c$$

which is the required solution.

PROBLEM SET 3.1

Solve the following differential equations :

1. $(x+2)\dfrac{dy}{dx} = x^2 + 4x - 9$

Ans. $y = \dfrac{x^2}{2} + 2x - 13\log(x+2) + c$

2. $\dfrac{dy}{dx} = \sin^3 x \cos^2 x + xe^x$

Ans. $y = -\dfrac{1}{3}\cos^3 x + \dfrac{1}{5}\cos^5 x + xe^x - e^x + c$

3. $\dfrac{dy}{dx} = \dfrac{3e^{2x} + 3e^{4x}}{e^x + e^{-x}}$

Ans. $y = e^{3x} + c$

4. $\dfrac{dy}{dx} = \tan^{-1} x$

Ans. $y = x\tan^{-1} x - \dfrac{1}{2}\log(1+x^2) + c$

5. $e^{\frac{dy}{dx}} = x + 1$

Ans. $y = x\log(x+1) - x + \log(x+1) + c$

6. $\dfrac{dy}{dx} - x\sin^2 x = \dfrac{1}{x\log x}$

Ans. $y = \dfrac{x^2}{4} - \dfrac{x\sin 2x}{4} - \dfrac{\cos 2x}{8} + \log(\log x) + c$

7. $\sqrt{a+x}\,dy + x\,dx = 0$

Ans. $y + \dfrac{2}{3}(a+x)^{3/2} - 2a\sqrt{a+x} = c$

3.3 DIFFERENTIAL EQUATIONS OF THE FORM $\dfrac{dy}{dx} = f(y)$

This type of equations can be solved by using following procedure:

(i) We have

$$\frac{dy}{dx} = f(y) \Rightarrow \frac{dx}{dy} = \frac{1}{f(y)}$$

$$\Rightarrow \qquad dx = \frac{1}{f(y)}\,dy$$

(ii) On integrating both sides, we obtain

$$\int dx = \int \frac{1}{f(y)}\,dy + c$$

Example 3.3 **Solve :**

(i) $\dfrac{dy}{dx} + \dfrac{1+y^2}{y} = 0$ (ii) $\dfrac{dy}{dx} + y = 1$

Solution : (i) We have

$$\frac{dy}{dx} + \frac{1+y^2}{y} = 0$$

$$\Rightarrow \quad \frac{dy}{dx} = -\frac{1+y^2}{y} \Rightarrow \frac{dx}{dy} = -\frac{y}{1+y^2} \Rightarrow dx = -\frac{y}{1+y^2}\,dy$$

On integrating both sides, we obtain

$$\int dx = -\int \frac{y}{1+y^2}\,dy + c \Rightarrow x = -\frac{1}{2}\log(1+y^2) + c$$

$$\Rightarrow \quad x + \frac{1}{2}\log(1+y^2) = c$$

which is the required solution

(ii) We have

$$\frac{dy}{dx} + y = 1 \Rightarrow \frac{dy}{dx} = 1 - y \Rightarrow \frac{dx}{dy} = \frac{1}{1-y}$$

$$\Rightarrow \quad dx = \frac{1}{1-y}\,dy$$

On integrating both sides, we get

$$\int dx = \int \frac{1}{1-y}\,dy + c$$

$$\Rightarrow \quad x = -\log(1-y) + c \Rightarrow x + \log(1-y) = c$$

which is the required solution.

3.4 EQUATIONS IN VARIABLE SEPARABLE FORM

The differential equations of the form $f(x)\,dx = g(y)\,dy$ are said to be the equations in variable separable form. To solve this type of equations we integrate both sides. The solution is given by

$$\int f(x)\,dx = \int g(y)\,dy + c$$

where c is a constant of integration.

Example 3.4 **Solve $y\,dx - x\,dy = 0$**

Solution : We have $y\,dx - x\,dy = 0$

$$\Rightarrow \quad y\,dx = x\,dy$$

$$\Rightarrow \quad \frac{dx}{x} = \frac{dy}{y}$$

On integration, $\log x = \log y + \log c$

$$\Rightarrow \quad \log \frac{x}{y} = \log c \quad \Rightarrow \quad \frac{x}{y} = c$$

$$\Rightarrow \quad x = cy$$

Example 3.5 **Solve** $-x^2 \dfrac{dy}{dx} + y + 1 = 0$

Solution : We have,

$$-x^2 \frac{dy}{dx} + y + 1 = 0 \Rightarrow -x^2 \frac{dy}{dx} = -(y+1)$$

$$\Rightarrow \quad \frac{dy}{y+1} = \frac{dx}{x^2}$$

On integration, we get

$$\log (y+1) = -\frac{1}{x} + \log c$$

$$\log\left(\frac{y+1}{c}\right) = -\frac{1}{x}$$

$$\Rightarrow \quad y + 1 = ce^{-1/x}$$

Example 3.6 $\cot y \, dx - \tan x \, dy = 0$

Solution : We have, $\cot y \, dx - \tan x \, dy = 0$

$$\Rightarrow \quad \cot y \, dx = \tan x \, dy$$

$$\Rightarrow \quad \frac{dx}{\tan x} = \frac{dy}{\cot y}$$

$$\Rightarrow \quad \frac{\cos x}{\sin x} dx = \frac{\sin y}{\cos y} dy$$

On integration, we get

$$\log \sin x = -\log \cos y + \log c$$

$$\log \sin x + \log \cos y = \log c$$

$$\sin x \cos y = c$$

Example 3.7 **Solve :**

$$(x+1)\frac{dy}{dx} = x(y^2+1)$$

Solution : $(x+1)\dfrac{dy}{dx} = x(y^2+1)$

$$\Rightarrow \quad \frac{x}{(x+1)} dx = \frac{dy}{y^2+1}$$

On integration, we get

$$x - \log(x+1) = \tan^{-1} y + c$$

Example 3.8 **Solve :**

(i) $\cos x(1 + \cos y)\, dx - \sin y\,(1 + \sin x)\, dy = 0$

(ii) $\sec^2 x \tan y \, dx + \sec^2 y \tan x \, dy = 0$

Solution : (i) We have

$$\cos x\,(1 + \cos y)\, dx - \sin y\,(1 + \sin x)\, dy = 0$$

Separating the variables, we get

$$\frac{\cos x}{1+\sin x}\,dx = \frac{\sin y}{1+\cos y}\,dy$$

Integrating both sides, we obtain

$$\int \frac{\cos x}{1+\sin x}\,dx = \int \frac{\sin y}{1+\cos y}\,dy$$

$\Rightarrow \quad \log(1+\sin x) = -\log(1+\cos y) + \log c$

$\Rightarrow \quad \log(1+\sin x) + \log(1+\cos y) = \log c$

$\Rightarrow \quad \log[(1+\sin x)(1+\cos y)] = \log c \qquad [\because \log m + \log n = \log mn]$

$\Rightarrow \quad (1+\sin x)(1+\cos y) = c$

which is the required solution.

(ii) We have

$$\sec^2 x \tan y\,dx + \sec^2 y \tan x\,dy = 0$$

Separating the variables, we get

$$\frac{\sec^2 x}{\tan x}\,dx = -\frac{\sec^2 y}{\tan y}\,dy$$

Integrating both sides, we obtain

$$\int \frac{\sec^2 x}{\tan x}\,dx = -\int \frac{\sec^2 y}{\tan y}\,dy$$

$\Rightarrow \quad \log \tan x = -\log \tan y + \log c$

$\Rightarrow \quad \log \tan x + \log \tan y = \log c$

$\Rightarrow \quad \log(\tan x \tan y) = \log c$

$\Rightarrow \quad \tan x \tan y = c$

which is the required solution.

Example 3.9 **Solve :**

(i) $\dfrac{dy}{dx} = \dfrac{1+y^2}{1+x^2}$ ***[UPTU 2000, 01, 02, 05]***

(ii) $\dfrac{dy}{dx} = 1 + x + y + xy$

Solution : (i) We have

$$\frac{dy}{dx} = \frac{1+y^2}{1+x^2}$$

Separating the variables, we get

$$\frac{dy}{1+y^2} = \frac{dx}{1+x^2}$$

Integrating both sides, we obtain

$$\int \frac{dy}{1+y^2} = \int \frac{dx}{1+x^2}$$

$\Rightarrow \quad \tan^{-1} y = \tan^{-1} x + \tan^{-1}$

$$\Rightarrow \quad \tan^{-1} y - \tan^{-1} x = \tan^{-1} c$$

$$\Rightarrow \quad \tan^{-1}\left(\frac{y-x}{1+xy}\right) = \tan^{-1} c$$

$$\Rightarrow \quad \frac{y-x}{1+xy} = c \Rightarrow y - x = c\,(1+xy) \Rightarrow y = x + c\,(1+xy)$$

which is the required solution

(ii) We have

$$\frac{dy}{dx} = 1 + x + y + xy$$

$$\Rightarrow \quad \frac{dy}{dx} = (1+x) + y(1+x) = (1+x)\,(1+y)$$

Separating the variables, we get

$$\frac{dy}{1+y} = \frac{dx}{1+x}$$

Integrating both sides, we obtain

$$\int \frac{dy}{1+y} = \int \frac{dx}{1+x}$$

$$\Rightarrow \quad \log\,(1+y) = \log\,(1+x) + \log c$$

$$\Rightarrow \quad \log \frac{1+y}{1+x} = \log c$$

$$\Rightarrow \quad \frac{1+y}{1+x} = c \Rightarrow 1 + y = c\,(1+x)$$

which is the required solution.

Example 3.10 **Solve the differential equation $(1 + e^{2x})\,dy + (1 + y^2)\,e^x\,dx = 0$ given that when $x = 0, y = 1$.**

Solution : We have

$$(1 + e^{2x})\,dy + (1 + y^2)\,e^x\,dx = 0$$

$$\Rightarrow \quad (1 + e^{2x})\,dy = -\,(1 + y^2)\,e^x\,dx$$

Separating the variables, we get

$$\frac{dy}{1+y^2} = -\frac{e^x}{1+e^{2x}}\,dx$$

Integrating both sides, we obtain

$$\int \frac{dy}{1+y^2} = -\int \frac{e^x}{1+e^{2x}}\,dx$$

$$\Rightarrow \quad \tan^{-1} y = -\tan^{-1}(e^x) + c \qquad \text{...(1)}$$

But at $x = 0, y = 1$. From (1), we get

$$\tan^{-1}(1) = -\tan^{-1}(e^0) + c$$

$$\Rightarrow \quad \frac{\pi}{4} = -\frac{\pi}{4} + c \Rightarrow c = \frac{\pi}{2}$$

Putting $c = \frac{\pi}{2}$ in (1), we get

$$\tan^{-1} y = -\tan^{-1}(e^x) + \frac{\pi}{2} \quad \text{or} \quad \tan^{-1} y + \tan^{-1}(e^x) = \frac{\pi}{2}$$

which is the required solution.

Example 3.11 **Solve :**

(i) $y - x\frac{dy}{dx} = a\left(y^2 + \frac{dy}{dx}\right)$

(ii) $\frac{dy}{dx} = e^{x-y} + x^2 e^{-y}$

[UPTU 2000]

Solution : (i) We have,

$$y - x\frac{dy}{dx} = a\left(y^2 + \frac{dy}{dx}\right)$$

$$\Rightarrow \quad y - x\frac{dy}{dx} = ay^2 + a\frac{dy}{dx} \Rightarrow y - ay^2 = (x+a)\frac{dy}{dx}$$

Separating the variables, we get

$$\frac{dx}{x+a} = \frac{dy}{y - ay^2}$$

Integrating both sides, we obtain

$$\int \frac{dx}{x+a} = \int \frac{dy}{y - ay^2} \Rightarrow \int \frac{dx}{x+a} = \int \frac{dy}{y(1-ay)}$$

$$\Rightarrow \quad \int \frac{dx}{x+a} = \int \left(\frac{1}{y} + \frac{a}{1-ay}\right) dy \qquad \text{[Using partial fractions]}$$

$$\Rightarrow \quad \log(x+a) = \log y - \log(1-ay) + \log c$$

$$\Rightarrow \quad \log(x+a) - \log y + \log(1-ay) = \log c$$

$$\Rightarrow \quad \log\frac{(x+a)(1-ay)}{y} = \log c \Rightarrow \frac{(x+a)(1-ay)}{y} = c$$

$$\Rightarrow \quad (x+a)(1-ay) = cy$$

which is the required solution.

(ii) We have,

$$\frac{dy}{dx} = e^{x-y} + x^2 e^{-y} = e^x e^{-y} + x^2 e^{-y} = (e^x + x^2)e^{-y}$$

Separating the variables, we get

$$\frac{dy}{e^{-y}} = (e^x + x^2)\,dx \Rightarrow e^y\,dy = (e^x + x^2)\,dx$$

Integrating both sides, we get

$$\int e^y\,dy = \int (e^x + x^2)\,dx \Rightarrow e^y = e^x + \frac{x^3}{3} + c$$

which is the required solution.

PROBLEM SET 3.2

Solve the following differential equations :

1. $x(1+y^2)\,dx - y\,(1+x^2)\,dy = 0$

Ans. $1+x^2 = c\,(1+y^2)$

2. $\sin^3 x \dfrac{dx}{dy} = \sin y$

Ans. $\cos y - \dfrac{3}{4}\cos x + \dfrac{1}{12}\cos 3x = c$

3. $\dfrac{dy}{dx} = -\sqrt{\dfrac{1-y^2}{1-x^2}}$

Ans. $y\sqrt{1-x^2} + x\sqrt{1-y^2} = c$

4. $(x^2 - yx^2)\,dy + (y^2 + x^2y^2)\,dx = 0$

Ans. $\log y + \dfrac{1}{y} = -\dfrac{1}{x} + x + c$

5. $\dfrac{dy}{dx} = e^{x+y}$

Ans. $e^x + e^{-y} = c$

6. $x\sqrt{1-y^2}\,dx + y\sqrt{1-x^2}\,dy = 0$

Ans. $\sqrt{1-x^2} + \sqrt{1-y^2} = c$

7. $x\cos y\,dy = (xe^x\log x + e^x)\,dx$

Ans. $\sin y = e^x\log x + c$

8. $(e^y+1)\cos x\,dx + e^y\sin x\,dy = 0$

Ans. $\sin x\,(e^y+1) = c$

9. $\sqrt{1+x^2+y^2+x^2y^2} + xy\dfrac{dy}{dx} = 0$

Ans. $\sqrt{1+y^2} + \sqrt{1+x^2} + \dfrac{1}{2}\log\left(\dfrac{\sqrt{1+x^2}-1}{\sqrt{1+x^2}+1}\right) = c$

10. $\dfrac{dy}{dx} = 1 - x + y - xy$

Ans. $\log(1+y) = x - \dfrac{x^2}{2} + c$

11. $y\sqrt{1+x^2} + x\sqrt{1+y^2}\dfrac{dy}{dx} = 0$

Ans. $\sqrt{1+y^2} + \sqrt{1+x^2} + \dfrac{1}{2}\log\left(\dfrac{\sqrt{1+x^2}-1}{\sqrt{1+x^2}+1}\right) + \dfrac{1}{2}\log\left(\dfrac{\sqrt{1+y^2}-1}{\sqrt{1+y^2}+1}\right) + c$

12. $\dfrac{dx}{x} = \tan y \, dy$

Ans. $x \cos y = a$

13. $3e^x \tan y \, dx + (1 - e^x) \sec^2 y \, dy = 0$

Ans. $(1 - e^x)^3 = c \tan y$

14. $y \sec^2 x + (y + 7) \tan x \dfrac{dy}{dx} = 0$

Ans. $(e^y + 2) \cos x = c$

15. $(e^y + 2) \sin dx - e^y \cos x \, dy = 0$

Ans. $e^y = e^x + \dfrac{x^3}{3} + c$

16. $\dfrac{dy}{dx} = e^{x-y} + x^2 e^{-y}$

Ans. $e^y = e^x + \dfrac{x^3}{3} + c$

3.5 EQUATIONS REDUCIBLE TO VARIABLE SEPARABLE FORM

The differential equations of the type $\dfrac{dy}{dx} = f(ax + by + c)$ can be solved by reducing them to the variable separable form by taking $ax + by + c = t \Rightarrow a + b\dfrac{dy}{dx} = \dfrac{dt}{dx}$.

Example 3.12 **Solve :**

(i) $\dfrac{dy}{dx} = \cos(x + y) + \sin(x + y)$

(ii) $\dfrac{dy}{dx} = (4x + y + 1)^2$, **if** $y(0) = 1$ *[UPTU 2002]*

Solution : (i) We have

$$\frac{dy}{dx} = \cos(x + y) + \sin(x + y) \qquad \ldots(1)$$

Let $x + y = t \Rightarrow 1 + \dfrac{dy}{dx} = \dfrac{dt}{dx} \Rightarrow \dfrac{dy}{dx} = \dfrac{dt}{dx} - 1$

From (1), we get

$$\frac{dt}{dx} - 1 = \cos t + \sin t \Rightarrow \frac{dt}{dx} = 1 + \cos t + \sin t$$

Separating the variables, we get

$$\frac{dt}{1 + \cos t + \sin t} = dx$$

Integrating both sides, we obtain

$$\int \frac{dt}{1 + \cos t + \sin t} = \int dx + c$$

$$\Rightarrow \quad \int \frac{dt}{1+\frac{1-\tan^2\frac{t}{2}}{1+\tan^2\frac{t}{2}}+\frac{2\tan\frac{t}{2}}{1+\tan^2\frac{t}{2}}} = x + c$$

$$\Rightarrow \quad \int \frac{\sec^2\frac{t}{2}}{2\left(1+\tan\frac{t}{2}\right)}\,dt = x + c$$

$$\Rightarrow \quad \log\left(1+\tan\frac{t}{2}\right) = x + c$$

$$\Rightarrow \quad \log\left(1+\tan\left(\frac{x+y}{2}\right)\right) = x + c$$

which is the required solution.

(ii) We have

$$\frac{dy}{dx} = (4x+y+1)^2 \quad \ldots(1)$$

Let $\quad 4x+y+1 = t \Rightarrow 4 + \frac{dy}{dx} = \frac{dt}{dx} \Rightarrow \frac{dy}{dx} = \frac{dt}{dx} - 4$

From (1), we get

$$\frac{dt}{dx} - 4 = t^2 \Rightarrow \frac{dt}{dx} = t^2 + 4$$

Separating the variables, we get

$$\frac{dt}{t^2+4} = dx$$

On integration, we obtain

$$\int \frac{dt}{t^2+4} = \int dx \Rightarrow \frac{1}{2}\tan^{-1}\left(\frac{t}{2}\right) = x + c$$

$$\Rightarrow \quad \frac{1}{2}\tan^{-1}\left(\frac{4x+y+1}{2}\right) = x + c \quad \ldots(2)$$

But at $x = 0,\ y = 1$

$\therefore$ From (2), we have

$$\frac{1}{2}\tan^{-1}\left(\frac{1+1}{2}\right) = 0 + c \Rightarrow \frac{1}{2}\tan^{-1}(1) = c \Rightarrow c = \frac{1}{2}\cdot\frac{\pi}{4} \Rightarrow c = \frac{\pi}{8}$$

From (2), we obtain

$$\frac{1}{2}\tan^{-1}\left(\frac{4x+y+1}{2}\right) = x + \frac{\pi}{8}$$

which is the required solution.

Example 3.13 **Solve :**

(i) $\mathbf{\cos(x+y)\, dy = dx}$ **(ii)** $\mathbf{\frac{dy}{dx} = \frac{(x-y)+3}{2(x-y)+5}}$

Solution : (i) Given

$$\cos(x+y)\, dy = dx \Rightarrow \frac{dy}{dx} = \frac{1}{\cos(x+y)} \qquad \ldots(1)$$

Let $x + y = t \Rightarrow 1 + \frac{dy}{dx} = \frac{dt}{dx} \Rightarrow \frac{dy}{dx} = \frac{dt}{dx} - 1$

From (1), we get

$$\frac{dt}{dx} - 1 = \frac{1}{\cos t} \Rightarrow \frac{dt}{dx} = \frac{1}{\cos t} + 1 = \frac{1+\cos t}{\cos t}$$

On separating the variables, we get

$$\frac{\cos t}{1+\cos t}\, dt = dx \Rightarrow \frac{\cos t\,(1-\cos t)}{1-\cos^2 t}\, dt = dx$$

$$\Rightarrow \frac{\cos t(1-\cos t)}{\sin^2 t}\, dt = dx \Rightarrow \left(\frac{\cos t}{\sin^2 t} - \frac{\cos^2 t}{\sin^2 t}\right) dt = dx$$

$$\Rightarrow (\cot t \cdot \operatorname{cosec} t - \cot^2 t)\, dt = dx$$

$$\Rightarrow (\cot t \cdot \operatorname{cosec} t - \operatorname{cosec}^2 t + 1)\, dt = dx \qquad [\because \cot^2 t = \operatorname{cosec}^2 t - 1]$$

On integrating, we have

$$\int \cot t \operatorname{cosec} t\, dt - \int \operatorname{cosec}^2 t\, dt + \int dt = \int dx + c$$

$$\Rightarrow -\operatorname{cosec} t + \cot t + t = x + c$$

$$\Rightarrow -\operatorname{cosec}(x+y) + \cot(x+y) + (x+y) = x + c \qquad [\because t = x + y]$$

$$\Rightarrow -\operatorname{cosec}(x+y) + \cot(x+y) + y = c$$

which is the required solution.

(ii) We have,

$$\frac{dy}{dx} = \frac{x-y+3}{2(x-y)+5} \qquad \ldots(1)$$

Let $x - y = t \Rightarrow 1 - \frac{dy}{dx} = \frac{dt}{dx} \Rightarrow \frac{dy}{dx} = 1 \; \frac{dt}{dx}$

From (1), we get

$$1 - \frac{dt}{dx} = \frac{t+3}{2t+5} \Rightarrow \frac{dt}{dx} = 1 - \frac{t+3}{2t+5} = \frac{2t+5-t-3}{2t+5}$$

$$\Rightarrow \frac{dt}{dx} = \frac{t+2}{2t+5}$$

Separating the variables, we have

$$\frac{2t+5}{t+2}\, dt = dx \Rightarrow \left(2 + \frac{1}{t+2}\right) dt = dx$$

Integrating both sides, we get

$$\int \left(2 + \frac{1}{t+2}\right) dt = \int dx + c$$

$\Rightarrow$ $2t + \log (t + 2) = x + c$

$\Rightarrow$ $2(x - y) \log (x - y + 2) = x + c$ $[\because t = x - y]$

which is the required solution.

PROBLEM SET 3.3

Solve the following differential equations :

1. $\left[\dfrac{x+y-a}{x+y-b}\right]\dfrac{dy}{dx} = \dfrac{x+y+a}{x+y+b}$

Ans. $(b - a) \log [(x + y)^2 - ab] = 2 (x - y + c)$

2. $\dfrac{dy}{dx} = 1 + \tan (y - x)$

Ans. $\sin (y - x) = e^{x + c}$

3. $(x + y)^2 \dfrac{dy}{dx} = a^2$

Ans. $(x + y) - a \tan^{-1}\left(\dfrac{x+y}{a}\right) = x + c$

4. $\sin^{-1}\left(\dfrac{dy}{dx}\right) = x + y$

Ans. $x = \tan (x + y) - \sec (x + y) + c$

5. $(x + y) (dx - dy) = dx + dy$

Ans. $\dfrac{1}{2}(y - x) + \dfrac{1}{2}\log (x + y) = c$

6. $\dfrac{dy}{dx} = (2x - y + 1)^2$

Ans. $2x - y + 1 - \sqrt{2} = c(2x - y + 1 + \sqrt{2})e^{-2\sqrt{2}x}$

7. $\dfrac{dy}{dx}(x + y + 1) = 1$

Ans. $x = ce^y - y - 2$

8. $(x - y)^2 \dfrac{dy}{dx} = a^2$

Ans. $y + c = \dfrac{a}{2}\log\left(\dfrac{x-y-a}{x-y+a}\right)$

9. $(x + y)^2 \dfrac{dy}{dx} = a^2$

Ans. $y - a \tan^{-1}\left[\dfrac{x+y}{a}\right] = c$

10. $(4x + y)^2 \dfrac{dx}{dy} = 1$

Ans. $\tan^{-1} \dfrac{4x + y}{2} = 2x + c$

3.6 HOMOGENEOUS DIFFERENTIAL EQUATIONS

3.6.1 Homogeneous Functions

If $f(\lambda x, \lambda y) = \lambda^n f(x, y)$ then the function $f(x, y)$ is known as a homogeneous function of degree n.

For example, $f(x, y) = x^3 + y^3 + 3x^2y + 2xy^2$ is a homogeneous function of degree 3.

Here,
$$f(\lambda x, \lambda y) = \lambda^3 x^3 + \lambda^3 y^3 + 3\lambda^2 x^2 \lambda y + 2\lambda x \lambda^2 y^2$$
$$= \lambda^3 (x^3 + y^3 + 3x^2y + 2xy^2) = \lambda^3 f(x, y)$$

3.6.2 Homogeneous Differential Equation

A differential equation of the form

$$\frac{dy}{dx} = \frac{f(x, y)}{g(x, y)}$$

where $f(x, y)$ and $g(x, y)$ are homogeneous functions of x and y of same degree is known as homogeneous differential equation of first order and first degree.

For example,

$$\frac{dy}{dx} = \frac{2xy}{x^2 + y^2}, \quad \frac{dy}{dx} = \frac{x^2 + y^2 + 2xy}{xy + y^2} \text{ etc.}$$

Working Rule

To solve homogeneous differential equations we use the following procedure.

1. Write the given differential equation in the form

$$\frac{dy}{dx} = \frac{f(x, y)}{g(x, y)}$$

2. Now, Put $y = vx$ and $\dfrac{dy}{dx} = v + x\dfrac{dv}{dx}$ in step 1 and cancel out x from the right hand side, we get the following equation

$$v + x\frac{dv}{dx} = \varphi(v)$$

3. Separate the variables and integrate both sides to obtain the solution in terms of v and x.

4. Replace v by $\dfrac{y}{x}$ to obtain the required solution.

Example 3.14 **Solve $x^2\, dy + y\,(x + y)\, dx = 0$.** *[UPTU 2002]*

Solution : Given $x^2\, dy + y\,(x + y)\, dx = 0$

$$\Rightarrow \quad \frac{dy}{dx} = -\frac{y(x + y)}{x^2} \qquad \ldots(1)$$

It is clear that each of the functions $y(x+y)$ and x^2 is homogeneous function of degree 2, therefore equation (1) is a homogeneous differential equation of degree 2.

Let $y = vx \Rightarrow \dfrac{dy}{dx} = v + x\dfrac{dv}{dx}$

From (1), we get

$$v + x\frac{dv}{dx} = -\frac{vx\,(x+vx)}{x^2} = -v\,(1+v) = -v - v^2$$

$$\Rightarrow \quad x\frac{dv}{dx} = -v - v^2 - v = -2v - v^2 = -(2v + v^2)$$

Separting the variables, we obtain

$$\frac{dv}{2v+v^2} = -\frac{dx}{x} \Rightarrow \frac{1}{v(2+v)}\,dv = -\frac{dx}{x}$$

$$\Rightarrow \quad \frac{1}{2}\left[\frac{1}{v} - \frac{1}{2+v}\right]dv = -\frac{dx}{x}$$

On integration, we obtain

$$\frac{1}{2}\int\left(\frac{1}{v} - \frac{1}{2+v}\right)dv = -\int\frac{dx}{x}$$

$$\Rightarrow \quad \frac{1}{2}\,[\log v - \log(2+v)] = -\log x + \log c$$

$$\Rightarrow \quad \frac{1}{2}\log\frac{v}{2+v} = \log\frac{c}{x} \Rightarrow \log\frac{v}{2+v} = \log\frac{c^2}{x^2}$$

$$\Rightarrow \quad \frac{v}{2+v} = \frac{c^2}{x^2}$$

Now, substituting $v = \dfrac{y}{x}$, we have

$$\frac{\frac{y}{x}}{2+\frac{y}{x}} = \frac{c^2}{x^2} \Rightarrow \frac{y}{2x+y} = \frac{c^2}{x^2} \Rightarrow yx^2 = c^2\,(2x+y)$$

which is the required solution.

Example 3.15 **Solve :**

$$\boldsymbol{x\,dy - y\,dx = \sqrt{x^2+y^2}\,dx}$$

Solution : Given

$$x\,dy - y\,dy = \sqrt{x^2+y^2}\,dx$$

$$\Rightarrow \quad \frac{dy}{dx} = \frac{\sqrt{x^2+y^2}+y}{x} \qquad \ldots(1)$$

Clearly, it is a homogeneous differential equation of degree 1.

Put $y = vx$ and $\dfrac{dy}{dx} = v + x\dfrac{dv}{dx}$ in (1), we have

$$v + x\frac{dv}{dx} = \frac{\sqrt{x^2 + v^2x^2} + vx}{x} = \sqrt{1+v^2} + v$$

$$\Rightarrow \quad x\frac{dv}{dx} = \sqrt{1+v^2} + v - v = \sqrt{1+v^2}$$

Separating the variables, we obtain

$$\frac{dv}{\sqrt{1+v^2}} = \frac{dx}{x}$$

Integrating both sides, we get

$$\int \frac{dv}{\sqrt{1+v^2}} = \int \frac{dx}{x} \Rightarrow \log(v + \sqrt{1+v^2}) = \log x + \log c$$

$$\Rightarrow \quad \log(v + \sqrt{1+v^2}) = \log cx \Rightarrow v + \sqrt{1+v^2} = cx \qquad \ldots(2)$$

Substituting $v = \frac{y}{x}$ in (2), we get

$$\frac{y}{x} + \sqrt{1 + \frac{y^2}{x^2}} = cx \Rightarrow y + \sqrt{x^2 + y^2} = cx^2$$

which is the required solution.

Example 3.16 **Solve :**

$$(1 + e^{x/y})\,dx + e^{x/y}\left(1 - \frac{x}{y}\right)dy = 0$$

Solution : The differential equation can be written as:

$$e^{x/y}\left(1 - \frac{x}{y}\right) + (1 + e^{x/y})\frac{dx}{dy} = 0 \qquad \ldots(1)$$

Putting $x = vy$ and $\frac{dx}{dy} = v + y\frac{dv}{dy}$ in (1), we get

$$e^v(1 - v) + (1 + e^v)\left(v + y\frac{dv}{dy}\right) = 0$$

$$\Rightarrow \quad e^v - ve^v + v + ve^v + y(1 + e^v)\frac{dv}{dy} = 0 \Rightarrow e^v + v + y(1 + e^v)\frac{dv}{dy} = 0$$

$$\Rightarrow \quad y(1 + e^v)\frac{dv}{dy} = -v - e^v = -(v + e^v)$$

Separating the variables, we get

$$\frac{1 + e^v}{v + e^v}dv = -\frac{dy}{y}$$

Integrating both sides, we obtain

$$\int \frac{1 + e^v}{v + e^v}dv = -\int \frac{dy}{y} \Rightarrow \log(v + e^v) = -\log y + \log c$$

$$\Rightarrow \quad \log(v + e^v) = \log\frac{c}{y} \Rightarrow v + e^v = \frac{c}{y} \qquad \ldots(2)$$

Substituting $v = \dfrac{x}{y}$ in (2), we get

$$\frac{x}{y} + e^{x/y} = \frac{c}{y} \Rightarrow x + y\, e^{x/y} = c$$

which is the required solution.

Example 3.17 **Solve the initial value problem**

$$(3xy + y^2)\, dx + (x^2 + xy)\, dy = 0,\; y(1) = 1$$

Solution : The given differential equation can be written as

$$\frac{dy}{dx} = -\frac{3xy + y^2}{x^2 + xy} \qquad \ldots(1)$$

Put $y = vx$ and $\dfrac{dy}{dx} = x\dfrac{dv}{dx}$ in (1), we get

$$v + x\frac{dv}{dx} = -\frac{3vx^2 + v^2x^2}{x^2 + x^2v} = \frac{3v + v^2}{1 + v}$$

$$\Rightarrow \qquad x\frac{dv}{dx} = -\frac{3v + v^2}{1 + v} - v$$

$$= \frac{-3v - v^2 - v(1 + v)}{1 + v} = \frac{-3v - v^2 - v - v^2}{1 + v}$$

$$= \frac{-4v - 2v^2}{1 + v} = \frac{-2v(2 + v)}{1 + v}$$

Separating the variables, we have

$$\frac{1 + v}{v(2 + v)}\, dv = -2\frac{dx}{x}$$

$$\Rightarrow \qquad \left(\frac{1}{v} + \frac{1}{2 + v}\right) dv = -4\frac{dx}{x}$$

Integrating both sides, we obtain

$$\int\left(\frac{1}{v} + \frac{1}{2 + v}\right) dv = -4\int\frac{dx}{x}$$

$$\Rightarrow \qquad \log v + \log(2 + v) = -4\log x + \log c$$

$$\Rightarrow \qquad \log v(2 + v) = \log\frac{c}{x^4} \Rightarrow v(2 + v) = \frac{c}{x^4} \qquad \ldots(2)$$

Substituting $v = \dfrac{y}{x}$ in (2), we get

$$\frac{y}{x}\left(2 + \frac{y}{x}\right) = \frac{c}{x^4} \Rightarrow y(2x + y) = \frac{c}{x^2} \Rightarrow c = x^2 y\,(2x + y) \qquad \ldots(3)$$

But at $x = 1, y = 1$. From (3), we get $c = 3$

$\therefore \qquad x^2 y\,(2x + y) = 3$

which is required solution.

Example 3.18 **Solve $(2xy + x^2)\dfrac{dy}{dx} = 3y^2 + 2xy$.**

Solution : We have, $(2xy + x^2)\dfrac{dy}{dx} = 3y^2 + 2xy$

$$\Rightarrow \quad \frac{dy}{dx} = \frac{3y^2 + 2xy}{2xy + x^2}$$

Let $\quad y = vx \Rightarrow \dfrac{dy}{dx} = v + x\dfrac{dv}{dx}$

On substituting it in the given differential equation, we get

$$v + x\frac{dv}{dx} = \frac{3v^2x^2 + 2xvx}{2x\,vx + x^2} = \frac{v^2 3x^2 + 2x^2 v}{2x^2 v + x^2}$$

$$v + x\frac{dv}{dx} = \frac{3v^2 + 2v}{2v + 1}$$

$$\Rightarrow \quad x\frac{dv}{dx} = \frac{3v^2 + 2v - 2v^2 - v}{2v + 1}$$

$$\Rightarrow \quad x\frac{dv}{dx} = \frac{v^2 + v}{2v + 1}$$

$$\Rightarrow \quad \frac{2v + 1}{v^2 + v}\,dv = \frac{dx}{x}$$

On integration, we get

$$\int \frac{2v + 1}{v^2 + v}\,dv = \int \frac{dx}{x} + \log c$$

$$\log(v^2 + v) = \log x + \log c$$

$$\Rightarrow \quad v^2 + v = cx$$

$$\Rightarrow \quad \frac{y^2}{x^2} + \frac{y}{x} = cx$$

$$\Rightarrow \quad y^2 + xy = cx^3$$

Example 3.19 **Solve the equation**

$$\frac{dy}{dx} = \frac{y}{x} + x\sin\frac{y}{x}$$

Solution : We have,

$$\frac{dy}{dx} = \frac{y}{x} + x\sin\frac{y}{x}$$

Let $\quad y = vx \Rightarrow \dfrac{dy}{dx} = v + x\dfrac{dv}{dx}$

On substituting it into the given differential equation, we get

$$v + x\frac{dv}{dx} = \frac{vx}{x} + x\sin\frac{vx}{x}$$

$$\Rightarrow \quad v + x\frac{dv}{dx} = v + x\sin v$$

$$\Rightarrow \quad x\frac{dv}{dx} = x \sin v$$

$$\Rightarrow \quad \frac{dv}{\sin v} = dx$$

On integration, we get

$$\int \operatorname{cosec} v \, dv = \int dx + c$$

$$\Rightarrow \quad \log \tan \frac{v}{2} = x + c$$

$$\Rightarrow \quad \log \tan \frac{y}{2x} = x + c$$

PROBLEM SET 3.4

Solve the following differential equations :

1. $x^2 \, dy + y\,(x + y)\, dx = 0$

Ans. $x^2 y = c\,(y + 2x)$

2. $x\frac{dy}{dx} = x + y$

Ans. $y = x \log x + cx$

3. $\frac{dy}{dx} = \frac{x+y}{x-y}$

Ans. $\tan^{-1}\left(\frac{y}{x}\right) = \frac{1}{2}\log\,(x^2 + y^2) + c$

4. $2x\,(x + y)\frac{dy}{dx} = 3y^2 + 4xy,\ y(1) = 1$

Ans. $y^2 + 2xy = 3x^3$

5. $[x\,(x^2 - y^2)^{-1/2} + e^{y/x}]\, x\frac{dy}{dx} = x + \left[x\,(x^2 - y^2)^{-1/2} + e^{y/x}\right] y,\ y(1) = 1$

Ans. $e^{y/x} + \sin^{-1}(y/x) = \log(x) + e + \frac{\pi}{2}$

6. $(x + y)\, dy + (x - y)\, dx = 0,\ y(1) = 1$

Ans. $\log\,(x^2 + y^2) + 2\tan^{-1}\left(\frac{y}{x}\right) = \log 2 + \frac{\pi}{2}$

7. $x\frac{dy}{dx}\cos\left(\frac{y}{x}\right) = y\cos\left(\frac{y}{x}\right) - x$

Ans. $\sin\left(\frac{y}{x}\right) = -\log x + c$

8. $x^2 \dfrac{dy}{dx} = x^2 - 2y^2 + xy$

Ans. $\dfrac{x + \sqrt{2}y}{x - \sqrt{2}y} = (cx)^{2\sqrt{2}}$

9. $[x \sqrt{x^2 + y^2} - y^2]\, dx + xy\, dy = 0$

Ans. $\sqrt{x^2 + y^2} = x \log \dfrac{c}{x}$

10. $x \cos \left(\dfrac{y}{x}\right) (y\, dx + x\, dy) = y \sin \left(\dfrac{y}{x}\right) \cdot (x\, dy - y\, dx)$

Ans. $\sec \dfrac{y}{x} = c\, xy$

11. $\dfrac{dy}{dx} + \dfrac{x - 2y}{2x - y} = 0$

Ans. $y - x = c\,(x + y)^3$

12. $\dfrac{dy}{dx} = \dfrac{3xy + y^2}{3x^2}$

Ans. $3x + y \log x + cy = 0$

13. $(x^2 + y^2)\; dy = xy\, dx$

Ans. $-\dfrac{x^2}{2y^2} + \log y = c$

14. $x^2y\, dx - (x^3 + y^3)\, dy = 0$

Ans. $-\dfrac{x^3}{3y^3} + \log y = c$

15. $\dfrac{dy}{dx} = \dfrac{y + \sqrt{x^2 + y^2}}{x}$

Ans. $y + \sqrt{x^2 + y^2} = cx^2$

16. $x \dfrac{dy}{dx} = y\,(\log y - \log x + 1)$

Ans. $\log \dfrac{y}{x} = cx$

3.7 EQUATIONS REDUCIBLE TO HOMOGENEOUS FORM

We have some differential equations which are non-homogeneous. These equations can be solved by reducing them into homogeneous form.

A differential equation of the form

$$\frac{dy}{dx} = \frac{a_1x + b_1y + c_1}{a_2x + b_2y + c_2} \qquad \ldots(1)$$

can be reduced to the homogeneous form as follows:

CASE 1 : If $\frac{a_1}{a_2} \neq \frac{b_1}{b_2}$, then we substitute $x = X + h$, $Y = y + k$ in the given differential equation, we get

$$\frac{dY}{dX} = \frac{a_1X + b_1Y + (a_1h + b_1k + c_1)}{a_2X + b_2Y + (a_2h + b_2k + c_2)} \quad ...(2)$$

Here, h and k are constant such that

$$a_1h + b_1k + c_1 = 0$$

and

$$a_2h + b_2k + c_2 = 0$$

Then, equation (2) simplifies to

$$\frac{dY}{dX} = \frac{a_1X + b_1Y}{a_2X + b_2Y}$$

which is a homogeneous equation in the variables X, Y. We solve this equation and substitute $X = x - h$ and $Y = y - k$ to get the general solution.

CASE 2 : If $\frac{a_1}{a_2} = \frac{b_1}{b_2} = r$ (say)

From (1), we have

$$\frac{dy}{dx} = \frac{r(a_2x + b_2y) + c_1}{a_2x + b_2y + c_2} \quad ...(3)$$

Put $a_2x + b_2y = t$ and $a_2 + b_2\frac{dy}{dx} = \frac{dt}{dx}$ in (3) to eliminate y from the equation. The new equation so obtained can be solved by separating the variables as discussed earlier.

Example 3.20 **Solve :**

(i) $\frac{dy}{dx} = \frac{y - x + 1}{y + x + 5}$ *[Raj 94]*

(ii) $(2x + y - 3)\,dy = (x + 2y - 3)\,dx$

Solution : (i) We have

$$\frac{dy}{dx} = \frac{y - x + 1}{y + x + 5} \quad ...(1)$$

Here $a_1 = -1, b_1 = 1, c_1 = 1$ and $a_2 = 1, b_2 = 1, c_2 = 5$

Clearly, $\frac{a_1}{a_2} \neq \frac{b_1}{b_2}$. Thus, given equation can be reduced to homogeneous form by taking new variables X and Y such that

$$x = X + h, \; y = Y + k$$

Then

$$\frac{dy}{dx} = \frac{dY}{dX}$$

From (1), We get

$$\frac{dY}{dX} = \frac{Y + k - X - h + 1}{Y + k + X + h + 5} = \frac{Y - X + (k - h + 1)}{Y + X + (k + h + 5)} \quad ...(2)$$

Choose h and k such that,

$$\left.\begin{array}{l} k - h + 1 = 0 \\ k + h + 5 = 0 \end{array}\right] \Rightarrow k = -3, h = -2$$

Now from (2), we have

$$\frac{dY}{dX} = \frac{Y - X}{Y + X} \qquad \ldots(3)$$

This is homogeneous equation. Put $Y = vX$ and $\frac{dY}{dX} = v + X\frac{dv}{dX}$ in (3), we obtain

$$v + X\frac{dv}{dX} = \frac{vX - X}{vX + X} = \frac{v-1}{v+1}$$

$$\Rightarrow \quad X\frac{dv}{dX} = \frac{v-1}{v+1} - v = \frac{v - 1 - v^2 - v}{v+1} = -\frac{1+v^2}{1+v}$$

Separating the variables, we get

$$\frac{1+v}{1+v^2}\,dv = -\frac{dX}{X}$$

$$\Rightarrow \quad \frac{1}{1+v^2}\,dv + \frac{v}{1+v^2}\,dv = -\frac{dX}{X}$$

On integrating, we get

$$\int \frac{1}{1+v^2}\,dv + \int \frac{v}{1+v^2}\,dv = -\int \frac{dX}{X}$$

$$\Rightarrow \quad \tan^{-1} v + \frac{1}{2}\log(1+v^2) = -\log X + \log c$$

$$\Rightarrow \quad \tan^{-1} v + \log\sqrt{1+v^2} = \log\frac{c}{X}$$

$$\Rightarrow \quad \tan^{-1} v = \log\frac{c}{X\sqrt{1+v^2}} \qquad \ldots(4)$$

Substituting $v = \frac{Y}{X}$ in (4), we have

$$\tan^{-1}\frac{Y}{X} = \log\frac{c}{\sqrt{X^2+Y^2}} \qquad \ldots(5)$$

Again substitute $Y = y + 3$ and $X = x + 2$ in (5), we get

$$\tan^{-1}\left(\frac{y+3}{x+2}\right) = \log\frac{c}{\sqrt{(x+2)^2 + (y+3)^2}}$$

which is the required solution.

(ii) Given $(2x + y - 3)\,dy = (x + 2y - 3)\,dx$

$$\Rightarrow \quad \frac{dy}{dx} = \frac{x + 2y - 3}{2x + y - 3} \qquad \ldots(1)$$

Here, $a_1 = 1,\ b_1 = 2,\ c_1 = -3$ and $a_2 = 2, b_2 = 1,\ c_2 = -3$

Clearly, $$\frac{a_1}{a_2} \neq \frac{b_1}{b_2}$$

Putting $x = X + h,\ y = Y + k$ in (1), we have

$$\frac{dY}{dX} = \frac{X + 2Y + (h + 2k - 3)}{2X + Y + (2h + k - 3)} \qquad ...(2)$$

Choose h and k, such that

$$\left.\begin{matrix} h + 2k - 3 = 0 \\ 2h + k - 3 = 0 \end{matrix}\right] \Rightarrow h = 1 \text{ and } k = 1$$

From (2), we have

$$\frac{dY}{dX} = \frac{X + 2Y}{2X + Y} \qquad ...(3)$$

This is a homogeneous equation.

Put $Y = vX$ and $\dfrac{dY}{dX} = v + X\dfrac{dv}{dX}$ in (3), we obtain

$$v + X\frac{dv}{dX} = \frac{X + 2vX}{2X + vX} = \frac{1 + 2v}{2 + v}$$

$$\Rightarrow \quad X\frac{dv}{dX} = \frac{1 + 2v}{2 + v} - v = \frac{1 + 2v - 2v - v^2}{2 + v} = \frac{1 - v^2}{2 + v}$$

Separating the variables, we get

$$\frac{2 + v}{1 - v^2}\,dv = \frac{dX}{X}$$

$$\Rightarrow \quad \frac{2 + v}{(1 - v)(1 + v)}\,dv = \frac{dX}{X}$$

$$\Rightarrow \quad \left(\frac{1}{2}\cdot\frac{1}{1 + v} + \frac{3}{2}\cdot\frac{1}{1 - v}\right)dv = \frac{dX}{X}$$

Integrating both sides, we get

$$\frac{1}{2}\int\frac{dv}{1 + v} + \frac{3}{2}\int\frac{dv}{1 - v} = \int\frac{dX}{X}$$

$$\Rightarrow \quad \frac{1}{2}\log(1 + v) - \frac{3}{2}\log(1 - v) = \log X + \log c$$

$$\Rightarrow \quad \log\left[\frac{1 + v}{(1 - v)^3}\right] = \log(c^2 X^2)$$

$$\Rightarrow \quad \frac{1 + v}{(1 - v)^3} = c^2X^2 \qquad ...(4)$$

Substitute $v = \dfrac{Y}{X}$ in (4), we get

$$\frac{1 + \dfrac{Y}{X}}{\left(1 - \dfrac{Y}{X}\right)^3} = c^2X^2 \Rightarrow \frac{X + Y}{(X - Y)^3} = c^2$$

$$\Rightarrow \quad X + Y = c^2(X - Y)^3 \qquad ...(5)$$

Again, substitute $X = x - 1$ and $Y = y - 1$ in (5), we get

$$x - 1 + y - 1 = c^2 (x - 1 - y + 1)^3 \Rightarrow x + y - 2 = c^2 (x - y)^3$$

which is the required solution.

Example 3.21 **Solve :**

(i) Given $\dfrac{dy}{dx} = \dfrac{x + 2y - 1}{x + 2y + 1}$

(ii) $\dfrac{dy}{dx} = \dfrac{3y + 2x + 4}{4x + 6y + 5}$

[Raj. 2003, 04; RJU 2007]

Solution : (i) Given

$$\frac{dy}{dx} = \frac{x + 2y - 1}{x + 2y + 1} \quad \text{...(1)}$$

Here, $a_1 = 1,\ b_1 = 2,\ c_1 = -1$ and $a_2 = 1,\ b_2 = 2,\ c_2 = 1$

Clearly, $\dfrac{a_1}{a_2} = \dfrac{b_1}{b_2} = 1$

Now substitute

$$x + 2y = v \Rightarrow 1 + 2\frac{dy}{dx} = \frac{dv}{dx} \Rightarrow \frac{dy}{dx} = \frac{1}{2}\left(\frac{dv}{dx} - 1\right)$$

From (1), we have

$$\frac{1}{2}\left(\frac{dv}{dx} - 1\right) = \frac{v - 1}{v + 1} \Rightarrow \frac{dv}{dx} - 1 = \frac{2v - 2}{v + 1}$$

$$\Rightarrow \quad \frac{dv}{dx} = \frac{2v - 2}{v + 1} + 1 = \frac{2v - 2 + v + 1}{v + 1} = \frac{3v - 1}{v + 1}$$

Separating the variables, we obtain

$$\frac{v + 1}{3v - 1}\, dv = dx \Rightarrow \left[\frac{1}{3} + \frac{4}{3(3v - 1)}\right] dv = dx$$

Integrating both sides, we get

$$\int \frac{1}{3}\, dv + \frac{4}{3}\int \frac{dv}{3v - 1} = \int dx$$

$$\Rightarrow \quad \frac{1}{3}v + \frac{4}{9}\log(3v - 1) = x + c \quad \text{...(2)}$$

Putting $v = x + 2y$ in (2), we have

$$\frac{1}{3}(x + 2y) + \frac{4}{9}\log(3x + 6y - 1) = x + c$$

which is the required solution.

(ii) Given $\dfrac{dy}{dx} = \dfrac{3y + 2x + 4}{4x + 6y + 5} = \dfrac{3y + 2x + 4}{2(3y + 2x) + 5}$ ...(1)

Here $\dfrac{a_1}{a_2} = \dfrac{b_1}{b_2} = \dfrac{1}{2}$

Put $3y + 2x = v$

$$\Rightarrow \quad 3\frac{dy}{dx} + 2 = \frac{dv}{dx} \Rightarrow \frac{dy}{dx} = \frac{1}{3}\left(\frac{dv}{dx} - 2\right)$$

From (1), we have

$$\frac{1}{3}\left(\frac{dv}{dx}-2\right)=\frac{v+4}{2v+5} \Rightarrow \frac{dv}{dx}-2=\frac{3v+12}{2v+5}$$

$$\Rightarrow \quad \frac{dv}{dx}=\frac{3v+12}{2v+5}+2 \Rightarrow \frac{dv}{dx}=\frac{3v+12+4v+10}{2v+5}=\frac{7v+22}{2v+5}$$

Separating the variables, we get

$$\frac{2v+5}{7v+22}\,dv=dx \Rightarrow \left[\frac{2}{7}-\frac{9}{7(7v+22)}\right]dv=dx$$

Integrating both sides, we obtain

$$\frac{2}{7}\int dv-\frac{9}{7}\int\frac{dv}{7v+22}=\int dx+c$$

$$\Rightarrow \quad \frac{2}{7}v-\frac{9}{49}\log(7v+22)=x+c \qquad \ldots(2)$$

Substitute $v=3y+2x$ in (2), we have

$$\frac{2}{7}(3y+2x)-\frac{9}{49}\log(21y+14x+22)=x+c$$

which is the required solution.

Example 3.22 **Solve :**

$$\frac{dy}{dx}=\frac{7x-3y-z}{7y-3x+3}$$

Solution : Here

$$\frac{a}{A}\neq\frac{b}{B} \qquad \left(\because \frac{7}{-3}\neq\frac{-3}{7}\right)$$

Let $x=X+h,\ y=Y+k$, then $\dfrac{dy}{dx}=\dfrac{dY}{dX}$

$$\therefore \quad \frac{dY}{dX}=\frac{7(X+h)-b(Y+k)-7}{7(Y+k)-3(X+h)+3}$$

$$\text{or} \quad \frac{dY}{dX}=\frac{7X-3Y+(7h-3k-7)}{7Y-3X+(7k-3h+3)} \qquad \ldots(1)$$

Choose h and k such that

$$7h-3k-7=0 \text{ and } 7h-5h+3=0$$

$$\Rightarrow \quad h=1 \text{ and } k=0 \qquad \ldots(2)$$

$$\therefore \quad \frac{dY}{dX}=\frac{7X-3Y}{7Y-3X} \qquad \ldots(3)$$

Putting $Y=vX$

$$\frac{dY}{dX}=v+X\frac{dv}{dX} \qquad \ldots(4)$$

From (3) and (4), we get

$$v+X\frac{dv}{dX}=\frac{7X-3vX}{7vX-3X}$$

or $$\frac{7v-3}{7(1-v^2)}\,dv = \frac{dX}{X}$$

or $$-\frac{1}{7}\left[\frac{5}{v+1}+\frac{2}{v-1}\right]dv = \frac{dX}{X}$$

Integrating

$$-\frac{1}{7}[5\log(v+1)+2\log(v-1)] = \log X + c\log C$$

or $$\log X^7 (v+1)^5 (v-1)^2 = \log C$$

or $$X^7 (v+1)^5 (v-1)^2 = C$$

or $$(Y+X)^5 (Y-X)^2 = C$$

or $$(y+x-1)^5 (y-x+1)^2 = C$$

Example 3.23 **Solve :**

$$\frac{dy}{dx} = \frac{2x-y+1}{x+2y+3}$$ ***(Raj. 1998 B.F. 2005)***

Solution : Here

$$\frac{a}{A} \neq \frac{b}{B} \qquad \left(\because \frac{2}{1} \neq \frac{-1}{2}\right)$$

Let $x = X + h$ and $y = Y + k$, then $\frac{dy}{dx} = \frac{dY}{dX}$

$$\therefore \quad \frac{dY}{dX} = \frac{2X-Y+(2h-k+1)}{X+2Y+(h+2k-3)} \quad \text{...(1)}$$

Now, choose h and k such that $2h-k+1=0$, $h+2k-3=0$

$$\Rightarrow \quad h = \frac{1}{5} \text{ and } k = \frac{7}{5} \quad \text{...(2)}$$

Eq. (1) can be written (using Eq. (2))

$$\frac{dY}{dX} = \frac{2X-Y}{X+2Y} \quad \text{...(3)}$$

which is homogeneous equation

Putting $Y = vX$

or $$\frac{dY}{dX} = v + X\frac{dv}{dX} \quad \text{...(4)}$$

From Eq. (3) and (4), we get

$$v + X\frac{dv}{dX} = \frac{2X-vX}{X+2vX}$$

or $$X\frac{dv}{dX} = -2\left(\frac{v^2+v-1}{2v+1}\right)$$

or $$\frac{(2v+1)\,dv}{v^2+v-1} = -2\frac{dX}{X}$$

Integrating

$$\log(v^2+v-1) = -2\log X + \log c$$

or $$v^2 + v - 1 = \frac{c}{X^2}$$

or $$\frac{Y^2}{X^2} + \frac{Y}{X} - 1 = \frac{c}{X^2}$$

or $$Y^2 + XY - X^2 = c$$

or $$\left(y - \frac{7}{5}\right)^2 + \left(y - \frac{7}{5}\right)\left(x - \frac{1}{5}\right) - \left(x - \frac{1}{5}\right)^2 = c$$

or $$(5y - 7)^2 + (5y - 7)(5x - 1) - (5x - 1)^2 = 25c$$

or $$xy + y^2 - 3y - (x^2 + x) = c$$

Example 3.24 **Solve :**

$$\frac{dy}{dx} = \frac{x + y - 1}{x - y + 1}$$

Solution : Given, equation can be written as

$$\frac{dy}{dx} = \frac{x + y - 1}{x - y + 1} \quad \text{...(1)}$$

Here, $$\frac{a}{A} \neq \frac{b}{B} \quad \left(\because \frac{1}{1} \neq \frac{1}{-1}\right)$$

Let $x = X + h$ and $y = Y + k$, then $\frac{dy}{dx} = \frac{dY}{dX}$

$$\frac{dY}{dX} = \frac{X + Y + (h + k - 1)}{X - Y + (h - k + 1)} \quad \text{...(2)}$$

Now, choose h and k such that $h + k - 1 = 0,\ h - k + 1 = 0$, we get

$$h = 0 \text{ and } k = 1.$$

Eq. (2) can be written

$$\frac{dY}{dX} = \frac{X + Y}{X - Y} \quad \text{...(3)}$$

which is homogeneous equation

Putting $Y = vX$

or $$\frac{dY}{dX} = v + X\frac{dv}{dX} \quad \text{...(4)}$$

From Eq. (3) and (4), we get

$$v + X\frac{dv}{dX} = \frac{X + vX}{X - vX}$$

or $$X\frac{dv}{dX} = \frac{1 + v}{1 - v} - v = \frac{1 + v^2}{1 - v}$$

or $$\frac{1 - v}{1 + v^2}\,dv = \frac{dX}{X}$$

Integrating, $$\tan^{-1} v - \frac{1}{2}\log(1 + v^2) = \log X + c$$

or $$\tan^{-1} v - \log X\sqrt{1 + v^2} = c$$

Substituting the values of v, X and Y, we get

$$\tan^{-1}\left(\frac{y-1}{x}\right) - \log\sqrt{x^2+(y-1)^2} = c$$

PROBLEM SET 3.5

Solving the following differential equations :

1. $\dfrac{dy}{dx} = \dfrac{2x-y+1}{x+2y-3}$

Ans. $(5y-7)^2 + (5x-1)(5y-1) - (5x-1)^2 = 25c$

2. $(3y-7x+7)\,dx + (7y-3x+3)\,dy = 0$

Ans. $(x-y-1)^2(x+y-1)^5 = c$

3. $(x+y-1)dy = (x+y)\,dx$

Ans. $2(y-x) - \log(2x+2y-1) = 4c$

4. $\dfrac{dy}{dx} = \dfrac{x+y+3}{2x+2y+2}$

Ans. $x + c = \dfrac{2}{3}(x+y) - \dfrac{5}{9}\log(3x+3y+4)$

5. $(x+y)(dx-dy) = dx+dy$

Ans. $(y-x) + \log(x+y) = c$

6. $(2x-2y+5)\,dy - (x-y+3)\,dx = 0$

Ans. $x - 2y + \log(x-y+2) = c$

7. $(2x+y-1)\,dy + (4x+2y-3)\,dx = 0$

Ans. $4x^2 + y^2 + 4xy - 6x - 2y = c$

8. $(x-2y)\,dy - (2x-4y-3)\,dx = 0$

Ans. $2x - y + \log(x-2y-2) = c$

9. $(y+x-2)\,dy - (y-x+1)\,dx = 0$

Ans. $2\tan^{-1}\left(\dfrac{2y-1}{2x-3}\right) + \log\left[\dfrac{1}{4}(2x-3)^2 + \dfrac{1}{4}(2y-1)^2\right] = c$

10. $(6x-4y+1)\,dy - (3x-2y+1)\,dx = 0$ ***[AMIETE, 2006]***

Ans. $16y = 8x + \log(24x+16y+5) + c$

11. $(2x+y-1)\,dy - (x-2y+5)\,dx = 0$

Ans. $\left(x+\dfrac{3}{5}\right)^2 - 4\left(x+\dfrac{3}{5}\right)\left(y-\dfrac{11}{5}\right) - \left(y-\dfrac{11}{5}\right)^2 = c$

12. $(x-y-2)\,dx = (2x-2y-6)dy$

Ans. $(x-y-4)^2 = ce^{2y-x}$

3.8 LINEAR DIFFERNETIAL EQUATIONS

A differential equation of the form

$$\frac{dy}{dx} + Py = Q \qquad ...(1)$$

where P and Q are either functions of x or constants is said to be the linear differential equation of first order.

For example, $\frac{dy}{dx} + 3xy = x^3$, $2\frac{dy}{dx} + \sin x \;\; y = \cos x$ etc. are linear differential equations.

Working Rule

1. Write the given differential equation in the form

$$\frac{dy}{dx} + Py = Q \text{ or } \frac{dx}{dy} + Px = Q$$

as the case may be.

2. Find integrating factor (I.F.) given by $e^{\int Pdx}$ or $e^{\int P\,dy}$.
3. The required solution is given by

$$y\,(\text{I.F.}) = \int Q(\text{I.F.})\,dx + c \quad \text{or} \quad x(\text{I.F.}) = \int Q(\text{I.F.})\,dy + c$$

as the case may be.

Example 3.25 **Solve**

$$\cos^2 x\,\frac{dy}{dx} + y = \tan x.$$

[UPTU 2002]

Solution : Given

$$\cos^2 x\,\frac{dy}{dx} + y = \tan x \;\Rightarrow\; \frac{dy}{dx} + \frac{1}{\cos^2 x}\,y = \frac{\tan x}{\cos^2 x}$$

$$\Rightarrow \quad \frac{dy}{dx} + \sec^2 x,\;\; y = \tan x \sec^2 x \qquad ...(1)$$

This is linear differential equation of the form

$$\frac{dy}{dx} + Py = Q$$

where $P = \sec^2 x$, $Q = \tan x \sec^2 x$

Now, $\quad \text{I.F.} = e^{\int Pdx} = e^{\int \sec^2 x\,dx} = e^{\tan x}$

The required solution of (1) is given by

$$y(\text{I.F.}) = \int Q(\text{I.F.})\,dx + c$$

$$\Rightarrow \quad ye^{\tan x} = \int \tan x \sec^2 x\, e^{\tan x}\,dx + c$$

Let $\quad \tan x = t \Rightarrow \sec^2 x\,dx = dt$

$$\Rightarrow \quad ye^{\tan x} = \int t\, e^t\, dt + c$$

$$= t\, e^t - \int 1\, e^t\, dt + c \qquad \text{[integration by parts]}$$

$$= te^t - e^t + c = e^t (t-1) + c$$

$$\Rightarrow \quad ye^{\tan x} = e^{\tan x} (\tan x - 1) + c$$

$$\Rightarrow \quad y = (\tan x - 1) + ce^{-\tan x}$$

Exmaple 3.26 **Solve :**

$$(x^2 - 1)\frac{dy}{dx} + 2xy = 1.$$

Solution : We have

$$(x^2 - 1)\frac{dy}{dx} + 2xy = 1$$

$$\Rightarrow \quad \frac{dy}{dx} + \frac{2x}{x^2-1} y = \frac{1}{x^2-1}$$

Here, $P = \dfrac{2x}{x^2-1}$, $Q = \dfrac{1}{x^2-1}$

Now, $\text{I.F.} = e^{\int P\, dx} = e^{\int \frac{2x}{x^2-1} dx} = e^{\log (x^2-1)} = x^2 - 1$

The required solution of the given differential equation is

$$y(\text{I.F.}) = \int Q(\text{I.F.})\, dx + c$$

$$\Rightarrow \quad y(x^2-1) = \int \frac{1}{x^2-1}(x^2-1)\, dx + c = \int dx + c = x + c$$

$$\Rightarrow \quad y\,(x^2-1) = x + c$$

Example 3.27 **Solve : $(1 + y^2)\, dx = (\tan^{-1} y - x)\, dy$** *[UPTU 2005]*

Solution : We have $(1 + y^2)\, dx = (\tan^{-1} y - x)\, dy$

$$\Rightarrow \quad (1+y^2)\frac{dx}{dy} = \tan^{-1} y - x$$

$$\Rightarrow \quad (1+y^2)\frac{dx}{dy} + x = \tan^{-1} y$$

$$\Rightarrow \quad \frac{dx}{dy} + \frac{1}{1+y^2} \cdot x = \frac{\tan^{-1} y}{1+y^2}$$

This is a linear differential equation of the form

$$\frac{dx}{dy} + Px = Q,$$

where $P = \dfrac{1}{1+y^2}$, $Q = \dfrac{\tan^{-1} y}{1+y^2}$

Now, $\quad$ I.F. $= e^{\int P\,dy} = e^{\int \frac{1}{1+y^2} dy} = e^{\tan^{-1} y}$

The required solution is

$$x(\text{I.F.}) = \int Q \cdot (\text{I.F.})\, dy + c$$

$$\Rightarrow \quad x\, e^{\tan^{-1} y} = \int \frac{\tan^{-1} y}{1+y^2}\, e^{\tan^{-1} y}\, dy + c$$

Let $\quad \tan^{-1} y = t$

$$\Rightarrow \quad \frac{1}{1+y^2}\, dy = dt = \int t\, e^t\, dt + c$$

$$= t\, e^t - \int e^t\, dt + c \qquad \text{[Integration by parts]}$$

$$\Rightarrow \quad xe^{\tan^{-1} y} = te^t - e^t + c = e^t\,(t-1) + c$$

$$\Rightarrow \quad xe^{\tan^{-1} y} = e^{\tan^{-1} y}\,(\tan^{-1} - 1) + c$$

Example 3.28 **Solve :**

$$(\boldsymbol{x^2 + 1})\,\frac{\boldsymbol{dy}}{\boldsymbol{dx}} + \boldsymbol{2xy} = \sqrt{\boldsymbol{x^2 + 4}}.$$

Solution : The given differential equation can be written as

$$\frac{dy}{dx} + \frac{2x}{x^2+1}\, y = \frac{\sqrt{x^2+4}}{x^2+1}$$

Here, $P = \dfrac{2x}{x^2+1}$, $Q = \dfrac{\sqrt{x^2+4}}{x^2+1}$

Now, $\quad$ I.F. $= e^{\int P\,dx} = e^{\int \frac{2x}{x^2+1} dx} = e^{\log (x^2+1)} = x^2 + 1$

The required solution is given by

$$y(\text{I.F.}) = \int Q(\text{I.F.})\, dx + c$$

$$\Rightarrow \quad y(x^2+1) = \int \frac{\sqrt{x^2+4}}{x^2+1}\,(x^2+1)\, dx + c$$

$$\Rightarrow \quad y(x^2+1) = \int \sqrt{x^2+4}\, dx + c$$

$$\Rightarrow \quad y(x^2+1) = \frac{1}{2}\, x\, \sqrt{x^2+4} + 2 \log\,(x + \sqrt{x^2+4}) + c$$

Example 3.29 **Solve :**

$$(x+1)\frac{dy}{dx} - y = e^x (x+1)^2.$$

Solution : We have

$$(x+1)\frac{dy}{dx} - y = e^x (x+1)^2$$

$$\Rightarrow \quad \frac{dy}{dx} - \frac{y}{x+1} = e^x (x+1)$$

The integrating factor (I.F.) $= e^{-\int \frac{dx}{x+1}}$

$$= e^{-\log(x+1)} = e^{\log(x+1)^{-1}} = \frac{1}{x+1}$$

$\therefore$ The solutioin is

$$y \cdot \frac{1}{x+1} = \int e^x (x+1)\left(\frac{1}{x+1}\right) dx + c$$

or

$$\frac{y}{x+1} = \int e^x \, dx + c$$

$$\Rightarrow \quad \frac{y}{x+1} = e^x + c$$

Example 3.30 $x(x-1)\frac{dy}{dx} - (x-2)y = x^2(2x-1)$

Solution : We have

$$\frac{dy}{dx} - \frac{x-2}{x(x-1)} y = \frac{x^2(2x-1)}{x(x-1)}$$

The integrating factor (I.F.) $= e^{\int -\frac{(x-1)}{x(x-1)} dx}$

$$= e^{\int \left(\frac{1}{x-1} - \frac{2}{x}\right) dx} = e^{(\log(x-1) - 2\log x)}$$

$$= e^{\log \frac{(x-1)}{x^2}} = \frac{x-1}{x^2}$$

$\therefore$ The solution is $y \cdot \left(\frac{x-1}{x^2}\right) = \int \frac{x^2(2x-1)}{x(x-1)} \cdot \frac{(x-1)}{x^2} dx + c$

$$= \int \frac{(2x-1)}{x} dx + c$$

$$\Rightarrow \quad y\frac{(x-1)}{x^2} = \int \left(2 - \frac{1}{x}\right) dx + c = 2x - \log x + c$$

Example 3.31 **If $\frac{dy}{dx} + 2y\tan x = \sin x$ and $y = 0$ for $x = \frac{\pi}{3}$.**

Solution : We have, $\frac{dy}{dx} + 2y\tan x = \sin x$

Integrating factor (I.F.) $= e^{2\int \tan x \, dx} = e^{2\log \sec x} = \sec^2 x$

$\therefore$ The solution is $y \cdot (\sec^2 x) = \int \sin x \cdot \sec^2 x \, dx + c$

$$\Rightarrow \quad y(\sec^2 x) = \int \sec x \tan dx + c$$

$$\Rightarrow \quad y(\sec^2 x) = \sec x + c$$

For $x = \frac{\pi}{3}$, $y = 0$ (given)

$$\therefore \quad 0 \cdot \sec^2 \frac{\pi}{3} = \sec \frac{\pi}{3} + c$$

$$\Rightarrow \quad 0 = 2 + c \Rightarrow c = -2$$

This the solution is

$$y \sec^2 x = \sec x - 2$$

$$y = \cos x - 2\cos^2 x$$

PROBLEM SET 3.6

Solve the following equations :

1. $\frac{dy}{dx} + 2y = \sin x$

Ans. $y = ce^{-2x} + \frac{1}{5}(2\sin x - \cos x)$

2. $\frac{dy}{dx} + y = \sin x$

Ans. $y = ce^{-x} + \frac{1}{2}(\sin x - \cos x)$

3. $\frac{dy}{dx} + y \cot x = \cos x$

Ans. $y \sin x = \frac{1}{2}\sin^2 x + c$

4. $\frac{dy}{dx} + \frac{2x}{1+x^2} y = \frac{1}{(1+x^2)^2}$

Ans. $y(1+x^2) = \tan^{-1} x + c$

5. $\frac{dy}{dx} + \frac{y}{2x} = 3x^2$

Ans. $y\sqrt{x} = \frac{6}{7}x^{7/2} + c$

6. $\frac{dy}{dx} + y \tan x = x^2 \cos^2 x$

Ans. $y \sec x = x^2 \sin x + 2x \cos x - 2\sin x + c$

7. $y^2 \frac{dx}{dy} + x - \frac{1}{y} = 0$

Ans. $x = \frac{y+1}{y} + ce^{1/y}$

8. $4\dfrac{dy}{dx} + 8y = 5e^{-3x}$

Ans. $ye^{2x} = -\dfrac{5}{4}e^{-x} + c$

9. $(x-a)\dfrac{dy}{dx} + 3y = 12(x-a)^3, \ x > a > 0$

Ans. $y = 2(x-a)^3 + \dfrac{c}{(x-a)^3}$

10. $\dfrac{dy}{dx} + \dfrac{1}{x}y = x^3 - 3$

Ans. $xy = \dfrac{x^5}{5} - \dfrac{3x^2}{2} + c$

11. $\dfrac{dy}{dx} + 2xy = 2e^{-x^2}$

Ans. $y = (2x + c)\,e^{-x^2}$

12. $(x+a)\dfrac{dy}{dx} - 3y = (x+a)^5$

Ans. $2y = (x+a)^5 + 2c\,(x+a)^3$

13. $x\log x\dfrac{dy}{dx} + y = 2\log x$

Ans. $y\log x = (\log x)^2 + c$

14. $\sec x\dfrac{dy}{dx} = y + \sin x$

Ans. $y = -\sin x - 1 + ce^{\sin x}$

15. $x\dfrac{dy}{dx} + 2y = x^2\log x$

Ans. $yx^2 = \dfrac{x^4}{4}\log x - \dfrac{x^4}{16} + c$

16. $y' + y\tan x = \cos x;\ y(0) = 0$

Ans. $y = x\cos x$

17. Find the value of a so that e^{ax^2} is an integrating factor of defferential equation $x(1-y)\,dx - dy = 0$.

Ans. $a\,\dfrac{1}{2}$.

18. Find the value of a so that e^{ay^2} is an integrating factor of the differential equation $(e^{y^2/2} - xy)\,dy - dx = 0$.

Ans. $a = \dfrac{1}{2}$

3.9 EQUATIONS REDUCIBLE TO THE LINEAR FORM

There are certain differential equations which are not linear, but their solution can be obtained by reducing them to linear form by using some mathematical substitutions.

3.9.1 Bernoulli Equation

A differential equation

$$\frac{dy}{dx} + P(x)\,y = Q(x)\,y^n \qquad \text{...(1)}$$

where n is any real number is said to be Bernoulli equation. This equation is not linear, but it can be made linear by dividing both sides of (1) by y^n

Working Rule

1. Write the differential equation in the form

$$\frac{dy}{dx} + Py = Qy^n$$

2. Divide both sides by y^n, we get

$$y^{-n}\frac{dy}{dx} + Py^{1-n} = Q$$

3. Put $y^{1-n} = t$, so that

$$(1-n)\,y^{-n}\frac{dy}{dx} = \frac{dt}{dx} \Rightarrow y^{-n}\frac{dy}{dx} = \frac{1}{1-n}\frac{dt}{dx}$$

By these substitutions we get a linear differential equation of first order and can be solved easily.

Example 3.32 **Solve:**

$$3\frac{dy}{dx} + \frac{2}{x+1}y = \frac{x^3}{y^2}$$

[UPTU 2005]

Solution : We have

$$3\frac{dy}{dx} + \frac{2}{x+1}y = \frac{x^3}{y^2}$$

$$\Rightarrow \quad 3\frac{dy}{dx} + \frac{2}{x+1}y = x^2\,y^{-2} \qquad \text{...(1)}$$

Divide both sides by y^{-2}, we have

$$3y^2\frac{dy}{dx} + \frac{2}{x+1}y^3 = x^2 \qquad \text{...(2)}$$

Put $y^3 = t \Rightarrow 3y^2\dfrac{dy}{dx} = \dfrac{dt}{dx}$

From (2), we get

$$\frac{dt}{dx} + \frac{2}{x+1}t = x^2 \qquad \text{...(3)}$$

which is linear differential equation of first order.

Here, $P = \dfrac{2}{x+1}$, $Q = x^2$

Now,

$$\text{I.F.} = e^{\int \frac{2}{x+1} dx} = e^{2 \log (x+1)} = e^{\log (x+1)^2} = (x+1)^2$$

The general solution is given by

$$t \cdot (\text{I.F.}) = \int Q \, (\text{I.F.}) \, dx + c$$

$$\Rightarrow \quad t (x+1)^2 = \int x^2 (x+1)^2 \, dx + c$$

$$\Rightarrow \quad t (x+1)^2 = \int x^2 (x^2 + 2x + 1) \, dx + c$$

$$\Rightarrow \quad t (x+1)^2 = \int (x^4 + 2x^3 + x^2) \, dx + c$$

$$\Rightarrow \quad t (x+1)^2 = \frac{x^5}{5} + \frac{2x^4}{4} + \frac{x^3}{3} + c$$

$$\Rightarrow \quad y^3 (x+1)^2 = \frac{x^5}{5} + \frac{x^4}{2} + \frac{x^3}{3} + c \qquad [\because \; t = y^3]$$

Example 3.33 **Solve :**

$$x \frac{dy}{dx} + y = xy^3$$

[UPTU 2002]

Solution : Given

$$x \frac{dy}{dx} + y = xy^3$$

$$\Rightarrow \quad \frac{dy}{dx} + \frac{1}{x} y = y^3 \qquad \text{...(1)}$$

Divide both sides by y^3, we get

$$y^{-3} \frac{dy}{dx} + \frac{1}{x} y^{-2} = 1 \qquad \text{...(2)}$$

Put $y^{-2} = t \;\Rightarrow\; -2y^{-3} \dfrac{dy}{dx} = \dfrac{dt}{dx}$

From (2), we get

$$-\frac{1}{2} \frac{dt}{dx} + \frac{1}{x} t = 1 \;\Rightarrow\; \frac{dt}{dx} - \frac{2}{x} t = -2 \qquad \text{...(3)}$$

which is a linear differential equation where

$$P = -\frac{2}{x}, \; Q = -2$$

$$\text{I.F.} = e^{\int P \, dx} = e^{\int -\frac{2}{x} dx} = e^{-2 \int \frac{1}{x} dx} = e^{-2 \log x}$$

$$= e^{\log x^{-2}} = x^{-2} = \frac{1}{x^2}$$

The general solution of (3) is given by

$$t(\text{I.F.}) = \int Q(\text{I.F.})\, dx + c$$

$$\Rightarrow \quad t\left(\frac{1}{x^2}\right) = \int -2\left(\frac{1}{x^2}\right) dx + c$$

$$\Rightarrow \quad \frac{t}{x^2} = -2\left[\frac{x^{-1}}{-1}\right] + c = \frac{2}{x} + c \Rightarrow t = 2x + cx^2$$

$$\Rightarrow \quad y^{-2} = 2x + cx^2 \qquad [\because\ t = y^{-2}]$$

$$\Rightarrow \quad \frac{1}{y^2} = 2x + cx^2$$

Example 3.34 **Solve :**

$$\frac{dy}{dx} + xy = xy^5.$$

[UPTU 2001]

Solution : Given

$$\frac{dy}{dx} + xy = xy^5 \qquad ...(1)$$

Divide both sides of (1) by y^5, we get

$$y^{-5}\frac{dy}{dx} + xy^{-4} = x \qquad ...(2)$$

Put $y^{-4} = t$

$$\Rightarrow \quad -4y^{-5}\frac{dy}{dx} = \frac{dt}{dx}$$

From (2), we get

$$-\frac{1}{4}\frac{dt}{dx} + xt = x \Rightarrow \frac{dt}{dx} - 4x\, t = -4x \qquad ...(3)$$

which is a linear differential equation, where $P = -4x,\ Q = -4x$

Now, $\quad \text{I.F.} = e^{\int P\, dx} = e^{\int -4x\, dx} = e^{-2x^2}$

The general solutin is given by

$$t(\text{I.F.}) = \int Q(\text{I.F.})\, dx + c$$

$$\Rightarrow \quad te^{-2x^2} = \int -4x\, e^{-2x^2}\, dx + c$$

$$\Rightarrow \quad te^{-2x^2} = e^{-2x^2} + c \Rightarrow t = 1 + ce^{2x^2}$$

$$\Rightarrow \quad y^{-4} = 1 + ce^{2x^2} \qquad [\because t = y^{-4}]$$

$$\Rightarrow \quad \frac{1}{y^4} = 1 + ce^{2x^2}$$

$$\Rightarrow \quad 1 = y^4(1 + ce^{2x^2})$$

Example 3.35 **Solve**

$$\frac{dy}{dx} + \frac{y}{x} = \frac{y^2}{x^2}$$

Solution : We have

$$\frac{dy}{dx} + \frac{y}{x} = \frac{y^2}{x^2}$$

$$\Rightarrow \quad \frac{1}{y^2}\frac{dy}{dx} + \frac{1}{xy} = \frac{1}{x^2}$$

Let $-\frac{1}{y} = t$

$$\Rightarrow \quad \frac{1}{y^2}\frac{dy}{dx} = \frac{dt}{dx}$$

$\therefore$ The given equation reduces to

$$\frac{dt}{dx} - \frac{t}{x} = \frac{1}{x^2}$$

$$\text{I.F.} = e^{-\int \frac{1}{x}dx} = e^{-\log x} = e^{\log\left(\frac{1}{x}\right)} = \frac{1}{x}$$

$\therefore$ The solution will be $t \cdot \frac{1}{x} = \int \frac{1}{x^2} \cdot \frac{1}{x}\, dx + c$

$$\Rightarrow \quad \frac{t}{x} = \int x^{-3}\, dx + c$$

$$\Rightarrow \quad \frac{t}{x} = \frac{x^{-2}}{-2} + c$$

$$\Rightarrow \quad \frac{-1}{xy} = \frac{1}{-2x^2} + c$$

$$\Rightarrow \quad \frac{1}{xy} = \frac{1}{2x^2} - c$$

Example 3.36 **Solve :**

$$x\frac{dy}{dx} + y \log y = xye^x$$

Solution : We have

$$x\frac{dy}{dx} + y \log y = xy\, e^x$$

Dividing it by xy, we get

$$\frac{1}{y}\frac{dy}{dx} + \frac{1}{x}\log y = e^x$$

Let $\log y = t$

$$\Rightarrow \quad \frac{1}{y}\frac{dy}{dx} = \frac{dt}{dx}$$

$\therefore$ The given equation becomes $\frac{dt}{dx} + \frac{t}{x} = e^x$

$$\text{I.F.} = e^{\int \frac{1}{x} dx} = e^{\log x} = x$$

∴ The solution is

$$t \cdot x = \int xe^x \, dx + c$$

$$t \cdot x = xe^x - e^x + c$$

$$\Rightarrow \quad x \log y = (x - 1) e^x + c$$

Example 3.37 **Solve :**

$$\frac{dy}{dx} + \frac{2xy}{3(1 + x^2)} = \frac{2xy^4}{3(1 + x^2)}$$

Solution : Given equation is

$$\frac{dy}{dx} + \frac{2xy}{3(1 + x^2)} = \frac{2xy^4}{3(1 + x^2)}$$

or

$$y^{-4} \frac{dy}{dx} + \frac{2xy^{-4}}{3(1 + x^2)} = \frac{2x}{3(1 + x^2)} \quad \text{...(1)}$$

Putting $y^{-3} = v$

So that,

$$-3y^{-4} \frac{dy}{dx} = \frac{dv}{dx}$$

∴ Eq. (1) becomes

$$-\frac{1}{3} \frac{dv}{dx} + \frac{3xv}{3(1 + x^2)} = \frac{2x}{3 1 + x^2)}$$

or

$$\frac{dv}{dx} - \frac{3xv}{1 + x^2} = \frac{-2x}{1 + x^2}$$

which is a linear Eq.

$$\text{I.F.} = e^{\int -\frac{2x}{1 + x^2} dx} = e^{-\log(1 + x^2)} = \frac{1}{1 + x^2}$$

∴ Solution is given by

$$\frac{v}{1 + x^2} = -\int \frac{2x}{(1 + x^2)^2} dx + c$$

or

$$\frac{v}{1 + x^2} = \frac{1}{1 + x^2} + c$$

or

$$\frac{v}{1 + x^2} = \frac{1}{(1 + x^2)} + c$$

or

$$y_{-3} = 1 + c(1 + x^2)$$

Example 3.38 **Solve :**

$$x \frac{dy}{dx} - (1 + x) y = xy^2$$

Solution : Given equation can be written as

$$y^{-2} \frac{dy}{dx} - \frac{(1 + x)}{x} y^{-1} = 1 \quad \text{...(1)}$$

Putting $y^{-1} = v$, so that $-y^{-2}\dfrac{dy}{dx} = \dfrac{dv}{dx}$

$\therefore$ Eq. (1) becomes

$$-\frac{dv}{dx} - \left(\frac{1}{x} + 1\right)v = 1$$

or
$$-\frac{dv}{dx} + \left(\frac{1}{x} + 1\right)v = 1$$

$\therefore$
$$I.F. = \rho^{\int\left(\frac{1}{x}+1\right)dx} = xe^x$$

$\therefore$ The solution is $xe^x \,.\, v = \int xe^x \, dx + c$

$$= xe^x - e^x + c$$

$$\frac{x}{y}e^x = xe^x - e^x + c$$

or
$$\frac{x}{y} = x - 1 + ce^{-x}$$

PROBLEM SET 3.7

Solve the following equations :

1. $\dfrac{dy}{dx} - y = y^2(\sin x + \cos x)$

Ans. $y = \dfrac{1}{ce^{-x} - \sin x}$

2. $y\dfrac{dy}{dx} + xy^2 = x$

Ans. $y^2 = 1 + ce^{-x^2}$

3. $x\dfrac{dy}{dx} + y = y^2$

Ans. $y = \dfrac{1}{cx + 1}$

4. $y^2\dfrac{dy}{dx} = x + y^3$

Ans. $y^3 = ce^{3x} - \left(x + \dfrac{1}{3}\right)$

5. $\dfrac{dy}{dx} + \dfrac{y}{x} = y^2$

Ans. $1 = xy\,(c - \log x)$

6. $\dfrac{dy}{dx} + \dfrac{1}{x} = \dfrac{e^y}{x^2}$

Ans. $\sec y = (1 + x) + ce^x$

7. $\dfrac{dy}{dx} - y \tan x = -y^2 \sec x$

Ans. $-\dfrac{1}{y} \sec x = c - \tan x$

8. $\dfrac{dy}{dx} - 2y \tan x = y^2 \tan^2 x$

Ans. $-\dfrac{1}{y} \sec^2 x = c + \dfrac{1}{3} \tan^3 x$

9. $(1 - x^2)\left(\dfrac{dy}{dx}\right) + xy = xy^2$

Ans. $cy = (1 - y)\sqrt{1 - x^2}$

10. $x\left(\dfrac{dy}{dx}\right) + y = y^2 \log x$

Ans. $\{\log (xe) + cx\}\, y = 1$

11. $\sec^2 y\left(\dfrac{dy}{dx}\right) + 2x \tan y = x^3$

Ans. $\tan y = \dfrac{1}{2}(x^2 - 1) + ce^{-x^2}$

12. $\dfrac{dy}{dx} + \dfrac{y}{x} \log y = \dfrac{y}{x^2} (\log y)^2$

Ans. $\dfrac{1}{x \log x} = \dfrac{1}{2x^2} + c$

13. $xy - \left(\dfrac{dy}{dx}\right) = y^3 e^{-x^2}$

Ans. $xy = c - \dfrac{1}{3}x^3 + e^{-x}$

14. $\dfrac{dy}{dx} + 2xy = y^2 e^{x^2} \log x$

Ans. $\dfrac{1}{y} = ce^{x^2} - x(\log x - 1)\, e^{x^2}$

15. $(x^3y^3 - xy)\, dx = dy$

Ans. $1 = y^2 (x^2 + 1) + cy^2 e^{x^2}$

16. $\dfrac{dy}{dx} = \dfrac{x^2 + y^2 + 1}{2xy}$

Ans. $y^2 = x^2 + cx - 1$

3.10 EXACT DIFFERENTIAL EQUATIONS

A differential equation of the form $M(x, y)\,dx + N(x, y)\,dy$ is said to be an exact differential equation if

$$\frac{\partial M}{\partial y} = \frac{\partial N}{\partial x}$$

This type of differential equation can be obtain by directly integrating its solution.

3.10.1 Necessary Condition for a Defferential Equation to be Exact

The Eq. $M\,dx + N\,dy = 0$ will be exact, if

$$M\,dx + N\,dy = du \qquad \ldots(1)$$

But
$$du = \frac{\partial u}{\partial x}dx + \frac{\partial u}{\partial y}dy \qquad \ldots(2)$$

$$\therefore \quad M\,dx + N\,dy = \frac{\partial u}{\partial x}dx + \frac{\partial u}{\partial y}dy \qquad \ldots(3)$$

Equating the coefficient of dx and dy in equation (1) and (2), we get

$$M = \frac{\partial u}{\partial x} \text{ and } N = \frac{\partial u}{\partial y}$$

$$\therefore \quad \frac{\partial M}{\partial y} = \frac{\partial^2 u}{\partial y\,\partial x} \text{ and } \frac{\partial N}{\partial x} = \frac{\partial^2 u}{\partial x\,\partial y}$$

Since,
$$\frac{\partial^2 u}{\partial y\,\partial x} = \frac{\partial^2 u}{\partial x\,\partial y}$$

$$\therefore \quad \frac{\partial M}{\partial y} = \frac{\partial N}{\partial x}$$

Working Rule

The exact differential equation can be solved by using the following steps.

Step I : $M(x, y)$ to be integrated with respect to x keeping y as a constant.

Step II : The terms of $N(x, y)$ not containg x, to be integrated with respect ro y.

Step III : The solution of the given exact differential equation will be :

Sum of Result I and II = constant.

Example 3.39 **Solve the differential equation**

$$(5x^4 + 3x^2y^2 - 2xy^3)\,dx + (2x^3y - 3x^2y^2 - 5y^4)\,dy = 0$$

Solution : We have, $M(x, y) = 5x^4 + 3x^2y^2 - 2xy^3$

$$N(x, y) = 2x^3y - 3x^2y^2 - 5y^4$$

$$\therefore \quad \frac{\partial M}{\partial y} = 6x^2y - 6xy^2$$

and
$$\frac{\partial N}{\partial x} = 6x^2y - 6xy^2$$

$$\because \quad \frac{\partial M}{\partial y} = \frac{\partial N}{\partial x}$$

So the given differential equation is an exact equation.

$\therefore$ The solution is

$$\int_{(y \text{ as constant})} M\,dx + \int (\text{terms of } N \text{ not containing x})\,dy = c$$

$$\Rightarrow \int_{(y \text{ as a constant})} (5x^4 + 3x^2y^2 - 2xy^3)\,dx + \int (-5y^4)\,dy = c$$

$$\Rightarrow x^5 + x^3y^2 - x^2y^3 - y^5 = c$$

Example 3.40 **Solve $3x(xy - 2)\,dx + (x^3 + 2y)\,dy = 0$**

Solution : Here, $M = 3x(xy - y)$ and $N = (x^3 + 2y)$

$$\frac{\partial M}{\partial y} = 3x^2 = \frac{\partial N}{\partial x}$$

Thus, the given equation is exact and its solution is

$$\int_{y \text{ constant}} M\,dx + \int (\text{terms of N not containing } x)\,dy = c$$

$$\int_{y \text{ cosntant}} 3x(xy - 2)\,dx + \int 2y\,dy = c$$

$$x^3y - 3x^2 + y^2 = c$$

Example 3.41 **Solve :**

$$\left(\frac{1}{x} + \frac{1}{x^2} - \frac{y}{x^2 + y^2}\right) dx + \left(ye^y + \frac{x}{x^2 + y^2}\right) dy = 0$$

Solution :

Here, $M = \frac{1}{x} + \frac{1}{x^2} - \frac{y}{x^2 + y^2} \Rightarrow \frac{\partial M}{\partial y} = \frac{y^2 - x^2}{(y^2 + x^2)^2}$

and $N = ye^y + \frac{x}{x^2 + y^2} \Rightarrow \frac{\partial N}{\partial x} = \frac{y^2 - x^2}{(y^2 + x^2)^2}$

$$\therefore \quad \frac{\partial M}{\partial y} = \frac{\partial N}{\partial x}$$

Thus, the given Eq. is exact and its solution is

$$\int_{y \text{ constant}} \left(\frac{1}{x} + \frac{1}{x^2} - \frac{y}{x^2 + y^2}\right) dx + \int ye^y\,dy = C$$

or $\log x - \frac{1}{x} - \tan^{-1}\frac{x}{y} + ye^y - e^y = C$

Example 3.42 **Solve :**

$$(2y \sin x \cos x - y + 2y^2 e^{xy^2})\,dx + (-x + \sin^2 x + 4xye^{xy^2})\,dy = 0.$$

Solution :

Here, $M = 2y \sin x \cos x - y + 2y^2 e^{xy^2}$

and $N = -x + \sin^2 x + 4xy\,e^{xy^2}$

$$\therefore \quad \frac{\partial M}{\partial y} = 2 \sin x \cos x - 1 + 4y\,e^{xy^2} + 4xy^3\,e^{xy^2}$$

$$\text{and} \quad \frac{\partial N}{\partial x} = 2 \sin x \cos x - 1 + 4y\,e^{xy^2} + 4xy^3 e^{xy^2}$$

Thus, the given equation is exact and its solution is

$$\int_{y\text{ constant}} (2y \sin x \cos x - y + 2y^2 e^{xy^2})\, dx = C$$

$$y \sin^2 x - xy + 2e^{xy^2} = C$$

Example 3.43 **Solve :**

$$\left[\boldsymbol{y\left(1+\frac{1}{x}\right)+\cos y}\right]\boldsymbol{dx + [x + \log x - \sin y]\, dy = 0}$$

Solution : Here, $M = y\left(1+\frac{1}{x}\right)+\cos y$

$$\Rightarrow \qquad \frac{\partial M}{\partial y} = 1 + \frac{1}{x} - \sin y$$

and $$N = x + \log x - x \sin y$$

$$\Rightarrow \qquad \frac{\partial N}{\partial x} = 1 + \frac{1}{x} - \sin y$$

$$\therefore \qquad \frac{\partial M}{\partial y} = \frac{\partial N}{\partial x}$$

Thus, the given equation is exact and its solution is

$$\int_{y\text{ cosntant}} \left[y\left(1+\frac{1}{x}\right)+\cos y\right] dx = c$$

or $$y\,[x + \log x] + x \cos y = c$$

Example 3.44 **Solve the following differential equation**

$$\boldsymbol{-(y^2 + y \sin x)\, dx + \left(\frac{1}{1+y^2} + \cos x - 2xy\right) dy = 0}$$

where $y(0) = 1$.

Solution : Here, $M = -(y^2 + y \sin x)$

$$\Rightarrow \qquad \frac{\partial M}{\partial y} = -(2y + \sin x)$$

and $$N = \left(\frac{1}{1+y^2} + \cos x - 2xy\right)$$

$$\Rightarrow \qquad \frac{\partial N}{\partial x} = -(\sin x + 2y)$$

$$\because \qquad \frac{\partial M}{\partial y} = \frac{\partial N}{\partial x}$$

Thus, the given equation is exact and its solution is

$$\int_{y\text{ constant}} [-(y^2 + y \sin x)]\, dx + \int \frac{1}{1+y^2}\, dy = c$$

or $$-xy^2 + y \cos x + \tan^{-1} y = c$$

Using the given condition $y(0) = 1$, *i.e.*, when $x = 0$, $y = 1$

$$\cos 0 + \tan^{-1} 1 = c$$

or $\quad c = \frac{\pi}{4} + 1$

Hence, the solution is

$$y \cos x - xy^2 + \tan^{-1} y = \frac{\pi}{4} + 1$$

Example 3.45 **Solve the following differential equation**

$$(y^2 \cos x - 3x^2y - 2x)\, dx + (2y \sin x - x^3 + \log y)\, dy = 0,$$

where $y(0) = e$.

Solution : Here, $M = y^2 \cos x - 3x^2y - 2x$

$\Rightarrow \quad \frac{\partial M}{\partial y} = 2y \cos x - 3x^2$

and $\quad N = 2y \sin x - x^3 + \log y$

$\Rightarrow \quad \frac{\partial N}{\partial x} = 2y \cos x - 3x^2$

$\therefore \quad \frac{\partial M}{\partial y} = \frac{\partial N}{\partial x}$

Thus, the given equation is exact and its solution

$$\int_{y \text{ constant}} (y^2 \cos x - 3x^2y - 2x)\, dx + \int \log y\, dy = c$$

$$y^2 \sin x - x^3y - x^2 + y \log y - y = c$$

Using the given condition $y(0) = e$, *i.e.*, when $x = 0$, $y = e$, we get

$$e^2 \sin 0 - 0 + e \log e - e = c$$

$\therefore \quad c = 0$

Hence, the solution is

$$y^2 \sin x - x^3y - x^2 + y \log y - y = 0$$

Example 3.46 **Solve $(x^2 + y^2 - m^2)\, x\, dx + (x^2 - y^2 - n^2)\, y\, dy = 0$.**

Solution : We have, $M = (x^2 + y^2 - m^2)\, x$

and $\quad N = (x^2 - y^2 - n^2)\, y$

$\therefore \quad \frac{\partial M}{\partial y} = 2xy$ and $\frac{\partial N}{\partial x} = 2xy$

$\because \quad \frac{\partial M}{\partial y} = \frac{\partial N}{\partial x}$

Thus, the given equation is exact and its solution is

$$\int_{y \text{ constant}} (x^2 + y^2 - m^2)\, x\, dx + \int y\, (-y^2 - n^2)\, dy = c$$

$$\frac{x^4}{4} + (-m^2)\frac{x^2}{2} - \frac{y^4}{2} - \frac{n^2y^2}{2} = c$$

or $\quad x^4 + 2x^2 (y^2 - m^2) - y^4 - 2n^2y^2 = 2c$

Example 3.47 **Solve $(c^2 - 2xy - y^2)\, dx = (x + y)^2\, dy$.**

Solution : Given equation can be written as

$$(c^2 - 2xy - y^2)\, dx - (x + y)^2\, dy = 0$$

Here, $M = c^2 - 2xy - y^2$ and $N = -(x + y)^2$

$$\therefore \quad \frac{\partial M}{\partial y} = -2(x + y) \text{ and } \frac{\partial N}{\partial x} = -2(x + y)$$

$$\because \quad \frac{\partial M}{\partial y} = \frac{\partial N}{\partial x}$$

Thus, the given equation is exact and its solution

$$\int_{y \text{ constant}} (c^2 - 2xy - y^2)\, dx - \int y^2\, dy = c_1$$

or $$c^2x - x^2y - xy^2 - \frac{y^3}{3} = c_1$$

or $$3c^2x - 3x^2y - 3xy^2 - y^3 = 3c_1$$

PROBLEM SET 3.8

Solve the following differentia equations :

1. $(x + y - 10)\, dx + (x - y - 2)\, dy = 0$

Ans. $\frac{x^2 - y^2}{2} + xy - 10x - 2y = c$

2. $(y^2 - x^2)\, dx = 2xy\, dy = 0$

Ans. $xy^2 - \frac{x^3}{3} = c$

3. $(y^2 e^{xy^2} + 4x^3)\, dx + (2xy\, e^{xy^2} - 3y^2)\, dy = 0$

Ans. $e^{xy^2} + x^4 - y^3 = C$

4. $\frac{y}{x}\, dx + (y \log 4x)\, dx = 0$

Ans. $2y \log x + y^2 = c$

5. $(x^2 - 2xy - y^2)\, dx - (x + y)\, dy = 0$

Ans. $x^3 - y^3 - 3x^2y - 3xy^2 = 3c$

6. $x\, dx + y\, dy + \frac{x\, dy - y\, dx}{x^2 + y^2}$

Ans. $x^2 + 2\tan^{-1}\left(\frac{x}{y}\right) + y^2 = c$

7. $(1 + e^{x/y})\, dx + e^{x/y}\left(1 - \frac{x}{y}\right) dy = 0$

Ans. $x + ye^{x/y} = c$

8. $e^y\, dx + (1 + xe^y)\, dy = 0$

Ans. $xe^y + y = c$

9. $2x\,(ye^{x^2} - 1)\,dx + e^{x^2}\,dy = 0$

Ans. $ye^{x^2} - x^2 = c$

10. $(a^2 - 2xy - y^2)\,dx - (x + y)^2\,dy = 0$

Ans. $a^2x - \dfrac{y^3}{3} - xy^2 - x^2y = c$

11. $(x^4 - 2xy^2 + y^4)\,dx - (2x^2y - 4xy^3 + \sin y)\,dy = 0$

Ans. $\dfrac{x^5}{5} - x^2y^2 + xy^4 + \cos y = c$

3.11 EQUATIONS REDUCIBLE TO EXACT DIFFERENTIAL EQUATION

Sometimes, there might be the situation, where a defferential equation is not exact, but it may become so by multiplying it with a suitable function. The function, that reduces the given differential equation to an exact equation is known as 'Integrating factor'.

The factor might be obtained by different methods.

Integrating factor by Inspection method : Following exact defferentials might be widely used to determine the integrating factor.

(i) $d\,(xy) = x\,dy + y\,dx$

(ii) $d\left(\dfrac{y}{x}\right) = \dfrac{x\,dy - y\,dx}{x^2}$

(iii) $d\left(\tan^{-1}\dfrac{x}{y}\right) = \dfrac{y\,dx - x\,dy}{x^2 + y^2}$

(iv) $d\left(\tan^{-1}\dfrac{y}{x}\right) = \dfrac{x\,dy - y\,dx}{x^2 + y^2}$

(v) $d\left(\log\dfrac{x}{y}\right) = \dfrac{y\,dx - x\,dy}{xy}$

(vi) $d\left(\log\dfrac{y}{x}\right) \dfrac{x\,dy - y\,dx}{xy}$

(vii) $d\left(\dfrac{1}{xy}\right) = -\dfrac{x\,dy + y\,dx}{x^2y^2}$

(viii) $d\left(\dfrac{x^2}{y}\right) = \dfrac{2xy\,dx' - x^2dy}{y^2}$

(ix) $d\left(\dfrac{y^2}{x^2}\right) = \dfrac{2x^2y\,dy - 2y^2x\,dx}{x^4}$

(x) $d\left(\frac{e^x}{y}\right) = \frac{ye^x d - e^x dy}{y^2}$

(xi) $d\,[\log(x^2 + y^2)] = \frac{2x\,dx + 2y\,dy}{x^2 + y^2}$

Example 3.48 **Solve the differential equation $(y^2e^x + 2xy)\,dx - x^2dy = 0$.**

Solution : We have,

$$M(x, y) = y^2e^x + 2xy$$

$$\Rightarrow \quad \frac{\partial M}{\partial y} = 2ye^x + 2x$$

$$N(x, y) = -x^2 \Rightarrow \frac{\partial N}{\partial x} = -2x$$

$$\because \quad \frac{\partial M}{\partial y} \neq \frac{\partial N}{\partial x}$$

So the given differential equations not exact.

The given equation many to written as

$$y^2e^xdx + 2xy\,dx - x^2dy = 0$$

On dividing this equation by y^2, we get

$$e^xdx + \frac{2xy\,dx - x^2\,dy}{y^2} = 0$$

The second term is the exact differential $d\left(\frac{x^2}{y}\right)$

$$\Rightarrow \quad e^x\,dx + d\left(\frac{x^2}{y}\right) = 0$$

On integration, we have

$$e^x + \frac{x^2}{y} = c$$

Example 3.49 **Solve the differential equation $(1 + xy)\,y\,dx + (1 - xy)x\,dy = 0$.**

Solution : Here, we have

$$M(x, y) = (1 + xy)\,y \Rightarrow \frac{\partial M}{\partial y} = 1 + 2xy$$

$$N(x, y) = (1 - xy) \Rightarrow \frac{\partial N}{\partial x} = 1 - 2xy$$

$$\because \quad \frac{\partial M}{\partial y} \neq \frac{\partial N}{\partial x}$$

The given equation is not exact.

The given equation may be written as

$$y\,dx + xy^2\,dx + x\,dy - x^2y\,dy = 0$$

or

$$y\,dx + x\,dy + xy^2\,dx - x^2y\,dy = 0$$

Dividing it by x^2y^2, we get

$$\frac{d\,(x,y)}{x^2y^2} + \frac{xy^2\,dx - x^2y\,dy}{x^2y^2} = 0$$

$$\Rightarrow \quad \frac{d\,(xy)}{(xy)^2} + \frac{dx}{x} - \frac{dy}{y} = 0$$

On integration, we have

$$\frac{(xy)^{-1}}{-1} + \log x - \log y = \log C$$

$$\Rightarrow \quad \log \frac{x}{cy} = \frac{1}{xy} \text{ or } \frac{x}{cy} = e^{1/xy}$$

Example 3.50 **Solve the defferential equation $y\,(2x^2y + e^x)\,dx - (e^x + y^3)\,dy = 0$.**

Solution : Here, we have

$$M\,(x, y) = y\,(2x^2y + e^x)\,dx$$

$$\Rightarrow \quad \frac{\partial M}{\partial y} = 4x^2y + e^x$$

$$N\,(x, y) = -\,(e^x + y^3) \Rightarrow \frac{\partial N}{\partial x} = -\,e^x$$

$$\because \quad \frac{\partial M}{\partial y} \neq \frac{\partial N}{\partial x}$$

$\therefore$ The given equation may be written is

$$(y\,e^x\,dx - e^x\,dy) + 2x^2y^2dx - y^3\,dy = 0$$

Dividing it by y^2, we get

$$\frac{y\,e^x\,dx - e^x\,dy}{y^2} + 2x^2\,dx - y\,dy = 0$$

$$\Rightarrow \quad d\left(\frac{e^x}{y}\right) + 2x^2\,dx - y\,dy = 0$$

On integration, we get

$$\frac{e^x}{y} + 2\,\frac{x^3}{3} - \frac{y^2}{2} = c$$

Example 3.51 **Solve the differential equation $x\,dy - y\,dx = (x^2 + y^2)\,dx$.**

Solution : Here, we have

$$M = -\,(y + x^2 + y^2)$$

$$\Rightarrow \quad \frac{\partial M}{\partial y} = -\,(1 + 2y)$$

$$N = x \Rightarrow \frac{\partial N}{\partial x} = 1$$

$$\because \quad \frac{\partial M}{\partial y} \neq \frac{\partial N}{\partial x}$$

$\therefore$ The given equation is not exact.

The given equation may be rewritten as

$$x\,dy - y\,dx = (x^2 + y^2)\,dx$$

On dividing it by $x^2 + y^2$, we get

$$\frac{x\,dy - y\,dx}{x^2 + y^2} = dx \;\Rightarrow\; d\left(\tan^{-1}\frac{y}{x}\right) = dx$$

On integration, we get

$$\tan^{-1}\frac{y}{x} = x + c$$

$$\frac{y}{x} = \tan(x + c)$$

PROBLEM SET 3.9

Solve the following differential equations :

1. $x\,dy - y\,dx + 4x^3 dx = 0$

Ans. $y + 2x^3 = cx$

2. $e^y\,dx + (xe^y + 2y)\,dy = 0$

Ans. $x\,e^y + y^2 = c$

3. $x\,dy - (y - x)\,dx = 0$

Ans. $\frac{y}{x} + \log x = c$

4. $(x^3 e^x - ay^2)\,dx + axy = 0$

Ans. $e^x + \frac{a}{2}\frac{y^2}{x^2} = c$

5. $y\,dy\,(x^2 + y^2 + l^2) + x\,dx\,(x^2 + y^2 - l^2) = 0$

Ans. $\frac{1}{2}(x^2 + y^2) + l^2(y^2 - x^2) = 6$

6. $x\,dx + y\,dy + 5\,(x^2 + y^2)\,dy = 0$

Ans. $x^2 + y^2 = ce^{-10y}$

Rule II : For the defferential equation $M\,dx + N\,dy$, if $\dfrac{\dfrac{\partial M}{\partial y} - \dfrac{\partial N}{\partial x}}{N}$ is a function of x along say $f(x)$, then I.F. $= e^{\int f(x)\,dx}$.

Example 3.52 **Solve $(x^2 + y^2 + 2x)\,dx + 2y\,dy = 0$.**

Solution : Here, we have

$$M = x^2 + y^2 + 2x \Rightarrow \frac{\partial M}{\partial y} = 2y$$

$$N = 2y \Rightarrow \frac{\partial N}{\partial x} = 2$$

$$\because \quad \frac{\dfrac{\partial M}{\partial y} - \dfrac{\partial N}{\partial x}}{N} = \frac{2y - 0}{2y} = 1 \text{ (may be treated as a function of } x)$$

$$\therefore \quad \text{I.F.} = e^{\int 1\, dx} = e^x$$

On multiplication the given equation by e^x, we get

$$e^x (x^2 + y^2 + 2x)\, dx + 2y\, e^x dy = 0$$

Now, $M = e^x (x^2 + y^2 + 2x) \Rightarrow \dfrac{\partial M}{\partial y} = 2ye^x$

$$N = 2ye^x \Rightarrow \frac{\partial N}{\partial x} = 2ye^x$$

Thus, the new equation is exact and the solution is

$$\int_{y \text{ a constant}} e^x (x^2 + y^2 + 2x)\, dx = c$$

$$\Rightarrow \quad x^2 e^x - \int 2xe^x\, dx + y^2 \int e^x\, dx + \int 2xe^x\, dx = c$$

$$\Rightarrow \quad (x^2 + y^2)\, e^x = c$$

Example 3.53 **Solve $(2x \log x - xy)\, dy + 2y\, dx = 0$.**

Solution : Here, we have

$$M = 2y \Rightarrow \frac{\partial M}{\partial y} = 2$$

$$N = 2x \log x - xy \Rightarrow \frac{\partial N}{\partial x} = 2 + 2 \log x - y$$

$$\because \quad \frac{\partial M}{\partial y} \neq \frac{\partial N}{\partial x}$$

$\therefore$ The given equationi is not exact

Here $\dfrac{\dfrac{\partial M}{\partial y} - \dfrac{\partial N}{\partial x}}{N} = -\dfrac{2 \log x + y}{2x \log x - xy}$

$$= -\frac{1}{x} = f(x) \text{ (say)}$$

Thus the I.F. $= e^{\int -\frac{1}{x} dx} = e^{-\log x} = \dfrac{1}{x}$

On multiplication, the given equqtion by $\dfrac{1}{x}$, we get

$$\frac{2y}{x}\, dx + (2 \log x - y)\, dy = 0$$

Obviously the new equation is exact. Hence, the solution is

$$\int_{(y \text{ a constant})} \frac{2y}{x}\, dx = \int y\, dy = c$$

$$\Rightarrow \quad 2y \log x - \frac{y^2}{2} = c$$

Rule III : For the defferential equation $M\,dx + N\,dy = 0$, if $\dfrac{\dfrac{\partial M}{\partial y} - \dfrac{\partial N}{\partial x}}{N}$ is a function of y only, say $g(y)$. Then I.F. $= e^{\int g(y)\,dy}$.

Example 3.54 **Solve $(3x^2 y^4 + 2xy)\,dx + (2x^3y^3 - x^2)\,dy = 0$.**

Solution : Here, we have

$$M = 3x^2y^4 + 2xy$$

$$\Rightarrow \quad \frac{\partial M}{\partial y} = 12x^2y^3 + 2x$$

$$N = 2x^3y^3 - x^2$$

$$\Rightarrow \quad \frac{\partial N}{\partial x} = 6x^2y^3 - 2x$$

$$\because \quad \frac{\partial M}{\partial y} \neq \frac{\partial N}{\partial x}$$

$\therefore$ The given equation is not exact.

$$\frac{\dfrac{\partial M}{\partial y} - \dfrac{\partial N}{\partial x}}{N} = \frac{6x^2y^2 - 2x - 12x^2y^3 - 2x}{3x^2y^2 + 2xy}$$

$$= \frac{-6x^2y^3 - 4x}{3x^2y^4 + 2xy} = -\frac{2x\,(3xy^3 + 2)}{xy\,(3xy^3 + 2)} = -\frac{2}{y}$$

$$\therefore \text{ The I.F.} = e^{-\int \frac{2}{y}\,dy}$$

$$= e^{-2\log y} = \frac{1}{y^2}$$

Multiplying the given equatioin by $\dfrac{1}{y^2}$, we get

$$\left(3x^2y^2 + \frac{2x}{y}\right)dx + \left(2x^3y - \frac{x^2}{y^2}\right)dy = 0$$

Which is an exact equation.

Thus the solution is

$$\int_{y\ a\ \text{cosntant}} \left(3x^2y^2 + \frac{2x}{y}\right)dx = C$$

$$\Rightarrow \quad x^3y^2 + \frac{x^2}{y} = C$$

$$\Rightarrow \quad x^3y^3 + x^2 = cy$$

Example 3.55 **Solve $(y^4 + 2y)\,dx + xy^3 + 2y^4 - 4x)\,dy = 0$.**

Solution : Here, we have

$$M = y^4 + 2y \Rightarrow \frac{\partial M}{\partial y} = 4y^3 + 2$$

$$N = xy^3 + 2y^4 - 4x \Rightarrow \frac{\partial N}{\partial x} = y^3 - 4$$

$$\because \quad \frac{\partial M}{\partial y} \neq \frac{\partial N}{\partial x}$$

$\therefore$ The given equationi is not exact

But $$\frac{\frac{\partial M}{\partial y} - \frac{\partial N}{\partial x}}{N} = \mathbf{2y^3 + 6}$$

$$= \frac{y^3 - 4 - 4y^3 - 2}{y^4 + 2y} = \frac{-3y^3 - 6}{y(y^3 + 2)}$$

$$= -\frac{3}{y}$$

$$\therefore \quad \text{I.F.} = e^{\int -\frac{3}{y} dy} = e^{-3 \log y} = \frac{1}{y^3}$$

On multiplying the given equatin by $\frac{1}{y^3}$, we get

$$\left(y + \frac{2}{y^2}\right) dx + \left(x + 2y - \frac{4x}{y^3}\right) dy = 0$$

Obviously this is an exact equation and its solution is

$$\int_{(y \text{ constant})} \left(y + \frac{2}{y^2}\right) dx + \int 2y\, dy = c$$

$$\left(y + \frac{2}{y^2}\right) x + y^2 = c$$

Rule IV : For the differential equation $M\,dx + N\,dy$, if $M(x, y)$ and $N(x, y)$ are of the form $y f_1(xy)$ and $f_2(xy)$. Then I.F. $= \frac{1}{Mx - Ny}$.

Example 3.56 **Solve $y(xy + 2x^2y^2)\,dx + x(xy - x^2y^2)\,dy = 0$.**

Solution : Here we have,

$$M(x, y) = y(xy + 2x^2y^2)$$

$$N(x, y) = x(xy - x^2y^2)$$

Obviously M and N are of the form $y f_1(xy)$ and $x f_2(xy)$ respectively.

$$\therefore \quad \text{I.F.} = \frac{1}{Mx - Ny} = \frac{1}{x^2y^2(1 + 2xy) - x^2y^2(1 - xy)} = \frac{1}{3x^3y^3}$$

On multiply the given equqtion by $\frac{1}{3x^3y^3}$, we get

$$\left(\frac{1}{3x^2y} + \frac{2}{3x}\right) dx + \left(\frac{1}{3x^2y^2} - \frac{1}{3y}\right) dy = 0$$

Obviously, this is an exact equation hence, the fsolution is

$$\int\left(\frac{1}{3x^2y}+\frac{2}{3x}\right)dx-\int\frac{1}{3y}\,dy=c$$

$$\Rightarrow \quad -\frac{1}{3xy}+\frac{2}{3}\log x-\frac{1}{3}\log y=c$$

or $$-\frac{1}{xy}+2\log x-\log y=c_1$$

Example 3.57 **Solve $(x^2y^2+xy+1)\,y\,dx+(x^2y^2-xy+1)\,x\,dy=0$.**

Solution : Here, we have

$$M=(x^2y^2+xy+1)\,y$$

$$\Rightarrow \quad \frac{\partial M}{\partial y}=3x^2y^2+2xy+1$$

$$N=(x^2y^2-xy+1)\,x$$

$$\Rightarrow \quad \frac{\partial N}{\partial x}=3x^2y^2-2xy+1$$

$$\because \quad \frac{\partial M}{\partial y}\neq\frac{\partial N}{\partial x}$$

$\therefore$ The given equation is not exact. But $M=y\,f_1(xy)$ and $N=x\,f_2(xy)$.

$$\therefore \quad \text{I.F.}=\frac{1}{Mx-Ny}=\frac{1}{x^3y^3+x^2y^2+xy-x^3y^3+x^2y^2-xy}$$

$$=\frac{1}{2x^2y^2}$$

On multiplying the given equqtion by $\frac{1}{2x^2y^2}$, we get

$$\frac{1}{2}\left(y+\frac{1}{x}+\frac{1}{x^2y}\right)dx+\frac{1}{2}\left(x-\frac{1}{y}+\frac{1}{xy^2}\right)dy=0$$

Which is obviously an exact equation and it solution is

$$\frac{1}{2}\int_{y\text{ constant}}\left(y+\frac{1}{x}+\frac{1}{x^2y}\right)dx+\frac{1}{2}\int\left(-\frac{1}{y}\right)dy=c$$

$$xy+\log x-\frac{1}{xy}-\log y=2c$$

$$xy-\frac{1}{xy}+\log\frac{x}{y}=c_1.$$

Rule V : If the given equation is of the type

$$x^my^n\,(a_1y\,dx+b_1x\,dy)+x^ly^k\,(a_2y\,dx+b_2\,x\,dy)=0,\ \text{the I.F.}=x^py^q,$$

where $\frac{m+p+1}{a_1}=\frac{n+q+1}{b_1}$ and $\frac{l+p+1}{a_2}=\frac{k+q+1}{b_2}$.

Example 3.58 **Solve $(2x^2y - 3y^4)\,dx + (3x^3 + 2xy^3)\,dy = 0$.**

Solution : The given equation may be written as

$$x^2\,(2y\,dx + 3x\,dy) + y^3\,(-\,3y\,dx + 2x\,dy) = 0$$

This equation matches with the above form, where

$$m = 2,\ n = 0,\ a_1 = 2,\ b_1 = 3,\ a_2 = -\,3,\ b_2 = 2,\ l = 0,\ k = 3$$

Here the integrating factor must be $x^p q^q$ such that

$$\frac{2+p+1}{2} = \frac{0+q+1}{3} \text{ and } \frac{0+p+1}{-3} = \frac{3+q+1}{2}$$

or $\quad 3\,(p+3) = 2\,(q+1)$ and $2\,(p+1) = -\,3\,(q+4)$

$\Rightarrow \quad 3p - 2q = -\,7$ and $2p + 3q = -\,14$

$\Rightarrow \quad p = -\dfrac{49}{13}$ and $q = -\dfrac{23}{13}$

$$\text{I.F.} = x^{-49/13}\,y^{-23/13}$$

On multiplying the given equation by I.F. $= x^{-49/13}\,y^{-23/13}$. The equation reduces to

$$(2x^{-23/13}\,y^{-15/13} - 3x^{-49/13}\,y^{29/13})\,dx + (3x^{10/13}\,y^{-28/13} + 2x^{-36/13}\,y^{11/13})\,dy = 0$$

It solution is

$$\int_{(y\text{ constant})} 2x^{-23/13}\,y^{-25/13} - 3x^{-49/13}\,y^{29/313})\,dx = c$$

or $\quad -\dfrac{13}{10}\cdot 2x^{-10/13}\,y^{-15/13} + 3\cdot\dfrac{13}{36}\,x^{-36/13}\,y^{29/13} = c$

or $\quad 5x^{-36/13}\,y^{29/13} - 12x^{-10/13}\,y^{-15/13} = c$

Rule VI : If the equation $M\,dx + N\,dy$ is an homogeneous equation in x and y. Then I.F. $= \dfrac{1}{Mx + Ny}$.

Example 3.59 **Solve $(x^2y - 2xy^2)\,dx - (x^3 - 3x^2y)\,dy = 0$.**

Solution : Here, we have

$$M = x^2y - 2xy^2 \Rightarrow \frac{\partial M}{\partial y} = x^2 - 4xy$$

$$N = 3x^2y - x^3 \Rightarrow \frac{\partial N}{\partial x} = 6xy - 3x^2$$

$$\because \quad \frac{\partial M}{\partial y} \neq \frac{\partial N}{\partial x}$$

$\therefore$ The given equation is not exact. But M and N are homogeneous functions of x and y and $\dfrac{1}{Mx + Ny} = \dfrac{1}{x^2y^2} \neq 0$.

Hence, I.F. $= \dfrac{1}{x^2y^2}$

Multiplying the given equation by I.F. the given equation reduces to

$$\left(\frac{1}{y}-\frac{2}{x}\right)dx+\left(\frac{3}{y}-\frac{x}{y^2}\right)dy=0$$

Obviously, this is an exact equation its solution is

$$\int_{y\text{ constant}}\left(\frac{1}{y}-\frac{2}{x}\right)dx+\int\frac{3}{y}\,dy=c$$

$$\frac{x}{y}-2\log x+3\log y=c$$

$$\frac{x}{y}+\log\frac{y^3}{x^2}=c$$

Example 3.60 **Solve :**

$$\frac{dy}{dx}=\frac{x^3+y^3}{xy^2}.$$

Solution : Here, we have

$$M=x^3+y^3 \Rightarrow \frac{\partial M}{\partial y}=3y^2$$

$$N=-xy^2 \Rightarrow \frac{\partial N}{\partial x}=-y^2$$

$$\because \quad \frac{\partial M}{\partial y}\neq\frac{\partial N}{\partial x}$$

∴ The given equation is not exact.

But the given equation is a homogeneous equation

$$\therefore \quad \text{I.F.}=\frac{1}{Mx+Ny}$$

$$=\frac{1}{(x^3+y^3)\,x+(-xy^2)\,y}=\frac{1}{x^4}$$

On multiplying the given by $\frac{1}{x^4}$, we get

$$\frac{1}{x^4}(x^3+y^3)\,dx-\frac{xy^2}{x^4}\,dy=0$$

$$\Rightarrow \quad \left(\frac{1}{x}+\frac{y^3}{x^4}\right)dx-\frac{y^2}{x^3}\,dy=0$$

It is an exact equation and its solution is

$$\int_{y\text{ constant}}\left(\frac{1}{x}+\frac{y^3}{x^4}\right)dx=c$$

$$\log x-\frac{y^3}{3x^3}=c$$

PROBLEM SET 3.10

Solve the following defferential equations :

1. $(x^3 + xy^4)\, dx + 2y^3 dy = 0$

   Ans. $e^{x^2}(x^2 - 1 + y^4) = c$

2. $(x^2 + y^2 + 1)\, dx - 2xy\, dy = 0$

   Ans. $x^2 - y^2 - 1 = cx$

3. $\left[2x \sinh \frac{y}{x} + 3y \cosh \frac{y}{x}\right] dx - 3x \cosh \frac{y}{x}\, dy = 0$

   Ans. $-3 \sinh \frac{y}{x} = cx^{2/3}$

4. $(x \sec^2 y - x^2 \cos y)\, dy = (\tan y - 3x^4)\, dx$

   Ans. $-\frac{1}{x} \tan y - x^3 + \sin y = c$

5. $(xy^3 + y)\, dx + 2(xy^2 + x + y^4)\, dy = 0$

   Ans. $3x^2y^4 + 6xy^3 + 2y^6 = c$

6. $(3x^2y^4 + 2xy)\, dx + (2x^3y^3 - x^2)\, dy = 0$

   Ans. $x^3y^2 + \frac{x^2}{y} = c$

7. $(y^4 + 2y)\, dx + (xy^3 + 2y^4 - 4x)\, dy = 0$

   Ans. $xy + y^2 + \frac{2x}{y^2} = c$

8. $(xy^2 - x^2)\, dx + (3x^2y^2 + x^2y - 2x^3 + y^2)\, dy$

   Ans. $4x^5y + 2x^4y^2 + \frac{4}{3}x^3y^3 + y^4x^3 = c$

9. $x(4y\, dx + 2x\, dy) + y^3(3y\, dx + 5x\, dy) = 0$

   Ans. $x^4y^2 + x^3y^5 = c$

10. $(20x^2 + 8xy + 4y^2 + 3y^3)\, y\, dx + 4(x^2 + xy + y^2 + y^3)\, x dy = 0$

    Ans. $4x^5y + 2x^4y^2 + \frac{4}{3}x^3y^3 + y^4x^2 = c$

11. $y(1 + xy)\, dx + x(1 - xy)\, dy = 0$

    Ans. $xy \log \frac{y}{x} = cxy - 1$

12. $(xy \sin xy + \cos xy)\, y\, dx + (xy \sin xy - \cos xy)\, x\, dy = 0$

    Ans. $x = cy \cos xy$

13. $x^2y\, dx - (x^3 + y^3)\, dy = 0$

    Ans. $y = ce^{x^3/3y^3}$

14. $(x^4y^4 + x^2y^2 + xy)\,y\,dx + (x^4y^4 - x^2y^2 + xy)\,x\,dy = 0$

Ans. $\frac{1}{2}x^2y^2 - \frac{1}{xy} + \log\frac{x}{y} = c$

15. $(y^2 + 2x^2y)\,dx + (2x^3 - xy)\,dy = 0$

Ans. $4x^{1/2}y^{1/3} + \frac{2}{3}x^{-3/2}y^{3/2} = c$

16. $(3x + 2y^2)\,y\,dx + 2x\,(2x + 3y^2)\,dy = 0$

Ans. $x^2y^4\,(x + y^2) = c$

17. $x\,(3y\,dx + 2x\,dy) + 8y^4\,(y\,dx + 3x\,dy) = 0$

Ans. $x^2y^2\,(x + 4y^4) = c$

18. $(2y\,dx + 3x\,dy) + 2xy\,(3y\,dx + 4x\,dy) = 0$

Ans. $x^2y^3 + 2x^2y^4 = c$

19. $y\log y\,dx + (x - \log y)\,dy = 0$

Ans. $2x\log y = c + (\log y)^2$

20. $(x - y^2)\,dx + 2xy\,dy = 0$

Ans. $y^2 = cx - x\log x$

21. $(y^3 - 3xy^2)\,dx + (2x^2y - xy^2)\,dy = 0$

Ans. $\frac{y}{x} + 3\log x - 2\log y = c$

22. $\left(\frac{y}{x}\sec y - \tan y\right)dx + (\sec y\log x - x)\,dy = 0$

Ans. $y\log x - x\sin y = c$

23. $(2xy^4e^y + 2xy^3 + y)\,dx + (x^2y^4e^y - x^2y^2 - 3x)\,dy = 0$

Ans. $x^2e^y + \frac{x^2}{y} + \frac{x}{y^3} = c$

24. $y\,(2xy + 1)\,dx + x\,(1 + 2xy - x^3y^3)\,dy = 0$

Ans. $y = c\,e^{-\frac{(3xy+1)}{x^3y^3}}$

3.12 LINEAR DIFFERENTIAL EQUATIONS OF HIGHER ORDER

3.12.1 Linear Combination of Functions

Let $f_1(x), f_2(x), \ldots, f_n(x)$ are n functions. Then $k_1f_1(x) + k_2f_2(x) + \ldots + k_rf_n(x)$, where $k_1, k_2, \ldots, k_n$ are constants is called a linear combination of the given functions.

3.12.2 Linear Dependence and Independence of Functions

A set of functions $f_1(x), f_2(x), \ldots, f_n(x)$ are said to be **linearly independent** if

$$k_1f_1(x) + k_2f_2(x) + \ldots + k_nf_n(x) = 0 \;\Rightarrow\; k_1 = k_2 = \ldots = k_n = 0$$

These functions are said to be **linearly dependent** if

$$k_1 f_1(x) + k_2 f_2(x) + \ldots + k_n f_n(x) = 0$$

$\Rightarrow$ $k_1, k_2, \ldots, k_n$ are not all zero.

In this case, one or more functions can be expressed as a linear combination of the remaining functions. *i.e.*, if $k_1 \neq 0$, then

$$f_1(x) = -\frac{1}{k_1}[k_2 f_2(x) + k_3 f_3(x) + \ldots + k_n f_n(x)]$$

3.12.3 Criterion for Linear Dependence or Independence of Functions

A very elegant procedure to test the linear independence or dependence of a given set of functions is the application of **wronskians**. The wronskian of n-functions $f_1(x), f_2(x), \ldots, f_n(x)$ is denoted by $w(f_1, f_2, \ldots, f_n)$ and is defined as

$$w(f_1, f_2, \ldots, f_n) = \begin{vmatrix} f_1 & f_2 & \ldots & f_n \\ f_1' & f_2' & \ldots & f_n' \\ \ldots & \ldots & \ldots & \ldots \\ f_1^{n-1} & f_2^{n-1} & \ldots & f_n^{n-1} \end{vmatrix}$$

1. The functions $f_1(x), f_2(x), \ldots, f_n(x)$ are linearly independent if $w \neq 0$
2. The functions $f_1(x), f_2(x) \ldots, f_n(x)$ are linearly dependent if $w = 0$

Example 3.61 **Show that the functions x, x^2, x^3 are linearly independent.**

Solution : Let $f_1(x) = x,\ f_2(x) = x^2$ and $f_3(x) = x^3$

The Wronskian of $f_1(x), f_2(x)$ and $f_3(x)$ is given by

$$w(f_1, f_2, f_3) = \begin{vmatrix} f_1 & f_2 & f_3 \\ f_1' & f_2' & f_3' \\ f_1' & f_2'' & f_3'' \end{vmatrix} = \begin{vmatrix} x & x^2 & x^3 \\ 1 & 2x & 3x^2 \\ 0 & 2 & 6x \end{vmatrix}$$

$$= x[12x^2 - 6x^2] - 1\,[6x^3 - 2x^3] + 0[3x^4 - 2x^4]$$

$$= 6x^3 - 4x^3 = 2x^3 \neq 0$$

Hence, the functions are linearly independent.

Example 3.62 **Find whether the functions $e^x \cos x$ and $e^x \sin x$ are linearly dependent or independent. If linearly independent, then determine a differential equation having the above functions as independent solutions.** *[UPTU 2002]*

Solution : Let $f_1(x) = e^x \cos x\ ,\ f_2(x) = e^x \sin x$

The Wronskian of $f_1(x)$ and $f_2(x)$ is given by

$$w(f_1, f_2) = \begin{vmatrix} f_1 & f_2 \\ f_1' & f_2' \end{vmatrix} = \begin{vmatrix} e^x \cos x & e^x \sin x \\ e^x \cos x - e^x \sin x & e^x \sin x + e^x \cos x \end{vmatrix}$$

$$= e^x \cos x\,(e^x \sin x + e^x \cos x) - e^x \sin x\,(e^x \cos x - e^x \sin x)$$
$$= e^{2x} \cos x \sin x + e^{2x} \cos^2 x - e^{2x} \sin x \cos x + e^{2x} \sin^2 x$$
$$= e^{2x} (\cos^2 x + \sin^2 x)$$
$$= e^{2x} \neq 0$$

Hence, the functions are linearly independent.

The required differential equation having $e^x \cos x$ and $e^x \sin x$ as independent solutions is given by

$$y = c_1 e^x \cos x + c_2 e^x \sin x$$
$$\Rightarrow \quad y' = c_1 (e^x \cos x - e^x \sin x) + c_2 (e^x \sin x + e^x \cos x)$$
$$= (c_1 e^x \cos x + c_2 e^x \sin x) - c_1 e^x \sin x + c_2 e^x \cos x$$
$$= y - c_1 e^x \sin x + c_2 e^x \cos x$$
$$\Rightarrow \quad y' - y = -c_1 e^x \sin x + c_2 e^x \cos x$$

Again $\quad y'' - y' = -c_1(e^x \sin x + e^x \cos x) + c_2 (e^x \cos x - e^x \sin x)$

$$\Rightarrow \quad y'' - y' = -c_1 e^x \sin x - c_1 e^x \cos x + c_2 e^x \cos x - c_2 e^x \sin x$$
$$= (-c_1 e^x \sin x + c_2 e^x \cos x) - (c_1 e^x \cos x + c_2 e^x \sin x)$$
$$= (y' - y) - y$$
$$\Rightarrow \quad y'' - 2y' + 2y = 0$$

which is the required differential equation.

PROBLEM SET 3.11

Examine whether the following set of functions are linearly independent.

1. e^x, xe^x

Ans. Linearly independent ($w = e^{2x}$)

2. $2x, 6x + 3, 3x + 2$

Ans. Linearly dependent

3. $\sin x, \sin 2x, \sin 3x$

Ans. Linearly independent ($w = -16 \sin^6 x$)

4. $1, \cos x, \sin x$

Ans. Linearly independent ($w = 1$)

5. $e^{-x}, \sinh x, \cosh x$

Ans. Linearly dependent

6. e^{ax}, e^{-ax}

Ans. Linearly independent ($w = -2a$)

3.13 LINEAR DIFFERENTIAL EQUATIONS OF nth ORDER WITH CONSTANT COEFFICIENTS

A linear differential equation of nth order is given by

$$a_0 \frac{d^n y}{dx^n} + a_1 \frac{d^{n-1} y}{dx^{n-1}} + \ldots + a_{n-1} \frac{dy}{dx} + a_n y = f(x) \quad \ldots(1)$$

where $a_0, a_1, \ldots, a_{n-1}, a_n$ are either functions of x or constant while, $f(x)$ is a function of x only.

A linear differential equation (1) of nth order is said to be the differential equation of nth order with constant coefficients if $a_0, a_1, \ldots, a_{n-1}, a_n$ are all constants

Let us introduce a symbolic operator D, where $D = \frac{d}{dx}$, $D^2 = \frac{d^2}{dx^2}, \ldots, D^n = \frac{d^n}{dx^n}$ then the given equation (1) can be written as

$$a_0 D^n y + a_1 D^{n-1} y + \ldots + a_{n-1} Dy + a_n y = f(x)$$

or $$(a_0 D^n + a_1 D^{n-1} + \ldots + a_{n-1} D + a_n)\, y = f(x)$$

or $$\phi(D) y = f(x) \quad \ldots(2)$$

where $$\phi(D) = a_0 D^n + a_1 D^{n-1} + \ldots + a_{n-1} D + a_n$$

Here, we have two type of differential equations

3.13.1 Homogeneous Linear Differential Equations

If $f(x) = 0$, then equation (2) is said to be **homogeneous linear differential equation** with consteant coefficients.

3.13.2 Non-Homogeous Linear Differential Equations

If $f(x) \neq 0$, then equation (2) is said to be **non-homogeneous linear differential equation** with constant coefficients.

3.14 SOLUTION OF HOMOGENEOUS LINEAR DIFFERENTIAL EQUATIONS OF nth ORDER WITH CONSTANT COEFFICIENTS

As we know that a linear differential equation of the form

$$(a_0 D^n + a_1 D^{n-1} + \ldots + a_{n-1} D + a_n)\, y = 0 \quad \ldots(1)$$

is said to be homogeneous linear differential equations, of nth order with constant coefficients.

Let $y = e^{mx}$ is the solution of (1), where m is a parameter, then from (1)

$$a_0 m^n e^{mx} + a_1 m^{n-1} e^{mx} + \ldots + a_{n-1} m e^{mx} + a_n e^{mx} = 0$$

$$\Rightarrow \quad e^{mx} [a_0 m^n + a_1 m^{n-1} + \ldots + a_{n-1} m + a_n] = 0$$

$$\Rightarrow \quad a_0 m^n + a_1 m^{n-1} + \ldots + a_{n-1} m + a_n = 0 \quad \ldots(2)$$

The equation (2) is called the **Auxilliary Equation (A.E.) of (1)**. The degree of this algebraic equation is same as the order of the differential equation. This equation has degree n. Hence, n roots.

NOTE : The A.E. of (1) is obtained by replacing D by m in $\varphi(D) = 0$.

3.14.1 Method for Finding the Complementary Function

The solution of equation (1) depends upon the nature of roots of the auxilliary equation (2). We have the following three cases:

Case 1 : When the roots are real and distinct.

If $m_1, m_2, ..., m_n$ are the roots of auxilliary equation, then

$$\text{C.F.} = c_1 e^{m_1 x} + c_2 e^{m_2 x} + c_3 e^{m_3 x} + ... + c_n e^{m_n x}$$

where $c_1, c_2, c_3, ..., c_n$ are arbitrary constants.

Case 2 : When the roots are real and equal

(i) when two roots are equal say $m_1 = m_2 = m \neq m_3$ etc.

$$\text{C.F.} = (c_1 + c_2 x) e^{mx} + c_3 e^{m_3 x} + ... + c_n e^{m_n x}$$

(ii) when three roots are equal say $m_1 = m_2 = m_3 = m \neq m_4$ etc.

$$\text{C.F.} = (c_1 + c_2 x + c_3 x^2) e^{mx} + c_4 e^{m_4 x} + ... + c_n e^{m_n x}$$

and so on.

Case 3 : When the roots are imaginary

If the roots are $\alpha \pm i\beta$, then the solution will be

$$\text{C.F.} = c_1 e^{(\alpha + i\beta)x} + c_2 e^{(\alpha - i\beta)x} + c_3 e^{m_3 x} + ... + c_n e^{m_n x}$$

$$= c_1 e^{\alpha x} [\cos \beta x + i \sin \beta x] + c_2 e^{\alpha x} [\cos \beta x - i \sin \beta x] + c_3 e^{m_3 x} + ... + c_n e^{m_n x}$$

$$= e^{\alpha x} [(c_1 + c_2) \cos \beta x + i (c_1 - c_2) \sin \beta x] + c_3 e^{m_3 x} + ... + c_n e^{m_n x}$$

$$= e^{\alpha x} [A \cos \beta x + B \sin \beta x] + c_3 e^{m_3 x} + ... + c_n e^{m_n x}$$

where $A = c_1 + c_2$ and $B = i(c_1 - c_2)$

Example 3.63 **Solve : $(D^3 + 6D^2 + 11D + 6)y = 0$** *[UPTU 2000]*

Solution : Given $(D^2 + 6D^2 + 11D + 6)y = 0$

A.E. is $m^3 + 6m^2 + 11m + 6 = 0$

$\Rightarrow \quad (m + 1)(m^2 + 5m + 6) = 0$

$\Rightarrow \quad (m + 1)(m + 2)(m + 3) = 0$

$\Rightarrow \quad m = -1, -2, -3$

Here, the roots of A.E. are real and distinct

$\therefore \quad C.F. = c_1 e^{-x} + c_2 e^{-2x} + c_3 e^{-3x}$

Hence, the required solution is

$$y = c_1 e^{-x} + c_2 e^{-2x} + c_3 e^{-3x}$$

Example 3.64 **Solve : $(D - 1)^2 (D^2 + 1)^2 y = 0$** *[UPTU 2001]*

Solution : Given $(D - 1)^2 (D^2 + 1)^2 y = 0$

A.E. is $(m - 1)^2 (m^2 + 1)^2 = 0$

$\Rightarrow \quad (m - 1)^2 = 0$ or $(m^2 + 1)^2 = 0$

$\Rightarrow \quad m = 1, 1$ or $m = \pm i, \ m = \pm i$

Here, $\quad m_1 = m_2 = 1$ and $m_3 = i = m_4, m_5 = -i = m_6$

Hence, $C.F. = (c_1 + c_2 x) e^x + (c_3 + c_4 x) \cos x + (c_5 + c_6 x) \sin x$

Thus, the required solutin is

$$y = (c_1 + c_2 x) e^x + (c_3 + c_4 x) \cos x + (c_5 + c_6 x) \sin x$$

Example 3.65 **Solve:**

$$\frac{d^2y}{dx^2} - 7\frac{dy}{dx} + 12y = 0$$

[UPTU 2000]

Solution : Given

$$\frac{d^2y}{dx^2} - 7\frac{dy}{dx} + 12y = 0 \Rightarrow (D^2 - 7D + 12)y = 0$$

A.E. is $m^2 - 7m + 12 = 0 \Rightarrow m^2 - 4m - 3m + 12 = 0$

$\Rightarrow$ $(m - 4)(m - 3) = 0 \Rightarrow m = 3, 4$

$\therefore$ $\text{C.F.} = c_1 e^{3x} + c_2 e^{4x}$

Hence, the required solution is $y = c_1 e^{3x} + c_2 e^{4x}$

Example 3.66 **Solve:**

$$\frac{d^3y}{dx^3} - 2\frac{d^2y}{dx^2} + 4\frac{dy}{dx} - 8y = 0$$

Solution : Given $\frac{d^3y}{dx^3} - 2\frac{d^2y}{dx^2} + 4\frac{dy}{dx} - 8y = 0$

$\Rightarrow$ $(D^3 - 2D^2 + 4D - 8)y = 0$

The A.E. is

$$m^3 - 2m^2 + 4m - 8 = 0 \Rightarrow (m - 2)(m^2 + 4) = 0$$

$\Rightarrow$ $m - 2 = 0$ or $m^2 + 4 = 0 \Rightarrow m = 2$ or $m = \pm 2i$

$\therefore$ $\text{C.F.} = c_1 e^{2x} + c_3 \cos 2x + c_4 \sin 2x$

Hence, the required solution is

$$y = c_1 e^{2x} + c_3 \cos 2x + c_4 \sin 2x$$

Example 3.67 **Solve :**

$$\frac{d^3y}{dx^3} + 6\frac{d^2y}{dx^2} + 12\frac{dy}{dx} + 8y = 0$$

Given $y(0) = 0$, $y'(0) = 0$ and $y''(0) = 2$.

Solution : We have

$$\frac{d^3y}{dx^3} + 6\frac{d^2y}{dx^2} + 12\frac{dy}{dx} + 8y = 0$$

$\Rightarrow$ $(D^3 + 6D^2 + 12D + 8)y = 0$

A.E. is $m^3 + 6m^2 + 12m + 8 = 0$

$\Rightarrow$ $(m + 2)(m^2 + 4m + 4) = 0$

$\Rightarrow$ $(m + 2)(m + 2)^2 = 0 \Rightarrow m = -2, -2, -2$

$\therefore$ $\text{C.F.} = (c_1 + c_2 x + c_3 x^2)e^{-2x}$

Hence, the required solution is

$$y = (c_1 + c_2 x + c_3 x^2)\, e^{-2x}$$

Given, at $x = 0, y = 0 \Rightarrow c_1 = 0$

At $x = 0, y'(0) = 0 \Rightarrow 0 = c_2$

Also at $x = 0, y''(0) = 2 \Rightarrow c_3 = 1$

Hence, $y = x^2 e^{-2x}$

Example 3.68 **Solve :**

$$\frac{d^2y}{dx^2} - 3\frac{dy}{dx} + 2y = 0$$

Solution : The equation can be written as

$$(D^2 - 3D + 2)\, y = 0$$

The auxiliary equation (A.E.) is given by

$$m^2 - 3m + 2 = 0$$

or $(m - 1)(m - 2) = 0$

or $m = 1, 2$

Hence, the solution is $y = c_1 e^x + c_2 e^{2x}$

Example 3.69 **Solve**

$$\frac{d^3y}{dx^3} - 6\frac{d^2y}{dx^2} + 11\frac{dy}{dx} - 6y = 0$$

Solution : Given Eq. can be written as

$$(D^3 - 6D^2 + 11D - 6)\, y = 0$$

Its A.E. is $m^3 - 6m^2 + 11m - 6 = 0$

or $(m - 1)(m - 2)(m - 3) = 0$

$\Rightarrow$ $m = 1, 2, 3$

Hence, the C.S. is $y = c_1 e^x + c_2 e^x + c_3 e^{3x}$

Example 3.70 **Solve :**

$$\frac{d^4y}{dx^4} + 2\frac{d^3y}{dx^3} + 3\frac{d^2y}{dx^2} + 2\frac{dy}{dx} + y = 0.$$

Solution : The eqn. can be written as

$$(D^4 + 2D^3 + 3D^2 + 2D + 1)\, y = 0$$

The auxiliary Eq. (A.E.) is given by

$$m^4 + 2m^3 + 3m^2 + 2m + 1 = 0$$

or $(m^2 + m + 1)^2 = 0$

$$\therefore \quad m = \frac{-1 \pm i\sqrt{3}}{2}, \frac{-1 \pm i\sqrt{3}}{2}$$

$$\therefore \quad y = e^{-\frac{1}{2}x}\left[(c_1 + c_2 x)\cos\left(\frac{\sqrt{3}}{2}x\right) + (c_3 + c_4 x)\sin\left(\frac{\sqrt{3}}{2}x\right)\right]$$

Example 3.71 **Solve :**

$$\text{(i)}\ \frac{d^4x}{dt^4} - a^4x = 0 \qquad \text{(ii)}\ \frac{d^4x}{dt^4} + a^4x = 0$$

Solution : (i) The eqn. can be written as

$$(D^4 - a^4)\,x = 0$$

where $D = \dfrac{d}{dt}$

The auxiliary eqn. (A.E.) is given by

$$(m^4 - a^4) = 0$$

or $$(m^2 - a^2)(m^2 + a^2) = 0$$

or $$m = a, -a, ia, -ia$$

$\therefore$ $$x = c_1e^{at} + c_2e^{-at} + c_3 \cos at + c_4 \sin at$$

(ii) The Eq. can be written as

$$(D^4 + a^4)\,x = 0$$

The auxiliry eqn. (A.E.) is given by

$$m^4 + a^4 = 0$$

or $$(m^2 + a^2)^2 - 2m^2a^2 = 0$$

or $$(m^2 + a^2) = \pm\, ma\sqrt{2}$$

or $$m^2 - ma\sqrt{2} + a^2 = 0$$

or $$m^2 + ma\sqrt{2} + a^2 = 0$$

or $$m = \frac{a\,(\sqrt{2} \pm i\sqrt{2})}{2},\ \frac{a\,(-\sqrt{2} \pm i\sqrt{2})}{2}$$

or $$m = \frac{a}{\sqrt{2}}(1 \pm i),\ -\frac{a}{\sqrt{2}}(1 \pm i)$$

$\therefore$ $$x = e^{(a/\sqrt{2})\,t}\left(c_1 \cos \frac{a}{\sqrt{2}}t + c_2 \sin \frac{a}{\sqrt{2}}t\right) + e^{-a/\sqrt{2}\,t}\left(c_3 \cos \frac{a}{\sqrt{2}}t + c_4 \sin \frac{a}{\sqrt{2}}t\right)$$

Example 3.72 **Solve :**

$$\frac{d^4y}{dx^4} + 2\frac{d^3y}{dx^3} + 3\frac{d^2y}{dx^2} + 2\frac{dy}{dx} + y = 0.$$

Solution : Given equation can be written as

$$(D^4 + 2D^3 + 3D^2 + 2D + 1)\,y = 0$$

A.E. is $m^4 + 2m^3 + 3m^2 + 2m + 1 = 0$

or $$(m^2 + m + 1)^2 = 0$$

$\therefore$ $$m = \frac{-1 \pm i\sqrt{3}}{2},\ \frac{-1 \pm i\sqrt{3}}{2}$$

Hence, the C.S. is

$$y = e^{-(1/2)\,x}\left[(c_1 + c_2x) \cos\left(\frac{\sqrt{3}}{2}x\right) + (c_3 + c_4x) \sin\left(\frac{\sqrt{3}}{2}x\right)\right]$$

Example 3.73 **Solve $y'' - 2y' - 10y = 0$, gives $y(0) = 4$, $y'(0) = 1$.**

Solution Given Eq. can be written as

$$(D^2 - 2D + 10)\, y = 0$$

Its A.E. is $m^2 - 2m + 10 = 0$

or $\quad m = 1 \pm 3i$

Hence, the C.S. is $y = e^x (c_1 \cos 3x + c_2 \sin 3x)$

Now $y(0) = 4 \Rightarrow y = 4$, when $x = 0$

$\therefore \quad y = e^0 (c_1 \cos 0 + c_2 \sin 0)$ or $4 = c_1$

$\therefore \quad y = e^x (4 \cos 3x + c_2 \sin 3x)$

$$y' = e^x (4 \cos 3x + c_2 \sin 3x) + e^x (-12 \sin 3x + 3c_2 \cos 3x)$$

$\because \quad y'(0) = 1 \Rightarrow y' = 1$ when $x = 0$

$$1 = 4 + 3c_2 \Rightarrow c_2 = -1$$

$\therefore$ C.S. is $y = e^x (4 \cos 3x - \sin 3x)$

PROBLEM SET 3.12

Solve the following differential equations :

1. $\dfrac{d^2y}{dx^2} - \dfrac{dy}{dx} - 12y = 0$

Ans. $y = c_1 e^{-3x} + c_2 e^{4x}$

2. $4\dfrac{d^2y}{dx^2} - 12\dfrac{dy}{dx} + 5y = 0$

Ans. $y = c_1 e^{x/2} + c_2 e^{5x/2}$

3. $(D^3 - D^2 - 12D)\, y = 0$

Ans. $y = c_1 + c_2 e^{4x} + c_3 e^{-3x}$

4. $(D^3 + 2D^2 - 5D - 6)\, y = 0$

Ans. $y = c_1 e^{2x} + c_2 e^{-x} + c_3 e^{-3x}$

5. $(D^3 - 2D^2 - 4D + 8)\, y = 0$

Ans. $y = c_1 e^{-2x} + (c_2 + c_3 x)\, e^{2x}$

6. $\dfrac{d^4y}{dx^4} - 4\dfrac{d^3y}{dx^3} + 8\dfrac{d^2y}{dx^2} - 8\dfrac{dy}{dx} + 4y = 0$

Ans. $y = e^x [(c_1 + c_2 x) \cos x + (c_3 + c_4 x) \sin x]$

7. $(D^4 + 2a^2 D^2 + a^4)\, y = 0$

Ans. $y = (c_1 + c_2 x) \cos ax + (c_3 + c_4 x) \sin ax$

8. $(D^4 + D^3 + 2D^2 + D + 3)\, y = 0$

Ans. $y = e^{-x}\{c_1 \cos \sqrt{2}x + c_2 \sin \sqrt{2}x + e^{x/2}\left\{c_3 \cos \frac{\sqrt{3}}{2}x + c_4 \sin \frac{\sqrt{3}}{2}x\right\}$

9. $(D^4 + 4)\, y = 0$

Ans. $y = (c_1 e^x + c_2 e^{-x}) \cos x + (c_3 e^x + c_4 e^{-x}) \sin x$

10. $\frac{d^3y}{dx^3} - 3\frac{d^2y}{dx^2} + 4y = 0$

Ans. $y = (c_1 + c_2 x)\, e^{2x} + c_3 e^x$

11. $\frac{d^4y}{dx^4} - 2\frac{d^3y}{dx^3} + 2\frac{d^2y}{dx^2} - 2\frac{dy}{dx} + y = 0$

Ans. $y = (c_1 + c_2 x)\, e^x + c_3 \cos x + c_4 \sin x$

12. $\frac{d^3y}{dx^3} - 4\frac{d^2y}{dx^2} + 5\frac{dy}{dx} - 2y = 0$

Ans. $y = (c_1 + c_2 x)\, e^x + c_3 e^{2x}$

13. $(D^2 + D + 1)^2 (D - 2)\, y = 0$

Ans. $y = e^{-(1/2)x}\left[(c_1 + c_2 x) \cos \frac{\sqrt{3}}{2}x + (c_3 + c_4 x) \sin \frac{\sqrt{3}}{2}x\right] + c_5 e^x$

14. $(D^4 + 8D^2 + 16)\, y = 0$

Ans. $y = (c_1 + c_2 x) \cos 2x + (c_3 + c_4 x) \sin 2x$

15. $\frac{d^2y}{dx^2} + 2\frac{dy}{dx} + 5y = 0$

Ans. $y = e^{-x}(c_1 \cos 2x + c_2 \sin 2x)$

16. $(D^3 - 13D + 12)\, y = 0$

Ans. $y = c_1 e^x + c_2 e^{3x} + c_3 e^{-4x}$

3.15 SOLUTION OF NON-HOMOGENEOUS LINEAR DIFFERENTIAL EQUATION OF nth ORDER WITH CONSTANT COEFFICIENTS

Consider the non-homogeneous linear differential equation of nth order with constant coefficients.

$$(a_0 D^n + a_1 D^{n-1} + \ldots + a_{n-1} D + a_n)\, y = f(x) \qquad \ldots(1)$$

Then
$$(a_0 D^n + a_1 D^{n-1} + \ldots + a_{n-1} D + a_n) y = 0 \qquad \ldots(2)$$

is the corresponding homogeneous equation of (1), obtained by putting the R.H.S. $f(x) = 0$ in (1)

3.15.1 General Solution (Complete Solution)

A general solution of the differential equation (1) consisting of two parts say

$$y = \text{C.F.} + \text{P.I.}$$

where, C.F. is the solution of differential equation (2) and P.I. is any particular solution of the non-homogeneous equation (1) containing no arbitrary constants.

3.16 METHOD OF OBTAINING PARTICULAR INTEGRAL (P.I.)

From equation (1), we have

$$(a_0 D^n + a_1 D^{n-1} + \dots + a_{n-1} D + a_n)y = f(x)$$

Let $\phi(D)y = f(x)$

where $\phi(D) = a_0 D^n + a_1 D^{n-1} + \dots + a_{n-1} D + a_n$ is a function of the differential operator D.

Then the particular integral (P.I.) is given by

$$\text{P.I.} = \frac{1}{\phi(D)} f(x) \qquad \dots(3)$$

3.16.1 When $f(x) = e^{ax}$

If $f(x) = e^{ax}$, then P.I. $= \frac{1}{\phi(D)} \cdot e^{ax} = \frac{1}{\phi(a)} e^{ax}$, provided $\phi(a) \neq 0$

If $\phi(a) = 0$, then P.I. $= x \cdot \frac{1}{\phi'(a)} e^{ax}$, provided $\phi'(a) \neq 0$

If $\phi'(a) = 0$, then P.I. $= x^2 \frac{1}{\phi''(a)} e^{ax}$, provided $\phi''(a) \neq 0$ and so on.

Example 3.74 **Solve : $(D^3 - 7D + 6)y = e^{2x}$** *[UPTU 2001]*

Solution : We have $(D^3 - 7D + 6)y = e^{2x}$

The A.E. is $m^3 - 7m + 6 = 0 \Rightarrow (m-1)(m^2 + m - 6) = 0$

$\Rightarrow \quad (m-1)(m-2)(m+3) = 0 \Rightarrow m = 1, 2, -3$

Complementary function is

$$\text{C.F.} = c_1 e^x + c_2 e^{2x} + c_3 e^{-3x}$$

Particular integral is

$$\text{P.I.} = \frac{1}{\varphi(2)} e^{2x}, \text{ where } \varphi(D) = D^3 - 7D + 6$$

At $D = 2$, $\varphi(2) = 0$. Now

$$\text{P.I.} = \frac{x}{\varphi'(2)} e^{2x}, \text{ where } \varphi'(D) = 3D^2 - 7$$

At $D = 2$, $\varphi'(2) = 5 \neq 0$

Hence, P.I. $= \frac{x}{5} e^{2x}$

General solution is

$$y = \text{C.F.} + \text{P.I.} = c_1 e^x + c_2 e^{2x} + c_3 e^{-3x} + \frac{x}{5} e^{2x}$$

Example 3.75 **Solve : $(D^2 - 5D + 6)y = e^{3x}$.** *[UPTU 2001]*

Solution : We have $(D^2 - 5D + 6)y = e^{3x}$

The A.E. is $m^2 - 5x + 6 = 0 \Rightarrow (m-2)(m-3) = 0 \Rightarrow m = 2, 3$

Complementary function is

$$\text{C.F.} = c_1e^{2x} + c_2e^{3x}$$

Particular integral is

$$\text{P.I.} = \frac{1}{\varphi(3)}\, e^{3x}$$

where $\varphi(D) = D^2 - 5D + 6$

At $D = 3,\ \varphi(3) = (3)^2 - 5(3) + 6 = 0$

Now $\quad \text{P.I.} = \dfrac{x}{\varphi'(3)}\, e^{3x}$

where $\varphi'(D) = 2D - 5$

At $\quad D = 3,\ \varphi'(3) = 2(3) - 5 = 6 - 5 = 1 \neq 0$

Hence, $\text{P.I.} = \dfrac{x}{1}\, e^{3x} = xe^{3x}$

General solution is

$$y = \text{C.F.} + \text{P.I.} = c_1e^{2x} + c_2e^{3x} + xe^{3x}$$

Example 3.76 **Solve :**

$$\frac{d^2y}{dx^2} - 5\,\frac{dy}{dx} + 6y = e^{4x}$$

[UPTU 2005]

Solution : Given

$$\frac{d^2y}{dx^2} - 5\,\frac{dy}{dx} + 6y = e^{4x} \;\Rightarrow\; (D^2 - 5D + 6)y = e^{4x}$$

The A.E. is

$$m^2 - 5m + 6 = 0 \Rightarrow (m-2)(m-3) = 0 \;\Rightarrow\; m = 2, 3$$

Complementary function is

$$\text{C.F.} = c_1e^{2x} + c_2e^{3x}$$

Particular Integral is

$$\text{P.I.} = \frac{1}{\varphi(4)}\, e^{4x}, \quad \text{where } \varphi(D) = D^2 - 5D + 6$$

At $D = 4$, $\varphi(4) = 16 - 20 + 6 = 2 \neq 0$

$\therefore \quad \text{P.I.} = \dfrac{1}{2}\, e^{4x}$

General solution is

$$y = \text{C.F.} + \text{P.I.} = c_1e^{2x} + c_2e^{3x} + \frac{1}{2}\, e^{4x}$$

Example 3.77 **Solve :**

$$\frac{d^2y}{dx^2} - \frac{dy}{dx} - 6y = e^x \cosh 2x$$

Solution : Given

$$\frac{d^2y}{dx^2} - \frac{dy}{dx} - 6y = e^x \cosh 2x$$

$$\Rightarrow \quad (D^2 - D - 6)y = e^x \left(\frac{e^{2x} + e^{-2x}}{2}\right) \qquad \left[\because \cosh 2x = \frac{e^{2x} + e^{-2x}}{2}\right]$$

$$\Rightarrow \quad (D^2 - D - 6)y = \frac{1}{2}(e^{3x} + e^{-x})$$

The A.E. is $m^2\ m - 6 = 0 \Rightarrow (m + 2)(m - 3) = 0 \Rightarrow m = -2, 3$

The complementary function is

$$\text{C.F.} = c_1 e^{-2x} + c_2 e^{3x}$$

The particular integral is

$$\text{P.I.} = \frac{1}{\varphi(D)} \frac{1}{2} [e^{3x} + e^{-x}]$$

where, $\varphi(D) = D^2 - D - 6$

$$= \frac{1}{2} \cdot \frac{1}{\varphi(3)} e^{3x} + \frac{1}{2} \cdot \frac{1}{\varphi(-1)} e^{-x}$$

$$= \frac{1}{2} \cdot \frac{x}{\varphi'(3)} e^{3x} + \frac{1}{2} \cdot \frac{1}{(-1)^2 - (-1) - 6} \cdot e^{-x} \qquad [\because \varphi(3) = 0]$$

$$= \frac{1}{2} \frac{x}{2(3) - 1} e^{3x} - \frac{1}{8} e^{-x}$$

$$= \frac{xe^{3x}}{10} - \frac{1}{8} e^{-x}$$

General solution is

Example 3.78 **Solve :**

$$\frac{d^3y}{dx^3} - 2\frac{d^2y}{dx^2} - 5\frac{dy}{dx} + 6y = e^{4x}.$$

Solution : The given Eq. can be written as

$$(D^3 - 2D^2 - 5D + 6)\, y = e^{4x}$$

The A.E. is $m^3 - 2m^2 - 5m + 6 = 0$ or $m = 1, -2, 3$

$$\therefore \quad \text{C.F.} = c_1 e^x + c_2 e^{-2x} + c_3 e^{3x}$$

and
$$\text{P.I.} = \frac{1}{(D^3 - 2D^2 - 5D + 6)} e^{4x}$$

$$= \frac{1}{(4)^3 - 2(4)^2 - 5(4) + 6} e^{4x} = \frac{1}{18} e^{4x}$$

Therefore the general solution

$$\therefore \quad \text{C.S. is } y = c_1 e^x + c_2 e^{-2x} + c_3 e^{3x} + \frac{1}{18} e^{4x}$$

Example 3.79 **Solve :**

$$\frac{d^2y}{dx^2} - 3\frac{dy}{dx} + 2y = e^x$$

Solution : The give Eq. can be written as

$$(D^2 - 3D + 2)\, y = e^x$$

The A.E. is $m^2 - 3m + 2 = 0 \Rightarrow m = 1, 2$

$\therefore$ $\quad$ C.F. $= c_1e^x + c_2e^{2x}$

and $\quad$ $$\text{P.I.} = \frac{1}{(D-1)(D-2)} e^x$$

$$= \frac{1}{(D-1)}\left[\frac{1}{(D-2)} e^x\right]$$

$$= \frac{1}{(D-1)}\left[\frac{1}{(1-2)} e^x\right] = \frac{1}{(D-1)} e^x = [xe^x]$$

Therefore the general solution is $y = c_1e^x + c_2e^{2x} - xe^x$.

Example 3.80 **Solve :**

$$\frac{d^3y}{dx^3} - 2\frac{d^2y}{dx^2} - 5\frac{dy}{dx} + 6y = (e^{2x} + 3)^2.$$

Solution : $\quad$ C.F. $= c_1e^x + c_2e^{-2x} + c_3e^{3x}$

and $$\text{P.I.} = \frac{1}{(D^3 - 2D^2 - 5D + 6)}(e^{2x} + 3)^2$$

$$= \frac{1}{(D-1)(D-3)(D+2)}(e^{4x} + 6e^{2x} + 9)$$

$$= \frac{1}{(D-1)(D-3)(D+2)} e^{4x} + \frac{1}{(D-1)(D-3)(D+2)} 6e^{2x} + \frac{1}{(D-1)(D-3)(D+2)} ge^{0x}$$

$$= \frac{1}{(3)(1)(6)} e^{4x} + \frac{6}{(1)(-1)(4)} e^{2x} + \frac{9}{(-1)(-3)(2)}$$

$$= \frac{e^{4x}}{18} - \frac{3}{2}e^{2x} + \frac{3}{2}$$

Therefore, the general solution is

$$y = c_1e^x + c_2e^{-2x} + c_3e^{3x} + \frac{e^{4x}}{18} - \frac{3}{2}e^{2x} + \frac{3}{2}$$

Example 3.81 **Find the P.I. of**

$$\frac{d^3y}{dx^3} - 6\frac{d^2y}{dx^2} + 11\frac{dy}{dx} - 6y = e^{-2x} + e^{-3x}.$$

Solution : $$\text{P.I.} = \frac{1}{(D^3 - 6D^2 + 11D - 6)}(e^{-2x} + e^{-2x})$$

$$= \frac{1}{(D^3 - 6D^2 + 11D - 6)} e^{-2x} + \frac{1}{(D^3 - 6D^2 + 11D - 6)} e^{-3x}$$

$$= \frac{1}{(-2)^3 - 6(-2)^2 + 11(-2) - 6} e^{-2x} + \frac{1}{(-3)^3 - 6(-3)^2 + 11(-3) - 6} e^{-3x}$$

$$= -\frac{1}{60} e^{-2x} - \frac{1}{120} e^{-3x}$$

$$y = \text{C.F.} + \text{P.I.} = c_1e^{-2x} + c_2e^{3x} + \frac{1}{10}xe^{3x} - \frac{1}{8}e^{-x}$$

PROBLEM SET 3.13

Solve the following differential equations :

1. $(D^2 - 2D + 1)\,y = e^{2x}$

Ans. $y = (c_1 + c_2x)\,e^x + e^{2x}$

2. $(3D^2 + D - 14)\,y = 16\,e^{3x}$

Ans. $y = c_1e^{2x} + c_2e^{-\frac{7}{3}x} + e^{3x}$

3. $\dfrac{d^2y}{dx^2} + 6\dfrac{dy}{dx} + 9y = 5e^{3x}$

Ans. $y = (c_1 + c_2x)\,e^{-3x} + \dfrac{5e^{3x}}{54}$

4. $(D^3 + 3D^2 + 3D + 1)\,y = e^{-x}$

Ans. $y = (c_1 + c_2x + c_3x^2)\,e^{-x} + \dfrac{x^3e^{-x}}{6}$

5. $(D^3 - 1)\,y = (e^x + 1)^2$

Ans. $y = c_1e^x + e^{-x/2}\left\{c_2\cos\dfrac{\sqrt{3}}{2}x + c_3\sin\dfrac{\sqrt{3}}{2}x\right\} + \dfrac{2}{3}xe^x + \dfrac{3^{2x}}{7} - 1$

6. $(D^3 - 7D + 6)\,y = e^{2x}$

Ans. $y = c_1e^x + c_2e^{2x} + c_3e^{-3x} + \dfrac{x}{5}e^{2x}$

7. $(D^2 + 4D + 3)\,y = e^{-3x}$

Ans. $y = c_1e^{-x} + c_2e^{-3x} - \dfrac{1}{2}xe^{-3x}$

8. $(D^3 + 1)\,y = 3 + e^{-x} + 5e^{2x}$

Ans. $y = c_1e^{-x} + e^{x/2}\left\{c_2\cos\dfrac{\sqrt{3}}{2}x + c_3\sin\dfrac{\sqrt{3}}{2}x\right\} + \dfrac{1}{3}xe^{-x} + \dfrac{5}{9}e^{2x} + 3$

9. $(D - 1)^2(D + 1)^2\,y = e^x$

Ans. $y = (c_1 + c_2x)\,e^x + (c_3 + c_4x)\,e^{-x} + \dfrac{1}{8}x^3e^x$

10. $(D^2 - 6D + 8)\,y = (e^{2x} - 1)^2$

Ans. $y = c_1e^{4x} + c_2e^{2x} + \dfrac{x}{2}e^{4x} + xe^{2x} + \dfrac{1}{8}$

11. $(D^5 - a^2D^3) = e^{mx}$

Ans. $y = c_1 + c_2x + c_3x^2 + c_4e^{ax} + c_5e^{-ax} + \dfrac{e^{mx}}{m^3(m^2 - a^2)}$

12. $\frac{d^2y}{dx^2} + 2\frac{dy}{dx} + 2y = \sinh x$

Ans. $y = e^{-x}(c_1 \cos x + c_2 \sin + \frac{1}{10}e^x - \frac{e^{-x}}{2}$

13. $\frac{d^2y}{dx^2} + 31\frac{dy}{dx} + 240y = 272\, e^{-x}$

Ans. $y = c_1 e^{-15x} + c_2 e^{-16x} + \frac{136}{105}e^{-x}$

14. $\frac{d^2y}{dx^2} - 4\frac{dy}{dx} + 5y = e^x - 2e^{2x} + 3e^{3x}$

Ans. $y = \left(c_1 \cos\frac{\sqrt{11}\,x}{3} + c_2 \sin\frac{\sqrt{11}}{3}x\right)e^{2x/3} + \frac{1}{4}e^x - \frac{2}{9}e^{2x} + \frac{3}{20}e^{3x}$

15. $(D^2 + 4D + 13)\,y = 18e^{-2x},\; y(0) = 0,\; y'(0) = 4$

Ans. $y = e^{-2x}\left[\frac{4}{3}\sin 3x - 2\cos x + 2e^{-2x}\right]$

3.16.2 When $f(x) = \sin(ax + b)$ or $\cos(ax + b)$

$$\text{P.I.} = \frac{1}{\varphi(D^2)}f(x) = \frac{1}{\varphi(-a^2)}f(x) \quad \text{provided } \varphi(-a^2) \neq 0$$

If $\varphi(-a^2) = 0$ then P.I. $= \frac{x}{\varphi'(-a^2)}f(x)$, provided $\varphi'(-a^2) \neq 0$

If $\varphi'(-a^2) = 0$, then P.I. $= \frac{x^2}{\varphi''(-a^2)}f(x)$, provided $\varphi''(-a^2) \neq 0$ and so on.

Working Rule

(i) Replace D^2 by $-a^2$, D^4 by $(-a^2)^2$ and so on in $\varphi(D)$.

(ii) If $\varphi(D)$ contains terms of odd powers $D, D^3, \ldots$ etc. even after application of the above result, multiply by D so as to get even powers of $D^2, D^4, \ldots$ etc. Interpreting D as differentiation and $\frac{1}{D}$ as integration.

Example 3.82 **Solve :**

$$\frac{d^2y}{dx^2} - \frac{dy}{dx} - 2y = \sin 2x$$

[UPTU 2000]

Solution : Given

$$\frac{d^2y}{dx^2} - \frac{dy}{dx} - 2y = \sin 2x$$

$\Rightarrow \quad (D^2 - D - 2)\,y = \sin 2x$

The A.E. is $m^2 - m - 2 = 0$

$\Rightarrow \quad (m - 2)(m + 1) = 0 \Rightarrow m = -1, 2$

Complementary function is

$$\text{C.F.} = c_1 e^{-x} + c_2 e^{2x}$$

Particular Integral is

$$\text{P.I.} = \frac{1}{D^2 - D - 2} \sin 2x$$

$$= \frac{1}{-2^2 - D - 2} \sin 2x$$

$$= \frac{1}{-6 - D} \sin 2x = \frac{-6 + D}{(-6 - D)(-6 + D)} \sin 2x$$

$$= \frac{-6 + D}{(-6)^2 - (D)^2} \sin 2x = \frac{-6 + D}{36 - D^2} \sin 2x$$

$$= \frac{-6 + D}{36 - (-4)} \sin 2x = \frac{-6 + D}{40} \sin 2x$$

$$= -\frac{6}{40} \sin 2x + \frac{D}{40} \sin 2x$$

$$= -\frac{6}{40} \sin 2x + \frac{1}{40} \cdot 2 \cos 2x$$

$$= -\frac{3}{20} \sin 2x + \frac{1}{20} \cos 2x$$

General Solution is

$$y = \text{C.F.} + \text{P.I.}$$

$$= c_1 e^{-x} + c_2 e^{2x} - \frac{3}{20} \sin 2x + \frac{1}{20} \cos 2x$$

Example 3.83 **Solve :**

$$\frac{d^2y}{dx^2} + 9y = \cos 2x$$ ***[UPTU 2001]***

Solution : Given

$$\frac{d^2y}{dx^2} + 9y = \cos 2x$$

$$\Rightarrow \quad (D^2 + 9)\, y = \cos 2x$$

The A.E. is $m^2 + 9 = 0 \Rightarrow m = \pm 3i$

The complementary function is

$$\text{C.F.} = c_1 \cos 3x + c_2 \sin 3x$$

Particular integral is

$$\text{P.I.} = \frac{1}{D^2 + 9} \cos 2x = \frac{1}{(-2^2) + 9} \cos 2x \qquad [\because a = 2]$$

$$= \frac{1}{-4 + 9} \cos 2x = \frac{1}{5} \cos 2x$$

General solution is

$$y = \text{C.F.} + \text{P.I.} = c_1 \cos 3x + c_2 \sin 3x + \frac{1}{5} \cos 2x$$

Example 3.84 **Solve :**

$$\frac{d^2y}{dx^2} - 4y = \cos^2 x$$

[UPTU 2002]

Solution : Given

$$\frac{d^2y}{dx^2} - 4y = \cos^2 x$$

$$\Rightarrow \quad \frac{d^2y}{dx^2} - 4y = \frac{1}{2}(1 + \cos 2x)$$

$$\Rightarrow \quad (D^2 - 4)\,y = \frac{1}{2} + \frac{1}{2}\cos 2x$$

The A.E. is $m^2 - 4 = 0 \Rightarrow m = \pm 2$

The complementary function is

$$\text{C.F.} = c_1e^{2x} + c_2\,e^{-2x}$$

Particular Integral is

$$\text{P.I.} = \frac{1}{D^2 - 4}\left[\frac{1}{2} + \frac{1}{2}\cos 2x\right]$$

$$= \frac{1}{2}\,\frac{1}{D^2 - 4}\,e^{0x} + \frac{1}{2}\,\frac{1}{D^2 - 4}\cos 2x$$

$$= \frac{1}{2}\,\frac{1}{0^2 - 4}\,e^{0x} + \frac{1}{2}\cdot\frac{1}{-2^2 - 4}\cos 2x$$

$$= -\frac{1}{8} - \frac{1}{16}\cos 2x$$

General solution is

$$y = \text{C.F.} + \text{P.I.}$$

$$= c_1e^{2x} + c_2\,e^{-2x} - \frac{1}{16}\cos 2x - \frac{1}{8}$$

Example 3.85 **Solve :**

$$\frac{d^2y}{dx^2} - 4y = e^x + \sin 2x$$

[UPTU 2005]

Solution : Given

$$\frac{d^2y}{dx^2} - 4y = e^x + \sin 2x$$

$$\Rightarrow \quad (D^2 - 4)\,y = e^x + \sin 2x$$

The A.E. is $m^2 - 4 = 0$

$$\Rightarrow \quad m^2 = 4$$

$$\Rightarrow \quad m = \pm 2$$

The complementary function is

$$\text{C.F.} = c_1e^{2x} + c_2e^{-2x}$$

Particular Integral is

$$\text{P.I.} = \frac{1}{D^2 - 4}\,[e^x + \sin 2x] = \frac{1}{D^2 - 4}\,e^x + \frac{1}{D^2 - 4}\sin 2x$$

$$= \frac{1}{(1)^2 - 4}\,e^x + \frac{1}{-2^2 - 4}\sin 2x$$

$$= \frac{1}{-3}\,e^x + \frac{1}{-8}\sin 2x$$

$$= -\frac{1}{3}\,e^x - \frac{1}{8}\sin 2x$$

General solution is

$$y = \text{C.F.} + \text{P.I.}$$

$$= c_1 e^{2x} + c_2 e^{-2x}\ \frac{1}{3}\,e^x - \frac{1}{8}\sin 2x$$

Example 3.86 **Solve :**

$$\boldsymbol{\frac{d^3y}{dx^3} - 3\,\frac{d^2y}{dx^2} + 4\,\frac{dy}{dx} - 2y = e^x + \cos x}$$

Solution : Given

$$\frac{d^3y}{dx^3} - 3\,\frac{d^2y}{dx^2} + 4\,\frac{dy}{dx} - 2y = e^x + \cos x$$

$$\Rightarrow \quad (D^3 - 3D^2 + 4D - 2)y = e^x + \cos x$$

The A.E. is $m^3 - 3m^2 + 4m - 2 = 0$

$$\Rightarrow \quad (m - 1)\,(m^2 - 2m + 2) = 0$$

$$\Rightarrow \quad m = 1,\ 1 \pm i$$

The complementary function is

$$\text{C.F.} = c_1 e^x + e^x\,(c_2 \cos x + c_3 \sin x)$$

Particular integral

$$\text{P.I.} = \frac{1}{D^3 - 3D^2 + 4D - 2}\,[e^x + \cos x]$$

$$= \frac{1}{D^3 - 3D^2 + 4D - 2}\,e^x + \frac{1}{D^3 - 3D^2 + 4D - 2}\cos x$$

$$= \frac{x}{3D^2 - 6D + 4}\,e^x + \frac{1}{D^2 D - 3D^2 + 4D - 2}\cos x$$

$$= \frac{x}{3(1)^2 - 6(1) + 4}\,e^x + \frac{1}{(-1^2)\,D - 3(-1^2) + 4D - 2}\cos x$$

$$= \frac{x}{1}\,e^x + \frac{1}{-D + 3 + 4D - 2}\cos x = xe^x + \frac{1}{1 + 3D}\cos x$$

$$= xe^x + \frac{1 - 3D}{(1 + 3D)\,(1 - 3D)}\cos x = xe^x + \frac{1 - 3D}{1 - 9D^2}\cos x$$

$$= xe^x + \frac{1 - 3D}{1 - 9(-1^2)}\cos x = xe^x + \frac{1 - 3D}{1 + 9}\cos x$$

$$= xe^x + \frac{1}{10}(\cos x - 3D\cos x) = xe^x + \frac{1}{10}(\cos x + 3\sin x)$$

General solution is

$$y = \text{C.F.} + \text{P.I.}$$

$$= c_1 e^x + e^x(c_2\cos x + c_3\sin x) + xe^x + \frac{1}{10}(\cos x + 3\sin x)$$

Example 3.87 **Solve : $(D^2 + 4)y = \cos 2x$**

Solution : Given $(D^2 + 4)y = \cos 2x$

A.E. is $m^2 + 4 = 0 \Rightarrow m = \pm 2i$

$$\text{CF} = c_1\cos 2x + c_2\sin 2x$$

$$\text{P.I.} = \frac{1}{D^2+4}\cos 2x = x\cdot\frac{1}{2D}\cos 2x$$

$$= \frac{x}{2}\int \cos 2x\, dx = \frac{x}{4}\sin 2x$$

General solution is

$$y = \text{C.F.} + \text{P.I.} = c_1\cos 2x + c_2\sin 2x + \frac{x}{4}\sin 2x$$

Example 3.88 **Solve :**

$$\frac{d^2y}{dx^2} - 5\frac{dy}{dx} + 6y = \sin 3x$$

Solution : The given equation can be written as

$$(D^2 - 5D + 6)\,y = \sin 3x$$

The A.E. is $m^2 - 5x + 6 = 0$ or $m = 2, 3$

$\therefore$ $\text{C.F.} = c_1 e^{2x} + c_2 e^{3x}$

$$\text{P.I.} = \frac{1}{(D^2 - 5D + 6)}\sin 3x$$

$$= \frac{1}{(-9) - 5D + 6}\sin 3x = \frac{1}{-5D-3}\sin 3x$$

$$= -(5D - 3)\frac{1}{(25D^2 - 9)}\sin 3x$$

$$= \frac{1}{234}(5D - 3)\sin 3x$$

$$= \frac{1}{234}(15\cos 3x - 3\sin 3x)$$

$\therefore$ C.S. is $y = c_1 e^{2x} + c_2 e^{3x} + \frac{1}{234}(15\cos 3x - 3\sin 3x)$

Example 3.89 **Find the P.I. of $(D^3 + D^2 + D + 1)\,y = \sin 2x + \cos 3x$**

Solution :

$$\text{P.I.} = \frac{1}{(D^3 + D^2 + D + 1)}(\sin 2x + \cos 3x)$$

$$= \frac{1}{D^2+1)(D+1)}(\sin 2x + \cos 3x)$$

$$= \frac{1}{(D^2+1)(D+1)}\sin 2x + \frac{1}{(D^2+1)(D+1)}\cos 3x$$

$$= -\frac{1}{3}\frac{1}{(D+1)}\sin 2x - \frac{1}{8}\frac{1}{(D+1)}\cos 3x$$

$$= -\frac{1}{3}\frac{(D-1)}{(D^2-1)}\sin 2x - \frac{1}{8}\frac{(D-1)}{(D^2-1)}\cos 3x$$

$$= \frac{1}{15}(D-1)\sin 2x + \frac{1}{80}(D-1)\cos 3x$$

$$\text{P.I.} = \frac{1}{15}(2\cos 2x - \sin 2x) - \frac{1}{80}(3\sin 3x + \cos 3x)$$

Example 3.90 **Solve :**

$$(D^4 + (a^2+b^2)D^2 + a^2b^2)\,y = \cos\frac{(a+b)\,x}{2}\cos\left(\frac{a-b}{2}\right)x$$

Solution : The given Eq. can be written as

$$(D^2+a^2)(D^2+b^2)\,y = \frac{1}{2}(\cos ax + \cos bx)$$

The A.E. is given by $(m^2+a^2)(m^2+b^2)=0$ or $m = \pm ia, \pm ib$

$\therefore$ C.F. $= (c_1\cos ax + c_2\sin ax) + (c_3\cos bx + c_4\sin bx)$

$$\text{P.I.} = \frac{1}{2}\frac{1}{(D^2+a^2)(D^2+b^2)}(\cos ax + \cos bx)$$

$$= \frac{1}{2}\frac{1}{(D^2+a^2)(D^2+b^2)}\cos ax + \frac{1}{2}\frac{1}{(D^2+a^2)(D^2+b^2)}\cos bx$$

$$= \frac{1}{2}\frac{1}{(b^2-a^2)}\frac{1}{(D^2+a^2)}\cos ax + \frac{1}{2}\frac{1}{(a^2-b^2)}\frac{1}{(D^2+b^2)}\cos bx$$

$$= \frac{1}{2}\cdot\frac{1}{(b^2-a^2)}\left\{\frac{x}{2a}\sin ax - \frac{x}{2b}\sin bx\right\}$$

$$= \frac{1}{4}\frac{x}{(b^2-a^2)}\left\{\frac{1}{a}\sin ax - \frac{1}{b}\sin bx\right\}$$

Therefore the general solution is

$$y = (c_1\cos ax + c_2\sin ax) + (c_3\cos bx + c_4\sin bx) + \frac{1}{4(b^2-a^2)}x\left[\frac{1}{a}\sin ax - \frac{1}{b}\sin bx\right]$$

Example 3.91 **Solve :**

$$(D-1)^2(D^2+1)^2\,y = \sin^2\left(\frac{x}{2}\right) + e^x.$$

Solution : The A.E. is $(m-1)^2(m^2+1)^2 = 0$ or $m = 1, 1 \pm i, \pm i$

$\therefore$ C.F. $= (c_1 + c_2x)\,e^x + (c_3 + c_4x)\cos x + (c_5 + c_6\,x)\sin x$

$$\text{P.I.} = \frac{1}{(D-1)^2 (D^2+1)^2}\left(\sin^2 \frac{x}{2} + e^x\right)$$

$$= \frac{1}{(D-1)^2 (D^2+1)^2} \frac{(1-\cos x)}{2} + \frac{1}{(D-1)^2 (D^2+1)^2} e^x$$

$$= \frac{1}{2}\left[\frac{1}{(D-1)^2 (D^2+1)^2} - \frac{1}{(D-1)^2 (D^2+1)^2} \cos x\right] + \frac{1}{(D-1)^2} \cdot \frac{1}{(1+1)^2} e^x$$

$$= \frac{1}{2} \frac{e^{0x}}{(0-1)^2 (0+1)^2} - \frac{1}{2} \frac{1}{(D^2-2D+1)(D^2+1)} \cos x + \frac{1}{4} \frac{x^2}{2!} e^x$$

$$= \frac{1}{2} - \frac{1}{2} \frac{1}{(-1-2D+1)(D^2+1)^2} \cos x + \frac{1}{8} x^3 e^x$$

$$= \frac{1}{2} + \frac{1}{4} \frac{1}{(D^2+1)^2}\left\{\frac{1}{D} \cos x\right\} + \frac{1}{8} x^2 e^x$$

$$= \frac{1}{2} + \frac{1}{4} \frac{1}{(D^2+1)^2} \sin x + \frac{1}{8} x^2 e^x$$

Now, $\displaystyle \frac{1}{(D^2+1)^2} \sin x = \text{I.P.} \frac{1}{(D^2+1)^2} e^{ix}$

$$= \text{I.P.} = \frac{1}{(D+i)(D-i)^2} e^{ix}$$

$$= \text{I.P.} \frac{x^2}{2!} \frac{e^{ix}}{(i+i)^2}$$

$$= \text{I.P.} \left(-\frac{1}{8} x^2 (\cos x + i \sin x)\right)$$

$$= -\frac{1}{8} x^2 \sin x$$

$\therefore$ $$\text{P.I.} = \frac{1}{2} - \frac{1}{32} x^2 \sin x + \frac{1}{8} x^2 e^x$$

So, C.S. is

$$y = (c_1 + c_2 x) e^x + (c_3 + c_4 x) \cos x + (c_5 + c_6 x) \sin x + \frac{1}{8} x^2 e^x - \frac{1}{32} x^2 \sin x + \frac{1}{2}$$

PROBLEM SET 3.14

Solve the following equations:

1. $(D^2 + 4)y = \sin 3x + \cos 2x$

Ans. $y = c_1 \cos 2x + c_2 \sin 2x - \frac{1}{5} \sin 3x + \frac{x}{4} \sin 2x$

2. $(D^2 + 5D - 6)\, y = \sin 4x \cdot \sin x$

Ans. $y = c_1 e^x + c_2 e^{-6x} + \frac{1}{30} [\sin 3x - \cos 3x] - \frac{1}{1586} [-25 \sin 5x + 31 \cos 5x]$

3. $\frac{d^2y}{dx^2} - 4\frac{dy}{dx} + y = \sin^2 x$

Ans. $y = c_1 e^{(2+\sqrt{3})x} + c_2 e^{(2-\sqrt{3})x} + \frac{1}{2}\left[1 - \frac{1}{73}(8 \sin 2x + 3 \cos 2x)\right]$

4. $\frac{d^2x}{dt^2} + 2\frac{dx}{dt} + 5x = \sin 2t$, when $t = 0, x = 3$ and $\frac{dx}{dt} = 0$

Ans. $x = e^{-t}\left[\frac{55}{17}\cos 2t + \frac{53}{34}\sin 2t\right] - \frac{1}{17}(4 \cos 2t - \sin 2t)$

5. $\frac{d^2x}{dt^2} + 2\frac{dx}{dt} + 3x = \sin t$

Ans. $x = e^{-t}[c_1 \cos \sqrt{2}\, t + c_2 \sin \sqrt{2}\, t] - \frac{1}{4}(\cos t - \sin t)$

6. $(D^2 + a^2)y = \sin ax$

Ans. $y = c_1 \cos ax + c_2 \sin ax - \frac{x}{2a}\cos ax$

7. $\frac{d^2y}{dx^2} + 4y = e^x + \sin 2x$

Ans. $y = c_1 \cos 2x + c_2 \sin 2x + \frac{1}{5}e^x - \frac{x}{4}\cos 2x$

8. $\frac{d^2y}{dx^2} + \frac{dy}{dx} - 2y = -6 \sin 2x - 18 \cos 2x, y(0) = 2, y'(0) = 2$

Ans. $y = -e^{-2x} + 3 \cos 2x$

9. $\frac{d^2y}{dx^2} - 4\frac{dy}{dx} + 3y = e^{2x} + \cos x$

Ans. $y = c_1 e^x + c_2 e^{3x} - e^{2x} + \frac{1}{10}(\cos x - 2 \sin x)$

10. $\frac{d^2y}{dx^2} + y = \sin 3x \cos 2x$

Ans. $y = c_1 \cos x + c_2 \sin x - \frac{1}{48}[\sin 5x + 12x \cos x]$

11. $(D^4 - 1)y = \sin x$

Ans. $y = c_1 \cos x + c_2 \sin x + c_3 e^x + c_4 e^{-x} + \frac{x}{4}\cos x$

12. $(D^3 - 2D^2 + 9D - 18)y = 6 \cos 3x + 48 \sin x + 3e^{2x}$

Ans. $y = c_1 e^{2x} + c_2 \cos 3x + c_3 \sin x + \frac{3}{13}xe^{2x} + x(\cos 3x - 2 \sin 3x)$

13. $(D^2 + 4)y = \sin^2 x + 4$

Ans. $y = c_1 \cos 2x + c_2 \sin 2x + \frac{9}{8} - \frac{1}{8}x \sin 2x$

14. $(D^2 + 9)y = \cos 2x + \sin 2x$

Ans. $y = c_1 \cos (3x + c_2) + \frac{1}{5} (\cos 2x + \sin 2x)$

15. $(D^2 + 1)^2 y = \cos x \cos 2x$

Ans. $y = (c_1 + c_2 x) \cos x + (c_3 + c_4 x) \sin x - \frac{1}{16} x^2 \cos x + \frac{1}{128} \cos 3x$

3.16.3 When $f(x) = x^m$, m being a positive integer

$$\text{P.I.} = \frac{1}{\varphi(D)} x^m = [\varphi(D)]^{-1} x^m$$

Expand $[\varphi(D)^{-1}]$ by the Binomial theorem in ascending powers of D as far as the result of operation on x^m is zero.

NOTE : 1. $D^n x^n = n!$

2. $D^{n+1} x^n = 0$
3. $(1 - x)^{-1} = 1 + x + x^2 + \ldots$
4. $(1 + x)^{-1} = 1 - x + x^2 - x^3 + \ldots$

Example 3.92 **Solve :**

$$\frac{d^2y}{dx^2} - 5\frac{dy}{dx} + 6y = x$$

[UPTU 2002]

Solution : Given

$$\frac{d^2y}{dx^2} - 5\frac{dy}{dx} + 6y = x \Rightarrow (D^2 - 5D + 6)y = x$$

The A.E. is

$$m^2 - 5m + 6 = 0 \Rightarrow (m - 2)(m - 3) = 0 \Rightarrow m = 2, 3$$

The complementary function is

$$\text{C.F.} = c_1 e^{2x} + c_2 e^{3x}$$

Particular Integral

$$\text{P.I.} = \frac{1}{D^2 - 5D + 6} x = \frac{1}{6\left[1 - \left(\frac{5D - D^2}{6}\right)\right]} x$$

$$= \frac{1}{6}\left[1 - \left(\frac{5D - D^2}{6}\right)\right]^{-1} x$$

$$= \frac{1}{6}\left[1 + \left(\frac{5D - D^2}{6}\right) + \left(\frac{5D - D^2}{6}\right)^2 + \ldots\right] x \quad \text{[Using Binomial theorem]}$$

$$= \frac{1}{6}\left[x + \frac{5}{6} Dx - \frac{1}{6} D^2 x + \frac{25}{36} D^2 x + \ldots\right]$$

$$= \frac{1}{6}\left[x + \frac{5}{6} - \frac{1}{6} 0 + \frac{35}{36} 0 + \ldots\right]$$

$$= \frac{1}{6}x + \frac{5}{36}$$

General solution is

$$y = \text{C.F.} + \text{P.I.}$$

$$= c_1 e^{2x} + c_2 e^{3x} + \frac{1}{6}x + \frac{5}{36}$$

Example 3.93 **Solve :** $(D-2)^2 y = 8(e^{2x} + \sin 2x + x^2)$ *[UPTU 2002]*

Solution : Given $(D-2)^2 y = 8(e^{2x} + \sin 2x + x^2)$

The A.E. is $(m-2)^2 = 0 \Rightarrow m = 2, 2$

The complementary function is

$$CF = (c_1 + c_2 x)\, e^{2x}$$

Particular integral is

$$\text{P.I.} = \frac{1}{(D-2)^2}[8(e^{2x} + \sin 2x + x^2)]$$

$$= \frac{8}{(D-2)^2}e^{2x} + \frac{8}{(D-2)^2}\sin 2x + \frac{8}{(D-2)^2}x^2$$

$$= \frac{8\cdot x}{2(D-2)}e^{2x} + \frac{8}{D^2 - 4D + 4}\sin 2x + \frac{8}{4\left(1-\frac{D}{2}\right)^2}x^2$$

$$= \frac{4x^2}{1}e^{2x} + \frac{8}{-2^2 - 4D + 4}\sin 2x + 2\left(1 - \frac{D}{2}\right)^{-2}x^2$$

$$= 4x^2 e^{2x} + \frac{8}{-4D}\sin 2x + 2\left[1 + 2\left(\frac{D}{2}\right) + \frac{2.3}{2}\frac{D^2}{4} + \frac{2.3.4\, D^3}{3.2.8} + \ldots\right]x^2$$

$$= 4x^2 e^{2x} + \frac{2}{2}\cos 2x + 2x^2 + 4x + 3$$

$$= 4x^2 e^{2x} + \cos 2x + 2x^2 + 4x + 3$$

General solution is

$$y = \text{C.F.} + \text{P.I.} = (c_1 + c_2 x)e^{2x} + 4x^2 e^{2x} + \cos 2x + 2x^2 + 4x + 3$$

Example 3.94 **Solve :**

$$\frac{d^3y}{dx^3} + 2\frac{d^2y}{dx^2} + \frac{dy}{dx} = e^{2x} + x^2 + x$$

Solution : The given Eq. can be written as

$$(D^3 + 2D^2 + D)\, y = e^{2x} + x^3 + x$$

The A.e. is $m^3 + 2m^2 + m = 0$

$\Rightarrow$ $m = 0, -1, -1$

$\therefore$ $\text{C.F.} = c_1 + (c_2 + c_3 x)\, e^{-x}$

and $\quad \text{P.I.} = \dfrac{1}{D\,(D+1)^2}\,(e^{2x} + x^2 + x)$

$$= \frac{1}{D\,(D+1)^2}\,e^{2x} + \frac{1}{D\,(D+1)^2}\,(x^2 + x)$$

$$= \frac{1}{2\,(2+1)^2}\,e^{2x} + \frac{1}{D}\,\{(1+D)^{-2}\,(x + x^2)\}$$

$$= \frac{1}{18}\,e^{2x} + \frac{1}{D}\,(1 - 2D + 3D^2 - 4D^3 + \ldots)\,(x + x^2)$$

$$= \frac{1}{18}\,e^{2x} + \left(\frac{1}{D} - 2 + 3D - 4D^2 + \ldots\right)(x + x^2)$$

$$= \frac{1}{18}\,e^{2x} + \left[\frac{x^2}{2} + \frac{x^3}{3} - 2x - 2x^2 + 3 + 6x - 8\right]$$

$$= \frac{1}{18}\,e^{2x} + \frac{x^3}{3} - \frac{3}{2}\,x^2 + 4x - 5$$

Therefore the complete solution is

$$y = c_1 + (c_2 + c_3 x)\,e^{-x} + \frac{1}{18}\,e^{2x} + \frac{x^3}{3} - \frac{3}{g}\,x^2 + 4x - 5$$

Example 3.95 **Solve**

$$(D^4 + 2D^3 - 3D^2)\,y = x^2 + 3e^{2x} + 4\sin x$$

Solution : The A.E. is

$$m^4 + 2m^3 - 3m^2 = 0$$

$$\Rightarrow \quad m = 0, 0, 1, -3$$

$$\therefore \quad \text{C.F.} = (c_1 + c_2 x) + c_3 e^{x} + c_4 e^{-3x}$$

and $\quad \text{P.I.} = \dfrac{1}{D^2\,(D^2 + 2D - 3)}\,(x^2 + 3e^{2x} + 4\sin x)$

$$= \frac{1}{D^2\,(D^2 + 2D - 3)}\,x^2 + \frac{3.1}{D^2\,(D^2 + 2D - 3)}\,e^{2x} + \frac{4\cdot 1}{D^2\,(D^2 + 2D - 3)}\,\sin x$$

$$= \frac{1}{D^2}\left(-\frac{1}{3}\right)\left(1 - \frac{(D^2 + 2D)}{3}\right)^{-1} x^2 + \frac{3}{4\,(4 + 4 - 3)}\,e^{2x} - \frac{2}{D - 2}\,\sin x$$

$$= \frac{1}{D^2}\left(-\frac{1}{3} - \frac{2}{9}\,D - \frac{7}{27}\,D^2 \ldots\right)x^2 + \frac{3}{20}\,e^{2x} - 2\,\frac{D + 2}{D^2 - 4}\,\sin x$$

$$= -\frac{1}{27}\left(\frac{3}{4}\,x^4 + 2x^3 + 7x^2\right) + \frac{3}{20}\,e^{2x} + \frac{2}{5}\,(\cos x + 2\sin x)$$

Therefore the complete solutions is

$$y = c_1 + c_2 x + c_3 e^{x} + c_4 e^{-3x} - \frac{x^2}{108}\,(3x^2 + 3x + 28) + \frac{3}{20}\,e^{2x} + \frac{2}{5}\,(\cos x + 2\sin x)$$

Example 3.96 **Solve $(D^2 - 4D + 4)\,y = x^2 + e^x + \cos 2x$.**

Solution : The A.E. is $m^2 - 4m + 4 = 0$ or $(m-2)^2 = 0$

$\therefore \quad m = 2, 2$

$$\therefore \qquad \text{C.F.} = (c_1 + c_2x)\,e^{2x}$$

$$\text{P.I.} = \frac{1}{D^2 - 4D + 4}x^2 + \frac{1}{D^2 - 4D + 4}e^x + \frac{1}{D^2 - 4D + 4}\cos 2x$$

$$= \frac{1}{4}\left(1 + \frac{D^2 - 4D}{4}\right)^{-1}(x^2) + \frac{e^x}{1 - 4 + 4} + \frac{11}{0 - 4 - 4D + 4}\cos 2x$$

$$= \frac{1}{4}\left(1 - \frac{(D^2 - 4D)}{4} + D^2 + \ldots\right)x^2 + e^x - \frac{1}{4}\frac{1}{D}\cos 2x$$

$$= \frac{1}{4}\left(1 + D + \frac{3}{4}D^2 + \ldots\right)x^2 + e^x - \frac{1}{8}\sin 2x$$

$$= \frac{1}{4}\left(x^2 + 2x + \frac{3}{2}\right) + e^x - \frac{1}{8}\sin 2x$$

$$\therefore \text{ C.S. is } y = (c_1 + c_2x)\,e^{2x} + \frac{1}{4}\left(x^2 + 2x + \frac{3}{2}\right) + e^x - \frac{1}{8}\sin 2x$$

PROBLEM SET 3.15

Solve the following differential equations :

1. $(D^3 - 13D + 12)\,y = x$

Ans. $y = c_1e^x + c_2e^{3x} + c_3e^{-4x} + \frac{1}{12}\left(x + \frac{13}{12}\right)$

2. $(D^2 - 4D + 3)\,y = 2x^2$

Ans. $y = c_1e^x + c_2e^{3x} + \frac{2}{3}\left(x^2 + \frac{8}{3}x + \frac{26}{9}\right)$

3. $(D^2 - 4)\,y = x^2$

Ans. $y = c_1e^{2x} + c_2e^{-2x} - \frac{1}{4}\left(x^2 + \frac{1}{2}\right)$

4. $(D^3 + 3D^2 + 2D)\,y = x^2$

Ans. $y = c_1 + c_2e^{-x} + c_3e^{-2x} + \frac{1}{12}(2x^3 - 9x^2 + 21x)$

5. $(D^2 + 2D + 2)\,y = x^2$

Ans. $y = e^{-x}\,(c_1\cos + c_2\sin x) + \frac{1}{2}(x-1)^2$

6. $(D^2 - 5D + 6)\, y = x$

Ans. $y = c_1 e^{2x} + c_2 e^{3x} + \frac{1}{6} + x + \frac{5}{36}$

7. $(D^3 - D^2 - D + 1)\, y = x$

Ans. $y = (c_1 + c_2 x)\, e^x + c_3 e^{-x} + x + 1$

8. $(D^3 + 2D^2 + 4D + 8)\, y = x$

Ans. $y = c_1 e^{-x} + c_2 \cos 2x + c_3 \sin 2x + \frac{1}{16}(2x - 1)$

9. $(D^4 - a^4)\, y = x^4$

Ans. $y = c_1 e^{ax} + c_2 e^{-ax} + c_3 \cos ax + c_4 \sin ax - \frac{1}{a^4} x^4 - \frac{24}{a^8}$

10. $(D^2 - 3D + 2)\, y = e^{2x} + x^2$

Ans. $y = c_1 e^x + c_2 e^{2x} + x e^{2x} + \frac{1}{2}\left(x^2 + 3x + \frac{7}{2}\right)$

11. $(D^2 + D - 2)\, y = x + \sin x$

Ans. $y = c_1 e^x + c_2 e^{-2x} - \frac{1}{10}(\cos x + 3 \sin x) - \frac{1}{4}(2x + 1)$

12. $(D^2 - 2D + 3)\, y = \cos x + x^2$

Ans. $y = e^x (c_1 \cos \sqrt{2}\, x + c_2 \sin \sqrt{2}\, x) + \frac{1}{4}(\cos x - \sin x) + \frac{1}{27}(9x^2 + 12x + 2)$

13. $(D^2 - 4D + 4)\, y = x^2 + \sin 2x + e^x$

Ans. $y = (c_1 + c_2 x)\, e^{2x} + \frac{1}{4}\left(x^2 + 2x + \frac{3}{2}\right) + \frac{1}{8} \cos 2x + e^x$

14. $(D^2 - 4D + 4)\, y = x^2 + \cos 2x + e^x$

Ans. $y = (c_1 + c_2 x)\, e^{2x} + \frac{1}{4}\left(x^2 + 2x + \frac{3}{2}\right) + 3^x - \frac{3}{2} \sin 2x$

3.16.4 When $f(x) = e^{ax} \cdot V$, where V is a Function of x

$$\text{P.I.} = \frac{1}{\varphi(D)} e^{ax} \cdot V$$

$$= e^{ax} \cdot \frac{1}{\varphi(D + a)} \cdot V$$

Here $\varphi(D + a)$ is obtained by replacing D by $D + a$ in $\varphi(D)$.

Example 3.97 **Solve :**

$$\frac{d^2y}{dx^2} - 2\frac{dy}{dx} + 2y = x + e^x \cos x$$

Solution : Given

$$\frac{d^2y}{dx^2} - 2\frac{dy}{dx} + 2y = x + e^x \cos x$$

$$\Rightarrow \qquad (D^2 - 2D + 2)y = x + e^x \cos x$$

A.E. is $m^2 - 2m + 2 = 0$

$$\Rightarrow \qquad m = \frac{2 \pm \sqrt{4-8}}{2} = \frac{2 \pm \sqrt{-4}}{2} = \frac{2 \pm 2i}{2} \Rightarrow m = 1 \pm i$$

$$\text{C.F.} = e^x (c_1 \cos x + c_2 \sin x)$$

Particular integral

$$\text{P.I.} = \frac{1}{D^2 - 2D + 2} [x + e^x \cos x]$$

$$= \frac{1}{D^2 - 2D + 2} x + \frac{1}{D^2 - 2D + 2} e^x \cos x$$

$$= \frac{1}{2\left[1 - \left(\frac{2D - D^2}{2}\right)\right]} x + e^x \cdot \frac{1}{(D+1)^2 - 2(D+1) + 2} \cos x$$

$$= \frac{1}{2}\left[1 - \left(\frac{2D - D^2}{2}\right)\right]^{-1} x + e^x \cdot \frac{1}{D^2 + 2D + 1 - 2D - 2 + 2} \cos x$$

$$= \frac{1}{2}\left[1 + \left(\frac{2D - D^2}{2}\right) + \left(\frac{2D - D^2}{2}\right)^2 + \ldots\right] x + e^x \cdot \frac{1}{D^2 + 1} \cos x$$

$$= \frac{1}{2}\left[x + Dx - \frac{D^2}{2} x + \ldots\right] + e^x \cdot \frac{x}{2D} \cos x$$

$$= \frac{1}{2} x + \frac{1}{2} + \frac{xe^x}{2} \cdot \sin x$$

General solution is

$$y = \text{C.F.} + \text{P.I.} = e^x (c_1 \cos x + c_2 \sin x) + \frac{1}{2} x + \frac{xe^x}{2} \sin x + \frac{1}{2}$$

Example 3.98 **Solve :**

$$\frac{d^2y}{dx^2} - 3\frac{dy}{dx} + 2y = xe^{3x} + \sin 2x$$

Solution : We have

$$\frac{d^2y}{dx^2} - 3\frac{dy}{dx} + 2y = xe^{3x} + \sin 2x$$

$$\Rightarrow \qquad (D^2 - 3D + 2)y = xe^{3x} + \sin 2x$$

A.E. is $m^2 - 3m + 2 = 0$

$$\Rightarrow \qquad (m-2)(m-1) = 0 \Rightarrow m = 1, 2$$

$$\text{C.F.} = c_1 e^x + c_2 e^{2x}$$

Particular Integral is

$$\text{P.I.} = \frac{1}{D^2 - 3D + 2} [xe^{3x} + \sin 2x]$$

$$= \frac{1}{D^2 - 3D + 2} xe^{3x} + \frac{1}{D^2 - 3D + 2} \sin 2x$$

$$= e^{3x} \cdot \frac{1}{(D+3)^2 - 3(D+3) + 2} x + \frac{1}{-2^2 - 3D + 2} \sin 2x$$

$$= e^{3x} \cdot \frac{1}{D^2 + 6D + 9 - 3D - 9 + 2} x + \frac{1}{-2 - 3D} \sin 2x$$

$$= e^{3x} \frac{1}{D^2 + 3D + 2} x - \frac{3D - 2}{(3D+2)(3D-2)} \sin 2x$$

$$= e^{3x} \frac{1}{2\left[1 + \frac{3D + D^2}{2}\right]} x - \frac{3D - 2}{9D^2 - 4} \sin 2x$$

$$= \frac{e^{3x}}{2}\left[1 + \left(\frac{3D + D^2}{2}\right)\right]^{-1} x - \frac{3D - 2}{9(-2^2) - 4} \sin 2x$$

$$= \frac{e^{3x}}{2}\left[1 - \left(\frac{3D + D^2}{2}\right) + \ldots\right] x - \frac{3D - 2}{-40} \sin 2x$$

$$= \frac{e^{3x}}{2}\left[x - \frac{3}{2}\right] + \frac{1}{40}(3D - 2) \sin 2x$$

$$= \frac{e^{3x}}{2}\left(x - \frac{3}{2}\right) + \frac{3}{20} \cos 2x - \frac{1}{20} \sin 2x$$

General solution is

$$y = \text{C.F.} + \text{P.I.} = c_1 e^x + c_2 e^{2x} + \frac{e^{3x}}{2}\left(x - \frac{3}{2}\right) + \frac{1}{20}(3 \cos 2x - \sin 2x)$$

Example 3.99 **Solve :**

$$(D^2 - 4D + 4)\, y = 8x^2 e^{2x} \sin 2x$$

Solution : We have $(D^2 - 4D + 4)y = 8x^2 e^{2x} \sin 2x$

A.E. is $m^2 - 4m + 4 = 0 \Rightarrow (m-2)^2 = 0 \Rightarrow m = 2, 2$

$$\text{C.F.} = (c_1 + c_2 x)\, e^{2x}$$

$$\text{P.I.} = \frac{1}{D^2 - 4D + 4} 8x^2 e^{2x} \sin 2x$$

$$= 8e^{2x} \frac{1}{(D+2)^2 - 4(D+2) + 4} x^2 \sin 2x$$

$$= 8e^{2x} \cdot \frac{1}{D^2 + 4D + 4 - 4D - 8 + 4} x^2 \sin 2x$$

$$= 8e^{2x} \frac{1}{D^2} x^2 \sin 2x$$

$$= 8e^{2x} \int\left[\int \underset{I}{x^2} \underset{II}{\sin 2x}\, dx\right] dx$$

$$= 8e^{2x} \int \left[x^2 \left(\frac{-\cos 2x}{2} \right) - \int 2x \left(\frac{-\cos 2x}{2} \right) dx \right] dx$$

$$= 8e^{2x} \int \left[\frac{-x^2}{2} \cos 2x + \int \underset{I}{x} \underset{II}{\cos 2x}\, dx \right] dx$$

$$= 8e^{2x} \int \left(\frac{-x^2}{2} \cos 2x + x \frac{\sin 2x}{2} - \int 1 \cdot \frac{\sin 2x}{2} dx \right) dx$$

$$= 8e^{2x} \int \left(\frac{-x^2}{2} \cos 2x + \frac{x}{2} \sin 2x + \frac{\cos 2x}{4} \right) dx$$

$$= 8e^{2x} \left[\frac{-x^2}{2} \left(\frac{\sin 2x}{2} \right) - \left(\frac{-2x}{2} \right) \left(\frac{-\cos 2x}{4} \right) + (-1) \left(\frac{-\sin 2x}{8} \right) + \frac{x}{2} \left(\frac{-\cos 2x}{2} \right) - \frac{1}{2} \left(\frac{-\sin 2x}{4} \right) + \frac{\sin 2x}{8} \right]$$

$$= e^{2x} [-2x^2 \sin 2x - 2x \cos 2x + \sin 2x - 2x \cos 2x + \sin 2x + \sin 2x]$$

$$= e^{2x} [-2x^2 \sin 2x - 4x \cos 2x + 3 \sin 2x]$$

$$= -e^{2x} [4x \cos 2x + (2x^2 - 3) \sin 2x]$$

General solution is

$$y = \text{C.F.} + \text{P.I.} = (c_1 + c_2 x) e^{2x} - e^{2x} [4x \cos 2x + (2x^2 - 3) \sin 2x]$$

Example 3.100 **A body excecutes damped forced vibrations given by the equation**

$$\frac{d^2x}{dt^2} + 2k \frac{dx}{dt} + b^2 x = e^{-kt} \sin \omega t$$

Solve the differential equation for both the cases when $w^2 \neq b^2 - k^2$ and when $w^2 = b^2 - k^2$.

Solution : We have

$$\frac{d^2x}{dt^2} + 2k \frac{dx}{dt} + b^2 x = e^{-kt} \sin wt$$

$$\Rightarrow \quad (D^2 + 2kD + b^2) x = e^{-kt} \sin wt \qquad \ldots(1)$$

A.E. is $m^2 + 2km + b^2 = 0 \Rightarrow m = \dfrac{-2k \pm \sqrt{4k^2 - 4b^2}}{2}$

$$\Rightarrow \quad m = -k \pm \sqrt{k^2 - b^2} \Rightarrow m = -k \pm i \sqrt{b^2 - k^2}$$

$$\text{C.F.} = e^{-kt} [c_1 \cos (\sqrt{b^2 - k^2})\, t + c_2 \sin (\sqrt{b^2 - k^2})\, t]$$

$$\text{P.I.} = \frac{1}{D^2 + 2kD + b^2} e^{-kt} \sin wt$$

$$= e^{-kt} \frac{1}{(D - k)^2 + 2k (D - k) + b^2} \sin wt$$

$$= e^{-kt} \frac{1}{D^2 - 2kD + k^2 + 2kD - 2k^2 + b^2} \sin wt$$

$$= e^{-kt} \frac{1}{D^2 + (b^2 - k^2)} \sin wt$$

$$\text{P.I.} = e^{-kt} \frac{1}{-w^2 + (b^2 - k^2)} \sin wt; \text{ if } w^2 \neq b^2 - k^2$$

and $$\text{P.I.} = -e^{-kt} \cdot \frac{t}{2w} \cos wt; \text{ if } w^2 = b^2 - k^2$$

Case 1 : If $w^2 \neq (b^2 - k^2)$, the general solution of (1) is given by

$$x = e^{-kt} [c_1 \cos (\sqrt{b^2 - k^2})t + c_2 \sin (\sqrt{b^2 - k^2})t] + \frac{e^{-kt}}{b^2 - k^2 - w^2} \sin wt$$

Case II : If $w^2 = (b^2 - k^2)$, the general solution of (1) is given by

$$x = e^{-kt} [c_1 \cos (\sqrt{b^2 - k^2})t + c_2 \sin (\sqrt{b^2 - k^2})t] - \frac{e^{-kt}\, t \cos wt}{2w}$$

Example 3.101 Solve :

$$\frac{d^2y}{dx^2} - 2\frac{dy}{dx} + y = xe^x \sin x$$

Solution : We have

$$\frac{d^2y}{dx^2} - 2\frac{dy}{dx} + y = xe^x \sin x$$

$$\Rightarrow \quad (D^2 - 2D + 1)y = xe^x \sin x$$

A.E. is $m^2 - 2m + 1 = 0 \Rightarrow (m - 1)^2 = 0 \Rightarrow m = 1, 1$

$$\text{C.F.} = (c_1 + c_2x)e^x$$

$$\text{P.I.} = \frac{1}{D^2 - 2D + 1} xe^x \sin x$$

$$= e^x \frac{1}{(D + 1)^2 - 2(D + 1) + 1} x \sin x$$

$$= e^x \frac{1}{D^2 + 2D + 1 - 2D - 2 + 1} x \sin x$$

$$= e^x \frac{1}{D^2} x \sin x$$

$$= e^x \frac{1}{D} \int x \sin x \, dx = e^x \frac{1}{D} [-x \cos x + \sin x]$$

$$= e^x \int (-x \cos x + \sin x) \, dx$$

$$= e^x (-x \sin x - \cos x - \cos x)$$

$$= -e^x (x \sin x + 2 \cos x)$$

General solution is

$$y = \text{C.F.} + \text{P.I.}$$

$$= (c_1 + c_2x)e^x - e^x (x \sin x + 2 \cos x)$$

Example 3.102 **Solve $(D^2 + 2)\, y = x^2 e^{3x}$.**

Solution : The A.E. is $m^2 + 2 = 0$

$\therefore\ m = \pm \sqrt{2}\, i$

$\therefore$ C.F. $= c_1 \cos \sqrt{2}\, x + c_2 \sin \sqrt{2}\, x$

and P.I. $= \dfrac{1}{D^2 + 2} (x^2 e^{3x})$

$$= e^{3x} \frac{1}{(D+3)^2 + 2} x^2 = e^{3x} \frac{1}{D^2 + 6D + 11} x^2$$

$$= \frac{1}{11} e^{3x} \left(1 + \frac{6D + D^2}{11}\right)^{-1} x^2$$

$$= \frac{1}{11} e^{3x} \left(1 - \frac{6D + D^2}{11} + \frac{36}{121} D^2 + \ldots\right) x^2$$

$$= \frac{1}{11} e^{3x} \left\{1 - \frac{6}{11} D + \frac{25}{121} D^2 + \ldots\right\} x^2$$

$$= \frac{1}{11} e^{3x} \left\{x^2 - \frac{12}{11} x + \frac{50}{121}\right\}$$

Therefore the complete solution is

$$y = c_1 \cos \sqrt{2} + c_2 \sin \sqrt{2}\, x + \frac{1}{11} e^{3x} \left\{x^2 - \frac{12}{11} x + \frac{50}{121}\right\}$$

Example 3.103 **Solve $(D^4 + D^2 + 1)\, y = ax^2 + be^{-x} \sin 2x$**

Solution : The A.E. is $m^4 + m^2 + 1 = 0$

$\therefore \quad m = \dfrac{-1 \pm \sqrt{1-4}}{2},\ m = \dfrac{1 \pm \sqrt{1-4}}{2}$

or $\quad m = -\dfrac{1}{2} \pm i \dfrac{\sqrt{3}}{2}, \dfrac{1}{2} \pm i \dfrac{\sqrt{3}}{2}$

$$\therefore \quad \text{C.F.} = e^{-x/2}\left(c_1 \cos \frac{\sqrt{3}}{2} x + c_2 \sin \frac{\sqrt{3}}{2} x\right) + e^{x/2}\left(c_3 \cos \frac{\sqrt{3}}{2} x + c_4 \sin \frac{\sqrt{3}}{2} x\right)$$

and $\quad$ P.I. $= \dfrac{1}{(D^2 + D + 1)(D^2 - D + 1)} (ax^2 + be^{-x} \sin 2x)$

$$= \frac{1}{D^4 + D^2 + 1} ax^2 + b \frac{1}{(D^2 + D + 1)(D^2 - D + 1)} e^{-x} \sin 2x$$

$$= (1 + D^2 + D^4)^{-1} (ax^2)$$

$$+ be^{-x} \frac{1}{[(D-1)^2 + (D-1) + 1]\,[(D-1)^2 - (D-1) + 1]} \sin 2x$$

$$= (ax^2 - 2a) + be^{-x} \frac{1}{(D^2 - D + 1)(D^2 - 3D + 3)} \sin 2x$$

$$= (ax^2 - 2a) + be^{-x} \frac{1}{(-4 - D + 1)(-4 - 3D + 3)} \sin 2x$$

$$= (ax^2 - 2a) + be^{-x} \frac{1}{3D^2 + 10D + 3} \sin 2x$$

$$= (ax^2 - 2a) + be^{-x} \frac{1}{3(-4) + 10D + 3} \sin 2x$$

$$= (ax^2 - 2a) + be^{-x} \frac{1}{10D - 9} \sin 2x$$

$$= (ax^2 - 2a) + be^{x} \frac{(10D + 9)}{-481} \sin 2x$$

$$= (ax^2 - 2a) - \frac{b}{481} e^{-x} (20 \cos 2x + 9 \sin 2x)$$

Therefore the complete solution is

$$y = e^{-x/2}\left(c_1 \cos \frac{\sqrt{3}}{2} + c_2 \sin \frac{\sqrt{3}}{2}\right) + e^{x/2}\left(c_3 \cos \frac{\sqrt{3}}{2} + c_4 \sin \frac{\sqrt{3}}{2}\right)$$

$$+ (ax^2 - 2a) - \frac{be^{-x}}{481} (20 \cos 2x + 9 \sin 2x)$$

Example 3.104 **Solve $(D^3 - 2D^2 - 19D + 20)\, y = xe^x + 2e^{-4x} \sin x$**

Solution : The A.E. is $m^3 - 2m^2 - 19m + 20 = 0$

or $(m - 1)(m + 4)(m - 5) = 0$

$\therefore$ $m = 1, -4, 5$

$\therefore$ C.F. $= c_1 e^x + c_2 e^{-4x} + c_3 e^{5x}$

and P.I. $= \dfrac{1}{D^3 - 2D^2 - 19D + 20} xe^x + \dfrac{2}{D^3 - 2D^2 - 19D + 20} e^{-4x} \sin x$

$$= e^x \frac{1}{(D+1)^3 - 2(D+1)^2 - 19(D+1) + 20} x$$

$$+ 2e^{-4x} \frac{1}{(D-4)^3 - 2(D-4) - 19(D-4) + 20} \sin x$$

$$= e^x \frac{1}{D^3 + D^2 - 20D} x + 2e^{-4x} \frac{1}{D^3 - 14D^2 + 45D} \sin x$$

$$= -\frac{e^x}{20} \frac{1}{D}\left(1 - \frac{D + D^2}{20}\right)^{-1} x + 2e^{-4x} \frac{1}{-D + 14 + 45D} \sin x$$

$$= -\frac{e^x}{20} \frac{1}{D}\left\{1 + \frac{D + D^2}{20} + \ldots\right\} x + 2e^{-4x} \frac{1}{14 + 44D} \sin x$$

$$= -\frac{e^x}{20} \cdot \frac{1}{D}\left\{x + \frac{1}{20}\right\} + 2e^{-4x} \frac{1}{2(7 + 22D)} \sin x$$

$$= \frac{e^x}{20}\left(\frac{x^2}{2} + \frac{1}{20} x\right) + e^{-4x} (7 - 22D) \frac{1}{49 - 484D^2} \sin x$$

$$= -\frac{e^x}{20}\left(\frac{x^2}{2}+\frac{1}{20}x\right)+\frac{e^{-4x}}{533}(7\sin x - 22\cos x)$$

Therefore the general solution si

$$y = c_1e^x + c_2e^{-4x} + c_3e^{5x} - \frac{3/x}{20}\left(\frac{x^2}{2}+\frac{1}{20}x\right)+\frac{e^{-4x}}{533}(7\sin x - 22\cos x)$$

Example 3.105 **Solve :**

$$\frac{d^2y}{dx^2} + y = x^2 \sin 2x.$$

Solution The given Eq. can be written as

$$(D^2+1)\,y = x^2 \sin 2x$$

The A.E. is $m^2 + 1 = 0$

$\therefore\ m = \pm i$

$$\text{C.F.} = c_1 \cos x + c_2 \sin x$$

and $\text{P.I.} = \dfrac{1}{D^2+1}x^2 \sin 2x = \text{imaginary part (I.P.) of } \dfrac{1}{D^2+1}x^3e^{2ix}$

$$= \text{I.P. of } e^{2ix}\frac{1}{(D+2i)^2+1}x^2$$

$$= \text{I.P. of } e^{2ix}\frac{1}{D^2+4iD-3}x^2$$

$$= \text{I.P. of } e^{2ix}\frac{1}{-3\left(1-\frac{4}{3}iD-\frac{D^2}{3}\right)}x^2$$

$$= \text{I.P. of } \frac{e^{2ix}}{-3}\left[1-\frac{(4iD+D^2)}{3}\right]^{-1}x^2$$

$$= \text{I.P. of } -\frac{e^{2ix}}{3}\left[1+\left(\frac{4iD+D^2}{3}\right)+\left(\frac{4iD+D}{3}\right)^2+\ldots\right]x^2$$

$$= \text{I.P. of } -\frac{e^{2ix}}{3}\left[1+\frac{4iD}{3}+\left(-\frac{13}{9}\right)D^2+\ldots\right]x^2$$

$$= \text{I.P. of } -\frac{e^{2ix}}{3}\left[x^2+\frac{4i}{3}(2x)-\frac{13}{9}(2)\right]$$

$$= \text{I.P. of } -\frac{1}{3}(\cos 2x + i \sin 2x)\left[\left(x^2-\frac{26}{9}\right)+\left(\frac{8x}{3}\right)i\right]$$

$$= -\frac{1}{3}\left[\frac{8x}{3}\cos 2x + \left(x^2-\frac{26}{9}\right)\sin 2x\right]$$

$$= -\frac{1}{27}[24x\cos 2x + (9x^2 - 26)\sin 2x]$$

So, the general solution is

$$y = c_1 \cos x + c_2 \sin x - \frac{1}{27}[24x\cos 2x + (9x^2 - 26)\sin 2x]$$

Example 3.106 Solve :

$$\left(\frac{d^2y}{dx^2} - 2\frac{dy}{dx} + y\right) = xe^x \sin x$$

Solution The given Eq. can be written as

$$(D^2 - 2D + 1)\, y = xe^x \sin x$$

The A.E. is $m^2 - 2m + 1 = 0$

$\therefore\ m = 1, 1$

$$\therefore \quad \text{C.F.} = (c_1 + {}_2x)\, e^x \text{ and}$$

$$\text{P.I.} = \frac{1}{(D-1)^2} xe^x \sin x$$

$$= e^x \frac{1}{D^2} x \sin x = e^x \left\{x \cdot \frac{1}{D^2}\sin x - \frac{2D}{D^4}\sin x\right\}$$

$$= e^x \{-x\sin x - 2\cos x\} = -e^x (x \sin x + 2\cos x)$$

Therefore the general solution is

$$y = (c_1 + c_2x)\, e^x - e^x (x \sin x + 2\cos x)$$

PROBLEM SET 3.16

Solve the following differential equations:

1. $(D^2 - 4D + 3)y = e^x \cos 2x + \cos 3x$

   Ans. $y = c_1e^x + c_2e^{3x} - \frac{e^x}{8}(\sin 2x + \cos 2x) - \frac{1}{30}(2\sin 3x + \cos 3x)$

2. $\frac{d^2y}{dx^2} - 2\frac{dy}{dx} + y = x^2 e^{3x}$

   Ans. $y = (c_1 + c_2x)e^x + (2x^2 - 4x + 3)\, e^{3x}$

3. $\frac{d^2y}{dx^2} - 4y = x \sinh x$

4. $\frac{d^2y}{dx^2} + 4y = 3x \sin x$

   Ans. $y = c_1 \cos 2x + c_2 \sin 2x + x \sin x - \frac{2}{3}\cos x$

5. $\dfrac{d^2y}{dx^2} - 4\dfrac{dy}{dx} + 4y = e^{2x}\cos^2 x$

Ans. $y = (c_1 + c_2x)\,e^{2x} + \dfrac{e^{2x}}{8}(2x^2 - \cos 2x)$

6. $\dfrac{d^3y}{dx^3} - \dfrac{d^2y}{dx^2} + 3\dfrac{dy}{dx} + 5y = e^x \cos 2x$

Ans. $y = c_1e^{-x} + (c_2\cos 2x + c_3\sin 2x)\,e^x - \dfrac{1}{16}xe^x(\cos 2x - \sin 2x)$

7. $\dfrac{d^2y}{dx^2} + 2\dfrac{dy}{dx} + y = e^{-x}x^{-2}$

Ans. $y = (c_1 + c_2x)e^{-x} - e^{-x}\log x$

8. $\dfrac{d^2y}{dx^2} - 3\dfrac{dy}{dx} + 2y = 2x^2e^{4x} + 5e^{3x}$

Ans. $y = c_1e^x + c_2e^{2x} + \dfrac{e^{4x}}{54}(18x^2 - 30x + 19) + \dfrac{5}{2}e^{3x}$

9. $(D^2 - 4)y = \cosh(2x - 1) + 3^x$

Ans. $y = c_1e^{2x} + c_2e^{-2x} + \dfrac{x}{4}\sinh(2x - 1) + \dfrac{3^x}{(\log e)^2 - 4}$

10. $(D - 2)^2y = \dfrac{1}{x^2}e^{2x}$

Ans. $y = (c_1 + c_2x)e^{2x} - e^{2x}\log x$

11. $(D^2 - 3D - 4)y = xe^{-2x}$

Ans. $y = c_1e^x + c_2e^{-4x} - \dfrac{1}{6}e^{-2x}\left(x - \dfrac{1}{6}\right)$

12. $(D^2 - 1)y = (1 + x^2)\,e^x$

Ans. $y = c_1e^x + c_2e^{-x} + \dfrac{e^x}{2}\left(\dfrac{x^3}{3} - \dfrac{x^2}{2} + \dfrac{3x}{2}\right)$

3.16.5 General Method of Finding the Particular Integral of Any Function $f(x)$

$$\text{P.I.} = \frac{1}{D - a}f(x)$$

$$= e^{ax}\int e^{-ax}f(x)\,dx$$

Working Rule

1. Resolve $\varphi(D)$ in linear factors of D.
2. Resolve into partial fractions.
3. Use the formula given above.

Example 3.107 Solve :

$$\frac{d^2y}{dx} + 9y = \sec 3x$$

[UPTU 2001]

Solution : Given

$$\frac{d^2y}{dx^2} + 9y = \sec 3x$$

$$\Rightarrow \quad (D^2 + 9)\, y = \sec 3x$$

A.E. is $m^2 + 9 = 0$

$$\Rightarrow \quad m^2 = -9$$

$$\Rightarrow \quad m = \pm 3i$$

$$\text{C.F.} = c_1 \cos 3x + c_2 \sin 3x$$

$$\text{P.I.} = \frac{1}{D^2 + 9} \sec 3x$$

$$= \frac{1}{(D + 3i)(D - 3i)} \sec 3x$$

$$= \frac{1}{6i}\left[\frac{1}{D - 3i} - \frac{1}{D + 3i}\right] \sec 3x$$

$$= \frac{1}{6i}\frac{1}{D - 3i} \sec 3x - \frac{1}{6i}\frac{1}{D + 3i} \sec 3x$$

$$= \frac{1}{6i} e^{3ix} \int e^{-3ix} \sec 3x\, dx - \frac{1}{6i} e^{-3ix} \int e^{3ix} \sec 3x\, dx$$

$$= \frac{1}{6i} e^{3ix} \int \frac{\cos 3x - i \sin 3x}{\cos 3x} dx - \frac{1}{6i} e^{-3ix} \int \frac{\cos 3x + i \sin 3x}{\cos 3x} dx$$

$$= \frac{1}{6i} e^{3ix} \int (1 - i \tan 3x)\, dx - \frac{1}{6i} e^{-3ix} \int (1 + i \tan 3x)\, dx$$

$$= \frac{1}{6i} \cdot e^{3ix}\left[x + \frac{i}{3} \log \cos 3x\right] - \frac{1}{6i} e^{-3ix}\left[x - \frac{i}{3} \log \cos 3x\right]$$

$$= \frac{x}{3}\left[\frac{e^{3ix} - e^{-3ix}}{2i}\right] + \frac{\log \cos 3x}{9}\left[\frac{e^{3ix} + e^{-3ix}}{2i}\right]$$

$$= \frac{x}{3} \sin 3x + \frac{1}{9} \cos 3x \log \cos 3x$$

General solution is

$$y = \text{C.F.} + \text{P.I.}$$

$$= c_1 \cos 3x + c_2 \sin 3x + \frac{x}{3} \sin 3x + \frac{1}{9} \cos 3x \cdot \log \cos 3x$$

Example 3.108 **Solve :**

$$\frac{d^2y}{dx^2} + n^2y = \tan nx$$

Solution : The given equation can be written as

$$(D^2 + n^2)\, y = \tan nx$$

The A.E. is $m^2 + n^2 = 0$ or $m = \pm\, in$

$\therefore$ C.F. $= c_1 \cos nx + c_2 \sin nx$

and P.I. $= \dfrac{1}{D^2 + n^2} \tan nx$

$$= \frac{1}{(D + in)\,(D - in)} \tan nx$$

$$= \frac{1}{2\,in}\left[\frac{1}{D - in} - \frac{1}{D + in}\right] \tan nx$$

$$= \frac{1}{2\,in}\left[\frac{1}{D - in} \tan nx - \frac{1}{D + in} \tan nx\right]$$

Now, $\dfrac{1}{D - in} \tan nx = e^{inx} \int \tan nx\, e^{-inx}\, dx$

$$= e^{inx} \int \tan nx\,(\cos nx - i \sin nx)\, dx$$

$$= e^{inx} \int \left(\sin nx - \frac{i \sin^2 nx}{\cos nx}\right) dx$$

$$= e^{inx} \int \left(\sin nx - i\,\frac{(1 - \cos^2 nx}{\cos nx}\right) dx$$

$$= e^{inx} \int [\sin nx - i\,(\sec nx - \cos nx]\, dx$$

$$= e^{inx} \left[\frac{-\cos nx}{n} - \left(\frac{i}{n}\right) \log\,(\sec nx + \tan nx) + \frac{i \sin nx}{n}\right]$$

$$= -\frac{1}{n}\, e^{inx}\, [\cos nx - i \sin nx + i \log\,(\sec nx + \tan nx)]$$

$$= -\frac{1}{n}\, e^{inx}\, [e^{-inx} + i \log\,(\sec nx + \tan nx)]$$

$$= -\frac{1}{n}\, [1 + e^{inx} \log\,(\sec nx + \tan nx)]$$

Similarly, $\dfrac{1}{D + in} \tan nx = -\dfrac{1}{n}\, [1 - ie^{-inx} \log\,(\sec nx + \tan nx)]$

$$\therefore \quad \text{P.I.} = \frac{1}{2in}\left[-\frac{1}{n}\{1 + ie^{inx} \log\,(\sec nx + \tan nx)\}\right.$$

$$\left. + \frac{1}{n}\{1 - ie^{inx} \log\,(\sec nx + \tan nx)\}\right]$$

$$= -\frac{1}{n^2}\log(\sec nx + \tan nx)\left(\frac{e^{inx}+e^{-inx}}{2}\right)$$

$$= -\frac{1}{n^2}\log(\sec nx + \tan x)\cos nx$$

Thus, the general solution is

$$y = c_1 \cos nx + c_2 \sin nx - \frac{1}{n^2}\cos nx \log(\sec nx + \tan nx)$$

PROBLEM SET 3.17

Solve the following differential equations:

1. $\frac{d^2y}{dx^2} + a^2y = \sec ax$

   Ans. $y = c_1 \cos ax + c_2 \sin ax + \frac{x}{a}\sin ax + \frac{1}{a^2}\cos ax \log \cos ax$

2. $\frac{d^2y}{dx^2} + y = \operatorname{cosec} x$

   Ans. $y = c_1 \cos x + c_2 \sin x - x \cos x + \sin x \log \sin x$

3. $\frac{d^2y}{dx^2} + 4y = \tan 2x$

   Ans. $y = c_1 \cos 2x + c_2 \sin 2x - \frac{1}{4}\cos 2x \log(\sec 2x + \tan 2x)$

4. $\frac{d^2y}{dx^2} - 2\frac{dy}{dx} + 2y = e^x \tan x$

   Ans. $y = e^x(c_1 \cos x + c_2 \sin x) - e^x \cos x \log(\sec x + \tan x)$

5. $\frac{d^2y}{dx^2} + y = \tan x$

   Ans. $y = c_1 \cos x + c_2 \sin x - \cos x \log(\sec x + \tan x)$

6. $\frac{d^2y}{dx^2} - y = \frac{2}{1+e^x}$

   Ans. $y = c_1 e^x + c_2 e^{-x} - 1 + e^x \log(e^{-x}+1) - e^{-x}\log(e^x+1)$

7. $\frac{d^2y}{dx^2} + y = \sec x \tan x$

   Ans. $y = c_1 \cos x + c_2 \sin x + x \cos x + \sin x \log \sec x - \sin x$

8. $(D^2 + 1)y = \sec^2 x$

   Ans. $y = c_1 \cos x + c_2 \sin x + \sin \log(\sec x + \tan x)$

3.17 HOMOGENEOUS LINEAR DIFFERENTIAL EQUATION (Equations Reducible To Linear Equations With Constant Coefficients)

Here we will study the linear differential equations with variable coefficient, which can be converted to linear differential equations with constant coefficents by making appropriate substitution.

3.17.1 Cauchy's Homogeneous Linear Equations

A differential equation of the form

$$x^n \frac{d^n y}{dx^n} + k_1 x^{n-1} \frac{d^{n-1} y}{dx^{n-1}} + \ldots k_{n-1} x \frac{dy}{dx} + k_n y = R(x)$$

where $R(x)$ is a function of x; is termed as Cauchy's Homogeneous equation.

This type of equations can be reduced to linear differential equations with constant coefficients by making the substitution.

$$x = e^z \text{ or } z = \log_e x$$

$$\Rightarrow \quad \frac{dy}{dx} = \frac{dy}{dz} \frac{dz}{dx} = \frac{1}{x} \frac{dy}{dz}$$

$$\Rightarrow \quad x \frac{dy}{dx} = \frac{dy}{dz} \text{ or } x \frac{dy}{dx} = Dy, \text{ where } D \equiv \frac{d}{dz}$$

Also
$$\frac{d^2y}{dx^2} = \frac{d}{dx}\left(\frac{1}{x} \frac{dy}{dz}\right) = -\frac{1}{x^2} \frac{dy}{dz} + \frac{1}{x} \frac{d^2y}{dz^2} \frac{dz}{dx}$$

$$= -\frac{1}{x^2} \frac{dy}{dz} + \frac{1}{x^2} \frac{d^2y}{dz^2}$$

$$= \frac{1}{x^2}\left[\frac{d^2y}{dz^2} - \frac{dy}{dz}\right]$$

or
$$x^2 \frac{d^2y}{dx^2} = \frac{d^2y}{dz^2} - \frac{dy}{dz} = D(D-1)y$$

Similarly $x^3 \frac{d^3y}{dx^3} = D(D-1)(D-2)y$ and so on substituting these values in equation (1), the given reduces to linear diferential equation with constant coefficients.

Example 3.109 Solve :

$$x^2 \frac{d^2y}{dx^2} - x \frac{dy}{dx} + y = \log x$$

Solution : Obviously, this is a Cauchy's Homogeneous equation.

Putting $x = e^z$ or $z = \log x$

$$\Rightarrow \quad x \frac{dy}{dx} = Dy \text{ and } x^2 \frac{d^2y}{dx^2} = D(D-1)y \text{ where } D \equiv \frac{d}{dz}$$

The given equation, thus reduces to

$$[D(D-1) - D + 1]\, y = z$$

$$\Rightarrow \quad [D-1]^2\, y = z$$

The Auxiliary equation is

$$\Rightarrow \quad m = 1, +1$$

$\therefore$ The C.F. $= c_1e^z + c_2ze^z = c_1x + c_2\, x \log x$

and P.I. $= \dfrac{1}{(D-1)^2} z = (1-D)^{-2} z$

$$= (1 + 2D + 3D^2 + \ldots)\, z$$

$$= z + 2 = \log x + 2$$

Hence, the complete solution is

$$y = c_1x + c_2x \log x + \log x + 2$$

Example 3.110 Solve :

$$x^2 \frac{d^2y}{dx^2} - 3x \frac{dy}{dx} + y = \log x \frac{\sin(\log x) + 1}{x}$$

Solution : Let $x = e^z$, *i.e.*, $z = \log x$

So that $x \dfrac{dy}{dx} = Dy,\ x^2 \dfrac{d^2y}{dx^2} = D(D-1)\, y$

Then the given eqution reduces to

$$[D(D-1) - 3D + 1]\, y = z \frac{\sin z + 1}{e^z}$$

$\Rightarrow$ $[D^2\ 4D + 1]\, y = e^{-z} z (\sin z + 1)$

A.E. is $m^2 - 4m + 1 = 0$

$\Rightarrow$ $m = 2 \pm \sqrt{3}$

$\therefore$ C.F. $= c_1e^{(2+\sqrt{3})\, z} + c_2e^{(2-\sqrt{3})\, z} = e^{2z}\, [c_1e^{\sqrt{3}z} + c_2e^{-\sqrt{3}z}]$

and P.I. $= \dfrac{1}{(D^2 - 4D + 1)} e^{-z} z (1 + \sin z)$

$$= e^{-z}\left[\frac{1}{(D-1)^2 - 4(D-1) + 1}\right] = z(1 + \sin z)$$

$$= e^{-z}\left[\frac{1}{D^2 - 6D + 6}\right] = z(1 + \sin z)$$

$$= e^{-z}\left[\frac{1}{D^2 - 6D + 6} z + \frac{1}{D^2 - 6D + 6} z \sin z\right]$$

Now $\dfrac{1}{(D^2 - 6D + 6)} z = \dfrac{1}{6}\left[1 - \dfrac{6D - D^2}{6}\right]^{-1} z = \dfrac{1}{6}\left[1 + \left(\dfrac{6D - D^2}{6}\right) + \ldots\right] z$

$$= \frac{1}{6}[z + 1] \text{ and}$$

$$= \frac{1}{(D^2 - 6D + 6)} z \sin z$$

Imaginary Part (I.P.) of $\dfrac{1}{D^2 - 6D + 6} e^{iz} z$

$$= \text{I.P. of } e^{iz} \frac{1}{(D+i)^2 - 6(D+i) + 6} z$$

$$= \text{I.P. of } e^{iz}.\frac{1}{D^2 + (2D-6)D + (5-6i)} z$$

$$= \text{I.P. of } \frac{e^{iz}}{(5 \cdot \quad)}\left[1 + \left[\frac{(2i-6)}{(5-6i)} D + \frac{D^2}{5-6i}\right]\right]^{-1} z$$

$$= \text{I.P. of } \frac{e}{5-6i}\left[1 - \frac{2i-6}{5-6i} D + \ldots\right] z$$

$$= \text{I.P. of } \frac{e^{iz}}{5-6i}\left(z - \frac{2i-6}{5-6i}\right)$$

$$= \text{I.P. of } \frac{e^{iz}}{5-6i}\left(z - \frac{2i-6}{5-6i}\right)$$

$$= \text{I.P. of } (\cos z + i \sin z)\ \frac{(5+6i)}{61}\left(z - \frac{2i-6}{5-6i}\right)$$

$$= \text{I.P. of } \frac{1}{61}\left[(5\cos z - 6\sin z) + i(5\sin z + 6\cos z)\right]\left(z + \frac{42+26i}{61}\right)$$

$$= \frac{26}{3721}(5\cos z - 6\sin z) + \frac{1}{61}(5\sin z + \cos z)\left(z + \frac{42}{61}\right)$$

$$= \frac{z}{61}(5\sin z + \cos z) + \frac{2}{3721}(27\sin z + 191\cos z)$$

$$\therefore \quad \text{P.I.} = e^{-z}\left[\frac{1}{6}(z+1) + \frac{1}{61}(5\sin z + 6\cos z) + \frac{2}{3721}(27\sin z + 191\cos z)\right]$$

Hence, the complete solution is

$$y = e^{2z}(c_1 e^{\sqrt{3}z} + c_2 e^{-\sqrt{3}z}) + e^{-t}\left[\frac{1}{6}(t+1) + \frac{1}{61}(5\sin t + 6\cos 5\right.$$

$$\left. + \frac{2}{3721}(27\sin t + 191\cos t)\right]$$

or $$y = x^2(c_1 x^{\sqrt{3}} + c_2 x^{-}\sqrt{3} + c_2 x^{-\sqrt{3}}) + \frac{1}{x}$$

$$\left[\frac{1}{6}(\log x + 1) + \frac{\log x}{61}\{5\sin(\log x) + 6\cos(\log x)\}\right.$$

$$\left. + \frac{2}{3721}\{27\sin\log x + 191\cos(\log x)\}\right]$$

Example 3.111 **Solve :**

$$x^2\frac{d^2y}{dx^2} + x\frac{dy}{dx} - y = x^2$$

Solution : Given Eq. is Cauchy's homogeneous linear equation. Putting $z = \log x$, *i.e.*, $x = e^z$, we get

$$x\frac{dy}{dx} = Dy,\ x^2\frac{d^2y}{dx^2} = D(D-1)y$$

where $D = \frac{d}{dz}$.

Substituting these values in the given equation

$$[D(D-1) + D - 1]\,y = e^{2z}$$

or $\quad (D^2 - 1)\,y = e^{2z}$

Auxiliary equation is

$$m^2 - 1 = 0$$

$$m = 1, -1$$

$$\text{C.F.} = c_1e^z + c_2e^{-z} = c_1x + \frac{c_2}{x}$$

and $\quad \text{P.I.} = \frac{1}{D^2 - 1)}\,e^{2z}$

$$= \frac{1}{4-1}\,e^{2z} = \frac{1}{3}\,e^{2z} = \frac{1}{3}\,x^2$$

Hence, the complete solution is

$$y = c_1x + \frac{c_2}{x} + \frac{1}{3}x^2$$

Example 3.112 Solve :

$$\boldsymbol{x^4\frac{d^3y}{dx^3} + 2x^3\frac{d^2y}{dx^2} - x^2\frac{dy}{dx} + xy = 1}$$

Solution : Given Eq. can be written as

$$x^3\frac{d^3y}{dx^3} + 2x^2\frac{d^2y}{dx^2} - x\frac{dy}{dx} + y = \frac{1}{x} \quad \text{...(i)}$$

which is a Cauchy's homogeneous linear equation.

Putting $z = \log x$, *i.e.*, $x = e^z$ in equation (i), we get

$$x\frac{dy}{dx} = Dy,\ x^2\frac{d^2y}{dx^2} = D(D-1)y$$

$$x^3\frac{d^3y}{dx^3} = D(D-1)(D-2)\,y$$

where $D = \frac{d}{dz}$

On substituting these values in the given equatin, it reduces to

$$[D(D-1)(D-2) + 2D(D-1) - (D-1)]\,y = e^{-z}$$

or $\quad (D-1)^2(D+1)\,y = e^{-z}$

A.e. is $(m-1)^2(m+1) = 0 \Rightarrow m = 1, 1, -1$

$\therefore \quad \text{C.F.} = (c_1 + c_2z)\,e^z + c_3e^{-z} = (c_1 + c_2\log x)\,x + c_3x^{-1}$

and $$\text{P.I.} = \frac{1}{(D-1)^2 (D+1)} e^{-z}$$

$$= \frac{1}{(D+1)(-1-1)^2} e^{-z} = \frac{1}{4} \frac{1}{(D+1)} e^{-z} = \frac{e^{-z}}{4} \cdot \frac{1}{D} 1$$

$$= \frac{e^{-z}}{4} \cdot z = \frac{1}{4} x^{-1} \log x$$

Hence, the complete solution is given by

$$\therefore \quad y = (c_1 + c_2 \log x) x + c_3 x^{-1} + \frac{1}{4} x^{-1} \log x$$

Example 3.113 **Solve :**

$$x^2 \frac{d^2y}{dx^2} + 3x \frac{dy}{dx} + y = \frac{1}{(1-x)^2}$$

[Raj. 2005,, RTU 2007]

Solution : Given Eq. is Cauchy's homogeneous linear Eq.

Putting, $z = \log x$, *i.e.*, $x = e^z$ in given Eq., we get

$$[D(D-1) + 3D + 1] y = \frac{1}{(1-e^z)^2} \quad \text{where } D = \frac{d}{dz}$$

or $$(D^2 + 2D + 1) y = \frac{1}{(1-e^z)^2}$$

A.E. is $m^2 + 2m + 1 = 0$ or $m = -1, -1$

$$\therefore \quad \text{C.F.} = (c_1 + c_2 z) e^{-z} = (c_1 + c_2 \log x) \frac{1}{x}$$

and $$\text{P.I.} = \frac{1}{(D+1)^2} \frac{1}{(1-e^z)^2} = \frac{1}{(D+1)} \left[\frac{1}{D+1} \left(\frac{1}{(1-e^z)^2} \right) \right]$$

$$= \frac{1}{(D+1)} \left[e^{-z} \int \frac{e^z}{(1-e^z)^2} dz \right]$$

$$= \frac{1}{(D+1)} \left[e^{-z} \frac{1}{1-e^z} \right] = e^{-z} \int \frac{e^{-z}}{1-e^z} dz$$

$$= e^{-z} \int \frac{1}{1-e^z} dz = e^{-z} \int \frac{e^{-z}}{e^{-z}-1} dz$$

$$= -e^{-z} \log(e^{-z} - 1) = -x^{-1} \log \left(\frac{1-x}{x} \right)$$

Hence, the compete solution is

$$y = (c_1 + c_2 \log x) x^{-1} - x^{-1} \log \left(\frac{1-x}{x} \right)$$

or $$y = \frac{1}{x} \left[c_1 + c_2 \log x - \log \left(\frac{1-x}{x} \right) \right]$$

PROBLEM SET 3.18

Solve the following differential equations :

1. $x^2 \dfrac{d^2y}{dx^2} + 4x \dfrac{dy}{dx} + 2y = e^x$

Ans. $y = c_1x^{-1} + c_2x^{-2} + \dfrac{e^x}{x^2}$

2. $x^2 \dfrac{d^2y}{dx^2} + 2x \dfrac{dy}{dx} - 20y = (x+1)^2$

Ans. $y = c_1x^4 + c_2x^{-5} - \dfrac{1}{9}\left[\dfrac{9}{14}x^2 + x + \dfrac{9}{20}\right]$

3. $x^2 \dfrac{d^2y}{dx^2} - x \dfrac{dy}{dx} + y = \log x$

Ans. $y = (c_1 + c_2 \log x)\, x + 2 + \log x$

4. $x^2 \dfrac{d^2y}{dx^2} - 4x \dfrac{dy}{dx} + 6y = xy$

Ans. $y = c_1x^2 + c_2x^3 + \dfrac{1}{2}x^4$

5. $x^3 \dfrac{d^3y}{dx^3} + 2x \dfrac{dy}{dx} - 2y = x^2 \log x + 3x$

Ans. $y = c_1x + x\,[c_2 \cos(\log x) + c_3 \sin(\log x)] + \dfrac{1}{2}x^3 (\log x - 2) + 3x \log x$

6. $x^2 \dfrac{d^2y}{dx^2} - 2y = x^2 + \dfrac{1}{x}$

Ans. $y = c_1x^2 + \dfrac{c_2}{x} + \dfrac{1}{3}\log x \left(x^2 - \dfrac{1}{x}\right)$

7. $x^3 \dfrac{d^3y}{dx^3} + 2x^2 \dfrac{d^2y}{dx^2} + 2y = 10\left(x + \dfrac{1}{x}\right)$

Ans. $y = \dfrac{c_1}{x} + [c_2 \cos(\log x) + c_3 \sin \log x]\, x + 5x + 10\, \dfrac{\log x}{x}$

8. $\dfrac{d^2y}{dx^2} - \dfrac{3}{x}\dfrac{dy}{dx} + \dfrac{3y}{x^2} = 2x - 5$

Ans. $y = ax + bx^3 + x^3 \log x + 5x^2$

9. $x^2 \dfrac{d^2y}{dx^2} + 7x \dfrac{dy}{dx} + 5y = x^5$

Ans. $y = \dfrac{c_1}{x} + \dfrac{c_2}{x^5} + \dfrac{1}{60}x^5$

10. $x^2 \dfrac{d^2y}{dx^2} - x \dfrac{dy}{dx} + y = 2 \log x$

Ans. $y = (c_1 + c_2 \log x)\, x + 2 (\log x + 2)$

11. $x^2 \dfrac{d^2y}{dx^2} - 3x \dfrac{dy}{dx} + 4y = 2x^2$

Ans. $y = (c_1 + c_2 \log x)\, x^2 + (x \log x)^2$

12. $x^2 \dfrac{d^2y}{dx^2} - x \dfrac{dy}{dx} - 3y = x^2 \log x$

Ans. $y = c_1 x^{-1} + c_2 x^3 - \dfrac{1}{3} x^2 \log x - \dfrac{2}{9} x^2$

13. $x^2 \dfrac{d^2y}{dx^2} + x \dfrac{dy}{dx} \;\; 4y = x^2$

Ans. $y = c_1 x^2 + c_2 x^{-2} + \dfrac{1}{4} x^2 \log x$

14. $x \dfrac{d^2y}{dx^2} + 2 \dfrac{dy}{dx} = 6x^3$

Ans. $y = c_1 + c_2 x + \dfrac{3}{10} x^4$

15. $x^2 \dfrac{d^2y}{dx^2} + 3x \dfrac{dy}{dx} + y = \dfrac{1}{(1-x)^2}$

Ans. $(c_1 + c_2 \log x)\, x^{-1} + x^{-1} \log \left[\dfrac{x}{(1-x)}\right]$

16. $x^3 \dfrac{d^3y}{dx^3} + 2x^2 \dfrac{d^2y}{dx^2} + 3x \dfrac{dy}{dx} - 3y = x^2 + x$

Ans. $y = (c_1 x + c_2 \cos (\sqrt{3} \log x + c_3) + \dfrac{1}{7} x^2 + \dfrac{1}{4} x \log x$

17. $x^2 \dfrac{d^2y}{dx^2} + 2x \dfrac{dy}{dx} = \log x$

Ans. $y = c_1 + c_2 x^{-1} + \dfrac{1}{2} (\log x)^2 - \log x$

3.17.2 Legendre's Linear Differential Equations

A linear adifferential equation with variable coefficients of the form

$$(ax+b)^n \frac{d^ny}{dx^n} + k_1 (ax+b)^{n-1} \frac{d^{n-1}y}{dx^{n-1}} + \dots k_{n-1} (ax+b) \frac{dy}{dx} + k_n y = R(x) \quad \text{...(i)}$$

is said to be legendre's linear differential equation of order n, where a, b and k_i's are constants and $R(x)$ is a function of x.

This type of equations can be reduced to the linear differential equation with constant coefficients by making the substitution

$$ax + b = e^z \text{ or } \log (ax + b) = z$$

$$\therefore \quad \frac{dy}{dx} = \frac{dy}{dz}, \frac{dz}{dx} = \frac{a}{ax+b}\frac{dy}{dz}$$

$$\Rightarrow \quad (ax+b)\frac{dy}{dx} = a\frac{dy}{dx} = a\,Dy$$

where $D = \dfrac{d}{dz}$

$$\text{and} \quad \frac{d^2y}{dx^2} = \frac{d}{dx}\left(\frac{a}{ax+b}\frac{dy}{dz}\right)$$

$$= \frac{a}{ax+b}, \frac{d^2y}{dz^2}, \frac{dz}{dx} - \frac{a^2}{(ax+b)^2}\frac{dy}{dz}$$

$$= \frac{a^2}{(ax+b)^2}\left[\frac{d^2y}{dz^2} - \frac{dy}{dz}\right]$$

$$\Rightarrow \quad (ax+b)^2\frac{d^2y}{dx^2} = a^2 D(D-1)\,y$$

In a similar mannar, we can write $(ax+b)^3 \dfrac{d^3y}{dx^3} = a^3 D(D-1)(D-2)$ and so on.

On subsitituting these values in equation (i), the equation reduces to linear differential equation with constants coefficients.

Example 3.114 Solve :

$$(1+x)^2\frac{d^2y}{dx^2} + (1+x)\frac{dy}{dx} + y = \sin 2\,[\log(1+x)].$$

Solution : The given equation is a Legendre's equation.

Putting $z = \log(1+x)$ or $1+x = e^z$, we get

$$(1+x)\frac{dy}{dx} = Dy,\ (1+x)^2\frac{d^2y}{dx^2} = D(D-1)\,y,\ D = \frac{d}{dz}$$

$\therefore$ The given equation reduces to

$$D(D-1)\,y + Dy + y = \sin 2z$$

$$\Rightarrow \quad (D^2+1)\,y = \sin 2z$$

A.E. is $m^2 + 1 = 0 \Rightarrow m = \pm i$

$$\therefore \quad \text{C.F.} = c_1 \cos z + c_2 \sin z$$

$$= c_1 \cos(\log(1+x)) + c_2 \sin(\log(1+x))$$

$$\text{and} \quad \text{P.I.} = \frac{1}{(D^2+1)}\sin 2z$$

$$= \frac{\sin 2z}{-4+1} = \frac{\sin 2z}{-3}$$

This the complete solution is

$$y = c_1 \cos z + c_2 \sin z - \frac{\sin 2z}{3}$$

$$\Rightarrow \quad y = c_1 \cos(\log(1+x)) + c_2 \sin(\log(1+x)) - \sin 2\log(1+x).$$

Example 3.115 **Solve :**

$$(3x+2)^2 \frac{d^2y}{dx^2} + 3(3x+2)\frac{dy}{dx} - 36y = 3x^2 + 4x + 1$$

Solution : The given equation is a Legendre's equation

Putting $z = \log(3x+2)$, *i.e.*, $(3x+2) = e^z$

$$\therefore \quad (3x+2)\frac{dy}{dx} = 3Dy,\ (3x+2)^2\frac{d^2y}{dx^2} = 3^2 D(D-1)y, \text{ where } D = \frac{d}{dz}$$

The given equation reduces to

$$[(3)^2 D(D-1) + 3(3D) - 36]\,y = 3\left(\frac{e^z - 2}{3}\right)^2 + 4\left(\frac{e^z-2}{3}\right) + 1$$

or $\quad (D^2 - 4)\,y = \dfrac{1}{27}(e^{2z} - 1)$

A.E. is $m^2 - 4 = 0 \Rightarrow m = \pm 2$

$$\therefore \quad \text{C.F.} = c_1 e^{2z} + c_2 e^{-2z} = c_1 (3x+2)^2 + c_2 (3x+2)^{-2}$$

and $\quad \text{P.I.} = \dfrac{1}{27}\cdot\dfrac{1}{D^2-4}(e^{2z} - 1)$

$$= \frac{1}{27}\left[\frac{1\cdot e^{2z}}{D^2-4} - \frac{1}{D^2-4}\right] = \frac{1}{27}\left[z\,\frac{1}{2D}e^{2z} - \frac{1}{0-4}\right]$$

$$= \frac{1}{27}\left[\frac{z}{4}e^{2z} + \frac{1}{4}\right] = \frac{1}{108}(ze^{2z} + 1)$$

$$= \frac{1}{108}[(3x+2)^2 \log(3x+2) + 1]$$

PROBLEM SET 3.19

Solve the following differential equations.

1. $(x+a)^2 \dfrac{d^2y}{dx^2} - 4(x+a)\dfrac{dy}{dx} + 6y = x$

Ans. $y = c_1 (x+a)^2 + c_2 (x+a)^3 + \dfrac{1}{6}(3x + 2a)$

2. $\{(x+1)^2 D^2 + 3(x+1)D + 4\}\,y = x^2$

Ans. $y = [c_1 + c_2 \log(1+x)](x+1)^2 + [\log(1+x)]^2 \dfrac{(x+1)^2}{2} - 2(x+1) + \dfrac{1}{4}$

3. $(x+3)^2 \dfrac{d^2y}{dx^2} - 4(x+3)\dfrac{dy}{dx} + 6y = x$

Ans. $y = c_1 (x+3)^3 + c_2 (x+3)^2 + \dfrac{(x+3)}{2} - \dfrac{1}{2}$

4. $(1 + 2x)^2 \frac{d^2y}{dx^2} - 6(1 + 2x)\frac{dy}{dx} + 16y = 8(1 + 2x)^2$

Ans. $y = \{c_1 + c_2 \log(2x + 1)\}(2x + 1)^2 + (2x + 1)^2 \{\log(2x + 1)^2\}$

5. $(2x + 3)^2 \frac{d^2y}{dx^2} - (2x + 3)\frac{dy}{dx} - 12y = 6x$

Ans. $y = c_1 (2x + 3)^a + c_2 (2x + 3)^b - \frac{3}{14}(2x + 3) + \frac{3}{4}$ where $a = \frac{3 + \sqrt{57}}{4}$, $b = \frac{3 - \sqrt{57}}{4}$

6. $(1 + x)^2 \frac{d^2y}{dx^2} + (1 + x)\frac{dy}{dx} + y = 2 \sin[\log(1 + x)]$

Ans. $y = c_1 \cos(\log(1 + x)) + c_2 \sin(\log 1 + x)) - \log(1 + x) \cos[\log(1 + x))$

7. $(1 + x)^2 \frac{d^2y}{dx^2} + (1 + x)\frac{dy}{dx} y = 4 \cos \log(1 + x)$

Ans. $y = c_1 \cos[\log(1 + x)] + c_2 \sin[\log(1 + x)] + 2 \log(1 + x) \sin[\log(1 + x)]$

8. $(x + 1)^4 \frac{d^4y}{dx^4} + 16(x + 1)^3 \frac{d^3y}{dx^3} + 104(x + 1^2 \frac{d^2y}{dx^2} + 8(x + 1)\frac{dy}{dx} - y = x^2 + 4x + 3$

Ans. $y = [c_1 + c_2 \log(x + 1)](x + 1)^{1/2} + [c_3 + c_4 \log(x + 1)]\, x\,(x + 1)^{-1/2} + \frac{(x + 1)^2 + 50(x + 1)}{225}$

3.18 SIMULTANEOUS DIFFERENTIAL EQUATIONS

In several applied mathematics problems, there are more than one dependent variables, each of which is a function of one independent variable. The formulation of such problems leads to a system of simultaneous linear differential equations with constant coefficients. Here, we consider the system of linear differential equations having two dependent variables and one independent variable with constant coefficients. The method of solving these equations is based on the process of elimination, as we solve algebraic simultaneous equation.

Example 3.116 Solve :

$$\frac{dx}{dt} - 7x + y = 0$$

$$\frac{dy}{dt} - 2x - 5y = 0$$

[UPTU 2001]

Solution : We have

$$\frac{dx}{dt} - 7x + y = 0$$

$$\frac{dy}{dt} - 2x - 5y = 0$$

$$\Rightarrow \quad (D - 7)x + y = 0 \quad \text{...(i)} \quad \left[\because D = \frac{d}{dt}\right]$$

$$-2x + (D - 5)y = 0 \quad \text{...(ii)}$$

Operating (ii) by $D - 7$ and (i) by 2, we get

$$2(D - 7)x + 2y = 0$$

$$-2(D+7)x + (D-7)(D-5)y = 0$$

On adding, we get

$$[2 + (D-7)(D-5)]\, y = 0 \;\Rightarrow\; (D^2 - 12D + 37)y = 0$$

A.E. is $m^2 - 12m + 37 \Rightarrow m = \dfrac{12 \pm \sqrt{144-148}}{2} \Rightarrow m = 6 \pm i$

$$\text{C.F.} = 6t\,(c_1 \cos t + c_2 \sin t)$$

$$\text{P.I.} = 0$$

$\therefore$ $\quad y = \text{C.F.} + \text{P.I.}$

$$= e^{6t}(c_1 \cos t + c_2 \sin t)$$

Again, operating (i) by $(D-5)$, we get

$$(D-5)(D-7)x + (D-5)y = 0$$

$$-2x + (D-5)y = 0$$

On subtracting, we obtain

$$[(D-5)(D-7) + 2]x = 0 \Rightarrow (D^2 - 12D + 37)x = 0$$

$\Rightarrow$ $\quad x = e^{6t}(c_3 \cos t + c_4 \sin t)$

Example 3.117 **Solve :**

$$\frac{dx}{dt} + 2\frac{dy}{dt} - 2x + 2y = 3e^t$$

[UPTU 2002]

$$3\frac{dx}{dt} + \frac{dy}{dt} + 2x + y = 4e^{2t}$$

Solution : We have

$$\frac{dx}{dt} + 2\frac{dy}{dt} - 2x + 2y = 3e^t$$

$$3\frac{dx}{dt} + \frac{dy}{dt} + 2x + y = 4e^{2t}$$

$\Rightarrow$ $\quad (D-2)x + 2(D+1)y = 3e^t \quad$ ...(i)

$\quad (3D+2)x + (D+1)y = 4e^{2t} \quad$ ...(ii)

Multiply (ii) by 2, get

$$(D-2)x + 2(D+1)y = 3e^t$$

$$(6D+4)x + 2(D+1)y = 8e^{2t}$$

On subtracting, we obtain

$$(D - 2 - 6D - 4)\,x = 3e^t - 8e^{2t}$$

$\Rightarrow$ $\quad (-5D-6)x = 3e^t - 8e^{2t} \;\Rightarrow\; (5D+6)x = -3e^t + 8e^{2t}$

A.E. is $5m + 6 = 0 \Rightarrow m = -\dfrac{6}{5}$

$$\text{C.F.} = c_1 e^{-\frac{6}{5}t}$$

$$\text{P.I.} = \frac{1}{5D+6}[-3e^t + 8e^{2t}]$$

$$= -3 \cdot \frac{1}{5D+6} e^t + 8 \cdot \frac{1}{5D+6} e^{2t}$$

$$= -\frac{3}{11}e^t + \frac{1}{2}e^{2t}$$

$$\therefore \quad x = e^{-\frac{6}{5}t} - \frac{3}{11}e^t + \frac{1}{2}e^{2t}$$

Operating $(3D + 2)$ on (i) and $(D - 2)$ on (ii), we get

$$(3D + 2)(D - 2)x + 2(3D + 2)(D + 1)y = 3(3D + 2)e^t$$

$$(3D + 2)(D - 2)x + (D - 2)(D + 1)y = 4(D - 2)e^{2t}$$

On subtracting, we obtain

$$[2(3D^2 + 5D + 2) - (D^2 - D - 2)]\, y = 3[3e^t + 2e^t] - 4[2e^{2t} - 2e^{2t}]$$

$$\Rightarrow \quad (5D^2 + 11D + 6)y = 15e^t$$

A.E. is $5m^2 + 11m + 6 = 0$

$$\Rightarrow \quad 5m^2 + 5m + 6m + 6 = 0 \Rightarrow (5m + 6)(m + 1) = 0$$

$$\Rightarrow \quad m = -\frac{6}{5}, -1$$

$$\text{C.F.} = c_2e^{-t} + c_3\, e^{\frac{-6}{5}t}$$

$$\text{P.I.} = \frac{1}{5D^2 + 11D + 6} \cdot 15e^t = \frac{15}{5(1)^2 + 11(1) + 6}\, e^t = \frac{15}{22}e^t$$

$$\Rightarrow \quad y = \text{C.F.} + \text{P.I.} = c_2e^{-t} + c_3e^{-\frac{6}{5}t} + \frac{15}{22}e^t$$

Hence, $x = c_1e^{-\frac{6}{5}t} - \frac{3}{11}e^t + \frac{1}{2}e^{2t}$

$$y = c_2e^{-t} + c_3e^{-\frac{6}{5}t} + \frac{15}{22}e^t$$

Example 3.118 Solve : $Dx + Dy + 3x = \sin t$

$Dx + y - x = \cos t$

Solution : We have,

$$(D + 3)x + Dy = \sin t \qquad \text{...(i)}$$

$$(D - 1)x + y = \cos t \qquad \text{...(ii)}$$

Operating D on (ii), we get

$$(D + 3)x + Dy = \sin t$$

$$D(D - 1)x + Dy = D\cos t = -\sin t$$

On subtracting, we get

$$(D + 3 - D^2 + D)x = \sin t + \sin t = 2\sin t$$

$$\Rightarrow \quad (-D^2 + 2D + 3)x = 2\sin t$$

$$\Rightarrow \quad (D^2 - 2D - 3)\, x = -2\sin t$$

A.E. is

$$m^2 - 2m - 3 = 0 \Rightarrow m = -1, 3$$

$$\therefore \quad \text{C.F.} = c_1e^{-t} + c_2\, e^{3t}$$

$$\text{P.I.} = \frac{1}{D^2 - 2D - 3}(-2\sin t) = -2\frac{1}{(-1) - 2D - 3}\sin t$$

$$= \frac{2}{2(D+2)}\sin t = \frac{D-2}{D^2-4}\sin t$$

$$= \frac{D-2}{-1-4}\sin t = -\frac{1}{5}(\cos t - 2\sin t)$$

$$\Rightarrow \quad x = \text{C.F.} + \text{P.I.}$$

$$= c_1e^{-t} + c_2e^{3t} - \frac{1}{5}(\cos t - 2\sin t)$$

From (ii), we have

$$(D-1)\left[c_1e^{-t} + c_2e^{3t} - \frac{1}{5}\cos t + \frac{2}{5}\sin t\right] + y = \cos t$$

$$\Rightarrow \quad -c_1e^{-t} + 3c_2e^{3t} + \frac{1}{5}\sin t + \frac{2}{5}\cos t - c_1e^{-t} - c_2e^{-3t} + \frac{1}{5}\cos t - \frac{2}{5}\sin t + y = \cos t$$

$$\Rightarrow \quad -2c_1e^{-t} + 2c_2e^{3t} - \frac{1}{5}\sin t + \frac{3}{5}\cos t + y = \cos t$$

$$\Rightarrow \quad y = 2c_1e^{-t} - 2c_2e^{3t} + \frac{1}{5}\sin t + \frac{2}{5}\cos t$$

Example 3.119 Solve the simultaneous differential equations

$$\frac{dx}{dt} + \frac{dy}{dt} - 2y = 2\cos t - 7\sin t$$

[UPTU 2005]

$$\frac{dx}{dt} - \frac{dy}{dt} + 2x = 4\cos t - 3\sin t$$

Solution : We have

$$Dx + (D-2)y = 2\cos t - 7\sin t \quad \text{...(i)}$$

$$(D+2)x - Dy = 4\cos t - 3\sin t \quad \text{...(ii)}$$

Operating D on (i) and $(D-2)$ on (ii), we get

$$D^2x + D(D-2)y = 2D\cos t - 7D\sin t$$

$$(D^2-4)x - D(D-2)y = 4(D-2)\cos t - 3(D-2)\sin t$$

On adding, we get

$$(D^2 + D^2 - 4)x = -2\sin t - 7\cos t - 4\sin t - 8\cos t - 3\cos t + 6\sin t$$

$$(2D^2 - 4)x = -18\cos t \Rightarrow (D^2 - 2)x = -9\cos t$$

A.E. is $m^2 - 2 = 0 \Rightarrow m = \pm\sqrt{2}$

$$\text{C.F.} = c_1e^{\sqrt{2}t} + c_2e^{-\sqrt{2}t}$$

$$\text{P.I.} = \frac{1}{D^2-2}(-9\cos t) = \frac{-9}{-1-2}\cos t = 3\cos t$$

$$x = \text{C.F.} + \text{P.I.}$$

$$= c_1e^{\sqrt{2}t} + c_2e^{-\sqrt{2}t} + 3\cos t$$

From (ii), we get

$$(D+2)(c_1e^{\sqrt{2}t} + c_2e^{-\sqrt{2}t} + 3\cos t) - Dy = 4\cos t - 3\sin t$$

$$\Rightarrow \quad \sqrt{2}c_1e^{\sqrt{2}t} - \sqrt{2}\, c_2\, e^{\sqrt{2}t} - 3\sin t + 2c_1e^{\sqrt{2}t} + 2c_2e^{-\sqrt{2}t} + 6\cos t$$

$- Dy = 4\cos t - 3\sin t$

$$\Rightarrow \quad (\sqrt{2}+2)c_1e^{\sqrt{2}t} - (\sqrt{2}-2)\, c_2e^{-\sqrt{2}t} - 3\sin t + 6\cos t - Dy = 4\cos t - 3\sin t$$

$$\Rightarrow \quad Dy = -(\sqrt{2}-2)c_2\, e^{-\sqrt{2}t} - (\sqrt{2}+2)c_1e^{\sqrt{2}t} - 2\cos t$$

On integrating, we obtain

$$y = \frac{\sqrt{2}-2}{-\sqrt{2}}\, c_2\, e^{-\sqrt{2}t} - \frac{(\sqrt{2}+2)}{\sqrt{2}}\, c_1e^{\sqrt{2}t} - 2\sin t + c_3$$

$$y = (\sqrt{2}-1)c_2e^{-\sqrt{2}t} - (\sqrt{2}+1)\, c_1e^{\sqrt{2}t} - 2\sin t + c_3$$

Example 3.120 Solve the system of equations

$$\frac{d^2x}{dt^2} - 3x - 4y = 0, \quad \frac{d^2y}{dx^2} + x + y = 0.$$

Solution : Given Eq. can be written as

$$(D^2 -)\, x - 4y = 0 \qquad \ldots(1)$$

$$x + (D^2 + 1)\, y = 0 \qquad \ldots(2)$$

Eliminating y from equation (1) and(2), we get

$$[(D^2+1)\,(D^2-3) + 4]\, x = 0$$

or $\quad (D^4 - 2D^2 + 1)\, x = 0$

or $\quad (D^2 - 1)^2\, x = 0$

The auxiliary equation (A.E.) is given by $(m^2-1)^2 = 0$ or $m = -1, -1, 1, 1$

$$\therefore \quad x = (c_1 + c_2t)\, e^{-t} + (c_3 + c_4t)\, e^t \qquad \ldots(3)$$

Differentiating equation (3), w.r.t. 't'

and $$\frac{dx}{dt} = -(c_1 + c_2t)\, e^{-t} - 2\, c_2e^{-t} + (c_3 + c_4e^t)\, e^t + 2c_4e^t \qquad \ldots(4)$$

From equation (1) and (4)

$$4y = D^2y - 3x = (c_1 + c_2t)\, e^{-t} - 2c_3e^{-t} + (c_3 + c_4t)\, e^t + 2c_4e^t$$
$$- 3\,(c_1 + c_2t)\, e^{-t} - 3\,(c_3 + c_4t)\, e^t$$

or $$y = -\frac{1}{2}\,(c_1 + c_3 + c_2t)\, e^{-t} + \frac{1}{2}\,(c_4 - c_3 - c_4t)\, e^t$$

The complete solution is given by Eqs. (3) and (5).

Example 3.121 Solve the system of equations

$$\frac{d^2x}{dt^2} - \frac{dy}{dt} = 2x + 2t, \quad \frac{dx}{dt} + 4\frac{dy}{dt} = 3y$$

Solution : The given equations can be written

$$(D^2 - 2)\, x - Dy = 2t \qquad \ldots(1)$$

$$Dx + (4D - 3)\, y = 0 \qquad \ldots(2)$$

Eliminating y from equations (1) and (2),

$$[(4D-3)\,(D^2-2) + D^2]\, x = (4D-3)\, 2t$$

or $\quad (2D^3 - D^2 - 4D + 3)\, x = 4 - 3t$

The auxilary Eq. (A.E.) is given by

$$2m^3 - m^2 - 4m + 3 = 0$$

or $$m = 1,\ 1,\ -\frac{3}{2}$$

$\therefore$ $$\text{C.F.} = (c_1 + c_2 t)\, e^t + c_3 e^{-3/2\,t} \text{ and P.I.} = \frac{1}{(2D^3 - D^2 - 4D + 3)}(4 - 3t)$$

$$= \frac{1}{3}\left(1 - \frac{4}{3}D - \frac{1}{3}D^2 + \frac{2}{3}D^3\right)^{-1}(4 - 3t)$$

$$= \frac{1}{3}\left[4 - 3t + \frac{4}{3}(-3)\right] = -t$$

$\therefore$ $$x = (c_1 + c_2 t)\, e^{-t} + c_3 e^{-3/2\,t} - 1 \qquad ...(3)$$

Differentiating Eq. (3), w.r.t. 't'

$$\frac{dx}{dt} = -(c_1 + c_2 t)\, e^{-t} + c_2 e^t - \frac{3}{2} c_3 e^{-3/2\,t}$$

and $$\frac{d^2x}{dt^2} = (c_1 + c_2 t)\, e^t + 2\, c_2 e^t + \frac{9}{4} c_3 e^{-3/2\,t}$$

Multiplying Eq. (1) by (4) and then adding to Eq. (2), we get

$$4\frac{d^2x}{dt^2} + \frac{dx}{dt} - 8x - 3y = 8t$$

or $$3y = 4\frac{d^2x}{dt^2} + \frac{dx}{dt} - 8x - 8t \qquad ...(4)$$

Put the value of x, $\frac{dx}{dt}$ and $\frac{d^2x}{dt^2}$ in Eq. (4)

$$3y = 4\,(c_1 + c_2 t)\, e^t + 8c_2 e^t + (c_3 e^{-3/2\,t} + (c_1 + c_2 t)\, e^t + c_2 e^t$$

$$- \frac{3}{2} c_e e^{-3/2\,t} - 1 - 8\,(c_1 + c_2 t)\, e^t - 8\, c_3\, e^{-3/2\,t} + 8t - 8t$$

or $$y = (3c_2 - c_1 - c_2 t)\, e^t - \frac{c_3}{6} e^{-3/2\,t} - \frac{1}{3}$$

Hence, the complete solutio is given by Eqs. (3) and (5).

Example 3.122 **Solve the system**

$$\frac{dx}{dt} + \frac{2}{t}(x - y) = 1; \quad \frac{dy}{dt} + \frac{1}{t}(x + 5y) = t$$

Solution : Given Eq. can be written as

$$t\frac{dx}{dt} + 2\,(x - y) = t \qquad ...(1)$$

$$t\frac{dy}{dt} + x + 5y = t^2 \qquad ...(2)$$

Differentiating Eq. (1) w.r.t. 't'

$$t\frac{d^2x}{dt^2} + 3\frac{dx}{dt} - 2\frac{dy}{dt} = 1 \quad ...(3)$$

Eliminating y and $\frac{dy}{dt}$ from Eq. (1), (2) and (3)

$$t^2\frac{d^2x}{dt^2} + 8t\frac{dx}{dt} + 12x = 2t^2 + 6t \quad ...(4)$$

Which is a homogeneous linear differential Eq. of order.

Put $t = e^z$ or $z = \log_e t$ in Eq. (4)

$$[D(D-1) + 8D + 12]\,x = 2e^{2z} + 6e^z \text{ where } D \equiv \frac{d}{dz}$$

or $$(D^2 + 7D + 12)\,x = 2e^{2z} + 6e^z \quad ...(5)$$

Which is linear differential Eq. with constant coefficient. The auxiliary Eq. (A.E.) is given by

$$m^2 + 7m + 12 = 0 \text{ or } m = -3, -4$$

$$\therefore \quad \text{C.F.} = c_1e^{-3z} + c_2c^{-4z} = c_1t^{-3} + c_2t^{-4}$$

$$\text{P.I.} = \frac{1}{D^2 + 7D + 12}(2e^{2z} + 6e^z)$$

$$= \frac{2e^{2z}}{(2)^2 + 7(2) + 12} + \frac{6e^z}{(1)^2 + 7(1) + 12}$$

$$= \frac{1}{15}e^{2z} + \frac{3}{10}e^z = \frac{1}{15}t^2 + \frac{3}{10}t$$

$$\therefore \quad x = c_1t^{-3} + c_2t^{-4} - 4c_2t^{-5} + \frac{2}{15}t + \frac{3}{10}t \quad ...(6)$$

Differentiating Eq. (6) w.r.t. 't'

$$\frac{dx}{dt} = -3c_1t^{-4} - 4c_2t^{-5} + \frac{2}{15}t + \frac{3}{10} \quad ...(7)$$

Put the value of x and $\frac{dx}{dt}$ in the equation (1), we get

$$2y = t\left(-3c_1t^{-4} - 4c_2t^{-5} + \frac{2}{15}t + \frac{3}{10}\right) + \left(c_1t^{-3} + c_2t^{-4} + \frac{1}{15}t^2 + \frac{3}{10}t\right) - t$$

or $$y = -\frac{c_1}{2}t^{-3} - c_2\,t^{-4} + \frac{2t^2}{15} - \frac{t}{20} \quad ...(8)$$

The complete solution is given by equations (6) and (8).

Example 3.123 **Solve the system of equations.**

$$\frac{dx}{dt} + 2y + \sin t = 0; \quad \frac{dy}{dt} - 2x - \cos t = 0$$

Solution : Given equations can be written as

$$Dx + 2y = -\sin t \quad ...(1)$$

$$-2x + Dy = \cos t \quad ...(2)$$

Multiplying Eq. (1) by 2 and operating on Eq. (2) with D and then adding, we get

$$4y + D^2y = -2\sin t - \sin t$$

or $\quad (D^2 + 4)\,y = -3\sin t$

The auxiliary Eq. (A.E.) is given by

$$m^2 + 4 = 0 \text{ or } m = \pm 2i$$

$$\text{C.F.} = c_1\cos 2t + c_2\sin 2t$$

$$\text{P.I.} = -3\,\frac{1}{D^2+4}\sin t$$

$$= -3\,\frac{1}{-1+4}\sin t = -\sin t$$

$\therefore \quad y = c_1\cos 2t + c_2\sin 2t - \sin t$

Differentiating Eq. (3) w.r.t. 't'

$$\frac{dy}{dt} = -2c_1\sin 2t + 2c_2\cos 2t - \cos t \qquad \text{...(4)}$$

From Eq. (2) and (4), we get

$$2x = Dy - \cos t = 2c_1\sin 2t + 2c_2\cos 2t - 2\cos t$$

or $\quad x = -c_1\sin 2t + c_2\cos 2t - \cos t$

So the complete solution is given by Eq. (3) and (5).

PROBLEM SET 3.20

Solve the following differential equations:

1. $\dfrac{dx}{dt} + 4x + 3y = t$

$\dfrac{dy}{dt} + 2x + 5y = e^t$

Ans. $x = c_1e^{-2t} + c_2e^{-7t} - \dfrac{31}{196} + \dfrac{5}{14}t - \dfrac{1}{8}e^t;\ y = -\dfrac{2}{3}c_1e^{-2t} + c_2e^{-7t} + \dfrac{9}{98} - \dfrac{1}{7}t + \dfrac{5}{24}e^t$

2. $\dfrac{dx}{dt} + y = \sin t\ \dfrac{dy}{dt} + x = \cos t.$

Given that $x = 2, y = 0$ when $t = 0$.

Ans. $x = e^t + e^{-t};\ y = \sin t - e^t + e^{-t}$

3. $\dfrac{dx}{dt} + \dfrac{2}{t}(x - y) = 1$

$\dfrac{dy}{dt} + \dfrac{1}{t}(x + 5y) = t$

Ans. $x = c_1t^{-4} + c_2t^{-3} + \dfrac{t^2}{15} + \dfrac{3t}{10};\ y = -c_1t^{-4} - \dfrac{1}{2}c_2t^{-3} + \dfrac{2t^2}{15} - \dfrac{t}{20}$

4. $\dfrac{d^2y}{dt^2} = x,$

$\dfrac{d^2x}{dt^2} = y$

Ans. $x = c_1e^t + c_2e^{-t} + c_3 \cos t + c_4 \sin t$; $y = c_1e^t + c_2e^{-t} - c_3 \cos t - c_4 \sin t$

5. $\dfrac{dx}{dt} - \dfrac{dy}{dt} + 2y = \cos 2t$

$\dfrac{dx}{dt} + \dfrac{dy}{dt} + 2x = \cos 2t$

Ans. $x = e^t (c_1 \cos t + c_2 \sin t) - \dfrac{\cos 2t}{2}$; $y = e^t (c_1 \sin t - c_2 \cos t) - \dfrac{\sin 2t}{2}$

6. $\dfrac{dy}{dt} + y = x + e^t$

$\dfrac{dx}{dt} + x = y + e^t$

Ans. $y = c_1 + c_2\, e^{-2t}$, $x = c_1 - c_2\, e^{-2t} + e^t$

7. $\dfrac{dx}{dt} + 2y + x = e^t$

$\dfrac{dy}{dt} + 2x + y = 3e^t$

Ans. $x = c_1e^t + c_2e^{-3t} - te^t$, $y = -c_1e^t + c_2e^{-3t} + te^t + e^t$

8. $\dfrac{d^2x}{dt^2} - \dfrac{dx}{dt} + \dfrac{dy}{dt} - y = 0$

$\dfrac{d^2y}{dt^2} - \dfrac{dy}{dt} + x = 0$

Ans. $x = c_1e^t + c_2 \cos t + c_3 \sin t,$

$y = c_1e^t + c_2 \sin t - c_3 \cos t$

9. $4\dfrac{dx}{dt} - \dfrac{dy}{dt} + 3x + 3y = \sin t$

$\dfrac{dx}{dt} + y = \cos t$, given that $x = 1, y = 0$ for $t = 0$

Ans. $x = 2e^{-t} - e^{-3t}$, $y = 2e^{-t} - 3e^{-3t} + \cos t$

10. $\dfrac{dx}{dt} + 2x + 3y = 0,$

$\dfrac{dy}{dt} + 3x + 2y = 2e^{2t}$

Ans. $x + c_1e^t + c_2e^{-5t} - \dfrac{6}{7}e^{2t}$, $y = -c_1e^t + c_2e^{-5t} + \dfrac{8}{7}e^{2t}$

11. $\frac{dx}{dt} = 7x - y,$

$\frac{dy}{dt} = 2x + 5y$

Ans. $x = e^{6t}(c_1 \cos t + c_2 \sin t),\ y = e^{6t}\{(c_1 - c_2)\cos t + (c_1 - c_2)\sin t\}$

12. $\frac{dx}{dt} + \frac{dy}{dt} + 2x + y = 0;$

$\frac{dy}{dt} + 5x + 3y = 0$

Ans. $x = c_1 \cos t + c_2 \sin t;\ y = \frac{1}{2}[(c_2 - 3c_1)\cos t - (c_1 + 3c_2)\sin t]$

13. $\frac{dx}{dt} - x + y = t^2;$

$\frac{dx}{dt} - \frac{dy}{dt} + x = t$

Ans. $x = c_1 \cos t + c_2 \sin t + 3t;\ y = (c_1 + c_2)\sin t + c_1 + c_2 \cos t + t^2 + 3t - 3$

14. $t^2 \frac{d^2x}{dt^2} + t\frac{dx}{dt} + 2y = 0;$

$t^2 \frac{d^2y}{dx^2} - 2x = 0$

Ans. $x = t(c_1 \cos \log t + c_2 \sin \log t) + \frac{1}{t}(c_3 \cos \log t + c_4 \sin \log t);$

$y = t(c_1 \sin \log t - c_2 \cos \log t) + \frac{1}{t}(-c_3 \sin \log t + c_4 \cos \log t)$

15. $\frac{dx}{dt} + 2x = t;$

$\frac{dy}{dt} - 2x = \frac{1}{4}$

Ans. $x = c_1 e^{-2t} + \frac{1}{2}t - \frac{1}{4};\ y = c_1 e^{-2t} + c_2 + \frac{1}{2}t^2 - \frac{1}{2}t + \log t$

3.19 METIOD OF VARIATION OF PARAMETERS

(a) For second order. The method discussed earlier are to dealt with differential equation with constant coefficients. But the method of variation of parometers is quite general, since

$$y'' + P(x)\,y' + Q(x)y = R(x),$$

where $P(x)$, $Q(x)$ and $R(x)$ are the continuous functions of x.

Let $y = y_h(x) + y_p(x)$ be the general solution where $y_h = c_1y_1 + c_2y_2$ be the complementary function.

Then $y_p(x) = u(x)\,y(x) + v(x)\,y_2(x)$ be particular integral (P.I.)

where $u = -\int \frac{y_2R(x)}{W}dx,\ v = \int \frac{y_1R(x)}{W}dx$

and W (Wronskion of y_1 and y_2) $= \begin{vmatrix} y_1 & y_2 \\ y_1' & y_2' \end{vmatrix}$

Example 3.124 Solve :

$$\frac{d^2y}{dx^2} + 4y = \tan 2x.$$

Solution : The given equation may be written as

$$(D^2 + 4)\,y = \tan 2x, \text{ where } D \equiv \frac{d}{dx}$$

A.E. is $m^2 + 4 = 0 \Rightarrow m = \pm 2i$

$\therefore$ C.F. $= c_1 \cos 2x + c_2 \sin 2x$

Here we have, $y_1 = \cos 2x$, $y_2 = \sin 2x$ and $R(x) = \tan 2x$

$$\therefore \quad W = \begin{vmatrix} y_1 & y_2 \\ y_1' & y_2' \end{vmatrix} = \begin{vmatrix} \cos 2x & \sin 2x \\ -2\sin 2x & 2\cos 2x \end{vmatrix} = 2$$

Thus,
$$\text{P.I.} = -y_1 \int \frac{y_2 R(x)}{W}\,dx + y_2 \int \frac{y_1 R(x)}{W}\,dx$$

$$= -\cos 2x \int \frac{\sin 2x \tan 2x}{2}\,dx + \sin 2x \int \frac{\cos 2x \tan 2x}{2}\,dx$$

$$= -\frac{1}{2}\cos 2x \int (\sec 2x - \cos 2x)\,dx + \frac{1}{2}\sin 2x \int \sin 2x\,dx$$

$$= -\frac{1}{4}\cos 2x\,[\log(\sec 2x + \tan 2x) - \sin 2x] - \frac{1}{4}\sin 2x \cos 2x$$

$$= -\frac{1}{4}\cos 2x \log(\sec x + \tan 2x)$$

Hence, the complete solution is

$$y = c_1 \cos 2x + c_2 \sin 2x - \frac{1}{4}\cos 2x \log(\sec 2x + \tan 2x)$$

Example 3.125 Solve $y'' + y = \sec x$.

Solution : The given equation may be written as

$$(D^2 + 1)\,y = \sec x$$

A.E. is $m^2 + 1 = 0 \Rightarrow m = \pm i$

$\therefore$ C.F. $= c_1 \cos x + c_2 \sin x$

Here we have, $y_1 = \cos x$, $y_2 = \sin x$ and $R(x) = \sec x$

$$\therefore \quad W = \begin{vmatrix} y_1 & y_2 \\ y_1' & y_2' \end{vmatrix} = \begin{vmatrix} \cos x & \sin x \\ -\sin x & \cos x \end{vmatrix} = 1$$

Thus,
$$\text{P.I.} = -y_1 \int \frac{y_2 R(x)}{W}\,dx + y_2 \int \frac{y_1 R(x)}{W}\,dx$$

$$= -\cos x \int \frac{\sin x \sec x}{1}\,dx + \sin x \int \frac{\cos x \sec x}{1}\,dx$$

$$= -\cos x \int \frac{\sin x}{\cos x}\,dx + \sin x \int dx$$

$$= + \cos x \, [\log \cos x] + x \sin x$$

Hence, the complete solution is

$$y = c_1 \cos x + c_2 \sin x + \cos x \, (\log \cos x) + x \sin x$$

PROBLEM SET 3.21

Solve the following differential equations by method of variation of parameters :

1. $\dfrac{d^2y}{dx^2} + y = \operatorname{cosec} x$

Ans. $y = c_1 \cos x + c_2 \sin x - x \cos x + \sin x \log \sin x$

2. $\dfrac{d^2y}{dx^2} + y = x \sin x$

Ans. $y = c_1 \cos x + c_2 \sin x + \dfrac{x}{2} \sin x - \dfrac{x^2}{4} \cos x$

3. $\dfrac{d^2y}{dx^2} + y = \tan x$

Ans. $y = c_1 \cos x + c_2 \sin x - \cos x \log (\sec x + \tan x)$

4. $\dfrac{d^2y}{dx^2} - 6 \dfrac{dy}{dx} + 9y = \dfrac{e^{3x}}{x^2}$

Ans. $y = (c_1 + c_2 x) \, e^{4x} - e^{3x} \log x$

5. $\dfrac{d^2y}{dx^2} - 2 \dfrac{dy}{dx} + 2y = e^x \tan x$

Ans. $y = e^x (c_1 \cos x + c_2 \sin x) - e^x \cos x \log (\sec x + \tan x)$

6. $\dfrac{d^2y}{dx^2} + 2 \dfrac{dy}{dx} + y = e^{-x} \cos x$

Ans. $y = c_1 e^{-x} + c_2 x e^{-x} - e^{-x} \cos x$

4

Vector Calculus

4.1 SCALAR AND VECTOR FIELDS

It is well known that a scalar quantity is concerned only with magnitude, and direction has no role. A single real number is enough to give a complete representation of a scalar quantity. Thus, it can be conclude that s scalar quantity is nothing but a real number.

A physical quantity expressed as a continuous function, that assumes one or more definite values at every point of a region in space is termed as a point function and the region under consideration is called a field. The point functions may be termed as scalar or vector point function according to the quantity considered is a scalar or vector respectively.

Example 4.1 The temperature at any point within or on the surface of earth at a certain time defines a scalar field.

Example 4.2 The gravitational potential of a system of masses in a median also defines a scalar field.

Example 4.3 (1) The acceleration of a moving particle defines a vector field.

(2) $F(x, y, z) = x^2 yi + yz^2 j + z^2 zk$ defines a vector field.

4.2 SCALAR FUNCTION AND VECTOR FUNCTION

If to each element of a subset S of real numbers, we can associate a unique real number $f(t)$ by a definite rule, then this rule defines a scalar function of the scalar variable t.

Since $f(t)$ is a scalar quantity therefore f is a scalar function.

Similarly one can define a vector function.

If a unique vector $f(t)$ can be associated to each element t of a subset S of real numbers, then this rule defines a vector function of the scalar variable t. We observe that $f(t)$ is a vector quantity and thus f is a vector function.

Exanple 4.4 It is clear that $r = a\cos t\,\hat{i} + a\sin t\,\hat{j} + 0\,\hat{k}$ represents a vector equation of circle. In a similar manner $r = at^2 i + 2atj + 0k$ is a vector equation of a parabola.

4.3 LIMIT AND CONTINUITY OF A VECTOR FUNCTION

4.3.1 Limit

A vector function $f(t)$ tends to a limit l for t tending to t_0 if for a small positive number $\in$. There exists a positive number δ such that $| f(t) - l | < \in$, whenever $0 < | t - t_0 | < \delta$.

It may be written as $\lim\limits_{t \to t_0} f(t) = l$.

Exanple 4.5 **Let $r(t) = 9t^2 + 2t + 7$ is a vector function then find $\lim\limits_{t \to 2} r(t)$.**

Solution : It is given that $r(t) = 9t^2 + 2t + 7$ is a vector function. Then as for definition of limit of a vector function

$$\lim_{t \to 2} r(t) = \lim_{t \to 2} 9t^2 + 2t + 7$$
$$= 36 + 4 + 7 = 47$$

4.3.2 Continuity

A vector function $f(t)$ is said to be continuous at a point $t = a$, if :

(i) $f(a)$ is defined.

(ii) For a small positive number $\in (> 0)$, there exists a positive number $\delta > 0$ such that $| f(t) - f(a) | < \in$, whenever $| t - a | < \delta$.

Also a vector function $f(t)$ is said to be continuous over an interval if it is continuous at every point of the given interval.

Note : (i) The necessary and sufficient condition for a vector function $f(t)$ to be continous at a point $t = a$ is $\lim\limits_{t \to a} f(t) = f(a)$.

(ii) Also a function $f(t) = f_1(t)\hat{i} + f_2(t)\hat{j} + f_3(t)\hat{k}$ is continuous at $t = a$ if and only if $f_1(t)$, $f_2(t)$ and $f_3(t)$ are continuous.

Theorem: For vector functions $f(t)$, $g(t)$ and a scalar function $\phi(t)$ of a scalar variable t.

(i) $\lim\limits_{t \to t_0} [f(t) \pm g(t)] = \lim\limits_{t \to t_0} f(t) \pm \lim\limits_{t \to t_0} (t)$

(ii) $\lim\limits_{t \to t_0} [f(t) \,.\, g(t)] = [\lim\limits_{t \to t_0} f(t)] \,.\, [\lim\limits_{t \to t_0} g(t)]$

(iii) $\lim\limits_{t \to t_0} [f(t) \times g(t)] = \lim\limits_{t \to t_0} f(t) \times \lim\limits_{t \to t_0} g(t)$

(iv) $\lim\limits_{t \to t_0} [\phi(t)\ f(t)] = [\lim\limits_{t \to t_0} \phi(\tau)]\ [\lim\limits_{t \to t_0} f(t)]$

(v) $\lim\limits_{t \to t_0} |f(t)| = [\lim\limits_{t \to t_0} f(t)]$

4.4 DIFFERENTIATION OF A VECTOR FUNCTION WITH RESPECT TO A SCALAR

Definition: Let $r(t) = f(t)$ be a vector function depending on a single variable t. Then $\delta r = f(t + \delta t) - f(t)$ and

$$\frac{\delta r}{\delta t} = \frac{f(t + \delta t) - f(t)}{\delta t}$$

If $\lim\limits_{\delta t \to 0} . \frac{\delta r}{\delta t} = \lim\limits_{\delta t \to 0} . \frac{f(t + \delta t) - f(t)}{\delta t}$ exists, then we denote this limit as $\frac{dr}{dt}$ and it is called the derivative of the vector function r, with respect to scalar t. A vector function $r = f(t)$ is said to be differentiable if $\frac{dr}{dt}$ exists.

Differentiation formulae : If $a(t)$, $b(t)$ and $c(t)$ are differentiable vector functions of a scalar t and $\phi\,(t)$ is a differentiable scalar fnction of t, then

1. $\frac{d}{dt}(a + b) = \frac{da}{dt} + \frac{db}{dt}$

2. $\frac{d}{dt}(a \times b) = a \times \frac{db}{dt} + \frac{da}{dt} \times b$

3. $\frac{d}{dt}(\phi a) = \phi \frac{da}{dt} + \frac{d\phi}{dt} a$

4. $\frac{d}{dt}[a, b, c] = \left[\frac{da}{dt}\, bc\right] + \left[a \frac{db}{dt} c\right] + \left[ab \frac{dc}{dt}\right]$

5. $\frac{d}{dt}(a \times (b \times c)) = \frac{da}{dt} \times (b \times c) + a \times \left(\frac{db}{dt} \times c\right) + a \times \left(b \times \frac{dc}{dt}\right)$

Theorems on derivative of a constant vector:

Theorem1. A necessary and sufficient condition that a vector $\vec{A}(t)$ be of constant magnitude only is $\frac{d[\vec{A}]}{dt} = 0$

Proof : (i) Condition is necessary

Let $\vec{A}(t)$ be a constant magnitude, *i.e.*, $|\vec{A}|$ = constant

So, $$\frac{d[\vec{A}]}{dt} = 0$$

Since $|\vec{A}|^2 = \vec{A} . \vec{A}$

$\therefore$ $$\frac{d(|\vec{A}|^2)}{dt} = \frac{d\,(\vec{A}.\vec{A})}{dt}$$

or $$2\,|\vec{A}|\,\frac{d\,|\vec{A}|}{dt} = 2\,\vec{A} . \frac{d\vec{A}}{dt}$$

or $\qquad \vec{A}.\dfrac{d\vec{A}}{dt}=0 \qquad \left[\because \dfrac{d|\vec{A}|}{dt}=0\right]$

Therefore, if $\vec{A}(t)$ be of constant magnitude then

$$\vec{A}.\frac{d\vec{A}}{dt}=0$$

(ii) condition is sufficient

Let $\qquad \vec{A}.\dfrac{d\vec{A}}{dt}=0$

Since $\qquad \dfrac{d}{dt}(\vec{A}.\vec{A})=\dfrac{d\,|\vec{A}|^2}{dt} \qquad ...(1)$

or $\qquad 2\vec{A}.\dfrac{d\vec{A}}{dt}=2\,|\vec{A}|\,\dfrac{d\,|\vec{A}|}{dt}$

or $\qquad \vec{A}.\dfrac{d\vec{A}}{dt}=|\vec{A}|\,\dfrac{d\,|\vec{A}|}{dt} \qquad ...(2)$

from (1) and (2), we get

$$|\vec{A}|\,\frac{d\,|\vec{A}|}{dt}=0$$

But $\qquad |\vec{A}|\neq 0$

$\therefore \qquad \dfrac{d\,|\vec{A}|}{dt}=0$

Therefore $|\vec{A}|$ constant, *i.e.* $\vec{A}(t)$ be of constant magnitude.

Theorem 2. The necessary and sufficient condition for a vector $\vec{A}(t)$ to be constant is $\dfrac{d\vec{A}}{dt}=\vec{0}$

Proof. Condition is necesary :

Let $\vec{A}(t)$ be a constant vector.

When a small change δt occurs in scalar t, then also there will be no change in $A(t)$.

So $\delta\vec{A}=\vec{0}$

Now $\qquad \dfrac{d\vec{A}}{dt}=\lim\limits_{\delta t\to 0}\dfrac{\delta\vec{A}}{\delta t}$

$$=\lim_{\delta t\to 0}\frac{\vec{0}}{\delta t}=\vec{0} \qquad [\because \delta\vec{A}=0]$$

Thus, the derivative of a constant vector is zero vector.

Condition is sufficient:

Let $\qquad \dfrac{d\vec{A}}{dt}=\vec{0}$

and let $\qquad \vec{A}(t)=A_1(t)\,\hat{i}+A_2(t)\,\hat{j}+A_3(t)\,\hat{k}$

$$\frac{d\vec{A}}{dt}=\frac{d}{dt}(A_1\hat{i}+A_2\hat{j}+A_3\hat{k})$$

or $\qquad \vec{0}=\dfrac{dA_1}{dt}+\dfrac{dA_2}{dt}+\dfrac{dA_3}{dt} \qquad \left[\because \dfrac{d\vec{A}}{dt}=\vec{0}\right]$

$$\frac{dA_1}{dt} = 0, \frac{dA_2}{dt} = 0 \text{ and } \frac{dA_3}{dt} = 0$$

Thus, A_1, A_2 and A_3 are constant.

Therefore, $\vec{A}$ is a constant vector.

Lemma: Prove that $\vec{A} \times \frac{d\vec{A}}{dt} = a^2 \hat{A} \times \frac{d\hat{A}}{dt}$

where $\hat{A}$ is a unit vector along $\vec{A}$, and $a = |\vec{A}|$.

Proof. $\vec{A} = a\hat{A}$

$$\vec{A} \times \frac{d\vec{A}}{dt} = a\hat{A} \times \frac{d(A\hat{A})}{dt}$$

$$= a\hat{A} \times \left(\frac{da\,\hat{A}}{dt} + \frac{a\,d\hat{A}}{dt}\right)$$

$$= a\frac{da}{dt}\hat{A} \times \hat{A} + a^2\hat{A}\frac{d\hat{A}}{dt}$$

$$\vec{A} \times \frac{d\vec{A}}{dt} = a^2\hat{A}\frac{d\hat{A}}{dt} \qquad [\because \hat{A} \times \hat{A} = 0]$$

Theorem 3. A necessary and sufficient condition that a vector $\vec{A}(t)$ be of constant direction only $\vec{A} \times \frac{d\vec{A}}{dt} = \vec{0}$.

Proof. condition is necessary:

Let $\vec{A}(t)$ be of constant direction only.

Therefore $\hat{A}$ is a constant vector.

$$\frac{d\vec{A}}{dt} = \vec{0} \qquad ...(1)$$

By above lemma, we have

$$\vec{A} \times \frac{d\vec{A}}{dt} = a^2\hat{A} \times \frac{d\hat{A}}{dt} \qquad ..(2)$$

From (1) and (2), we have

$$\vec{A} \times \frac{d\vec{A}}{dt} = \vec{0}$$

So, $\vec{A}(t)$ be a constant direction only, then $\vec{A} \times \frac{d\vec{A}}{dt} = \vec{0}$

Condition is sufficient:

Let $$\vec{A} \times \frac{d\vec{A}}{dt} = \vec{0} \qquad ...(1)$$

From above lemma,

$$\vec{A} \times \frac{d\vec{A}}{dt} = a^2\hat{A} \times \frac{d\hat{A}}{dt} \qquad ...(2)$$

From (1) and (2)

$$a^2\hat{A} \times \frac{d\hat{A}}{dt} = \vec{0}$$

or $$\hat{A} \times \frac{d\hat{A}}{dt} = \vec{0} \qquad ...(3) \qquad [\because a \neq 0]$$

But $\hat{A}$ is a unit vector,

$$\therefore \quad \hat{A} \times \frac{d\hat{A}}{dt} = \vec{0} \qquad ...(4)$$

$[\because$ The magnitude of a vector $\vec{r}$ is a constant then $\vec{r} \cdot \frac{d\vec{r}}{dt} = 0]$

From equations (3) and (4), the vectors $\hat{A}$ and $\frac{d\hat{A}}{dt}$ are both parallel and vertical to each other and it is possible only when $\hat{A} \cdot \frac{d\hat{A}}{dt} = \vec{0}$, *i.e.*, $\hat{A}$ si a constant vector.

Since the direction of and are same, therefore direction of is also constant.

Thus, if then the direction of is constant.

Example 4.6 **If $\vec{r} = \sin t\,\hat{i} + \cos t\,\hat{j} + t\hat{k}$, find**

(i) $\frac{d\vec{r}}{dt}$ **(ii)** $\left|\frac{d\vec{r}}{dt}\right|$ **(iii)** $\frac{d^2r}{dt^2}$ **(iv)** $\left|\frac{d^2r}{dt^2}\right|$

Solution : As we know that for a vector $\vec{r} = r_1(t)\hat{i} + r_2(t)\hat{j} + r_3(t)\hat{k}$

$$\frac{d\vec{r}}{dt} = \frac{dr}{dt}\hat{i} + \frac{dr_2}{dt}\hat{j} + \frac{dr_3}{dty}\hat{k}$$

and
$$\frac{d^2\vec{r}}{dt^2} = \frac{d^2r_1}{dt^2}\hat{i} + \frac{d^2r_2}{dt^2}\hat{j} + \frac{d^2r_3}{dt^2}\hat{k}$$

Thus, the given vector $\vec{r} = \sin t\,\vec{i} + \cos t\,\hat{j} + t\,\hat{k}$

(i) $\frac{d\vec{r}}{dt} = \cos t\,\hat{i} - \sin t\,\hat{j} + \hat{k}$

(ii) $\left|\frac{d\vec{r}}{dt}\right| = \sqrt{\cos^2 t + \sin^2 t + 1} = \sqrt{2}$

(iii) $\frac{d\vec{r}}{dt} = -\sin t\,\hat{i} - \cos t\,\hat{j} + 0\,.\,\hat{k}$

(iv) $\left|\frac{d^2\vec{r}}{dt^2}\right| = \sqrt{\sin^2 t + \cos^2 t + 0} = \sqrt{1} = 1$

Example 4.7 **If $\vec{r} = (\cos a\hat{t})i + (\sin at)\hat{j}$, where a is a constant, show that**

$$\vec{r} \times \frac{d\vec{r}}{dt} = a\hat{k}$$

Solution : As per definition of $\frac{d\vec{r}}{dt}$, we have

$$\frac{d\vec{r}}{dt} = (-a \sin a\hat{t})i + (a \cos at)\hat{j}$$

$$\therefore \quad \vec{r} \times \frac{d\vec{r}}{dt} = [(\cos at)\hat{i} + (\sin at)\hat{j}] \times [(-a \sin at)\hat{i} + (a \cos at)\hat{j})]$$

$$= \begin{vmatrix} \hat{i} & \hat{j} & \hat{k} \\ \cos at & \sin at & 0 \\ -a\sin at & a\cos at & 0 \end{vmatrix} = a\hat{k}$$

Example 4.8 **If $\frac{d\vec{a}}{dt} = \vec{c} \times \vec{a}$, $\frac{d\vec{b}}{dt} = \vec{c} \times \vec{b}$, show that**

$$\frac{d}{dt}(\vec{a} \times \vec{b}) = \vec{c} \times (\vec{a} \times \vec{b}).$$

Solution : Since we know that

$$\frac{d}{dt}(\vec{a} \times \vec{b}) = \vec{a} \times \frac{d\vec{b}}{dt} + \frac{d\vec{a}}{dt} \times \vec{b}$$

$$= \vec{a} \times (\vec{c} \times \vec{b}) + (\vec{c} \times \vec{a}) \times \vec{b} \qquad \left(\because \frac{d\vec{b}}{dt} = \vec{c} \times \vec{b} \text{ and } \frac{d\vec{a}}{dt} = \vec{c} \times \vec{a}\right)$$

$$= [(\vec{a}.\vec{b})\vec{c} - (\vec{a}.\vec{c})\vec{b}] - [(\vec{b}.\vec{a})\vec{c} - (\vec{b}.\vec{c})\vec{a}]$$

$$= (\vec{a}.\vec{b})\vec{c} - (\vec{a}.\vec{c})\vec{b} - (\vec{a}.\vec{b})\vec{c} + (\vec{b}.\vec{c})\vec{a}$$

$$= (\vec{b}.\vec{c})\vec{a} - (\vec{a}.\vec{c})\vec{b}$$

$$= \vec{c} \times (\vec{a} \times \vec{b})$$

Example 4.9 **Find $\frac{d\vec{v}(t)}{dt}$ for the following function $v(t) = (\cos t + t^2)(t\hat{i} + \hat{j} + 2\hat{k})$.**

Solution : As per the definition of $\frac{d\vec{v}}{dt}$, we have

$$\frac{d\vec{v}}{dt} = \left[\frac{d}{dt}(\cos t + t^2)\right][t\hat{i} + \hat{j} + 2\hat{k}] + [\cos t + t^2]\frac{d}{dt}[t\hat{i} + \hat{j} + 2\hat{k}]$$

$$= (-\sin t + 2t)(t\hat{i} + \hat{j} + 2\hat{k}) + (\cos t + t^2)(\hat{i})$$

$$= (3t^2 - t\sin t + \cos t)\hat{i} + (2t - \sin t)\hat{j} + (2t - \sin t)\hat{k}$$

4.5 TANGENT TO A CURVE

Let the curve C is given by $r(t)$ and the point P and Q are represented by t and $t + \delta t$ respectively. Let L be the staight line joining points P and Q, then the direction of line L is given by the following vector.

$$\frac{1}{\Delta t}[r(t + \Delta t) - r(t)]$$

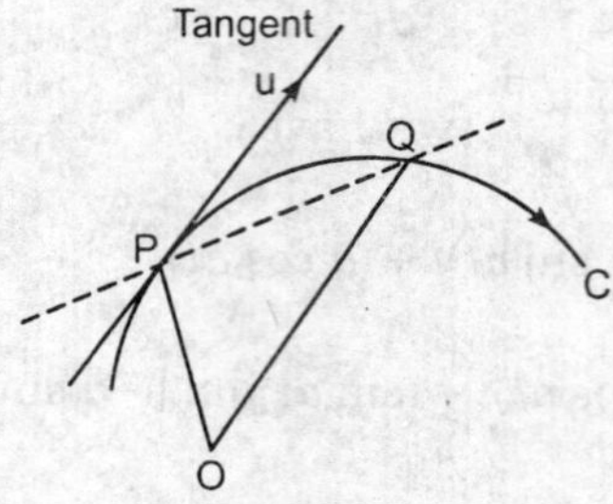

Fig. 4.1 Tangent of a curve

Also if limit of this vector exists then it becomes the derivative (provided $r(t)$ is differentiable) *i.e.*,

$$r'(t) = \lim_{\Delta t \to 0} \frac{1}{\Delta t} [r(t + \Delta t) - r(t)]$$

If $r'(t) \neq 0$; we call $r'(t)$ a tangent vector of C at P because it shows the direction of tangent. The unit vector corresponding to the tangent vector is given by

$$u = \frac{1}{|r'|} r'$$

Partial Derivative of Vectors : Suppose A is a vector depending on more than one scalar variable. Let $A = f(x, y, z,)$ *i.e.* A be a function of three scalar variables x, y and z. The partial derivative of A with respect to x is difined as

$$\frac{\partial A}{\partial x} = \lim_{\delta x \to 0} \frac{f(x + \delta x, y, z) - f(x, y, z)}{\delta x}$$

If this limit exists, we can say that the $\frac{\partial A}{\partial x}$ is the ordinary dervativee of A with respect to x provided the other variable y and z regarded as constants. Similarly, we may define the partial derivatives $\frac{\partial A}{\partial y}$ and $\frac{\partial A}{\partial z}$.

If $A = f(x, y, z)$ the total differential dA is given by

$$dA = \frac{\partial A}{\partial x} dx + \frac{\partial A}{\partial y} dy + \frac{\partial A}{\partial z} dz$$

The Vector differential operator (∇) **:** The Vector differential operator ∇ (read as del or nabla) is defined as

$$\nabla = \frac{\partial}{\partial x} \hat{i} + \frac{\partial}{\partial y} \hat{j} + \frac{\partial}{\partial z} \hat{k}$$

$$= \hat{i} \frac{\partial}{\partial x} + \hat{j} \frac{\partial}{\partial y} + \hat{k} \frac{\partial}{\partial z}$$

and operates distributively.

The vector operator ∇ possesses properties like ordinary vectors. The symbols $\frac{\partial}{\partial x}, \frac{\partial}{\partial y}, \frac{\partial}{\partial z}$ can be treated as its cotonents along, i, j, k.

Example 4.10 **Find the tangent vector to the curve whose parametric representation is $x = 2 \cos t$, $y = 2 \sin t$, $z = t$, $-\pi \leq t \leq \pi$. Hence find the unit tangent vector.**

Solution : The position vector of a point on the curve is

$\vec{r} = 2 \cos t \, \hat{i} + 2 \sin t \, \hat{j} + t\hat{k}$

$\because$ The tangent vector is

$$\vec{r} = \frac{d\vec{r}}{dt}$$

$$\therefore \quad \frac{d\vec{r}}{dt} = -2 \sin t \, \hat{i} + 2 \cos \hat{j} + \hat{k} \qquad \text{(The tangent vector)}$$

$$\because \quad \left| \frac{d\vec{r}}{dt} \right| = \sqrt{4 \sin^2 t + 4 \cos^2 t + 1}$$

$= \sqrt{4 + 1}$

$= \sqrt{5}$

Thus, the unit tangent vector is

$$\hat{r'} = \frac{-2 \sin t \,\hat{i} + 2 \cos \hat{j} + \hat{k}}{\sqrt{5}}$$

4.6 ARC LENGTH

To define are legth, first let us discuss the length of curve.

4.6.1 Length of Curve

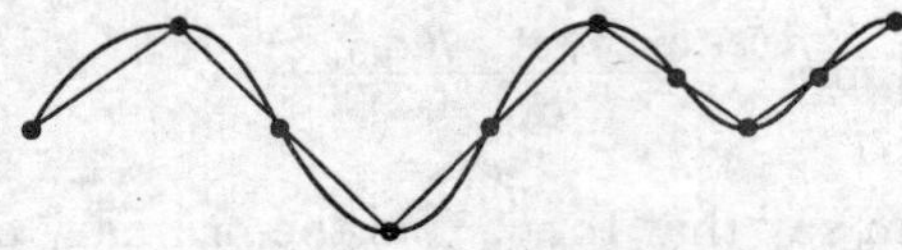

Fig. 4.2 Length of a curve

The length l of a curve is the limit of the lengths of broken lines of n chords for $n = 1, 2, \ldots$ larger n.

Let $r(t), a \le t \le b$ represents the cureve C. The interval $a \le t \le b$ is divided as

$$t_0(= a), t_1, \ldots., t_{n-1} (= b)$$

where $t_0 < t_1 < t_2 < \ldots < t_n$ for each $n = 1, 2, \ldots.$ In accordance with the property of intetrv $[a,b]$,the given line can be broken in to chords with end points r $(t_0) \ldots r(t_n)$.

This division of length is done arbitrarily but so that the greatest $| \Delta t_m | = | t_m - t_{m-1} |$ approaches 0 as $n \to \infty$. Using Pythagorean theorem one can find the length $l_1, l_2, \ldots$ of these chords. If $r(t)$ possesses a continuous derivative $r'(t)$, it can be proved that the sequence $l_1, l_2. \ldots$ is convergent (*i.e.*, has a linit). Also it is independent of the particular choice of the subdivision of curve C. This limit of sequence can be obtained as below :

$$l = \int_a^b r'. r' \, dt \qquad \ldots(1)$$

$$\left(r' = \frac{dr}{dt} \right)$$

This l is known as length of curve C and the curve C is called ractifiable.

4.6.2 Arc Length of a Curve

It has been proved that the length l of a curve C is constant and is a positive number. But if the fixed upper limit of integral in equation (1) is replaced by a variable t, the integral becomes a function of t and is denoted as $s(t)$ and called the are length of C. Thus

$$ss(t) = \int_a^t \sqrt{r'. r'} \, dt$$

Example 4.11 **Find the length of the following curve.**

$$\boldsymbol{r(t) = a\cos\hat{t} + a\sin t\hat{j} + ct\,\hat{k},\ 0 \le t \le 2\pi.}$$

Solution: It is given that

$$r(t) = x(t)\,\hat{i} + y(t)\,\hat{j} + z(t)\,\hat{k}$$
$$= a\cos t\, i + a\sin t\, j + ctk$$

$\therefore$ $x'(t) = -a\,sin\,t,\ y'(t) = a\cos t$ and $z'(t) = c$

Therefore, l= arc length $= \int_0^{2\pi} [a^2 \sin^2 t + a^2\cos^2 t + c^2]^{1/2}\, dt$

$$= \int_0^{2\pi} [a^2 + c^2]^{1/2}\, dt$$
$$= [t\,(a^2 + c^2)^{1/2}]_0^{2\pi}$$
$$= 2\pi\,(a^2 + c^2)^{1/2}$$

Example 4.12 **Find the length of the curve**

$$\boldsymbol{r(t) = t\,\mathrm{sin}t\,\hat{i} + t\cos t\,\hat{j}\,\sqrt{3}\,t\,\hat{k}\,.\ 0 \le t \le \pi}$$

Solution : It is given that

$$r(t) = x\,(t)\,\hat{i} + y\,(t)\,\hat{j} + z\,(t)\,\hat{k} = t\sin t\,\hat{i} + \cos t\,\hat{j} + \sqrt{3}\,t\,\hat{k}.$$

Thus $x(t) = \sin t,\ y(t) = t\cos t$ and $z\,(t) = \sqrt{3}\,t$ and

$x'(t) = \sin t + t\cos t,\ y'(t) = \cos t - t\sin t.\ z'(t) = \sqrt{3}$

Therefore, l= length of arc

$$= \int_0^{\pi} [(\sin t + t\cos t)^2 + (cos\,t - t\sin t)^2 + (\sqrt{3})^2]\, dt$$

$$= \int_0^{\pi} [\sin^2 t + t^2\cos^2 t + 2t\sin t\cos t + \cos^2 t + t^2\sin^2 t - 2t\sin t\cos t + 3]\, dt$$

$$= \int_0^{\pi} [1 + t^2 + 3]\, dt = \int_0^{\pi} (t^2 + 4)\, dt$$

$$= \left[\frac{t^3}{3} + 4t\right]_0^{\pi} = \frac{\pi^3}{3} + 4\pi$$

4.7 VELOCITY AND ACCELERATION

Let a particle is moving along the curve $\vec{r}(t)$ and at any time t, the particle is at a point P whose position vector is $\vec{r}$. If the particle reached to the point Q, whose position vector is $\vec{r} + \delta\vec{r}$ in time interval δt.

$\therefore$ $\delta\vec{r}$ is the displacement of the particle in the interval δt.

$\therefore$ The average velocity of the particle in the interval δt is $\frac{\delta r}{\delta t}$.

If $\vec{v}$ represents the velocity vector of the particle at the point P.

Then, $$\vec{v} = \lim_{\delta t \to 0} \frac{\delta\vec{r}}{\delta t} = \frac{d\vec{r}}{dt}$$

$\because$ $\dfrac{d\vec{r}}{dt}$ is the vector in the direction of tangent at P.

$\therefore$ $\vec{v}$, the velocity vector is always in the direction of the tangent.

Further if $\delta\vec{v}$ is the change in velocity during the time interval δt. Then $\dfrac{\delta v}{\delta t}$ represents the vector, that gives an average accelaration of the particle. If the acceleration vector of the particle is denoted by $\vec{a}$, then

$$\vec{a} = \lim_{\delta t \to 0} \frac{\delta\vec{v}}{\delta t} = \frac{d\vec{v}}{dt}$$

$$\Rightarrow \quad \vec{a} = \frac{d^2\vec{r}}{dt^2} \qquad \left(\because \ \vec{v} = \frac{d\vec{r}}{dt}\right)$$

Example 4.13 **The position vector of a moving particle is given by $\vec{r}(t) = t^3\hat{i} + t^2\hat{j} + t^4\hat{k}$. Determine the velocity speed and acceleration of the particle.**

Solution : We are given the position vector of the particle as

$\vec{r}(t) = t^3\hat{i} + t^2\hat{j} + t^4\hat{k}$

$\therefore$ $\quad \vec{r}'(t) = 3t^2\hat{i} + 2t\hat{j} + 4t^3\hat{k}$

and $\quad \hat{r}''(t) = 6t\hat{i} + 2\hat{j} + 12t^2\hat{k}$

Thus, the velocity $\vec{v}(t) = \vec{r}'(t)$

$$= 3t^2\hat{i} + 2t\hat{j} + 4t^3\hat{k}$$

$$\text{Speed} = |\ \vec{v}'(t)\ |$$

$$= \sqrt{9t^4 + 4t^2 + 16t^6}$$

$$\text{Acceleration} = \vec{a}(t)$$

$$= 6t\hat{i} + 2\hat{j} + 12t^2\hat{k}$$

Example 4.14 **Find the velocity and acceleration of a particle whose position vector is $(\sin t - \cos t)\hat{i} + (\sin t + \cos t)\hat{j} + t^2\hat{k}$.**

Solution : The position vector of the moving particle is given is

$$\vec{r}(t) = (\sin t - \cos t)\,\hat{i} + (\sin t + \cos t)\,\hat{j} + t^2\hat{k}$$

$$\therefore \quad \frac{d\vec{r}}{dt} = (\cos t + \sin t)\,\hat{i} + (\cos t - \sin t)\,\hat{j} + 2t\hat{k}$$

$$\text{and} \quad \frac{d^2\vec{r}}{dt^2} = (-\sin t + \cos t)\,\hat{i} + (-\sin t - \cos t)\,\hat{j} + 2\hat{k}$$

Thus, the velocity of the particle $= \vec{v} = \dfrac{d\vec{r}}{dt}$

$$= (\cos t + \sin t)\hat{i} + (\cos t - \sin t)\hat{j} + 2t\hat{k}$$

and the acceleration $= \hat{a}(t) = \dfrac{d^2r}{dt^2}$

$$= (-\sin t + \cos t)\hat{i} - (\sin t + \cos t)\hat{j} + 2\hat{k}$$

PROBLEM SET 4.1

1. If $\vec{r}(t) = (t+1)\hat{i} + (t^2+t+1)\hat{j} + (t^3+t^2+t+1)\hat{k}$, find $\dfrac{d\vec{r}}{dt}, \dfrac{d^2\vec{r}}{dt^2}$.

Ans. $\dfrac{d\vec{r}}{dt} = \hat{i} + (2t+1)\hat{j} + (3t^2+2t+1)\hat{k}$, $\dfrac{d^2\vec{r}}{dt} = 2\hat{j} + (6t+2)\hat{k}$.

2. If $\vec{a}, \vec{b}$ are constant vectors, w is a constant and $\vec{r}$ is vector function of the scalar variable t given by $\vec{r} = \vec{a}\cos wt + \vec{b}\sin wt$, show that

(i) $\dfrac{d^2\vec{r}}{dt^2} + w^2\vec{r} = 0$

(ii) $\vec{r} \times \dfrac{d\vec{r}}{dt} = w\vec{a} \times \vec{b}$

3. If $\vec{u}(t) = 5t^2\hat{i} + t\hat{j} + t^3\hat{k}$ and $f(t) = \sin t$, find $\dfrac{d}{dt}[f(t) \,.\, u(t)]$.

Ans. $(5t^2 \cos t + 10t \sin t)\hat{i} + (t\cos t + \sin t)\hat{j} + (t^3 \cos t + 3t^2 \sin t)\hat{k}$

4. If $\vec{u}(t) = [\sin(2t)\hat{i} - \cos(2t)\hat{j} + t\hat{k}]$ and $\vec{v}(t) = [\cos(2t)\hat{i} - \sin(2t)\hat{j} + t^2\hat{k}]$. Find

$$\frac{d}{dt}[u(t) \,.\, v(t)].$$

Ans. $4\cos 4t + 3t^2$

5. Find the equation of the tangent line to the following curves at the indicated point.

(i) $\vec{r}(t) = t\hat{i} + 2t^2\hat{j} + 3t^3\hat{k}\,;\ t = 2$

(ii) $\vec{r}(t) = \sin t\,\hat{i} + \cos t\,\hat{j} + t\,\hat{k}\,;\ t = \dfrac{\pi}{4}$

(iii) $\vec{r}(t) = t\hat{i} + e^t\hat{j} + \hat{k}\,;\ t = 1$

Ans. (i) $\hat{i} + 8\hat{j} + 36\hat{k}$, (ii) $\dfrac{1}{\sqrt{2}}\hat{i} + \dfrac{1}{\sqrt{2}}\hat{j} + \hat{k}$, (iii) $\hat{i} + e\hat{j}$

6. If $\vec{r}(t)$ is a vector function of a scalar variable t and $\vec{a}$ is a constant vector, then differentiate the following with respect to t.

(i) $\vec{r} \times \vec{a}$

(ii) $\vec{r} \times \dfrac{d\vec{r}}{dt}$

(iii) $\dfrac{\vec{r} \times \vec{a}}{\vec{r} \,.\, \vec{a}}$

Ans. (i) $\dfrac{d\vec{r}}{dt} \times \vec{a}$, (ii) $\vec{r} \times \dfrac{d^2\vec{r}}{dt^2}$, (iii) $\dfrac{\dfrac{d\vec{r}}{dt} \times \vec{a}}{\vec{r} \,.\, \vec{a}} - \dfrac{\dfrac{d\vec{r}}{dt} \,.\, \vec{a}}{(\vec{r} \,.\, \vec{a})^2}(\vec{r} \times \vec{a})$

7. If $\vec{r} = e^{at}\vec{a} + e^{-at}\vec{b}$, where $\vec{a}$ and $\vec{b}$ are constant vectors, show that

$$\frac{d^2\vec{r}}{dt^2} - a^2\vec{r} = 0.$$

8. Find the length of the following curves :

(i) $\vec{r}(t) = \frac{t^2}{2}\hat{i} + \frac{t^3}{3}\hat{k}$; $0 \le t \le 2$

(ii) $\vec{r}(t) = t\,\hat{i} + \frac{t^2}{2}\hat{j}$; $0 \le t \le 1$

(iii) $\vec{r}(t) = \cos\hat{i} + \sin t\,\hat{j} + 3t\,\hat{k}$; $-2\pi \le t \le 2\pi$

(iv) $r(t) = a\cos t\,\hat{i} + a\sin t\,\hat{j}$; $0 \le t \le 2\pi$

(v) $r(t) = a\cos^3 t\,\hat{i} + a\sin^3 t\,\hat{j}$; $0 \le t \le \frac{\pi}{2}$

Ans. (i) $\left(\frac{5\sqrt{5}-1}{3}\right)$ (ii) $\frac{[\sqrt{2} + \log(1+\sqrt{2})]}{2}$ (iii) $4\pi\sqrt{10}$ (iv) $2\pi a$ (v) $\frac{3a}{2}$

9. A particle is moving along the curve $x = 4\cos t$, $y = 4\sin t$, $z = 6t$, find the velocity and acceleration at $t = 0$ and $t = \frac{\pi}{2}$.

Ans. Veclocity $= 4\hat{j} + 6\hat{k}$, acceleration $= -4\hat{i}$ and velocity $= -4\hat{i} + 6\hat{k}$, acceleration $= -4\hat{j}$.

10. Find the velocity and acceleration for a particle moving along the curve $\vec{r}(t) = e^{te} + 2\cos 3t\,\hat{j} + 2\sin 3t\,\hat{k}$.

Ans. $\vec{v}(t) = -e^{-t}\,\hat{i} - 6\sin 3t\,\hat{j} + 6\cos 3t\,\hat{k}$, $\vec{a}(t) = e^{-t}\,\hat{i} - 18\cos 3t\,\hat{j} - 18\sin 3t\,\hat{k}$

4.8 GRADIENT OF A SCALAR FIELD

Let $f(x, y, z)$ be defined and differentiable at each point (x, y, z) in a certain region of space (*i.e.*, defines a differentiable scalar field). Then the gradient of $f(x, y, z)$ is denoted and defined by

$$\text{grad } f = \frac{\partial f}{\partial x}\hat{i} + \frac{\partial f}{\partial y}\hat{j} + \frac{\partial f}{\partial z}\hat{k} = \left(\frac{\partial}{\partial x}\hat{i} + \frac{\partial}{\partial y}\hat{j} + \frac{\partial}{\partial z}\hat{k}\right) f = \nabla f$$

Thus, the gradient of scalar field f can be obtained, operator ∇ (del) cver the field f.

It should be noted that ∇f is a vector whose three successive components are $\frac{\partial f}{\partial x}, \frac{\partial f}{\partial y}$ and $\frac{\partial f}{\partial z}$. Thus, the gradient of a scalar field is vector field. If f is a scalar point function, then ∇f is a vector point function.

Example 4.15 **Find the gradient of $f(x, y) = x^2 + y^2 - 4xy$ at (1, 1).**

Solution : We have $f(x, y) = x^2 + y^2 - 4xy$ and the point (1, 1)

The gradient of $f(x, y) = \nabla f(x, y)$

$$= \nabla f(x, y) = \hat{i}\frac{\partial f}{\partial x} + f\frac{\partial f}{\partial y}$$

$$= \hat{i}\,(2x - 4y) + \hat{j}(2y - 4x)$$

At (1, 1) grad $f = \hat{i}(2-4) + \hat{j}(2-4)$

$$= -2\hat{i} - 2j$$

4.9 IMPORTANT RESULTS INVOLVING GRADIENT

1. Gradient of the sum of two scalar point functions : If f and g are two scalar point functions, then

$$\text{grad } (f+g) = \text{grad } f + \text{grad } g \text{ or } \nabla(f+g) = \nabla f + \nabla g$$

Proof : We have grad $(f+g) = \nabla(f+g)$

$$= \left(\hat{i}\frac{\partial}{\partial x} + \hat{j}\frac{\partial}{\partial y} + \hat{k}\frac{\partial}{\partial z}\right)(f+g)$$

$$= \hat{i}\frac{\partial}{\partial x}(f+g) + \hat{j}\frac{\partial}{\partial y}(f+g) + \hat{k}\frac{\partial}{\partial z}(f+g)$$

$$= \hat{i}\frac{\partial f}{\partial x} + \hat{i}\frac{\partial g}{\partial x} + \hat{j}\frac{\partial f}{\partial y} + \hat{j}\frac{\partial g}{\partial y} + \hat{k}\frac{\partial f}{\partial z} + \hat{k}\frac{\partial g}{\partial z}$$

$$= \left(\hat{i}\frac{\partial f}{\partial x} + \hat{j}\frac{\partial f}{\partial y} + \hat{k}\frac{\partial f}{\partial z}\right) + \left(\hat{i}\frac{\partial g}{\partial x} + \hat{j}\frac{\partial g}{\partial y} + \hat{k}\frac{\partial g}{\partial z}\right)$$

$$= \left(\hat{i}\frac{\partial}{\partial x} + \hat{j}\frac{\partial}{\partial y} + \hat{k}\frac{\partial}{\partial z}\right)f + \left(\hat{i}\frac{\partial}{\partial x} + \hat{j}\frac{\partial}{\partial y} + \hat{k}\frac{\partial}{\partial z}\right)g = \nabla f + \nabla g$$

Thus, grad $(f+g)$ = grad f + grad g

Similarly, we can prove that

$$\nabla(f-g) = \nabla f - \nabla g.$$

2. Gradient of a constant : The necessary and sufficient condition for a scalar point function f to be constant is that $\nabla f = 0$.

Proof : If $f(x, y)$ is constant then $\frac{\partial f}{\partial x} = 0, \frac{\partial f}{\partial y} = 0, \frac{\partial f}{\partial z} = 0$

Therefore, grad $f = \hat{i}\frac{\partial f}{\partial x} + \hat{j}\frac{\partial f}{\partial y} + \hat{k}\frac{\partial f}{\partial z} = 0\hat{i} + 0\hat{j} + 0\hat{k} = 0$. Hence the condition is necessary.

Conversely : Let grad $f = 0$. Then $\hat{i}\frac{\partial f}{\partial x} + \hat{j}\frac{\partial f}{\partial y} + \hat{k}\frac{\partial f}{\partial z} = 0$

Therefore, $\frac{\partial f}{\partial x} = 0, \frac{\partial f}{\partial y} = 0, \frac{\partial f}{\partial z} = 0$

$\therefore$ f must be independent of x, y, z and f must be constant.

Hence, the condition is sufficient.

3. Gradient of the Product of two scalar point function : If f and g are scalar point functions, then,

$$\text{grad } fg = f \text{ grad } g + g \text{ grad } f \text{ or } \nabla(fg) = f\nabla g + g\Delta f$$

Proof : We have $\nabla(fg) = \left(\hat{i}\frac{\partial}{\partial x} + \hat{j}\frac{\partial}{\partial y} + \hat{k}\frac{\partial}{\partial z}\right)(fg)$

$$= \hat{i}\frac{\partial}{\partial x}(fg) + \hat{j}\frac{\partial}{\partial y}(fg) + \hat{k}\frac{\partial}{\partial z}(fg)$$

$$= \hat{i}\left(f\frac{\partial g}{\partial x} + g\frac{\partial f}{\partial x}\right) + \hat{j}\left(f\frac{\partial g}{\partial y} + g\frac{\partial f}{\partial y}\right) + \hat{k}\left(f\frac{\partial g}{\partial z} + g\frac{\partial f}{\partial z}\right)$$

$$= f\left(\hat{i}\frac{\partial g}{\partial x}+\hat{j}\frac{\partial g}{\partial y}+\hat{k}\frac{\partial g}{\partial z}\right)+g\left(\hat{i}\frac{\partial f}{\partial x}+\hat{j}\frac{\partial f}{\partial y}+\hat{k}\frac{\partial f}{\partial z}\right)=f\nabla g+g\nabla f$$

or grad $fg = f$ grad $g + g$ grad f

4. Gradient of the quotient of two scalar functions : If f and g are two scalar point functions, then

$$\nabla\left(\frac{f}{g}\right)=\frac{g\nabla f-f\nabla g}{g^2}$$

Proof : We have $\nabla\left(\frac{f}{g}\right)=\left(\hat{i}\frac{\partial}{\partial x}+\hat{j}\frac{\partial}{\partial y}+\hat{k}\frac{\partial}{\partial z}\right)\left(\frac{f}{g}\right)$

$$=\hat{i}\frac{\partial}{\partial x}\left(\frac{f}{g}\right)+\hat{j}\frac{\partial}{\partial y}\left(\frac{f}{g}\right)+\hat{k}\frac{\partial}{\partial z}\left(\frac{f}{g}\right)$$

But, $$\frac{\partial}{\partial x}\left(\frac{f}{g}\right)=\frac{g\frac{\partial f}{\partial x}-f\frac{\partial g}{\partial x}}{g^2},\ \frac{\partial}{\partial y}\left(\frac{f}{g}\right)=\frac{g\frac{\partial f}{\partial y}-f\frac{\partial g}{\partial y}}{g^2},\ \frac{\partial}{\partial z}\left(\frac{f}{g}\right)=\frac{g\frac{\partial f}{\partial z}-f\frac{\partial g}{\partial z}}{g^2}$$

$\therefore$ $$\nabla\left(\frac{f}{g}\right)=\frac{1}{g^2}\left\{\hat{i}\left(g\frac{\partial f}{\partial x}-f\frac{\partial g}{\partial x}\right)+\hat{j}\left(g\frac{\partial f}{\partial y}-f\frac{\partial g}{\partial y}\right)+\hat{k}\left(g\frac{\partial f}{\partial z}-f\frac{\partial g}{\partial z}\right)\right\}$$

$$=\frac{1}{g^2}\left\{g\left(\hat{i}\frac{\partial f}{\partial x}+\hat{j}\frac{\partial f}{\partial y}\hat{k}\frac{\partial f}{\partial z}\right)-f\left(\hat{i}\frac{\partial g}{\partial x}+\hat{j}\frac{\partial g}{\partial y}+\hat{k}\frac{\partial g}{\partial z}\right)\right\}$$

$$=\nabla\left(\frac{f}{g}\right)=\frac{1}{g^2}(g\nabla f-f\nabla g)$$

Example 4.16 **If $f(x, y, z) = 4xy - y^2z^3$. Find grad f at the point (2, 1, –2).**

Solution : We have

$$\text{grad } f=\nabla f=\left(\hat{i}\frac{\partial}{\partial x}+\hat{j}\frac{\partial}{\partial y}+\hat{k}\frac{\partial}{\partial z}\right)(4xy-y^2z^2)$$

$$=\hat{i}\frac{\partial}{\partial x}(4xy-y^2z^3)+\hat{j}\frac{\partial}{\partial y}(4xy-y^2z^3)+\hat{k}\frac{\partial}{\partial z}(4xy-y^2z^3)$$

$$=\hat{i}(4y)+\hat{j}(4x-2yz^3)+\hat{k}(-3y^2z^2)$$

Putting $x = 2, y = 1, z = -2$, we get

$$\nabla f=4(1)\hat{i}+\{4(2)-2(1)(-2)^3\}\hat{j}-3(1)^2(-2)^2\hat{k}$$

$$=4\hat{i}+24\hat{j}-12\hat{k}$$

Example 4.17 **If $\vec{r}=xi+yj+zk$, then show that**

(i) $\nabla(\vec{a}.\vec{r})=\vec{a}$, where $\vec{a}$ is constant vector

(ii) $\nabla\left(\frac{1}{r}\right)=-\frac{\vec{r}}{r^3}$

(iii) $\nabla r^n = nr^{n-2}\vec{r}$

Solution : (i) Let $\vec{a}=a_1\hat{i}+a_2\hat{j}+a_3\hat{k}$, where a_1, a_2 and a_3 are constants, then

$$\vec{a}\cdot\vec{r} = (a_1\hat{i} + a_2\hat{j} + a_3\hat{k})\cdot(x\hat{i} + y\hat{j} + z\hat{k})$$

$$\vec{a}\cdot\vec{r} = a_1x + a_2y + a_3z$$

$$\nabla(\vec{a}\cdot\vec{r}) = \left(\frac{\partial}{\partial x}\hat{i} + \frac{\partial}{\partial y}\hat{j} + \frac{\partial}{\partial z}\hat{k}\right)(a_1x + a_2y + a_3z)$$

$$= \hat{i}\frac{\partial}{\partial x}(a_1x + a_2y + a_3z) + \hat{j}\frac{\partial}{\partial y}(a_1x + a_2y + a_3z) + \hat{k}\frac{\partial}{\partial z}(a_1x + a_2y + a_3z)$$

$$= a_1\hat{i} + a_2\hat{j} + a_3\hat{k}$$

$$\nabla(\vec{a}\cdot\vec{r}) = \vec{a}$$

(ii)
$$\nabla\left(\frac{1}{r}\right) = \left(\frac{\partial}{\partial x}\hat{i} + \frac{\partial}{\partial y}\hat{j} + \frac{\partial}{\partial z}\hat{k}\right)\left(\frac{1}{r}\right)$$

$$= \hat{i}\frac{\partial}{\partial x}\left(\frac{1}{r}\right) + \hat{j}\frac{\partial}{\partial x}\left(\frac{1}{r}\right) + \hat{k}\frac{\partial}{\partial x}\left(\frac{1}{r}\right)$$

$$= \hat{i}\left(\frac{1}{r^2}\frac{\partial r}{\partial x}\right) + \hat{j}\left(\frac{1}{r^2}\frac{\partial r}{\partial y}\right) + \hat{k}\left(\frac{1}{r^2}\frac{\partial r}{\partial z}\right)$$

$$= -\frac{1}{r^2}\left[\frac{\partial r}{\partial x}\hat{i} + \frac{\partial r}{\partial y}\hat{j} + \frac{\partial r}{\partial z}\hat{k}\right] \quad ...(1)$$

To evaluate $\frac{\partial r}{\partial x}$, $\frac{\partial r}{\partial y}$ and $\frac{\partial r}{\partial z}$

Let $\vec{r} = x\hat{i} + y\hat{j} + z\hat{k}$; $r = |\vec{r}| = \sqrt{x^2 + y^2 + z^2}$

$$r^2 = x^2 + y^2 + z^2 \quad ...(2)$$

Differentiating (2) partially w.r.t., x we get

$$2r\frac{\partial r}{\partial x} = 2x$$

$$\Rightarrow \quad \frac{\partial r}{\partial x} = \frac{x}{r},\ \frac{\partial r}{\partial y} = \frac{y}{r},\ \frac{\partial r}{\partial z} = \frac{z}{r}$$

Put the values of $\frac{\partial r}{\partial x}$, $\frac{\partial r}{\partial y}$ and $\frac{\partial r}{\partial z}$ in (1)

$$\nabla\left(\frac{1}{r}\right) = -\frac{1}{r^2}\left[\frac{x}{r}\hat{i} + \frac{y}{r}\hat{j} + \frac{z}{r}\hat{k}\right]$$

$$= -\frac{1}{r^3}[x\hat{i} + y\hat{j} + z\hat{k}]$$

$$\nabla\left(\frac{1}{r}\right) = -\frac{1}{r^3}\vec{r}$$

(iii)
$$\nabla r^n = \left(\hat{i}\frac{\partial}{\partial x} + \hat{j}\frac{\partial}{\partial y} + \hat{k}\frac{\partial}{\partial z}\right)r^n = \hat{i}\frac{\partial}{\partial x}r^n + \hat{j}\frac{\partial}{\partial y}r^n + \hat{k}\frac{\partial}{\partial y}r^n$$

$$= \hat{i}\left(nr^{n-1}\frac{\partial r}{\partial x}\right) + \hat{j}\left(nr^{n-1}\frac{\partial r}{\partial y}\right) + \hat{k}\left(nr^{n-1}\frac{\partial r}{\partial z}\right)$$

Put the values of $\frac{\partial r}{\partial x}$, $\frac{\partial r}{\partial y}$ and $\frac{\partial r}{\partial z}$

$$\nabla r^n = \hat{i}\left(nr^{n-1}\frac{x}{r}\right) + \hat{j}\left(nr^{n-1}\frac{y}{r}\right) + \hat{k}\left(nr^{n-1}\frac{z}{r}\right)$$

$$= nr^{n-2}\,(x\hat{i} + y\hat{j} + z\hat{k})$$

$$\nabla r^n = nr^{n-2}\,\vec{r}$$

Example 4.18 **If $f = x + y + z$, $g = x^2 + y^2 + z^2$, $h = yz + zx + xy$, prove that grad f, grad g and grad h are coplanar vectors.**

Solution :

$$\text{grad } f = \left(\hat{i}\frac{\partial}{\partial x} + \hat{j}\frac{\partial}{\partial y} + \hat{k}\frac{\partial}{\partial z}\right)(x + y + z) = \hat{i} + \hat{j} + \hat{k}$$

$$\text{grad } g = \left(\hat{i}\frac{\partial}{\partial x} + \hat{j}\frac{\partial}{\partial y} + \hat{k}\frac{\partial}{\partial z}\right)(x^2 + y^2 + z^2) = 2x\hat{i} + 2y\hat{j} + 2z\hat{k}$$

$$\text{grad } h = \left(\hat{i}\frac{\partial}{\partial x} + \hat{j}\frac{\partial}{\partial y} + \hat{k}\frac{\partial}{\partial z}\right)(yz + zx + xy)$$

$$= \hat{i}(z + y) + \hat{j}(z + x) + \hat{k}(y + x)$$

$$\text{grad } f\,.\,(\text{grad } g \times \text{grad } h) = \begin{vmatrix} 1 & 1 & 1 \\ 2x & 2y & 2z \\ z+y & z+x & y+x \end{vmatrix}$$

$$= 2\begin{vmatrix} 1 & 1 & 1 \\ x & y & z \\ z+y & z+x & y+x \end{vmatrix} \qquad [\text{applying } R_2 \to R_2 + R_1]$$

$$= 2\begin{vmatrix} 1 & 1 & 1 \\ x+y+z & y+z+x & z+y+x \\ z+y & z+x & y+x \end{vmatrix}$$

$$= 2(x + y + z)\begin{vmatrix} 1 & 1 & 1 \\ 1 & 1 & 1 \\ y+z & z+x & y+x \end{vmatrix} = 0$$

Hence, grad f, grad g and grad h are coplanar vectors.

Example 4.19 **Find a unit vector normal to the surface $xy^2 + 2yz = 8$ at the point (3, –2, 1).**

Solution : It is given that $f(x, y, z) = xy^2 + 2yz - 8$

$$\therefore \quad \frac{\partial f}{\partial x} = y^2\,.\,\frac{\partial f}{\partial y} = 2xy + 2z \text{ and } \frac{\partial f}{\partial z} = 2y$$

$$\therefore \quad \nabla f = \hat{i}\frac{\partial f}{\partial x} + \hat{j}\frac{\partial f}{\partial y} + \hat{k}\frac{\partial f}{\partial z}$$

$$= y^2\hat{i} + (2xy + 2z)\hat{j} + 2y\hat{k}$$

At the point (3, –2, 1), we obtain the normal vector as

$$\nabla f(3, -2, 1) = 4\hat{i} - 10\hat{j} - 4\hat{k}$$

$\therefore$ The unit normal vector at (3, –2, 1) $= \dfrac{4\hat{i} - 10\hat{j} - 4\hat{k}}{\sqrt{16 + 100 + 16}} = \dfrac{2\hat{i} - 5\hat{j} + 2\hat{k}}{\sqrt{33}}$

Example 4.20 **Find the normal vector and equation of the tangent plane to the surface $z = (x^2 + y^2)^{1/2}$ at (3, 4, 5).**

Solution : Let $f(x, y, z) = z - \sqrt{x^2 + y^2} = 0$ be the given surface.

$\therefore$ The normal vector is given by

$$\nabla f = \left(\hat{i}\frac{\partial}{\partial x} + \hat{j}\frac{\partial}{\partial y} + \hat{k}\frac{\partial}{\partial z}\right)(z - \sqrt{x^2 + y^2})$$

$$= -\frac{x}{\sqrt{x^2 + y^2}}\hat{i} - \frac{y}{\sqrt{x^2 + y^2}}\hat{j} + \hat{k} \qquad (z \neq 0)$$

$$= -\frac{x}{z}\hat{i} - \frac{y}{z}\hat{j} + \hat{k}$$

At (3, 4, 5), the normal vector is given by

$$[\nabla f]_{(3,4,5)} = -\frac{3}{5}\hat{i} - \frac{4}{5}\hat{j} + \hat{k}$$

Also the tangent plane at the point (3, 4, 5) is given by

$$-\frac{3}{5}(x - 3) - \frac{4}{5}(y - 4) + (z - 5) = 0$$

or $\quad 3x + 4y - 5z = 0$

Example 4.21 **Find the angle between the surface $x^2 + y^2 + z^2 = 9$ and $z = x^2 + y^2 - 3$ at the point (2, –1, 2).**

Solution : Let $\quad f = x^2 + y^2 + z^2 - 9 = 0$

and $\quad \phi = x^2 + y^2 - z - 3 = 0$

$$\nabla f = \hat{i}\frac{\partial}{\partial x}(x^2 + y^2 + z^2 - 9) + \hat{j}\frac{\partial}{\partial y}(x^2 + y^2 + z^2 - 9) + \hat{k}\frac{\partial}{\partial z}(x^2 + y^2 + z^2 - 9)$$

$$= 2x\hat{i} + 2y\hat{j} + 2z\hat{k}$$

$$(\nabla f)_{(2, -1, 2)} = 4\hat{i} - 2\hat{j} + 4\hat{k}$$

$$\nabla\phi = \hat{i}\frac{\partial}{\partial x}(x^2 + y^2 - z - 3) + \hat{j}\frac{\partial}{\partial y}(x^2 + y^2 - z - 3) + \hat{k}\frac{\partial}{\partial z}(x^2 + y^2 - z - 3)$$

$$= 2x\hat{i} + 2y\hat{j} - \hat{k}$$

$$(\nabla\phi)_{(2, -1, 2)} = 4\hat{i} - 2\hat{j} - \hat{k}$$

$$\therefore \quad \cos\theta = \left[\frac{\nabla f \cdot \nabla\phi}{|\nabla f|\ |\nabla\phi|}\right] = \frac{(4\hat{i} - 2\hat{j} + 4\hat{k}) \cdot (4\hat{i} - 2\hat{j} - \hat{k})}{\sqrt{16 + 4 + 16} \cdot \sqrt{16 + 4 + 1}}$$

$$= \frac{16 + 4 - 4}{\sqrt{36}\ \sqrt{21}} = \frac{8\sqrt{21}}{63} = \frac{8}{3\sqrt{21}}$$

$$\therefore \quad \theta = \cos^{-1}\left(\frac{8}{3\sqrt{21}}\right)$$

Example 4.22 **Find a unit vector normal to the surface $x^3 + y^3 + 3xyz = 3$ at the point (1, 2, –1).**

Solution : Let $f = x^3 + y^3 + 3xyz - 3$

$$\frac{\partial f}{\partial x} = 3x^2 + 3yz, \quad \frac{\partial f}{\partial y} = 3y^2 + 3xz, \quad \frac{\partial f}{\partial z} = 3xy$$

$$\nabla f = \hat{i}\frac{\partial f}{\partial x} + \hat{j}\frac{\partial f}{\partial y} + \hat{k}\frac{\partial f}{\partial z}$$

$$\nabla f = (3x^2 + 3yz)\hat{i} + (3y^2 + 3xz)\hat{j} + 3xy\hat{k}$$

At the point (1, 2, –1)

$$\nabla f = -3\hat{i} + 9\hat{j} + 6\hat{k}$$

Which is a vector normal to the given surface at (1, 2, –1). Hence until vector normal to the given surface at (1, 2, –1).

$$= \frac{\nabla f}{|\nabla f|} = \frac{-3\hat{i} + 9\hat{j} + 6\hat{k}}{\sqrt{(-3)^2 + (9)^2 + (6)^2}}$$

$$= \frac{-3\hat{i} + 9\hat{j} + 6\hat{k}}{3\sqrt{14}} = \frac{1}{\sqrt{14}}(-\hat{i} + 3\hat{j} + 2\hat{k})$$

4.10 GEOMETRICAL INTERPRETATION OF GRADIENT

If a surface $f(x, y, z) = c$ is drawn through any point P such that at each point on the surface, the function has the same value as at P, then such a surface is called a level surface through P.

For example, if $f(x, y, z)$ represents potential at the point (x, y, z) the equipotential surface $f(x, y, z) = c$ is a level surface.

Note that, in general two such surfaces will not intersect because any point passes through one and only one level surface.

Let, $f = c$ and $f + \delta f = c$, be two level surface through neighbouring points P and Q and let $\vec{r}$ and $\vec{r} + \delta\vec{r}$ be the position vectors of point P and Q referred to some origin O. Then

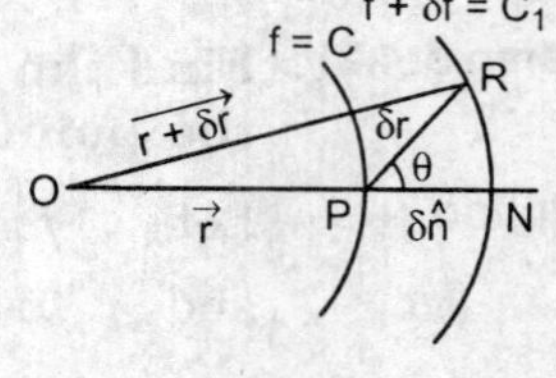

Fig. 4.3

$$\overrightarrow{PQ} = \overrightarrow{\delta r}$$

Now as $Q \to P$ the line PQ tends to the tangent at P to the level surface $f = c$ and hence δr lies in the tangent plane to this surface at P.

$$df = \frac{\partial f}{\partial x}dx + \frac{\partial f}{\partial y}dy + \frac{\partial f}{\partial z}dz$$

$$= \left(\hat{i}\frac{\partial f}{\partial x} + \hat{j}\frac{\partial f}{\partial y} + \hat{k}\frac{\partial f}{\partial z}\right) . (\hat{i}\,dx + \hat{j}\,dy + \hat{k}\,dz)$$

$$= \nabla f . dr \qquad [\because r = x\hat{i} + y\hat{j} + z\hat{k}]$$

$$= 0 \qquad [\because f \text{ is constant}]$$

Thus, ∇f is a vector perpendicular to dr.

Hence, ∇f is a vector normal to the level surface $f = c$ at the point P.

4.11 DIRECTIONAL DERIVATIVE

Let $f(x, y, z)$ be a scalar field in a region R and P be any point in this region. Also if Q is a variable point in the neighbourhood of P in the direction of a given unit vector $\vec{b}$ and s be the distance between P and R. Then the directional derivative of $\vec{f}$ at P in the direction of $\hat{b}$ is denoted as $\frac{df}{ds}$ and defined as

$$\frac{df}{ds} = \lim_{t \to t_0} \frac{f(Q) - f(P)}{s}$$

Theorem : The directional derivative of a scalar field f at a point $P(x, y, z)$ in the position vector of a unit vector $\hat{a}$ is given by $\frac{df}{ds} = \nabla f \cdot \hat{a}$.

Proof : Let $f(x, y, z)$ be a scalar field in the region R and $r = xi + yj + zk$, denote the position vector of any point $P(x, y, z)$ in this region. If the distance of P from a fixed point in the direction of $\hat{a}$ is denoted by s, then ds denotes infinitesimal small element at P in the direction of $\hat{a}$.

Thus, $\frac{dr}{ds}$ is a unit vector at P in the direction of $\vec{a}$,

i.e., $$\frac{dr}{ds} = \hat{a}$$

Also since $$r = xi + yj + zk$$

$\therefore$ $$\frac{dr}{ds} = \frac{dx}{ds}\hat{i} + \frac{dy}{ds}\hat{j} + \frac{dz}{ds}k = \hat{a}$$

Now, $$\nabla f \cdot \hat{a} = \left(\hat{i}\frac{\partial f}{\partial x} + \hat{j}\frac{\partial f}{\partial y} + \hat{k}\frac{\partial f}{\partial z}\right) \cdot \left(\frac{dx}{ds}\hat{i} + \frac{dy}{ds}\hat{j} + \frac{dz}{ds}\hat{k}\right)$$

$$= \frac{\partial f}{\partial x}\frac{dx}{ds} + \frac{\partial f}{\partial y}\frac{dy}{ds} + \frac{\partial f}{\partial z}\frac{dz}{ds}$$

$$= \frac{df}{ds} = \text{directional derivative of } f \text{ at } P \text{ in the direction of } \hat{a}$$

4.12 DIRECTIONAL DERIVATIVE OF A SCALAR POINT FUNCTION

Let $f(x, y, z)$ be a scalar field at a point P. Let $\hat{a}$ be the point vector along the direction $\vec{PQ}$ and $PQ = ds$. Then the directional derivative of f at P in the direction of $\vec{PQ}$ is denoted and defined by

$$\frac{\partial f}{\partial s} = \hat{a} \cdot \nabla f = \hat{a} \cdot \nabla f$$

Note : The directional derivative of f is maximum in the direction of the normal to the given surface, *i.e.*, in the direction of ∇f. The maximum value of this directional derivative is $|\nabla f|$.

Example 4.23 **Find the directional derivative of $\phi(x, y, z) = (x^2 + y^2 + z^2)^{-1/2}$ at the point $P(3, 1, 2)$ in the direction of the vector $yz\hat{i} + xz\hat{j} + xy\hat{k}$.**

Solution : If $\hat{a}$ be the unit vector in the direction of the vector $yz\hat{i} + zx\hat{j} + xy\hat{k}$, then

$$\hat{a} = \frac{yz\hat{i} + xz\hat{j} + xy\hat{k}}{\sqrt{y^2z^2 + x^2z^2 + x^2y^2}}$$

It can be observed that

$$\text{grad } \phi = -\frac{(x\hat{i} + y\hat{j} + z\hat{k})}{(x^2 + y^2 + z^2)^{3/2}}$$

At the point (3, 1, 2) grad $-\frac{1}{(14)^{3/2}}(3\hat{i} + \hat{j} + 2\hat{k})$

Therefore, required directional derivative is

$$\text{grad } \phi \cdot \hat{a} = -\frac{(3\hat{i} + \hat{j} + 2\hat{k})}{(14)^{3/2}} \cdot \frac{(yz\hat{i} + xz\hat{j} + xy\hat{k})}{\sqrt{y^2z^2 + x^2z^2 + x^2y^2}}$$

$$= -\frac{3yz + xz + 2xy}{(14)^{3/2}\sqrt{y^2z^2 + x^2z^2 + x^2y^2}}$$

$$= -\frac{9}{49\sqrt{14}} \text{ at the point } (3, 1, 2)$$

$$= \frac{9}{49\sqrt{14}} \text{ (omitting the negative sign)}$$

Example 4.24 **Find the direction derivative of $f(x, y, z) = x^2yz + 4xz^2$ at the point (1, –2, –1) in the direction of the vector $2\hat{i} - \hat{j} - 2\hat{k}$.**

Solution : It is given that

$$f(x, y, z) = x^2yz + 4xz^2$$

$$\text{grad } f = (2xyz + 4z^2)\,\hat{i} + x^2z\,\hat{j} + (x^2y + 8xz)\,\hat{k}$$

$$= 8\hat{i} - \hat{j} - 10\hat{k} \text{ the point } (1, -2, -1).$$

Let $\hat{a}$ be the unit vector in the direction of the vector $2i - j - 2k$, then

$$\hat{a} = \frac{2\hat{i} - \hat{j} - 2\hat{k}}{\sqrt{4 + 1 + 4}} = \frac{2}{3}\hat{i} - \frac{1}{3}\hat{j} - \frac{2}{3}\hat{k}$$

Therefore, the required directional derivative is

$$\frac{dr}{ds} = \text{grad } f \cdot \hat{a} = (8\hat{i} - \hat{j} - 10\hat{k}) \cdot \left(\frac{2}{3}\hat{i} - \frac{1}{3}\hat{j} - \frac{2}{3}\hat{k}\right)$$

$$= \frac{16}{3} + \frac{1}{3} + \frac{20}{3} = \frac{37}{3}$$

Theorem : Grad f is a vector in the direction of which the maximum value of the directional derivative of f, *i.e.*, $\frac{dr}{ds}$ occurs.

Proof : We know that the directional derivative of f in the direction of $\hat{a}$ is given by

$$\frac{df}{ds} = \nabla f \cdot \hat{a} = \left(\frac{dr}{dn}\hat{n}\right) \cdot \hat{a} \qquad \left(\because \nabla f = \frac{dr}{dn}\hat{n}\right)$$

$$= \frac{df}{dn}(\hat{n} \cdot \hat{a}) = \frac{df}{dn}\cos\theta$$

Thus, $\frac{df}{ds}$ will be maximum, where $\cos\theta$ is maximum, *i.e.*, when $\cos\theta = 1$.
But $\cos\theta = 1$, when the angle between $\hat{a}$ and $\hat{n}$ is zero, which concludes that $\hat{a}$ is along the unit normal vector $\hat{n}$.
Therefore, the directional derivative is maximum along the normal to the surface and the maximum value of the directional derivative is

$$= \frac{df}{dn} = |\,grad\, f\,|.$$

Example 4.25 **In what direction from the point (1, 1, –1) the directional derivative of $f = x^2 - 2y^2 + 4z^2$ is maximum ? Also find the value of this maximum directional derivative.**

Solution : We can find that $\quad \text{grad } f = 2x\hat{i} - 4y\hat{j} + 8z\hat{k}$

at ((1, 1, –1) we get $\quad \text{grad } f = 2\hat{i} - 4\hat{j} - 8\hat{k}$

Since the directional derivative of f is maximum in the direction of grad f.
Thus the maximum value of the directional derivative

$$= |\,\text{grad } f\,| = |\,2\hat{i} - 4\hat{j} - 8\hat{k}\,|$$
$$= \sqrt{4 + 16 + 64}$$
$$= \sqrt{84} = 2\sqrt{21}$$

Example 4.26 **Find the directional derivative of the function $f = x^2 - y^2 + 2z^2$ at the point $P(1, 2, 3)$ in the direction of the line PQ where Q is the piont 5, 0, 4). In what direction it will be maximum ? Find also the magnitude of this maximum.** *(UPTU 2001)*

Solution : It is given that

$$f = x^2 - y^2 + 2z^2$$
$$\nabla f = \hat{i}\frac{\partial f}{\partial x} + \hat{j}\frac{\partial f}{\partial y} + \hat{k}\frac{\partial f}{\partial z}$$
$$\frac{\partial f}{\partial x} = 2x, \quad \frac{\partial f}{\partial y} = -2y, \quad \frac{\partial f}{\partial z} = 4z$$
$$\nabla f = 2x\hat{i} - 2y\hat{j} + 4z\hat{k}$$
$$\overrightarrow{PQ} = \overrightarrow{OQ} - \overrightarrow{OP} = (5\hat{i} + 4\hat{k}) - (\hat{i} + 2\hat{j} + 3\hat{k}) = 4\hat{i} - 2\hat{j} + \hat{k}$$

If $\hat{n}$ is a unit vector in the direction $\overrightarrow{PQ}$, then

$$\hat{n} = \frac{4\hat{i} - 2\hat{j} + \hat{k}}{\sqrt{(4)^2 + (-2)^2 + 1}} = \frac{4\hat{i} - 2\hat{j} + \hat{k}}{\sqrt{21}}$$

$\therefore$ Directional derivative of f in the direction

$$\overrightarrow{PQ} = (\nabla f) \cdot \hat{n}$$
$$= (2\hat{i} - 4\hat{j} + 12\hat{k}) \cdot \frac{1}{\sqrt{21}}(4\hat{i} - 2\hat{j} + \hat{k})$$
$$= \frac{1}{\sqrt{21}}[2(4) + (-4))(-2) + 12(1)]$$

$$= \frac{1}{\sqrt{21}}(8+8+12) = \frac{28}{\sqrt{21}}$$

$$= \frac{4}{3}\sqrt{21}$$

The directional derivative of f is maximum in the direction of the normal to the given surface, *i.e.*, in the direction of $\nabla f = 2\hat{i} - 4\hat{j} + 12\hat{k}$.

The maximum value of this directional derivative $= |\nabla f|$

$$= \sqrt{(2)^2 + (-4)^2 + (12)^2} = \sqrt{164} = 2\sqrt{41}$$

Example 4.27 **Find the directional derivative $f = (x^2 + y^2 + z^2)^{-1/2}$ at the point $P(3, 1, 2)$ in the dirction of the vector $yz\hat{i} + zx\hat{j} + xy\hat{k}$.**

(UPTU Special Exam. 2001)

Solution : $f = (x^2 + y^2 + z^2)^{-1/2}$

$$\nabla f = \hat{i}\frac{\partial}{\partial x}(x^2+y^2+z^2)^{-1/2} + \hat{j}\frac{\partial}{\partial y}(x^2+y^2+z^2)^{-1/2} + \hat{k}\frac{\partial}{\partial z}(x^2+y^2+z^2)^{-1/2}$$

$$= \hat{i}\left\{-\frac{1}{2}(x^2+y^2+z^2)^{-3/2}(2x)\right\} + \hat{j}\left\{-\frac{1}{2}(x^2+y^2+z^2)^{-3/2}(2y)\right\}$$

$$+ \hat{k}\left\{-\frac{1}{2}(x^2+y^2+z^2)^{-3/2}(2z)\right\}$$

$$\nabla f = \frac{-1}{(x^2+y^2+z^2)^{3/2}}[x\hat{i} + y\hat{j} + z\hat{k}]$$

At the point $P(3, 1, 2) = \frac{-(3\hat{i} + \hat{j} + 2\hat{k})}{(9+1+4)^{3/2}} = \frac{-(3\hat{i} + \hat{j} + 2\hat{k})}{14\sqrt{14}}$

Let $\hat{a}$ be the unit vector in the given direction, then

$$\hat{a} = \frac{yz\hat{i} + zx\hat{j} + xy\hat{k}}{\sqrt{y^2z^2 + z^2x^2 + x^2y^2}}$$

$\hat{a}$ at the point $(3, 1, 2) = \frac{2\hat{i} + 6\hat{j} + 3\hat{k}}{\sqrt{49}} = \frac{2\hat{i} + 6\hat{j} + 3\hat{k}}{7}$

$\therefore$ Directional derivative of f in the given direction

$$= \hat{a}\,.\,(\nabla f) = -\left(\frac{2\hat{i} + 6\hat{j} + 3\hat{k}}{7}\right)\cdot\left(\frac{3\hat{i} + \hat{j} + 2k}{14\sqrt{14}}\right)$$

$$= -\frac{(2)(3) + (6)(1) + (3)(2)}{7 \times 14\sqrt{14}} = -\frac{6+6+6}{7 \times 14\sqrt{14}}$$

$$= -\frac{18}{7 \times 14\sqrt{14}} = -\frac{9}{49\sqrt{14}}$$

$$= \frac{9}{49\sqrt{14}}$$

Example 4.28 **For the function $\phi = \frac{y}{x^2 + y^2}$, find the magnitude of the directional derivative making an angle 30° with the positive x-axis at the point (0, 1).**

Solution : The given function is $\phi = \dfrac{y}{x^2+y^2}$

$$\therefore \quad \frac{\partial\phi}{\partial x} = \frac{-2xy}{(x^2+y^2)^2}\,;\ \frac{\partial\phi}{\partial y} = \frac{(x^2+y^2)-2y^2}{(x^2+y^2)^2} = \frac{x^2-y^2}{(x^2+y^2)} \text{ and } \frac{\partial\phi}{\partial z} = 0$$

Thus, $$grad\, f = \hat{i}\frac{\partial\phi}{\partial x} + \hat{j}\frac{\partial\phi}{\partial y}$$

$$= \frac{-2xy}{(x^2+y^2)^2}\hat{i} + \frac{x^2-y^2}{(x^2+y^2)^2}\hat{j}$$

$$grad\,\phi \text{ at } (0, 1) = -\hat{j}$$

Let $\hat{a}$ be the unit vector along the line making an angle 30° with the positive x-axis at the point (0,1), then

$$\hat{a} = \cos 30^\circ\, \hat{i} + \sin 30^\circ\, .\hat{j}$$

Hence, the directional derivative is given by $\hat{a}\,.\,grad\,\phi$

$$= (\cos 30\, \hat{i} + \sin 30\, j)\,.\,(-j) = \sin 30 = -\frac{1}{2}$$

Example 4.29 **Find the directional derivative of $f = x^2 - 2y^2 + 4z^2$ at (1, 1, –1) in the direction of the vector $2i + j - k$. Also find the direction of the maximum directional derivative at (1, 1, –1) and its maximum value.**

Solution :

$$\text{grad } f = \hat{i}\frac{\partial}{\partial x}(x^2-2y^2+4z^2) + \hat{j}\frac{\partial}{\partial y}(x^2-2y^2+4z^2) + \hat{k}\frac{\partial}{\partial z}(x^2-2y^2+4z^2)$$

$$= 2x\hat{i} - 4y\hat{j} + 8z\hat{k}$$

$$(\text{grad } f)_{(1,1,-1)} = 2\hat{i} - 4\hat{j} - 8\hat{k}$$

Directional derivative in the direction of the vector $2\hat{i} + \hat{j} - \hat{k}$

$$= (2\hat{i} - 4\hat{j} - 8\hat{k})\,.\,\frac{(2\hat{i}+\hat{j}-\hat{k})}{\sqrt{4+1+1}} = \frac{8}{\sqrt{6}}$$

Since, the directional derivative of f is maximum in the direction of grad f, *i.e.*, along $2(\hat{i} - 2\hat{j} - 4\hat{k})$.

Hence, the value of this maximum directional derivative at (1, 1, –1).

$$= |\,(\text{grad } f)_{(1,1,-1)}\,|$$

$$= |\,2\hat{i} - 4\hat{j} - 8\hat{k}\,| = 2\sqrt{21}.$$

Example 4.30 **Find the values of constant a, b and c so that the directional derivative of $f = axy^2 + byz + cz^2x^3$ at (1, 2, –1) has a maximum magnitude 64 in a direction parallel to the z-axis.**

Solution :

$$\text{grad } f = \hat{i}\frac{\partial}{\partial x}(axy^2+byz+cz^2x^3) + \hat{j}\frac{\partial}{\partial y}(axy^2+byz+cz^2x^3)$$

$$+ \hat{k}\frac{\partial}{\partial z}(axy^2+byz+cz^2x^3)$$

$$= (ay^2 + 3cx^2z^2)\,\hat{i} + (2axy + bz)\,\hat{j} + (by + 2czx^3)\,\hat{k}$$

$$\therefore \quad (\text{grad } f)_{(1,2,-1)} = (4a+3c)\,\hat{i} + (4a-b)\,\hat{j} + (2b-2c)\,\hat{k} \qquad ...(1)$$

Since, the max. directional is along the normal is value is equal to $|\nabla f|$. We know that ∇f is a vector along the normal.

$\therefore$ $(\text{grad } f)_{(1,2,-1)}$ represents the max. directional derivative and this is given to be along the z-axis.

$\therefore$ $$4a + 3c = 0,\ 4a - b = 0 \text{ and } 2b - 2c = 64$$

$\Rightarrow$ $$a = 6,\ b = 24 \text{ and } c = -8.$$

Example 4.31 **If the directional derivative of $\phi = ax^2y + by^2z + cz^2x$ at the point (1, 1, 1) has maximum magnitude 15 in the direction parallel to the line $\frac{x-1}{2} = \frac{y-3}{-2} = \frac{z}{1}$, find the value of a, b and c.** *(UPTU 2002)*

Solution : It is given that

$$\phi = ax^2y + by^2z + cz^2x$$

$$grad\ \phi = \hat{i}\frac{\partial\phi}{\partial x} + \hat{j}\frac{\partial\phi}{\partial y} + \hat{k}\frac{\partial\phi}{\partial z}$$

$$= (2axy + cz^2)\,\hat{i} + (ax^2 + 2byz)\,\hat{j} + (by^2 + 2czx)\,\hat{k}$$

$$grad\ \phi \text{ at } (1, 1, 1) = (2a + c)\,\hat{i} + (a + 2b)\,\hat{j} + (b + 2c)\,\hat{k} \quad ...(1)$$

Now, the directional derivative is maximum along the normal to the surface, *i.e.*, along $grad\ \phi$ and it's maximum value is $|\ grad\ \phi\ |$

Also $$|\ grad\ \phi\ | = \sqrt{(2a+c)^2 + (a+2b)^2 + (b+2c)^2}$$

$\Rightarrow$ $$(2a+c)^2 + (a+2b)^2 + (b+2c)^2 = 225 \quad ...(2)$$

But we are given that the directional derivative is maximum in the direction in the parallel to the line $\frac{x-1}{2} = \frac{y-3}{-2} = \frac{z}{1}$, *i.e.*, parallel to the vector $2\hat{i} - 2\hat{j} + \hat{k}$.

Hence, $2\hat{i} - 2\hat{j} + \hat{k}$ is parallel to equation (1)

Therefore, $$\frac{2a+c}{2} = \frac{a+2b}{-2} = \frac{b+2c}{1}$$

$\Rightarrow$ $$2a + c = -a - 2b \text{ or } 3a + 2b + c = 0 \quad ...(3)$$

and $$2b + a = -2b - 4c \text{ or } a + 4b + 4c = 0 \quad ...(4)$$

From equation (3) and (4) by cross multiplication, we get

$$-11a = 4b \text{ and } -10b = 11c$$

$\Rightarrow$ $$\frac{a}{4} = \frac{b}{-11} \text{ and } \frac{b}{-11} = \frac{c}{10}$$

$$\frac{a}{4} = -\frac{b}{11} = \frac{c}{10} = \lambda \text{ (say)} \quad ...(5)$$

From equation (2)

$\Rightarrow$ $$(8\lambda + 10\lambda)^2 + (\lambda - 22\lambda)^2 + (-11\lambda + 20\lambda)^2 = 225$$

$\Rightarrow$ $$324\lambda^2 + 324\lambda^2 + 81\lambda^2 = 225$$

$\Rightarrow$ $$729\lambda^2 = 225 \text{ or } \lambda^2 = \frac{225}{729}$$

$\Rightarrow$ $$\lambda = \pm\frac{15}{27}$$

$\Rightarrow$ $$\lambda = \pm\frac{5}{9}$$

$$\Rightarrow \qquad a = \pm\frac{20}{9}, \quad b = \pm\frac{55}{9}, \quad c = \pm\frac{50}{9} \qquad \text{[from equation (5)]}$$

Example 4.32 **Find the direction derivative of $\phi = 5x^2y - 5y^2z + \frac{5}{2}z^2x$ at the point (1, 1, 1) in the direction of the line $\frac{x-1}{2} = \frac{y-3}{-2} = \frac{z}{1}$.** *(UPTU 2004)*

Solution : It is given that

$$\phi = 5x^2y - 5y^2z + \frac{5}{2}z^2x$$

$$\therefore \qquad grad\,\phi = \hat{i}\frac{\partial\phi}{\partial x} + \hat{j}\frac{\partial\phi}{\partial y} + \hat{k}\frac{\partial\phi}{\partial z}$$

$$= \left(10xy + \frac{5}{2}z^2\right)\hat{i} + (5x^2 - 10yz)\,\hat{j} + (-5y^2 + 5zx)\,k$$

$$\therefore \qquad grad\,\phi \text{ at } (1, 1, 1) = \left(10 + \frac{5}{2}\right)\hat{i} + (5 - 10)\,\hat{j} + (-5 + 5)\,\hat{k}$$

$$= \frac{25}{2}\hat{i} - 5\hat{j}$$

Let $\hat{a}$ be the unit vector in the given direction

$$\therefore \qquad \hat{a} = \frac{2\hat{i} - 2\hat{j} + \hat{k}}{\sqrt{4 + 4 + 1}}$$

or

$$\hat{a} = \frac{2\hat{i} - 2\hat{j} + \hat{k}}{3}$$

$\therefore$ Directional derivative $= (grad\,\phi)\cdot\hat{a}$

$$= \left(\frac{25}{2}\hat{i} - 5\hat{j}\right)\cdot\left(\frac{2\hat{i} - 2\hat{j} + \hat{k}}{3}\right) = \left(\frac{25}{3} + \frac{10}{3}\right) = \frac{35}{3}$$

Example 4.33 **Find the directional derivative of $\frac{1}{r}$ in the direction of $\vec{r}$ where $\vec{r} = x\hat{i} + y\hat{j} + z\hat{k}$.**

Solution : Here, $f(x, y, z) = \frac{1}{r} = \frac{1}{\sqrt{x^2 + y^2 + z^2}}$

We have

$$grad\,f = \frac{\partial f}{\partial x}\hat{i} + \frac{\partial f}{\partial y}\hat{j} + \frac{\partial f}{\partial z}\hat{k}$$

$$\frac{\partial}{\partial x}(x^2 + y^2 + z^2)^{-1/2}\,\hat{i} + \frac{\partial}{\partial y}(x^2 + y^2 + z^2)^{-1/2}\,\hat{j} + \frac{\partial}{\partial z}(x^2 + y^2 + z)^{-1/2}\,\hat{k}$$

$$= -\frac{1}{2}(x^2 + y^2 + z^2)^{-3/2}\cdot 2x\hat{i} - \frac{1}{2}(x^2 + y^2 + z^2)^{-3/2}\cdot 2y\,\hat{j} - \frac{1}{2}(x^2 + y^2 + z^2)^{-3/2}\cdot 2z\hat{k}$$

$$= -\frac{(x\hat{i} + y\hat{j} + z\hat{k})}{(x^2 + y^2 + z^2)^{3/2}}$$

If $\hat{a}$ be the unit vector in the direction of the vector $\vec{r} = x\hat{i} + y\hat{j} + z\hat{k}$, then

$$\hat{a} = \frac{(x\hat{i} + y\hat{j} + z\hat{k})}{(x^2 + y^2 + z^2)^{1/2}}$$

$\therefore$ Directional derivative in the given direction = $(grad\ f) \cdot \hat{a}$

$$= -\frac{(x\hat{i} + y\hat{j} + z\hat{k})}{(x^2 + y^2 + z^2)^{3/2}} \cdot \frac{(x\hat{i} + y\hat{j} + z\hat{k})}{(x^2 + y^2 + z^2)^{1/2}} = -\frac{x^2 + y^2 + z^2}{(x^2 + y^2 + z^2)^2}$$

$$= \frac{-1}{x^2 + y^2 + z^2}$$

4.13 GRADIENT IN POLAR CO-ORDINATE

Let P be a point with position vector $\vec{r}$. Let $\hat{e}_r$ and $\hat{e}_\theta$ be the unit vector along and perpendicular to $\vec{r}$, repectively. If s be the distance in the direction of r increasing then $ds = dr$.

Also directional derivative of f at P along $\hat{e}_r$ is

$$\frac{\partial f}{\partial s} = \frac{\partial f}{\partial r} = \hat{e}_r \cdot \nabla f \qquad \text{...(1)}$$

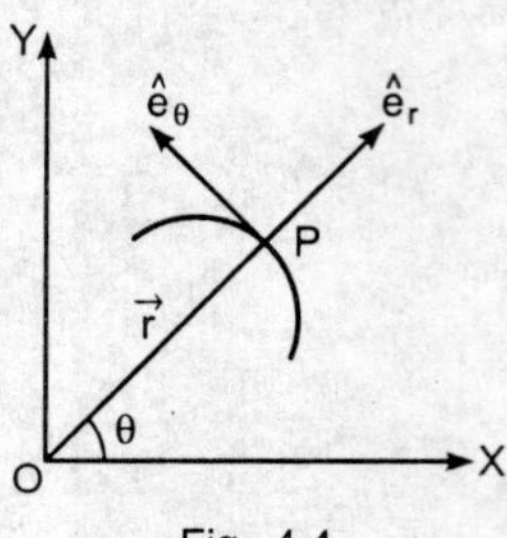

Fig. 4.4

Now it takes distance s in the direction perpendicular to $\vec{r}$, then

$$ds = r\,d\theta$$

Directional derivative of f at P along $\hat{e}_\theta$ is

$$\frac{\partial f}{\partial s} = \frac{\partial f}{r\partial\theta} = \hat{e}_{hat} \cdot \nabla f \qquad \text{...(2)}$$

Again componentof ∇f along and perpendicular to r are $\hat{e}_r$. ∇f fand $\hat{e}_\theta$ ∇f respectively.

Therefore, $\nabla f = (\hat{e}_r \cdot \nabla f) \cdot \hat{e}_r + (\hat{e}_\theta \cdot \nabla f) \cdot \hat{e}_\theta$

From Eq. (1) and (2)

$$\nabla f = \frac{\partial f}{\partial r} \hat{e}_r + \frac{\partial f}{r\partial\theta} \hat{e}_\theta \qquad \text{...(3)}$$

Equation (3) represents gradient in polar coordinate

Example 4.34 **Find $\nabla\ |\vec{r}|^2$.**

Solution : $\nabla\ |\vec{r}|^2 = \nabla\ r^2$

$$= \frac{\partial}{\partial r}(r^2)\ \hat{r} = 2r\ \hat{r} = 2\vec{r} \qquad (\because\ r\hat{r} = \vec{r})$$

Example 4.35 **Evaluate $grad\ e^{r^2}$.**

Solution : $\nabla\ e^{r^2} = \frac{\partial}{\partial r}(e^{r^2} \cdot \hat{r})$

$$= e^{r^2} \cdot 2r \cdot \hat{r} = 2e^{r^2}\ \vec{r} \qquad (\because\ r\hat{r} = \vec{r})$$

Example 4.36 **If $f(x, y, z) = \frac{1}{2}\log(x^2 + y^2)$, prove that $grad = f = \vec{r} \cdot \hat{k}$.**

Example 4.37 **If $f = r\cos\theta + \tan\theta$, find grad f in polar coordinates.**

Solution : $grad\ f = \frac{\partial f}{\partial r} \hat{e}_r + \frac{1}{r} \cdot \frac{\partial f}{z} \partial\theta\ \hat{e}_\theta$

$$= \frac{\partial}{\partial r}(r\cos\theta + \tan\theta)\,\hat{e}_r + \frac{1}{r}\cdot\frac{\partial}{\partial\theta}(r\cos\theta + \tan\theta)\,\hat{e}_\theta$$

$$grad\, f = \cos\theta\,\hat{e}_r + \frac{1}{r}(-r\sin\theta + \sec^2\theta)\,\hat{e}_\theta$$

Example 4.37 **Find the directional derivative of $\frac{1}{r}$ in the direction $\vec{r}$, where $r = x\hat{i} + y\hat{j} + z\hat{k}$.**

Solution : $f = \frac{1}{r}$

$$\nabla f = \nabla\left(\frac{1}{r}\right) = \frac{\partial}{\partial r}\left(\frac{1}{r}\right)\hat{r} = -\frac{1}{r^2}\hat{r} = \frac{-\vec{r}}{r^3} \qquad [r\,\hat{r} = \vec{r}]$$

Let $\hat{a}$ be the unit vector in the direction of $\vec{r}$, then

$$\hat{a} = \frac{\vec{r}}{|\vec{r}|} = \frac{\vec{r}}{r}$$

$\therefore$ Directional derivative $= \nabla f\cdot\hat{a} = \left(-\frac{\vec{r}}{r^3}\right)\cdot\left(\frac{\vec{r}}{r}\right) = -\frac{1}{r^4}(\vec{r}\cdot\vec{r})$

$$= \frac{-r^2}{r^4} = \frac{-1}{r^2}$$

PROBLEM SET 4.2

1. Find grad ϕ when ϕ is given by

(a) $\phi = x^2 + xyz$

(b) $\phi = x^2 + z^3 + 2xy$

(c) $\phi = \log(x^2 + y^2 + z^2)$

Ans. (a) $(2x + yz)\hat{i} + (xz)\hat{j} + xy\hat{k}$ (b) $(2x + 2y)\hat{i} + 2x\hat{j} + 3z^2\hat{k}$ (c) $\frac{2(x\hat{i} + y\hat{j} + z\hat{k})}{x^2 + y^2 + z^2}$

2. Compute the gradient of the scalor field $f(x, y, z) = x\sin(yz) + y\sin(xz) + z\sin(xy)$ at $\left(0, \frac{\pi}{4}, 1\right)$.

Ans. $\hat{i}\frac{(\pi + \sqrt{2})}{2}$

3. Find the gradient of the following fields :

(a) $f(x, y) = xy^2 - 3xy$ at (1, 2)

(b) $f(x, y, z) = x^2y^2 + xy^2 - z^2$ at (3, 1, 1)

(c) $f(x, y, z) = z^2 - x^2 + y^2$ at $(2, 1, \sqrt{3})$

Ans. (a) $-6\hat{i} + \hat{j}$ (b) $7\hat{i} + 24\hat{j} - 2\hat{k}$ (c) $2(2\hat{i} - \hat{j} - \sqrt{3}\hat{k}$

4. Find the normal vector and units normal vector to the following curves at the indicated points.

(a) $y^2 = 16x$ at (4, 8)

(b) $z = xy$ at (− 1, − 2, 2).

Ans. (a) $16(\hat{i} - \hat{j})$; $\frac{(\hat{i} - \hat{j})}{\sqrt{2}}$ (b) $-(2\hat{i} + \hat{j} + \hat{k})$; $\frac{-(2\hat{i} + \hat{j} + \hat{k}}{\sqrt{6}}$.

5. Find a unit vector normal to the surface $x^3 + y^3 + 3xyz = 3$ at (1, 2, – 1).

Ans. $-\dfrac{3\hat{i} + 9\hat{j} + 6\hat{k}}{3\sqrt{14}}$

6. Find a unit vector normal to the surface $xy^2z^3 = 4$ at the piont (– 1, 2, – 1).

Ans. $\dfrac{-\hat{i} + \hat{j} - 3k}{\sqrt{11}}$

7. Find the directional derivative of the function

(a) $f(x, y, z) = x^2y + y^3z$ at the point (2, – 1, 2) in the direction of the vector $= 2\hat{i} + \hat{j} + \hat{k}$

(b) $f(x, y, z) = 4xz^3 - 3x^2yz^2$ at (2, – 1, 2) along z axis.

(c) $f(x, y, z) = xy^2 = yz^3$ at the point (2, – 1, 1) in the direction of the normal to the normal to the surface $x \log z - y^2 + 4 = 0$ at (– 1, 2, 1).

Ans. (a) $\dfrac{10}{\sqrt{16}}$ (b) 144 (c) $\dfrac{15}{\sqrt{17}}$

8. In what direction from (3, 1, – 2) is the direcitonal derivative of $f = x^2y^2z^4$ maximum and what is its magnitude. ***(Rohtak 2003)***

9. Find the directional derivative of scalar function $f(x, y, z) = xyz$ at point $P(3, 1, 3)$ in the direction of the outward drawn normal of the surface $z = xy$ through the point P.

Ans. $\dfrac{27}{\sqrt{11}}$

10. Find a vector that gives the direction of maximum rate of increase and find the maximum rate for the following :

(a) $e^{2y} \cos x$ at $\left(\dfrac{\pi}{4}, 0\right)$

(b) $6xyz$ at (– 1, 2, 1)

(c) $x^2y^2z^2 + xz^2 + x^2y$ at (1, 2, – 1)

Ans. (a) $\dfrac{(-\hat{i} + 2\hat{j})}{\sqrt{2}}$; $\sqrt{5/2}$ (b) $6(2\hat{i} - \hat{j} - 2\hat{k})$; 18 (c) $13\hat{i} + 5\hat{j} - 10\hat{k}$; $\sqrt{294}$

11. Prove that $\nabla r^{-3} = -3r^{-5}\vec{r}$.

12. Find the angle between the surface $x \log z = -y^2 - 1$ and $x^2y = 2 - z$ of the ponit (1, 1, 1).

Ans. $\cos^{-1}\left(-\dfrac{1}{30}\right)$

4.14 DIVERGENCE OF A VECTOR POINT FUNCTION

The divergence of a differentiable vector point function $\vec{A}$ is denoted by div. $\vec{A}$ and defined as

$$div\,\vec{A} = \nabla \cdot \vec{A} = \left(\hat{i}\frac{\partial}{\partial x} + \hat{j}\frac{\partial}{\partial y} + \hat{k}\frac{\partial}{\partial z}\right) \cdot \vec{A}$$

$$= \hat{i}\cdot\frac{\partial \vec{A}}{\partial x} + \hat{\mathbf{j}}\cdot\frac{\partial \vec{A}}{\partial y} + \hat{k}\cdot\frac{\partial \vec{A}}{\partial z}$$

The divergence of a vector point function is a scalar point function.

If $\vec{A} = A_1\hat{i} + A_2\hat{j} + A_3\hat{k}$, then

$$div\, \vec{A} = \left(\hat{i}\frac{\partial}{\partial x} + \hat{j}\frac{\partial}{\partial y} + \hat{k}\frac{\partial}{\partial z}\right)\cdot(A_1\hat{i} + A_2\hat{j} + A_3\hat{k})$$

$$= \frac{\partial A_1}{\partial x} + \frac{\partial A_2}{\partial y} + \frac{\partial A_3}{\partial z} \qquad (\because \hat{i}\cdot\hat{i} = \hat{j}\cdot\hat{j} = \hat{k}\cdot\hat{k} = 1)$$

$$(\because \hat{i}\cdot\hat{j} = \hat{j}\cdot\hat{k} = \hat{k}\cdot\hat{i} = 0)$$

Example 4.38 **Find divergence of the vector $\vec{A} = (xyz)\,\hat{i} + (3x^2y)\,\hat{j} + (xz^2 - y^2z)\hat{k}$ at the piont (2, – 1, 1).**

Solution

$$div\, \vec{A} = \nabla\cdot\vec{A} = \frac{\partial}{\partial x}(xyz) + \frac{\partial}{\partial y}(3x^2y) + \frac{\partial}{\partial z}(xz^2 - y^2z)$$

$$= yz + 3x^2 + 2xz - y^2$$

$$= -1 + 3\times 4 + 2\times 2 - 1 = 14 \text{ at } (2, -1, 1)$$

4.15 SOLENOIDAL VECTOR

If $\vec{v}$ denote the veocity of a fluid in nature, the fluid is said to be incompresible, *i.e.*, $div\, \vec{v} = 0$. But in electromagnetic theory; if $div\, \vec{v} = 0$, then the vector field $\vec{v}$ is said to be solenoidal.

Example 4.39 **Show that the vector $3y^2z^2\hat{i} + x^2z^3\hat{j} + x^3y^2\hat{k}$ is solenoidal.**

Solution : The given vector is $3y^2z^2\hat{i} + x^2z^3\hat{j} + x^3y^2\hat{k} = \vec{V}$ (say)

$$\therefore \quad \text{Div}\, \vec{V} = \nabla\cdot\vec{V}$$

$$= \left(\hat{i}\frac{\partial}{\partial x} + \hat{j}\frac{\partial}{\partial y} + \hat{k}\frac{\partial}{\partial z}\right)\cdot(3y^2z^2\hat{i} + x^2xz^3\hat{j} + x^3y^2\hat{k})$$

$$= \frac{\partial}{\partial x}(3y^2z^2) + \frac{\partial}{\partial y}(x^2z^2) + \frac{\partial}{\partial z}(x^2z^2)$$

$$= 0 + 0 + 0 = 0$$

Thus, the given vector V is solenoidal.

Example 4.40 **Show that the vector field $\vec{A} = (x + 3y)\,\hat{i} + (y - 3z)\,\hat{j} + (x - 2z)\,\hat{k}$ is solenoidal.**

Solution : $\vec{A}$ is solenoidal if $div\, \vec{A} = 0$.

$$\vec{A} = (x + 3y)\,\hat{i} + (y - 3z)\,\hat{j} + (x - 2z)\,\hat{k}$$

$$\therefore \quad \text{div}\, \vec{A} = \nabla\cdot\vec{A} = \frac{\partial}{\partial x}(x + 3y) + \frac{\partial}{\partial y}(y + 3z) + \frac{\partial}{\partial z}(x - 2z)$$

$$= 1 + 1 - 2 = 0$$

Hence, $\vec{A}$ is solenoidal.

Example 4.41 **Show that div grad $r^n = n(n + 1)\, r^{n-2}$.**

Solution :

$$\text{grad } r^n = \nabla r^n = \left(\hat{i}\frac{\partial}{\partial x} + \hat{j}\frac{\partial}{\partial y} + \hat{k}\frac{\partial}{\partial z}\right) r^n$$

$$= \hat{i}nr^{n-1}\frac{\partial r}{\partial x} + \hat{j}nr^{n-1}\frac{\partial r}{\partial y} + \hat{k}\,nr^{n-1}\frac{\partial r}{\partial z}$$

$$= nr^{n-1}\left[\hat{i}\frac{x}{r} + \hat{j}\frac{y}{r} + \hat{k}\,\frac{z}{r}\right]$$

$$= nr^{n-2}\,[x\hat{i} + y\hat{j} + z\hat{k}]$$

Now div grad $r^n = \nabla\cdot[\nabla r^n]$

$$= \left(\hat{i}\frac{\partial}{\partial x} + \hat{j}\frac{\partial}{\partial y} + \hat{k}\,\frac{\partial}{\partial z}\right)\cdot\,[nr^{n-2}\,(x\hat{i} + y\hat{j} + z\hat{k})]$$

$$= n\left[\frac{\partial}{\partial x}\,(r^{n-2}\,x) + \frac{\partial}{\partial y}\,(r^{n-2}\,y) + \frac{\partial}{\partial z}\,(r^{n-2}\,z)\right]$$

$$= n\left[\left\{r^{n-2} + (n-2)\,xr^{n-3}\frac{\partial r}{\partial x}\right\} + \left\{r^{n-2} + (n-2)\,yr^{n-3}\frac{\partial r}{\partial y}\right\}\right.$$

$$\left. + \left\{r^{n-2} + (n-2)\,zr^{n-3}\frac{\partial r}{\partial z}\right\}\right]$$

$$= n\left[3r^{n-2} + (n-2)\,xr^{n-3}\,\frac{x}{r} + (n-2)\,yr^{n-3}\,\frac{y}{r} + (n-2)\,zr^{n-3}\,\frac{z}{r}\right]$$

$$= 3nr^{n-2} + n(n-2)\,r^{n-4}\,[x^2 + y^2 + z^2]$$

$$= 3nr^{n-2} + n(n-2)\,r^{n-2}$$

$$= r^{n-2}\,[3n + n^2 - 2n] = n\,(n+1)\,r^{n-2}$$

Example 4.42 **Show that** $\nabla^2 f(\vec{r}) = \dfrac{2}{r}\,f(\vec{r}) + f'(\vec{r})$**.**

Solution :

$$\nabla^2 f(\vec{r}) = \left(\frac{\partial^2}{\partial x^2} + \frac{\partial^2}{\partial y^2} + \frac{\partial^2}{\partial z^2}\right) f(r)$$

$$= \frac{\partial^2 f(\vec{r})}{\partial x^2} + \frac{\partial^2 f(\vec{r})}{\partial y^2} + \frac{\partial^2 f(\vec{r})}{\partial z^2}$$

$$= \frac{\partial}{\partial x}\left[\frac{\partial}{\partial x}\,f(\vec{r})\right] + \frac{\partial}{\partial y}\left[\frac{\partial}{\partial y}\,f(\vec{r})\right] + \frac{\partial}{\partial z}\left[\frac{\partial}{\partial z}\,f(\vec{r})\right]$$

$$= \frac{\partial}{\partial x}\left[f'\,(\vec{r})\frac{\partial r}{\partial x}\right] + \frac{\partial}{\partial y}\left[f'\,(\vec{r})\frac{\partial r}{\partial y}\right] + \frac{\partial}{\partial z}\left[f'\,(\vec{r})\frac{\partial r}{\partial z}\right]$$

$$= \frac{\partial}{\partial x}\left[f'\,(\vec{r})\frac{x}{r}\right] + \frac{\partial}{\partial y}\left[f'\,(\vec{r})\frac{y}{r}\right] + \frac{\partial}{\partial z}\left[f'\,(\vec{r})\frac{z}{r}\right]$$

$$= \Sigma\,\frac{\partial}{\partial x}\left[\frac{f'(\vec{r})}{r}\,x\right]$$

$$= \Sigma\left[\frac{f'(\vec{r})}{r} + f''(\vec{r})\frac{x}{r}\frac{\partial r}{\partial x} + x f'(r)\left(-\frac{1}{r^2}\frac{\partial r}{\partial x}\right)\right]$$

$$= \Sigma\left[\frac{f'(\vec{r})}{r} + \frac{x^2}{r^2} f''(\vec{r}) - \frac{x^2}{r^3} f'(\vec{})\right]$$

$$= \frac{3f'(\vec{r})}{r} + \frac{1}{r^2}(x^2+y^2+z^2) f''(\vec{r}) - \frac{1}{r^3}(x^2+y^2+z^2) f'(\vec{r})$$

$$= \frac{3f'(\vec{r})}{r} + \frac{f''(r)\, r^2}{r^2} - \frac{f'(r)\, r^2}{r^3}$$

$$= \frac{2}{r} f'(r) + f''(r)$$

Example 4.43 **If $\vec{r} = x\hat{i} + y\hat{j} + z\hat{k}$ and $r = |\vec{r}|$, show that $div\left(\frac{\vec{r}}{r^3}\right) = 0$.**

Solution : We have $\vec{r} = x\hat{\mathbf{i}} + y\hat{j} + z\hat{k}$ and $r = \sqrt{x^2+y^2+z^2}$

$$\therefore \quad \text{Div}\left(\frac{\vec{r}}{r^3}\right) = \nabla \cdot \frac{\vec{r}}{r^3}$$

$$= \left(\hat{i}\frac{\partial}{\partial x} + \hat{j}\frac{\partial}{\partial y} + \hat{k}\frac{\partial}{\partial z}\right)\cdot\left(\frac{x}{r^3}\hat{i} + \frac{y}{r^3}\hat{j} + \frac{z}{r^3}\hat{k}\right)$$

$$= \frac{\partial}{\partial x}\left(\frac{x}{r^3}\right) + \frac{\partial}{\partial y}\left(\frac{y}{r^3}\right) + \frac{\partial}{\partial z}\left(\frac{z}{r^3}\right)$$

$$= \frac{3}{r^3} - \frac{3}{r^4}\left[x\frac{\partial r}{\partial x} + y\frac{\partial r}{\partial y} + z\frac{\partial r}{\partial z}\right]$$

But, since $r^2 = x^2 + y^2 + z^2$, we get

$$\frac{\partial r}{\partial x} = \frac{x}{r}, \quad \frac{\partial r}{\partial y} = \frac{y}{r}, \quad \frac{\partial r}{\partial z} = \frac{z}{r}$$

$$\therefore \quad \text{Div}\left(\frac{\vec{r}}{r^3}\right) = \frac{3}{r^3} - \frac{3}{r^5}(x^2+y^2+z^2) = \frac{3}{r^3} - \frac{3}{r^3} = 0$$

Example 4.44 **Show that $f = ax^2 + by^2 + cz^2$ satisfies Laplace's equation $\nabla^2 f = 0$, provided $a + b + c = 0$.**

Solution :

$$\nabla^2 f = \left(\frac{\partial^2}{\partial x^2} + \frac{\partial^2}{\partial y^2} + \frac{\partial^2}{\partial z^2}\right) sub\,(ax^2 + by^2 + cz^2)$$

$$= \frac{\partial^2}{\partial x^2}(ax^2+by^2+cz^2) + \frac{\partial^2}{\partial y^2}(ax^2+by^2+cz^2) + \frac{\partial^2}{\partial z^2}(ax^2+by^2+cz^2)$$

$$= \frac{\partial}{\partial x}(2ax) + \frac{\partial}{\partial y}(2by) + \frac{\partial}{\partial z}(2cz)$$

$$= 2a + 2b + 2c = 2(a+b+c) = 0, \quad \text{if } a+b+c = 0$$

Example 4.45 **If $\vec{a}$ and $\vec{b}$ are constant vectors, $\vec{r} = x\hat{i} + y\hat{j} + z\hat{k}$, prove that**

$$\vec{a} \cdot \nabla \left[\vec{b} \cdot \nabla \left(\frac{1}{r} \right) \right] = \frac{3\,(\vec{a} \cdot \vec{r})\,(\vec{b} \cdot \vec{r})}{r^5} - \frac{\vec{a} \cdot \vec{b}}{r^3}$$

Solution ; Let $\vec{a} = a_1\hat{i} + a_2\hat{j} + a_3\hat{k}$ and $\vec{b} = b_1\hat{i} + b_2\hat{j} + b_3\hat{k}$

$$\vec{b} \cdot \nabla \left(\frac{1}{r} \right) = \left(b_1 \frac{\partial}{\partial x} + b_2 \frac{\partial}{\partial y} + b_3 \frac{\partial}{\partial z} \right) (x^2 + y^2 + z^2)^{-1/2}$$

$$= b_1 \frac{\partial}{\partial x} (x^2 + y^2 + z^2)^{-1/2} + b_2 \frac{\partial}{\partial y} (x^2 + y^2 + z^2)^{-1/2} + b_3 \frac{\partial}{\partial z} (x^2 + y^2 + z^2)^{-1/2}$$

$$= -\frac{1}{2} (x^2 + y^2 + z^2)^{-3/2}\, 2b_1x - \frac{1}{2} (x^2 + y^2 + z^2)^{-3/2}\, 2b_2y - \frac{1}{2} (x^2 + y^2 + z^2)^{-3/2}\, 2b_3z$$

$$= -(x^2 + y^2 + z^2)^{-3/2} (b_1x + b_2y + b_3z)$$

Now

$$(\vec{a} \cdot \nabla) \left\{ \vec{b} \cdot \nabla \left(\frac{1}{r} \right) \right\} = \left(a_1 \frac{\partial}{\partial x} + a_2 \frac{\partial}{\partial y} + a_3 \frac{\partial}{\partial} \right) \{ -(x^2 + y^2 + z^2)^{-3/2} (b_1x + b_2y + b_3z) \}$$

$$= -\left(-\frac{3}{2} \right) 2\, (x^2 + y^2 + z^2)^{-5/2} \{ (a_1x + a_2y + a_3z)\, b_1x + b_2y + b_z) \}$$

$$- (x^2 + y^2 + z^2)^{-3/2} \left(a_1 \frac{\partial}{\partial x} + a_2 \frac{\partial}{\partial z} + a_3 \frac{\partial}{\partial z} \right) (b_1x + b_2y + b_3z)$$

$$= 3r^{-5} (a_1x + a_2y + a_3z)(b_1x + b_2y + b_3z) - r^{-3} \{ a_1b_1 + a_2b_2 + a_3b_3 \}$$

$$= \frac{3}{r^5} (\vec{a} \cdot \vec{r})(\vec{b} \cdot \vec{r}) - \frac{1}{r^3} (\vec{a} \cdot \vec{b}) \text{ Ans.}$$

4.16 PHYSICAL INTERPRETATION OF DIVERGENCE

Let us consider the motion of a flux moving with velocity $V = v_x\hat{i} + v_y\hat{j} + v_z\hat{k}$ at any point $P(x, y, z)$ consider a small parallelopiped with edges. $\delta_x, \delta_y, \delta_z$ parallel to x-axis and z-axis respectively. Then the loss of fluid per unit volume per unit time in this small perallelopiped is given approximately by

$$\text{div } V = \nabla \cdot V$$

We have $V = v_x\hat{i} + v_y\hat{j} + v_z\hat{k}$

The amount of fluid entering the face PC in unit time.

$$= v_y \delta_z \delta_x$$

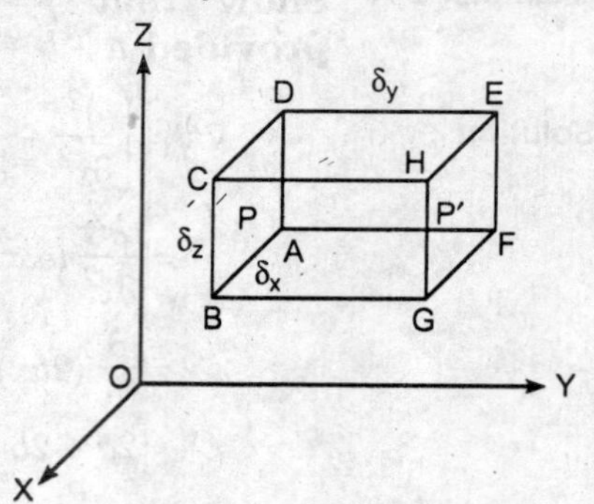

Fig. 4.5

And the amount of fluid leaving the face *P/F* in unit time.

$$= v_y + \delta_y \delta_z \delta_x$$

$$= \left(v_y + \frac{\partial v_y}{\partial y}\delta y\right)\delta z \delta x$$

Therefore, the net loss of the amount of fluid due to flow across these two faces

$$= \frac{dv_y}{\partial y}\delta x\ \delta y\ \delta z$$

In a similar way, we can find the loss of fluid from other two pairs of faces. Therefore, the total loss of the amount of fluid inside the parallelopiped per unit time

$$= \left(\frac{\partial v_x}{\partial x} + \frac{\partial v_y}{\partial y} + \partial\frac{v_z}{\partial z}\right)\delta x\ \delta y\ \delta z$$

Thus, the rate of loss of fluid per unit volume.

$$= \frac{\partial v_x}{\partial x} + \frac{\partial v_y}{\partial y} + \frac{\partial v_z}{\partial z} = div\ V$$

4.17 Curl of A Vector Point Function

Let we have a vector function $\vec{A} = A_1\hat{i} + A\hat{j} + A_3\hat{k}$, then the curl of a vector $\vec{A}$ is denoted by curl $\vec{A}$ and defined as

$$\text{curl } \vec{A} = \nabla \times \vec{A}$$

$$= \left(\frac{\partial}{\partial x}\hat{i} + \frac{\partial}{\partial z}\hat{j} + \frac{\partial}{\partial z}\hat{k}\right) \times (A_1\hat{i} + A_2\hat{j} + A_3\hat{k})$$

$$= \begin{vmatrix} \hat{i} & \hat{j} & \hat{k} \\ \frac{\partial}{\partial x} & \frac{\partial}{\partial y} & \frac{\partial}{\partial z} \\ A_1 & A_2 & A_3 \end{vmatrix}$$

$$= \hat{i}\left(\frac{\partial A_3}{\partial y} - \frac{\partial A_2}{\partial z}\right) + \hat{j}\left(\frac{\partial A_1}{\partial z} - \frac{\partial A_3}{\partial x}\right) + \hat{k}\left(\frac{\partial A_2}{\partial x} - \frac{\partial A_1}{\partial y}\right)$$

The curl of a vector point function is a vector point function.

Example 4.46 **If $\vec{A} = x\hat{i} + y\hat{j} + z\hat{k}$. Find the curl of the vector $\vec{A}$.**

Solution Curl $\vec{A} = \nabla \times \vec{A}$

$$= \begin{vmatrix} \hat{i} & \hat{j} & \hat{k} \\ \frac{\partial}{\partial x} & \frac{\partial}{\partial y} & \frac{\partial}{\partial z} \\ x & y & z \end{vmatrix} = \hat{i}\left[\frac{\partial}{\partial y}(z) - \frac{\partial}{\partial y}(y)\right] + \hat{j}\left[\frac{\partial}{\partial z}(x) - \frac{\partial}{\partial x}(z)\right] + ki\left[\frac{\partial}{\partial x}y - \frac{\partial}{\partial y}x\right]$$

$$= \hat{i}(0) + \hat{j}(0) + \hat{k}(0)$$

$$= \vec{0}$$

Example 4.47 **If $u\vec{F} = \nabla v$, where u, v are scalar fields and $\vec{F}$ is a vector field, show that $F, curl\ F = 0$.**

Solution : It is given that

$$\vec{F} = \frac{1}{u} \nabla v$$

$$\therefore \quad curl\ \vec{F} = \dot{\nabla} \times \left(\frac{1}{u} \nabla v\right)$$

$$= \nabla \frac{1}{u} \times \nabla v + \frac{1}{u} \nabla \times (\nabla v)$$

$$= \nabla \frac{1}{u} \times \nabla v \qquad (\because \nabla \times \nabla v = 0)$$

Hence, $F\ curl\ F = \frac{1}{u} \nabla v \cdot \left(\nabla \frac{1}{u} \times \nabla v\right) = 0$

($\because$ It is a triple product having two vectors identify)

4.18 IRROTATIONA VECTOR

A vector f is said to be irrotational if $\nabla \times f = 0$.

Example 4.48 **Show that vector field $\vec{F} = (x^2 - y^2 + x)\,\hat{i} - (2xy + y)\,\hat{j}$ is irrotational.**

Solution The given vector field is

$$\vec{F} = (x^2 - y^2 + x)\,\hat{i} - (2xy + y)\,\hat{j}$$

$$\text{Here, } \nabla \times F = \begin{vmatrix} \hat{i} & \hat{j} & \hat{k} \\ \frac{\partial}{\partial x} & \frac{\partial}{\partial y} & \frac{\partial}{\partial z} \\ x^2 - y^2 + x & -(2xy + y) & 0 \end{vmatrix}$$

$$= \hat{i}\left(0 + \frac{\partial}{\partial z}(2xy + y)\right) - \hat{j}\left(0 - \frac{\partial}{\partial z}(x^2 - y^2 + x)\right)$$

$$+ \hat{k}\left(-\frac{\partial}{\partial x}(2xy + y) - \frac{\partial}{\partial y}(x^2 - y^2 + x)\right)$$

$$= \hat{i} \cdot 0 - \hat{j}\,0 + k\,(-2y + 2y)$$

$$= 0$$

Hence, the given vector field $\vec{F}$ is irrotational.

Example 4.49 **A vector $\vec{r}$ is defined by $\vec{r} = x\hat{i} + y\hat{j} + z\hat{k}$. If $|\vec{r}| = r$, then show that the vector $r^n \vec{r}$ is irrotational.**

Solution : We know that a vector $\vec{r}$ is irrotational if

$$Curl\ \vec{r} = 0$$

$$\Rightarrow \quad \nabla \times \vec{r} = 0$$

$$\therefore \quad \nabla \times \vec{r} = \nabla \times r^n\, \vec{r}$$

$$= \left(\hat{i}\frac{\partial}{\partial x} + \hat{j}\frac{\partial}{\partial y} + \hat{k}\frac{\partial}{\partial z}\right) \times (x^2+y^2+z^2)^{n/2}(x\hat{i}+y\hat{j}+z\hat{k})$$

$$= \begin{vmatrix} \hat{i} & \hat{j} & \hat{k} \\ \frac{\partial}{\partial x} & \frac{\partial}{\partial y} & \frac{\partial}{\partial z} \\ (x^2+y^2+z^2)^{n/2}x & (x^2+y^2+z^2)^{n/2}y & (x^2+y^2+z^2)^{n/2}z \end{vmatrix}$$

$$= \left[\frac{n}{2}(x^2+y^2+z^2)^{n/2-1}(2yz) - (x^2+y^2+z^2)^{n/2-1}(2yz)\right]\hat{i}$$

$$+ \left[\frac{n}{2}(x^2+y^2+z^2)^{n/2-1}(2xz) - (x^2+y^2+z^2)^{n/2-1}(2xz)\right]\hat{j}$$

$$+ \left[\frac{n}{2}(x^2+y^2+z^2)^{n/2-1}(2xy) - (x^2+y^2+z^2)^{n/2-1}(2xy)\right]\hat{k}$$

$$= 0$$

Example 4.50 **A fluid motion is given by**

$$\vec{v} = (y \sin z - \sin x)\hat{i} + (x \sin z + 2yz)\hat{j} + (xy \cos z + y^2)\hat{k},$$

show that it is irrotational, find the velocity potnetial.

Solution : We know that $\vec{v}$ is irrotational if *curl* $\vec{v} = 0$

$$\therefore \quad Curl\ \vec{v} = \nabla \times \vec{v} = \left(\hat{i}\frac{\partial}{\partial x} + \hat{j}\frac{\partial}{\partial y} + k\frac{\partial}{\partial z}\right)$$

$$\times (y \sin z - \sin x)\hat{i} + (x \sin z + 2yz)\hat{j} + (xy \cos z + y^2)\hat{k}$$

$$= \begin{vmatrix} \hat{i} & \hat{j} & \hat{k} \\ \frac{\partial}{\partial x} & \frac{\partial}{\partial y} & \frac{\partial}{\partial z} \\ y \sin z - \sin z & x \sin z + 2yz & xy \cos z + y^2 \end{vmatrix}$$

$$= (x \cos z + 2y - x \cos z - 2y)\hat{i} - (y \cos z - y \cos z)\hat{j} + (\sin z - \sin z)\hat{k}$$

$$= 0$$

Hence the motion is irrotational.

So, $\vec{v} = \nabla\phi$, where ϕ is called velocity potential

Also $d\phi = \frac{\partial\phi}{dx}dx + \frac{\partial\phi}{dy}dy + \frac{\partial\phi}{dz}dz$

$$= \left(\hat{i}\frac{\partial\phi}{dx} + \hat{j}\frac{\partial\phi}{dy} + \hat{k}\frac{\partial\phi}{dz}\right)\cdot(\hat{i}\,dx + \hat{j}\,dy + \hat{k}\,dz) = \nabla\phi \,.\, d\vec{r} = \vec{v}\,.\,dr$$

$$= [(y \sin z - \sin x)\hat{i} + (x \sin z + 2yz)\hat{j} + (xy \cos z + y^2)\hat{k}]\,.\,[\hat{i}\,dx + \hat{j}\,dy + \hat{k}dz]$$

$$= (y \sin zdx + xdy \sin z + xy \cos zdz) - \sin x\,dx + (2yzdy + y^2dz)$$

$$= d(xy \sin z) + d(\cos x + d(y^2z)$$

$$\Rightarrow \phi = xy \sin z + \cos x + y^2 z + c$$

4.19 PHYSICAL INTERPRETATION OF CURL OF VECTOR

Consider the motion of a rigid body about a fixed axis through O. Let $\vec{w}$ be its angulr velocity where,

$$\vec{w} = w_1\hat{i} + w_2\hat{j} + w_3\hat{k}$$

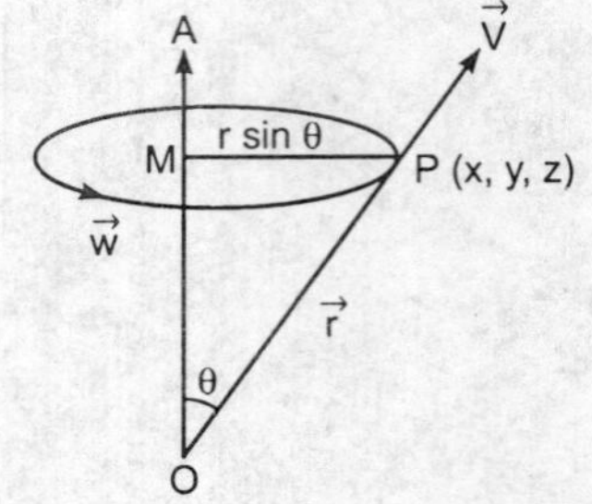

Fig. 4.6

Let $P\,(x, y, z)$be a point with position vector $\bar{r} = x\hat{i} + y\hat{j} + z\hat{k}$ and OP makes an angle θ with the fixed aixs OA.

The point P is descrbing a circle with centre M and radius, $PM = r \sin \theta$. Thus, the linear velocity of the body is $wr \sin \theta$. This velocity is perpendicular to the plane POM. Therefore,

$\vec{v} = wr \sin \theta\, \hat{n}$, where $\hat{n}$ is the unit vector perpendicular to the plane of $\vec{w}$ and $\vec{v}$.

$\therefore\ \vec{v} = \vec{w} \times \vec{r}$(by definition of cros product)

$$= \begin{vmatrix} \hat{i} & \hat{j} & \hat{k} \\ w_1 & w_2 & w_3 \\ x & y & z \end{vmatrix}$$

$$= (w_2 z - w_3 y)\,\hat{i} - (w_1 z - w_3 x)\,\hat{j} + (w_1 y - w_2 x)\,\hat{k}$$

Also, $\quad \text{curl } \bar{v} = \nabla \times \bar{v}$

$$= \begin{vmatrix} \hat{i} & \hat{j} & \hat{k} \\ \dfrac{\partial}{\partial x} & \dfrac{\partial}{\partial y} & \dfrac{\partial}{\partial z} \\ (w_2 z - w_3 y) & (w_2 x + w_1 z) & (w_1 y - w_2 x) \end{vmatrix}$$

$$= (w_1 + w_1)\,\hat{i} + (w_2 + w_2)\,\hat{j} + (w_3 + w_3)\,\hat{k}$$

$$= 2\,(w_1\hat{i} + w_2\hat{j} + w_3\hat{k})$$

Thus, it concludes that, if a rigid body is in motion, the curl of its linear velocity at any point gives twice its angular velocity, *i.e.*,

$$\text{curl } \vec{v} = 2\vec{w}$$

4.20 IMPORTANT VECTOR IDENTITIES

(i) Prove that $\nabla \cdot (\nabla \times \vec{a}) = 0$ or div curl $\vec{a} = 0$

Proof : Let $\vec{a} = a_1\,\hat{i} + a_2\hat{j} + a_3\hat{k}$

$$\nabla \times \vec{a} = \begin{vmatrix} \hat{i} & \hat{j} & \hat{k} \\ \dfrac{\partial}{\partial x} & \dfrac{\partial}{\partial y} & \dfrac{\partial}{\partial z} \\ a_1 & a_2 & a_3 \end{vmatrix}$$

$$= \left(\frac{\partial a_3}{\partial y} - \frac{\partial a_2}{\partial z}\right)\hat{i} + \left(\frac{\partial a_1}{\partial z} - \frac{\partial a_3}{\partial x}\right)\hat{j} + \left(\frac{\partial a_2}{\partial x} - \frac{\partial a_1}{\partial y}\right)\hat{k}$$

Now, $\nabla \cdot (\nabla \times (\vec{a}) = \left(\hat{i}\frac{\partial}{\partial x} + \hat{j}\frac{\partial}{\partial y} + \hat{k}\frac{\partial}{\partial z}\right) \cdot \left[\left(\frac{\partial a_3}{\partial y} - \frac{\partial a_2}{\partial y}\right)\hat{i} + \left(\frac{\partial a_1}{\partial z} - \frac{\partial a_3}{\partial x}\right)\hat{j} + \left(\frac{\partial a_2}{\partial x} - \frac{\partial a_1}{\partial y}\right)\hat{k}\right]$

$$= \Sigma \frac{\partial}{\partial x}\left[\frac{\partial a_3}{\partial y} - \frac{\partial a_2}{\partial z}\right] = \Sigma\left(\frac{\partial^2 a_3}{\partial x \partial y} - \frac{\partial^2 a_2}{\partial x \partial z}\right)$$

$$= \frac{\partial^2 a_3}{\partial x \partial y} - \frac{\partial^2 a_2}{\partial x \partial z} + \frac{\partial^2 a_1}{\partial y \partial z} - \frac{\partial^2 a_3}{\partial y \partial x} + \frac{\partial^2 a_2}{\partial z \partial x} - \frac{\partial^2 a_2}{\partial z \partial y}$$

$$= 0$$

(ii) Prove that $\nabla \times (\vec{a} \times \vec{b}) = (\vec{b} \cdot \nabla)\,\vec{a} - \vec{b}\,(\nabla \cdot \vec{a}) - (\vec{a} \cdot \nabla)\,\vec{b} \times \vec{a}\,(\nabla \cdot \vec{b})$

or Curl $(\vec{a} \times \vec{b}) = (b \cdot \nabla)\,\vec{a} - \vec{b}\,div\,\vec{a} - (\vec{a} \cdot \nabla)\,div\,\vec{b}$

Proof : $\nabla \times (\vec{a} \times \vec{b}) = \left(\hat{i}\frac{\partial}{\partial x} + \hat{j}\frac{\partial}{\partial y} + \hat{k}\frac{\partial}{\partial z}\right) \times (\vec{a} \times \vec{b})$

$$= \Sigma\, \hat{i}\frac{\partial}{\partial x} \times (\vec{a} \times \vec{b})$$

$$= \Sigma\, \hat{i} \times \left(\vec{a} \times \frac{\partial \vec{b}}{\partial x} + \frac{\partial \vec{a}}{\partial x} \times \vec{b}\right)$$

$$= \Sigma\left[i \times \left(\vec{a} \times \frac{\partial \vec{b}}{\partial x}\right)\right] + \Sigma\left[\hat{i} \times \left(\frac{\partial \vec{a}}{\partial x} \times \vec{b}\right)\right]$$

$$= \Sigma\left(i \cdot \frac{\partial \vec{b}}{\partial x}\right)\vec{a} - \Sigma\,(i \cdot \vec{a})\frac{\partial \vec{b}}{\partial x} + \Sigma\,(i \cdot \vec{b})\frac{\partial \vec{a}}{\partial x} - \Sigma\left(i \cdot \frac{\partial \vec{a}}{\partial x}\right)\vec{b}$$

$$= (\nabla \cdot \vec{b})\,\vec{a} - (\vec{a} \cdot \nabla)\,\vec{b} + (\vec{b} \cdot \nabla)\,\vec{d} - (\nabla \cdot \vec{a})\,\vec{b}$$

or $\nabla \times (\vec{a} \times \vec{b}) = (\vec{b} \cdot \nabla)\,\vec{a} - \vec{b}\,(\nabla \cdot \vec{a}) - (\vec{a} \cdot \nabla)\,\vec{b} + \vec{a}\,(\nabla \cdot \vec{b})$

(iii) Prove that $\nabla \cdot (\phi\vec{a}) = (\nabla\,\phi) \cdot \vec{a} + \phi\,(\nabla \cdot \vec{a})$

Proof : $\nabla \cdot (\phi\,\vec{a}) = \left(\hat{i}\frac{\partial}{\partial x} + \hat{j}\frac{\partial}{\partial y} + \hat{k}\frac{\partial}{\partial z}\right) \cdot (\phi\,\vec{a})$

$$= \Sigma\,\hat{i}\frac{\partial}{\partial x} \cdot (\phi\,\vec{a})$$

$$= \Sigma\,\hat{i} \cdot \left(\phi\frac{\partial \vec{a}}{\partial x} + \vec{a}\frac{\partial \phi}{\partial x}\right)$$

$$= \Sigma\,\hat{i} \cdot \left(\phi\frac{\partial \vec{a}}{\partial x}\right) + \Sigma\,\hat{i} \cdot \left(\vec{a}\frac{\partial \phi}{\partial x}\right)$$

$$= \Sigma\,\phi\left(\hat{i} \cdot \frac{\partial \vec{a}}{\partial x}\right) + \Sigma\frac{\partial \phi}{\partial x}(\hat{i} \cdot \vec{a})$$

$$= \phi\,(\nabla \cdot \vec{a}) + (\nabla\,\phi) \cdot \vec{a}$$

(iv) Prove that div $(\vec{a}\times\vec{b}) = \vec{b}\cdot(\nabla\times\vec{a}) - \vec{a}\cdot(\nabla\times\vec{b})$

Proof : div $(\vec{a}\times\vec{b}) = \nabla\cdot(\vec{a}\times\vec{b}) = \left(\hat{i}\frac{\partial}{\partial x} + \hat{j}\frac{\partial}{\partial y} + \hat{k}\frac{\partial}{\partial z}\right)\cdot(\vec{a}\times\vec{b})$

$$= \hat{i}\cdot\frac{\partial}{\partial x}(\vec{a}\times\vec{b}) + \hat{j}\cdot\frac{\partial}{\partial y}(\vec{a}\times\vec{b}) + \hat{k}\frac{\partial}{\partial z}(\vec{a}\times\vec{b})$$

$$= \Sigma\, i\cdot\left(\vec{a}\times\frac{\partial\vec{b}}{\partial x} + \frac{\partial\vec{a}}{\partial x}\times\vec{b}\right)$$

$$= -\Sigma\left(\hat{i}\times\frac{\partial\vec{b}}{\partial x}\cdot\vec{a}\right) + \Sigma\left(\hat{i}\times\frac{\partial\vec{a}}{\partial x}\cdot\vec{b}\right)$$

$[\because \vec{a}\cdot\vec{b}\times\vec{c} = \vec{a}\times\vec{b}\cdot\vec{c} = -\vec{d}\times\vec{c} = -\vec{d}\times\vec{c}\cdot\vec{b}$ etc.]

$$= \Sigma\left(\hat{i}\times\frac{\partial\vec{a}}{\partial x}\cdot\vec{b}\right) - \left(\hat{i}\times\frac{\partial\vec{b}}{\partial x}\cdot\vec{a}\right)$$

$$= \left(\hat{i}\times\frac{\partial\vec{a}}{\partial x} + \hat{j}\times\frac{\partial\vec{a}}{\partial y} + \hat{k}\times\frac{\partial\vec{a}}{\partial z}\right)\cdot\vec{b} - \left(\hat{i}\times\frac{\partial\vec{b}}{\partial x} + \hat{j}\times\frac{\partial\vec{b}}{\partial y} + \hat{k}\times\frac{\partial\vec{b}}{\partial z}\right)\cdot\vec{a}$$

$$= (\nabla\times\vec{a})\cdot\vec{b} - (\nabla\times\vec{b})\cdot\vec{a}$$

or div $(\vec{a}\times\vec{b}) = \vec{b}\cdot(\nabla\times\vec{a}) - \vec{a}\cdot(\nabla\times\vec{b})$

or div $(a\times\vec{b}) = \vec{b}\cdot(\text{curl }\vec{a}) - \vec{a}\cdot(\text{curl }\vec{b})$

(v) Prove div grad $\phi = \nabla^2\phi$

Proof : div $\nabla\phi = \nabla\cdot(\nabla\phi)$

$$= \left(\hat{i}\frac{\partial}{\partial x} + \hat{j}\frac{\partial}{\partial y} + \hat{k}\frac{\partial}{\partial z}\right)\cdot\left(\hat{i}\frac{\partial\phi}{\partial x} + \hat{j}\frac{\partial\phi}{\partial y} + \hat{k}\frac{\partial\phi}{\partial z}\right)$$

$$= \frac{\partial}{\partial x}\left(\frac{\partial\phi}{\partial x}\right) + \frac{\partial}{\partial y}\left(\frac{\partial\phi}{\partial y}\right) + \frac{\partial}{\partial z}\left(\frac{\partial\phi}{\partial z}\right)$$

(vi) Prove that grad $(\vec{a}\vec{b}) = \nabla(\vec{a}\cdot\vec{b}) = (\vec{b}\cdot\nabla)\vec{a} + (\vec{a}\cdot\nabla)\vec{b} + \vec{b} + \text{curl }\vec{a} + \vec{a}\times\text{curl }\vec{b}$

Proof : We take right hand side

$$= (\vec{b}\cdot\nabla)\vec{a} + (\vec{a}\cdot\nabla)\vec{v} + b\times\text{curl }\vec{a} + \vec{a}\times\text{curl }\vec{b}$$

$$= (\vec{b}\cdot\nabla)\vec{a} + (\vec{a}\cdot\nabla)\vec{b} + \vec{b}\times\left[\Sigma\,\hat{i}\frac{\partial}{\partial x}\times\vec{a}\right] + \vec{a}\times\left[\Sigma\,\hat{i}\frac{\partial}{\partial x}\times\vec{b}\right]$$

$$= (\vec{b}\cdot\nabla)\vec{d} + (\vec{a}\cdot\nabla)\vec{b} + \vec{b}\times\left[\Sigma\,\hat{i}\times\frac{\partial\vec{a}}{\partial x}\right] + \vec{a}\times\left[\Sigma\,\hat{i}\frac{\partial\vec{b}}{\partial x}\right]$$

$$= (\vec{b}\cdot\nabla)\vec{a} + (\vec{a}\cdot\nabla) + \Sigma\,\hat{i}\left(\vec{b}\cdot\frac{\partial\vec{a}}{\partial x}\right) - \Sigma(\vec{b}\cdot\hat{i})\frac{\partial\vec{a}}{\partial x}$$

$$= \Sigma\,\hat{i}\left(\vec{a}\cdot\frac{\partial\vec{b}}{\partial x}\right) - \Sigma(\vec{a}\cdot\hat{i})\frac{\partial\vec{b}}{\partial x}$$ [Using $\vec{a}\times(\vec{b}\times\vec{c}) = (\vec{a}\cdot\vec{c})\vec{b} - (\vec{a}\cdot\vec{b})\vec{c}$]

$$= (\vec{b}\cdot\nabla)\,\vec{a} + (\vec{a}\cdot\nabla)\,\vec{b} + \Sigma\,\hat{i}\left(\vec{b}\cdot\frac{\partial\vec{a}}{\partial x}\right) - \vec{b}\cdot\left(\Sigma\,\hat{i}\,\frac{\partial}{\partial x}\right)\vec{a} + \Sigma\,\hat{i}\left(\vec{a}\cdot\frac{\partial\vec{b}}{\partial x}\right) - \vec{a}\cdot\left(\Sigma\,\hat{i}\,\frac{\partial}{\partial x}\right)\vec{b}$$

$$= \Sigma\,\hat{i}\left(\vec{b}\cdot\frac{\partial\vec{a}}{\partial x}\right) + \Sigma\,\hat{i}\left(\vec{a}\cdot\frac{\partial\vec{b}}{\partial x}\right) + (\vec{b}\cdot\nabla)\,\vec{a} + (\vec{a}\cdot grd)\,\vec{b} - \vec{b}\cdot\left(\Sigma\,\hat{i}\,\frac{\partial}{\partial xz}\right)\mathbf{a} - \vec{a}\cdot\left(\Sigma\,\hat{i}\,\frac{\partial}{\partial x}\right)\vec{b}$$

$$= \Sigma\,\hat{i}\left\{\left(\vec{b}\cdot\frac{\partial\vec{a}}{\partial x}\right) + \left(\vec{a}\cdot\frac{\partial\vec{b}}{\partial x}\right)\right\} + (\vec{b}\cdot\nabla)\,\vec{a} + (\vec{a}\cdot\nabla)\,\vec{b} - \vec{b}\cdot\nabla\,\vec{a} - \vec{a}\cdot\nabla\vec{b}$$

$$= \Sigma\,\hat{i}\,\frac{\partial}{\partial x}\,(\vec{b}\cdot\vec{a}) = \nabla\,(\vec{b}\cdot\vec{a}) = \text{grad}\,(\vec{a}\cdot\vec{b}) = \text{L.H.S.}$$

(vii) Prove that $\nabla\times(\nabla\times\vec{a}) = \nabla\,(\nabla\cdot\vec{a}) - \nabla^2\vec{a}$

Proof : Let $\vec{a} = a_1\,\hat{i} + a_2\hat{j} + a_3\hat{k}$

$$\therefore \qquad \nabla\times\vec{a} = \begin{vmatrix} \hat{i} & \hat{j} & \hat{k} \\ \frac{\partial}{\partial x} & \frac{\partial}{\partial y} & \frac{\partial}{\partial z} \\ a_1 & a_2 & a_3 \end{vmatrix}$$

$$= \hat{i}\left[\frac{\partial a_3}{\partial y} - \frac{\partial a_2}{\partial z}\right] + \hat{j}\left[\frac{\partial a_1}{\partial z} - \frac{\partial a_3}{\partial x}\right] + \hat{k}\left[\frac{\partial a_2}{\partial x} - \frac{\partial a_1}{\partial y}\right]k$$

$$\text{So} \qquad \nabla\times(\nabla\times\vec{a}) = \begin{vmatrix} \hat{i} & \hat{j} & \hat{k} \\ \frac{\partial}{\partial x} & \frac{\partial}{\partial y} & \frac{\partial}{\partial z} \\ \frac{\partial a_3}{\partial y} - \frac{\partial a_2}{\partial z} & \frac{\partial a_1}{\partial z} - \frac{\partial a_3}{\partial x} & \frac{\partial a_1}{\partial x} - \frac{\partial a_1}{\partial y} \end{vmatrix}$$

$$= \Sigma\,\hat{i}\left[\frac{\partial}{\partial y}\left(\frac{\partial z_2}{\partial x} - \frac{\partial a_1}{\partial y}\right) - \frac{\partial}{\partial z}\left(\frac{\partial a_1}{\partial z} - \frac{\partial a_3}{\partial x}\right)\right]$$

$$= \Sigma\,\hat{i}\left[\frac{\partial^2 a_2}{\partial y\,\partial x} - \frac{\partial^2 a_1}{\partial y^2} - \frac{\partial^2 a_1}{\partial z^2} + \frac{\partial^2 a_3}{\partial z\,\partial x}\right]$$

$$= \Sigma\,\hat{i}\left\{\frac{\partial}{\partial x}\left[\frac{\partial a_2}{\partial y} + \frac{\partial a_3}{\partial z}\right] - \left[\frac{\partial^2 a_1}{\partial y^2} + \frac{\partial^2 a_1}{\partial z^2}\right]\right\}$$

Adding $\frac{\partial}{\partial x}\left(\frac{\partial a_1}{\partial x}\right)$ in first term and subtracting if from second term

$$= \Sigma\,\hat{i}\,\frac{\partial}{\partial x}\left[\frac{\partial a_1}{\partial x} + \frac{\partial a_2}{\partial y} + \frac{\partial a_3}{\partial z}\right] - \Sigma\,i\left[\frac{\partial^2 a_1}{\partial x^2} + \frac{\partial^2 a_1}{\partial y^2} + \frac{\partial^2 a_1}{\partial z^2}\right]$$

$$= \Sigma\,\hat{i}\,\frac{\partial}{\partial x}\,(\nabla\cdot\vec{a}z0 - \nabla^2\,\Sigma\,(a_1\,\hat{i})$$

$$= \nabla\,(\nabla\cdot\vec{d}) - \nabla^2\,\vec{a}$$

Example 4.51 **For the vector field $\vec{V} = (x^2 - y^2 + 2xz)\,\hat{i} + (xz - xy + yz)\,\hat{j} + (z^2 + x^2)\,\hat{k}$, find curl $\vec{V}$. Show that the vectors given by curl $\vec{V}$ at $P(1, 2, -3)$ and $Q(2, 3, 12)$ are orthogonal.**

Solution : We know that

$$\text{Curl } \vec{V} = \nabla \times \vec{V}$$

$$= \left(\hat{i}\frac{\partial}{\partial x} + \hat{j}\frac{\partial}{\partial y} + \hat{k}\frac{\partial}{\partial z}\right) \times [(x^2 - y^2 + 2xz)\,\hat{i} + (xz - xy + yz)\,\hat{j} + (z^2 + x^2)\,\hat{k}]$$

$$= \begin{vmatrix} \hat{i} & \hat{j} & \hat{k} \\ \dfrac{\partial}{\partial x} & \dfrac{\partial}{\partial y} & \dfrac{\partial}{\partial z} \\ x^2 - y^2 + 2xz & xz - xy + yz & z^2 + x^2 \end{vmatrix}$$

$$= -(x + y)\hat{i} - (2x - 2x)\,\hat{j} + (z - y + 2y)\,\hat{k}$$

$$= -(x + y)\,\hat{i} + (y + z)\,\hat{k}$$

Curl $\vec{V}$ at $P(1, 2, -3) = -(1 + 2)\,\hat{i} + (2, -3)\,\hat{k} = -3\hat{i} - \hat{k}$

Curl $\vec{V}$ at $Q(2, 3, 12) = -(2 + 3)\,\hat{i} + (3 + 12)\,\hat{k} = -5\hat{i} + 15\hat{k}$

$\because \quad (-3\hat{i} - \hat{k}) \,.\, (-5\hat{i} + 15\hat{k}) = 15 - 15 = 0$

So, Curl $\vec{V}$ at P and Q are perpendicular.

PROBLEM SET 4.3

1. Find divergence and curl of a vector function
$$\vec{F} = (x^2 - y^2)\,\hat{i} + 2xy\,\hat{j} + (y^2 - xy)\,\hat{k}$$
Ans. $4x$ and $(2y - x)\,\hat{i} + y\hat{j} + 4y\hat{k}$

2. Find divergence and curl of $\vec{F}$, where $\vec{F} = \text{grad } (x^3 + y^3 + z^3 - 3xyz)$
Ans. $6\,(x + y + z)$ and 0

3. Find div (grad ϕ); if $\phi = 2x^3y^2z^4$
Ans. $12xy^2z^4 + 4x^3z^4 + 24x^3y^2z^2$

4. Find div and curl of vector $\vec{v} = xe^{-y}\,\hat{i} + 2ze^{-y}\,\hat{j} + xy^2\,\hat{k}$ and verify that div (curl $\vec{V}$) = 0.
Ans. $(1 - 2z)e^{-y}$, $[2\hat{i}(xy - e^{-y}) - \hat{j}y^2 + \hat{k}\,xe^{-y}]$

5. For the scalar field $f(x, y, z) = e^{x + y + z}$, complete grad f and verify that curl (grad f) = 0.
Ans. $(\hat{i} + \hat{j} + \hat{k})\,e^{x + y + z}$

6. If $\vec{R} = x\hat{i} + y\hat{j} + z\hat{k}$, show that
(i) div $\vec{R} = 3$ and (ii) curl $\vec{R} = 0$

7. If $F = (x + y + z)\,\hat{i} + \hat{j} + (-x - y)\,\hat{k}$, prove that F. curl $F = 0$. ***(UPTU 2000)***

8. If $u = x^2yz$, $v = xy - 3z^2$. Find
(i) $\nabla\,(\nabla u \cdot \nabla v)$ and (ii) $\nabla \cdot (\nabla u \times \nabla v)$.
Ans. (i) $2\,(y^3 + 3x^2y - 6xz^2)\,z\hat{i} + 2\,(3xy^2 + x^3 - 6x^2y)\,z\hat{j} + 2\,(xy^2 + x^3 - 3x^2y)\,y\hat{k}$ (ii) 0

9. Show that the vector $3y^4z^2\hat{i} + 4x^3z^2\hat{j} + 3x^2y^2\hat{k}$ is solenoidal.

10. Find the value of α for which the vector field,
$\vec{V} = \alpha(x + y)\hat{i} + 4y\hat{j} + 3\hat{k}$ is solenoidal.
Ans. $\alpha = -4$

11. Show that a fluid motion given by $\vec{V} = (y+z)\hat{i} + (z+x)\hat{j} + (x+y)\hat{k}$ is irrotational. Also find the velocity potential. ***(UPTU 2004)***
Ans. $xy + yz + zx$

12. Show that the vector $\vec{v} = e^{xy}(y\hat{i} + x\hat{j}) + 2e^z\hat{k}$ is irrotational and find a scalar function $f(x, y, z)$ such that $v = grad\, f$.
Ans. $e^{xy} + 2e^z + c$

13. Show that the vector $\vec{v} = \cos(x^2 + y^2 + z^2)(x\hat{i} + y\hat{j} + z\hat{k})$ is irrotational, find a scalar function $\phi(x, y, z)$ such that $\vec{v} = grad\, \phi$.
Ans. $\frac{1}{2}\sin(x^2 + y^2 + z^2) + c$

14. Find the constants a, b and c such that
$\vec{v} = (3x + ay + z)\hat{i} + (2x - y + bz)\hat{j} + (x + cy + z)\hat{k}$ is irrotational.
Ans. $a = 2$, $b = c$, c is arbitrary.

15. For a vector $V = x^2\hat{i} + xye^x\hat{j} + \sin z\,\hat{k}$, show that $\nabla \cdot (\nabla \times V) = 0$.

16. Prove that $\text{div}\,(f\nabla g) - \text{div}\,(g\nabla f) = f\nabla^2 g - g\nabla^2 f$. ***(UPTU 2001)***

17. Find constants a, b and c so that the vector
$\vec{F} = (x + 2y + az)\hat{i} + (bx - 3y - z)\hat{j} + (4x + cy + 2z)\hat{k}$ is irrotational.
Ans. $a = 4$, $b = 2$, $c = -1$

18. Prove that curl grad $r^n = 0$

19. If $\vec{a}$ is a constant vector and $\vec{r} = x\hat{i} + y\hat{j} + z\hat{k}$, prove that
(i) grad $(\vec{a} \cdot \vec{r}) = \vec{a}$
(ii) div $(\vec{a} \times \vec{r}) = 0$
(iii) curl $(\vec{a} \times \vec{r}) = 2\vec{a}$ and
(iv) curl $[(\vec{a} \cdot \vec{r})\vec{r}] = \vec{a} \times \vec{r}$

20. For a solenoidal vector $\vec{F}$, show that curl curl curl curl $F = \nabla^4 F$.

4.21 INTEGRATION OF VECTOR FUNCTIONS

The integration may be defined as the reverse process of differentiation. If $f(t)$ and $F(t)$ are two vector functiosn of a single scalar t, such that

$$\frac{d}{dt}(F(t)) = f(t)$$

Then $F(t)$ is called the indefinite integral of $f(t)$ with respect to t.

$$\int f(t)\,dt = F(t) + C$$

The function, which is to be integrated is called the integrand.

4.22 LINE INTEGRAL

The concept of line integral is the generalization of definite integral in calculus. In definite integral a function, $f(x)$ is to be integrated from $x = a$ to $x = b$ along x-axis. But in line integral a vector function is to be integrated along a curve C in space.

Definition : Let $r(t) = x(t)\,\hat{i} + y(t)\hat{j} + z(t)\,\hat{k}$ be the position vector of (x, y, z) and C be a curve joining points P_1 and P_2. Then, the line integral of the vector function $F(r)$ over the curve C is defined by.

$$\int_C F(r) \cdot dr = \int_{P_1}^{P_2} F(r(t)) \cdot \frac{dr}{dt}\, dt$$

If $F(x\, y, z) = F_1\hat{i} + F_2\hat{j} + F_3\hat{k}$ be a position vector function defined and continuous inside C. Then the integral becomes

$$\int_C F \cdot dr = \int_C (F_1 dx + F_2 dy + F_3 dz)$$

4.23 GENERAL PROPERTIES OF THE LINE INTEGRAL

Similar to the properties of integral in calculus we can obtain corresponding formula for line integral.

(i) $\int_C kF{\cdot}dr = k\int_C F{\cdot}dr$ where k is constant

(ii) $\int_C (F + G){\cdot}dr = \int_C F{\cdot}dr + \int_C G{\cdot}dr$

(iii) $\int_C F{\cdot}dr = \int_{C_1} F{\cdot}dr + \int_{C_2} F{\cdot}dr$

Here the path C is subdivided in two arcs C_1 and C_2 of same orientation as C.

4.24 SOME MORE RESULTS ON INTEGRATION

1. $\int \left(\frac{dr}{dt} \cdot s + r \cdot \frac{ds}{dt}\right) dt = r{\cdot}s + c$

2. $\int \left(2r \cdot \frac{dr}{dt}\right) dt = r^2 + c$

3. $\int \left(r \times \frac{d^2r}{dt^2}\right) dt = r \times \frac{dr}{dt} + c$

Here, the constant of integration is a vector quantity.

Example 4.52 **Evaluate $\int_1^3 (a{\cdot}b \times c)\, dt$**

where $a = t\,\hat{i} - 3\hat{j} + 2t\hat{k},\ b = \hat{i} - 2\hat{j} + 2\hat{k},\ c = 3\hat{i} + t\hat{j} - \hat{k}$

Solution : We can obtain $a{\cdot}b \times c = [a, b, c]$

$$= \begin{vmatrix} t & -3 & 2t \\ 1 & -2 & 2 \\ 3 & t & -1 \end{vmatrix}$$

$$= \begin{vmatrix} t & -3+2t & 0 \\ 1 & 0 & 0 \\ 3 & t+t & -7 \end{vmatrix}$$

Applying $C_2 \to C_2 + 2C_1$, $C_3 \to C_3 - 2C_1$

$$= \begin{vmatrix} t & -3+2t & 0 \\ 1 & 0 & 0 \\ 3 & t+0 & -7 \end{vmatrix}$$

$$= 7\,(2t-3)$$

Hence, $\int_1^3 (a \cdot b \times c)\, dt = \int_1^3 7\,(2t-3)\, dt$

$$= 7\int_1^3 (2t-3)\, dt$$

$$= 7\,[t^2 - 3t]_1^3$$

$$= 7\,[9 - 9 - 1 + 3] = 14$$

Example 4.53 **Evaluate**

$$\int_0^1 (e^t\hat{i} + e^{-2t}\hat{j} + t\hat{k})\, dt$$

Solution : We have

$$\int_0^1 (e^t\hat{i} + e^{-2t}\hat{j} + t\hat{k})\, dt = \left[e^t\hat{i} + \frac{e^{-2t}}{-2}\hat{j} + \frac{t^2}{2}\hat{k} \right]_0^1$$

$$= e\hat{i} - \frac{e^{-2}}{2}\hat{j} + \frac{1}{2}\hat{k} - i + \frac{1}{2}\hat{j}$$

$$= (e-1)\,\hat{i} - \frac{1}{2}(e^{-2} - 1)\hat{j} + \frac{1}{2}\hat{k}$$

Example 4.54 **Evaluate $\int_C F \cdot dr$, where $\vec{F} = x^2 i + y^3 j$ and curve C is the arc of the parabola $y = x^2$ in xy-plane from (0, 0) to (1, 1).**

Solution : It is given that the curve C is the parabola $y = x^2$ from (0, 0) to (1, 1). We know that the parametric form of the parabola is $x = t$ and $y = t^2$.
if r is the position vector of any point (x, y) on C, then

$$r = ti + t^2 j \text{ and } \frac{dr}{dt} = i + 2tj$$

Also in terms of t, $F = t^2 i + t^6 j$
At the point (0, 0), $t = 0$ and at (1, 1), $t = 1$

$$\therefore \quad \int_C F \cdot dr = \int_C \left(F \cdot \frac{dr}{dt} \right) dt$$

$$= \int_0^1 (t^2 i + t^6 j) \cdot (i + 2tj) dt$$

$$= \int_0^1 (t^2 + 2t^7)\, dt = \left[\frac{t^3}{3} + \frac{2t^8}{8}\right]_0^1$$

$$= \frac{1}{3} + \frac{2}{8} = \frac{7}{12}$$

4.25 WORK DONE BY A FORCE

Let a force $\vec{F}$ is acted upon a particle m and the particle is displaced along a given path C in space. If r denotes the position vector of a point on C, then $\frac{dr}{dt}$ is a unit vector along the tangent to C at the ponit r. Since, the component of force F along tangent to C is $F \cdot \frac{dr}{dt}$. Therefore, the workdone by force F during the small displacement dt of a particle along C is

$$\left[\vec{F} \cdot \frac{dr}{dt}\right] dt \text{ or } F \cdot dr$$

and the total workdone by F in this displacement along C can be given by the line integral

$$W = \int_C F \cdot dr$$

Example 4.55 **Find the total work done $\int_C \vec{F} \cdot dr$, where $F = 3xy\hat{i} - y^2\hat{j}$ and curve C is the arc of the parabola $y = 2x^2$ in the xy plane from (0, 0) to (1, 2).**

Solution : The curve C is given as parabola $y = 2x^2$ from (0, 0) to (1, 2).

We know that in xy-plane $r = xi + yj$

$\therefore \quad dr = dx\hat{i} + dy\hat{j}$

Also $\quad \vec{F} = 3xy\hat{i} - y^2\hat{j}$

$\therefore \quad \vec{F} \cdot dr = (3xy\hat{i} - y^2\hat{j}) \cdot (dx\hat{i} + dy\hat{j}) = 3xy\, dx - y^2\, dy$

Now, the curve C is $y = 2x^2$.

Therefore, $dy = 4x\, dx$

$$\therefore \quad \int_C \vec{F} \cdot dr = \int_0^1 [3x \cdot 2x^2\, dx - 4x^4 \cdot 4x\, dx]$$

$$= \int_0^1 [6x^3 - 16x^5]\, dx$$

$$= \left[\frac{6x^4}{4} - 16\frac{x^6}{6}\right]_0^1 = \left[\frac{3}{2} - \frac{8}{3}\right]$$

$$= \frac{9 - 16}{6} = \frac{-7}{6}$$

$\therefore$ Work done is $\frac{7}{7}$ unit.

Example 4.56 **Find the workdone in moving a particle in a force field $\vec{F} = 3x^2\vec{i} + (2xz - y)\vec{j} + z\vec{k}$ along curve defined by $x^2 = 4y, 3x^2 = 8z$ from $x = 0$ to $x = 2$.**

Solution : In the curve $x^2 = 4y, 3x^2 = 8z$ from $x = 0$ to $x = 2$, we substitute $x = t$ in $x^2 = 4y$ and $3x^3 = 8z$. Then the parametric equations of C are $x = t, y = \frac{t^2}{4}$, $z = \frac{3t^3}{8}$ and t varies from 0 to 2.

$\therefore$ The required workdone.

$$= \int_C F \cdot dr$$

$$= \int_C (3x^2\hat{i} + (2xz - y)\hat{j} + 2\hat{k}) \cdot (\hat{i}dx + \hat{j}dy + \hat{k}dz)$$

$$= \int_C 3x^2dx + (2xz - y)\, dy + 2dz$$

$$= \int_0^2 3t^2dt + \left(2 \cdot t \cdot \frac{3t^3}{8} - \frac{t^2}{4}\right)\left(\frac{t}{2}\,dt\right) + \frac{3t^3}{8}\left(\frac{9t^2}{8}\right)dt$$

$$= \int_0^2 \left(3t^2 + \frac{3}{8}t^5 - \frac{t^3}{8} + \frac{27}{64}t^4\right) dt$$

$$= \int_0^2 \left(3t^2 - \frac{t^3}{8} + \frac{51}{64}t^5\right) dt$$

$$= \left[t^3 - \frac{t^4}{32} + \frac{51}{64} \cdot \frac{t^6}{6}\right]_0^2$$

$$= \left[8 - \frac{1}{2} + \frac{51}{6}\right] = \left[\frac{48 - 3 + 51}{6}\right]$$

$$= \frac{96}{6} = 16 \text{ units.}$$

Example 4.57 **Find the total workdone in moving a particle in a force field given by $\vec{F} = 3xy\hat{i} - 5z\hat{j} + 10x\hat{k}$ along the curve $x = t^2 + 1, y = 2t^2, z = t^3$ from $t = 1$ to $t = 2$.**

Solution : The curve C is given as

$$x = t^2 + 1,\ y = 2t^2 \text{ and } z = t^2 \text{ from } t = 1 \text{ to } t = 2$$

Also the force is

$$\vec{F} = 3xy\hat{i} - 5z\hat{j} + 10x\hat{k}$$

$$\therefore \quad \vec{F} \cdot dr = (3xy\hat{i} - 5z\hat{j} + 10x\hat{k}) \cdot (dx\hat{i} + dy\hat{j} + dz\hat{k})$$

$$= 3xydx - 5z\,dy + 10x\,dz$$

Therefore, the total workdone is

$$\int_C \vec{F} \cdot dr = \int_1^2 3(t^2+1)(2t^2)(2t \cdot dt) - 5(t^3)(4t \cdot dt) + 10(t^2+1)(3t^2)\, dt$$

$$= \int_1^2 [12(t^5+t^3)\, dt - 20t^4\, dt + 30(t^4+t^2)\, dt]$$

$$= 12\left[\frac{t^6}{6}+\frac{t^4}{4}\right]_1^2 - 20\left[\frac{t^5}{5}\right]_1^2 + \left[\frac{t^5}{5}+\frac{t^3}{3}\right]_1^2$$

$$= 12\left[\frac{64}{6}+\frac{16}{4}-\frac{1}{6}-\frac{1}{4}\right] - 20\left[\frac{32}{5}-\frac{1}{5}\right] + 30\left[\frac{32}{5}+\frac{8}{3}-\frac{1}{5}-\frac{1}{3}\right]$$

$$= [171] - 124 + 2\,[128]$$

$$= 171 - 124 + 256$$

$$= 303$$

4.26 Line Integral Independent of Path

In earlier section we have discussed the line integrals.

$$\int_C \vec{F} \cdot dr = \int_C (F_1 dx + F_2 dy + F_3 dz)$$

In this section we are concerned with the same line integrals. But here the integrals depend only at the end points, not on the path joining these points. This type of line integrals are said to be independent of path in a domain D in space.

Theorem : A line integral $\int_C \vec{F} \cdot dr = \int_C F_1 dx + F_2 dy + F_3 dz$ with continuous F_1, F_2, F_3 in a domain D in space is independent of path in D iff $F = F_1\hat{i} + F_2\hat{j} + F_3\hat{k}$ is the gradient of any function f in D.

i.e.,

$$F = \text{grad}\, f.$$

Example 4.58 **Show that the integral $\int_C \vec{F} \cdot dr = \int_C 2xdx + 2ydy + 4zdz$ is independent of path in any domain in space and find its value if C has the initial ponit A (0, 0, 0) and terminal point (2, 2, 2).**

Solution : By inspection we obseve that

$$\vec{F} = 2x\hat{i} + 2y\hat{j} + 4z\hat{k} = \text{grad}\, f$$

where $f = x^2 + y^2 + 2z^2$

Therefore, the given integration is independent of path and we may choose a convenient straight path, *i.e.*,

$$C : r(t) = ti + tj + tk, \quad 0 \le t \le 2 \text{ and get}$$

$$\frac{dr}{dt} = i + j + k$$

$$\therefore \quad \vec{F} \cdot \frac{dr}{dt} = 2t + 2t + 4t$$

$$= 8t$$

and $\int_C 2xdx + 2ydy + 4zdz = \int_0^2 \left(F \cdot \frac{dr}{dt}\right) dt$

$$= \int_0^2 8t dt$$

$$= 16$$

Theorem : The integral $\int_C F \cdot dr = \int_C F_1 dx + F_2 dy + F_3 dz$ is independent of path in a domain D if and only if its value around every closed path in D is zero.

Example 4.59 **(Theorem) : Let F_1, F_2, F_3 in the line integral**

$$\int_C \vec{F} \cdot dr = \int_C (F_1 dx + F_2 dy + F_3 dz)$$

be continuous and have continuous first order partial derivatives in a domain D in space. Then the given integral is independent of path in D if

$$\textbf{curl } F = 0$$

Example 4.60 **Show that the line integral**

$$\int_C 2xyz^2 dx + (x^2z^2 + z \cos yz)\, dy + (2x^2yz + y \cos yz)\, dz$$

is independent of the path of integration.

Solution : We have

$$= \int_C 2xyz^2 dx + (x^2z^2 + z \cos yz)\, dy + (2x^2yz + y \cos yz)\, dz$$

$$= \int_C [2xyz^2 \hat{i} + (x^2z^2 + z \cos yz)\hat{j} + (2x^2yzz + y \cos yz)\hat{k}] \cdot (\hat{i}dx + \hat{j}dy + \hat{k}dz)$$

$$= \int F \cdot dr$$

$\Rightarrow$ $F = 2xyz^2\hat{i} + (x^2z^2 + z \cos yz)\hat{j} + (2x^2yz + y \cos yz)\hat{k}$

The line integral is independent of path if

$$\text{curl } \vec{F} = 0$$

$\therefore$ $\text{curl } F = \nabla \times \vec{F}$

$$= \begin{vmatrix} \hat{i} & \hat{j} & \hat{k} \\ \frac{\partial}{\partial x} & \frac{\partial}{\partial y} & \frac{\partial}{\partial z} \\ 2xyz^2 & x^2z^2 + z \cos yz & 2x^2yz + y \cos yz \end{vmatrix}$$

$$= (2x^2z + \cos yz - yz \sin yz - 2x^2z - \cos yz + yz \sin yz)\hat{i}$$

$$- (4xyz - 4xyz)\hat{j} + (2xz^2 - 2xz^2)\hat{k} = 0$$

Hence, the given integral is independent of path.

PROBLEM SET 4.4

1. Evaluate $\int_0^2 (2t\,\hat{i} + e^t\,\hat{j} + t^3\,\hat{k})\,dt$.

Ans. $4\hat{i} + (e^2 - 1)\,\hat{j} + 4\hat{k}$

2. Evaluate the line integral $\int_C \vec{V} \cdot d\vec{r}$, where

(i) $\vec{v} = e^x\hat{i} + xe^{xy}\hat{j} + \hat{k},\ r(t) = t\hat{i} + t^2\hat{j} + t^3\hat{k},\ 0 \le t \le 2$

(ii) $\vec{v} = x\hat{i} + (\sin y)\,\hat{j} + \hat{k};$ C is given by $x = t^2, y = t,\ z = 2t,\ 0 \le t \le 1$.

Ans. (i) $\dfrac{2e^8 + 3e^2 + 19}{3}$ (ii) $\dfrac{7}{2} - \cos 1$

3. Evaluate the line integral $\int_C f(x, y)\,dx + g(x, y)\,dy$, where

(i) $f(x, y) = y,\ g(x, y) = x;$ C is $y = 2x^2$ from (0, 0) to (2, 8)

(ii) $f(x, y) = xy, g(x, y) = x + 2y;$ C is $y = x - 1$ from (1, 0) to (2, 1)

(iii) $f(x, y) = 2xy^2, g(x, y) = 2x + 3y;$ C is parabola $x = y^2$ from (1, – 1), to (4, 2).

(iv) $f(x, y) = y, g(x, y) = -2x;$ C is given by $x = 3\cos t, y = 2\sin t, 0 \le t \le \pi$

Ans. (i) 16 (ii) $\dfrac{10}{3}$ (iii) $\dfrac{105}{2}$ (iv) -9π.

4. Given that $r(t) = 5t^2\hat{i} + t\hat{j} + t^3\hat{k}$; evaluate $\int_1^2 \vec{r} \cdot \dfrac{d\vec{r}}{dt}\,dt$

Ans. $\dfrac{315}{2}$.

5. The acceleration of a particle at any time t is $e^t\,\hat{i} + e^{2t}\,\hat{j} + \hat{k}$. Find velocity v, given that $v = \hat{i} + \hat{j}$ at $t = 0$.

Ans. $v = 6\sin 2t\,\hat{i} + 4(\cos 2t - 1)\,\hat{j} + 8t^2\,\hat{k}$

6. Evaluate $\int_C \vec{F} \cdot dr$, where $F = (2x + y)\,\hat{i} + (3y - x)\,\hat{j}$ and C is the curve in the xy plane containing of the straight lines from (0, 0) to (2, 0) and then to (3, 2).

Ans. 11

7. For $F = y\hat{i} - x\hat{j}$ evaluate $\int_C \vec{F} \cdot d\vec{r}$ from (0, 0) to (1, 1) along the following paths C :

(a) the parabola $y = x^2$

(b) the straight lines from (0, 0) to (1, 0) and then to (1, 1)

(c) The straight line joining (0, 0) and (1, 1).

Ans. (a) $-\dfrac{1}{3}$ (b) 0

8. Find the work done by a force $\vec{F} = (5xy - 6x^2)\,\hat{i} + (2y - 4x)\,\hat{j}$, which displaces a particle from the point (1, 1) to (2, 8) along the curve $y = x^3$.

Ans. 35

9. Evaluate $\int_C \vec{F} \cdot d\vec{r}$, where $F = yz\hat{i} + zx\hat{j} - xy\hat{k}$ and C is given by $x = t, y = t^2, z = t^3$ from $P(0, 0, 0)$ to $(2, 4, 8)$.
Ans. 0

10. Evalute $\int_C \vec{F} \cdot d\vec{r}$, where $\vec{F} = xy\hat{i} = yz\hat{j} + zx\hat{k}$ and C is the arc of the curve

$$r = (a \cos \theta)\,\hat{i} + (a \sin \theta)\,\hat{j} + a\theta\hat{k} \text{ from } \theta = 0 \text{ to } \theta = \frac{\pi}{2}.$$

Ans. $a^3\left[\frac{5}{8}\pi - \frac{4}{3}\right]$

11. If $\vec{F} = (2x^2 + y^2)\,\hat{i} + (3y - 4x)\,\hat{j}$, evaluate $\int \vec{F} \cdot dr$ around the triangle ABC whose vetices are $A\ (0, 0)$, $B(2, 0)$ and $C(2, 1)$.
Ans. $-\frac{14}{5}$

12. Evaluate $\int (xdy - ydx)$ around the circle $x^2 + y^2 = 1$

13. If $A = (3x^2 + 6y)\,\hat{i} - 14yz\hat{j} + 20xz^2\hat{k}$, evaluate $\int \vec{A} \cdot dr$ from $(0, 0, 0)$ to $(1, 1, 1)$ along the path $x = t, y = t^2, z = t^3$
Ans. $\frac{23}{3}$

14. If $F = 2y\hat{i} - z\hat{j} + x\hat{k}$, evalute $\int_C \vec{F} \cdot dr$ along the curve $x = \cos t, y = \sin t, z = 2$ from $t = 0$ to $t = \frac{\pi}{2}$.
Ans. $-(\pi + 1)$

15. If $\vec{F} = (2y + 3)\hat{i} + xz\hat{j} + (yz - x)\,\hat{k}$, evaluate. $\int_C \vec{F} \cdot dr$, where C is the path consisting of the straight line C from $(0, 0, 0)$ to $(0, 0, 1)$ then to $(0, 1, 1)$ and then to $(2, 1, 1)$.
Ans. 10

16. Find the workdone when a force $\vec{F} = (x^2 - y^2 + x)\,\hat{i} - (2xy + y)\,\hat{j}$, moves a particle in xy-plane from $(0, 0)$ to $(1, 1)$ along the parabola $y^2 = x$.
Ans. $-\frac{2}{3}$.

17. A vector field is given by $\vec{F} = (\sin y)\,\hat{i} + x\,(1 + \cos y)\,\hat{j}$. Evaluate, the line integral of a circular path given by $x^2 + y^2 = a^2, z = 0$
Ans. πa^2

18. Show that the integral

$$\int_{(1,\,2)}^{(3,\,4)} (xy^2 + y^3)\,dx + (x^2y + 3xy^2)\,dy$$

is independent of path joining the points (1, 2) and (3, 4).
Hence, evaluate the integral.

4.27 Surface Integral

Any integral to be evaluted over a surface is called a surface integral. Let S be a surface of finite area and $f(x, y, z)$ is a single valued function of position defined over S. The area S is subdivided into n elements of areas

$$\delta S_1, \delta S_2, \ldots, \delta S_n.$$

Let an arbitrary path P_k with coordinates (x_k, y_k, z_k) is choosen from each δS_k.

We define $\qquad f(P_k) = f(x_k, y_k, z_k)$

Consider the sum $\sum_{k=1}^{n} f(P_k)\, \delta S_k$

Now, taking limit of this sum as $n \to \infty$ in such a way that the largest of the area tends to zero. If this limit exists is called the surface integral of $f(x, y, z)$ over S and denoted by

$$\iint_S F(x, y, z)\, dS$$

Z
S
O
Y
X

Fig. 4.7

Note : It is observed that to evaluate surface integrals it is convenient to express then in form of double integrals taken over the orthogonal projection of the surface S on one of the coordinates planes. But, it should be taken into consideration that any line perpendicular to the coordinate plane choosen meets the surface S in not more than one point. If this condition is not satisfied by the sruface then it can be subdivided into surfaces, which satisfy this condition.

Let us assume that surface S is in such a way that any line perpendicular to xy-plane meets S in one point only. Then the equation of surface can be written in the form of $z = h(x, y)$.

Suppose, R be the orthogonal projection of S on the xy-plane. If θ be the acute angle which the normal $\hat{n}$ at $P(x, y)$ to the surface S makes with z-axis, then

$$\cos\theta\, dS = dx\, dy$$

where ds is the small element of area of surface S at the point P

Hence, $\qquad dS = \dfrac{dx\, dy}{\cos\theta} = \dfrac{dx\, dy}{|\hat{n}\cdot\hat{k}|},$

where k is the unit vector along z-axis.

Therefore, $\qquad ds = \dfrac{dx\, dy}{\cos\theta} = \dfrac{dx\, dy}{|\hat{n}\cdot\hat{k}|}.$

Example 4.61 **Evaluate $\iint_S \vec{F}\cdot\hat{n}\, dS$, where $\vec{F} = yz\hat{i} + zx\hat{j} + xy\hat{k}$ and S is the surface of the sphere $x^2 + y^2 + z^2 = a^2$ which lies in the first octant.** *(UPTU 2005)*

Solution : We know that a vector normal to the surface S is given by

$$\nabla(x^2 + y^2 + z^2) = \left(\hat{i}\frac{\partial}{\partial x} + \hat{j}\frac{\partial}{\partial y} + \hat{k}\frac{\partial}{\partial y}\right)(x^2 + y^2 + z^2)$$

$$= 2x\hat{i} + 2y\hat{j} + 2z\hat{k}$$

Thus a unit vector normal to the surface S at any point (x, y, z) is

$$\hat{n} = \frac{(2x\hat{i} + 2y\hat{j} + 2z\hat{k})}{\sqrt{4x^2 + 4y^2 + 4z^2}} = \frac{(x\hat{i} + y\hat{j} + z\hat{k})}{\sqrt{x^2 + y^2 + z^2}}$$

$$= \frac{x\hat{i} + y\hat{j} + z\hat{k}}{a} \qquad (\because x^2 + y^2 + z^2 = a^2 \text{ on the surface } S)$$

Now, $$\iint_S \vec{F}\cdot\hat{n}\, dS = \iint_R \vec{F}\cdot\hat{n} \frac{dx\, dy}{|\hat{n}\cdot\hat{k}|}$$

where R is the projection of S on xy-plane. It is clear that region R is bounded by x-axism y-axis and circle $x^2 + y^2 = a^2; z = 0$.

We have,

$$\vec{F}\cdot\hat{n} = (yz\hat{i} + zx\hat{j} + xy\hat{k}) \cdot \frac{(x\hat{i} + y\hat{j} + z\hat{k}}{a}$$

$$= \frac{xyz + xyz + xyz}{a} = \frac{3xyz}{a}$$

Also, $\hat{n}\cdot\hat{k} = (x\hat{i} + y\hat{j} + z\hat{k}) \cdot \hat{k} = z$

$\therefore$ $|\hat{n}\cdot\hat{k}| = z$

Therefore, $$\iint_S \vec{F}\cdot\hat{n}\, dS = \iint_R \frac{3xyz}{a}\frac{1}{z}\, dx\, dy$$

$$= \iint_R \frac{3xy}{a}\, dx\, dy$$

$$= \frac{3}{a}\int_0^a\int_0^{\sqrt{a^2 - x^2}} xy\, dx\, dy = \frac{3}{a}\int_0^a \left[\frac{y^2}{2}\right]_0^{\sqrt{a^2 - x^2}} dx$$

$$= \frac{3}{a}\int_0^a \frac{(a^2x - x^3)}{2}\, dx = \frac{3}{2a}\int_0^a (a^2x - x^3)\, dx$$

$$= \frac{3}{2a}\left[a^2\frac{x^2}{2} - \frac{x^4}{4}\right]_0^a = \frac{3}{2a}\left[\frac{a^4}{2} - \frac{a^4}{4}\right]$$

$$= \frac{3}{8}a^3$$

Example 4.62 **Show that $\iint_S (\vec{F}\cdot\hat{n})\, dS = \frac{3}{2}$, where $\vec{F} = 4xz\hat{i} - y^2\hat{j} + yz\hat{k}$ and S is the surface of the cube bounded by the planes, $x = 0, x = 1, y = 1, z = 0$ $z = 1$.**

Solution : We are given that S is the surface of the cube $x = 0, x = 1, y = 0, y = 1$ and $z = 0, z = 1$ as shown in figure 4.8.

It is clear that

$$\iint_S \vec{F}\cdot\hat{n}\, dS = \iint_{OABC} \vec{F}\cdot\hat{n}\, dS + \iint_{DEFG} \vec{F}\cdot\hat{n}\, dS + \iint_{OAFG} \vec{F}\cdot\hat{n}\, dS + \iint_{BCDE} \vec{F}\cdot\hat{n}\, dS$$

$$+ \iint_{ABEF} \vec{F}\cdot\hat{n}\, dS + \iint_{OCDG} \vec{F}\cdot\hat{n}\, dS \qquad \ldots(1)$$

(i) $$\iint_{OABC} \vec{F}\cdot\hat{n}\, dS = \iint_{OABC} \frac{(4xz\hat{i} - y^2\hat{j} + yz\hat{k})\cdot(-\hat{k})\, dx\, dy}{|-\hat{k}\cdot\hat{k}|}$$

$$= -\iint_{OABC} yz\, dx\, dy = -\int_0^1\int_0^1 yz\, dx\, dy$$

$$= 0 \qquad (\because\ z = 0)$$

(ii) $$\iint_{DEFG} \frac{(4xz\hat{i} - y^2\hat{j} + yz\hat{k})\cdot\hat{k}}{|\hat{k}\cdot\hat{k}|}\, dx\, dy = \int_0^1\int_0^1 yz\, dx\, dy$$

$$= \int_0^1\int_0^1 y\cdot(1)\, dx\, dy \qquad (\because\ z = 1)$$

$$= \int_0^1\int_0^1 y\, dx\, dy = \int_0^1 \left(\frac{y^2}{2}\right)_0^1 dx$$

$$= \frac{1}{2}\,[x]_0^1 = \frac{1}{2}$$

(iii) $$\iint_{OAFG} \frac{(4xz\hat{i} - y^2\hat{j} + yz\hat{k})\cdot(-\hat{j})\, dx\, dz}{|-\hat{j}\cdot\hat{j}|} = \iint_{OAFG} (-y^2)\, dx\, dz = 0 \qquad (\because y = 0)$$

(iv) $$\iint_{BCDE} (4xzi - y^2j + yzk)\cdot j\, dx\, dz = \iint_{BCDE} (-y^2)\frac{dx\, dz}{|j\cdot j|} = -\int_0^1\int_0^1 dx\, dz$$

$$= -\int_0^1 dx \int_0^1 dz = -\int_0^1 dx = -[x]_0^1 = -1 \qquad (\because\ y = 1)$$

(v) $$\iint_{ABEF} (4zxi - y^2j + yzk)\cdot\frac{i\, dy\, dz}{|i\cdot i|} = \iint_{ABEF} 4xz\, dy\, dz$$

$$= 4\int_0^1\int_0^1 z\, dy\, dz \qquad (\because\ x = 1)$$

$$= 4\int_0^1 \left[\frac{z^2}{2}\right]_0^1 dy = 2\,[y]_0^1 = 2$$

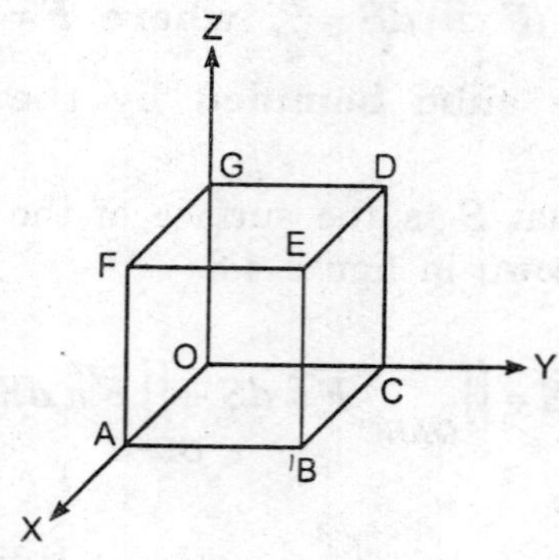

Fig. 4.8

(vi) $$\iint_{OCDG} \frac{(4xz\hat{i} - y^2\hat{j} + yz\hat{k}) \cdot (-\hat{i})\, dy\, dz}{|-\hat{i}\cdot\hat{i}|} = -\iint_{OCDG} 4xz \cdot dy\, dz$$

$$= 0 \qquad (\because x = 0)$$

Therefore, from Eq. (1) we get

$$\iint_S \vec{F}\cdot\hat{n}\, ds = 0 + \frac{1}{2} + 0 - 1 + 2 + 0 = \frac{3}{2}$$

Example 4.63 **Evaluate $\iint_S (yz\hat{i} + zx\hat{j} + xy\hat{k}), dS$, where S is the surface of the sphere $x^2 + y^2 + z^2 = a^2$ in the first octant.**

Solution : We have to evalute

$$\iint_S (yz\hat{i} + xz\hat{j} + xy\hat{k})\, dS = \iint_S (yz\hat{i} + xz\hat{j} + xy\hat{k}) \cdot (dy\, dz\hat{i} + dz\, dx\,\hat{j} + dx\, dy\,\hat{k})$$

$$= \iint_S (yz\, dy\, dz + xz\, dz\, dx + xy\, dx\, dy)$$

$$= \int_0^a \int_0^{\sqrt{a^2 - z^2}} yz\, dy\, dz + \int_0^a \int_0^{\sqrt{a^2 - x^2}} zx\, dz\, dx + \int_0^a \int_0^{\sqrt{a^2 - y^2}} xy\, dx\, dy$$

$$= \int_0^a z\left(\frac{y^2}{2}\right)_0^{\sqrt{a^2 - z^2}} + \int_0^a x\left[\frac{z^2}{2}\right]_0^{\sqrt{a^2 - x^2}} dx + \int_0^a y\left(\frac{x^2}{2}\right)_0^{\sqrt{a^2 - y^2}} dy$$

$$= \frac{1}{2}\left(\frac{a^2 z^2}{2} - \frac{z^4}{4}\right)_0^a + \frac{1}{2}\left(\frac{a^2 x^2}{2} - \frac{x^4}{4}\right)_0^a + \frac{1}{2}\left(\frac{a^2 y^2}{2} - \frac{y^4}{2}\right)_0^a$$

$$= \frac{1}{2}\left(\frac{a^4}{4}\right) + \frac{1}{2}\left(\frac{a^4}{4}\right) + \frac{1}{2}\left(\frac{a^4}{4}\right) = \frac{3}{8}a^4$$

Example 4.64 **Evaluate the surface integral $\iint_S F\cdot\hat{n}\, dS$, where $F = 6z\hat{i} + 6\hat{j} + 3y\hat{k}$ and S is the perform of the plane $2x + 3y + 4z = 12$, lying in first octant.**

Solution : Let $f(x, y, z) = 2x + 3y + 4z - 12 = 0$ be the surface of integration.

Then grad $f = 2\hat{i} + 3\hat{j} + 4\hat{k}$

$$\Rightarrow \qquad \hat{n} = \frac{\text{grad}\, f}{|\text{grad}\, f|} = \frac{(2\hat{i} + 3\hat{j} + 4\hat{k})}{\sqrt{29}}$$

Let we consider the projection of S over xy-plane, this projection is obviously a triagnle bounded by $x = 0, y = 0$ and $2x + 3y = 12$.

Thus $$\iint_S \vec{F}\cdot\hat{n}\, dS = \iint_R \vec{F}\cdot\hat{n}\, \frac{dx\, dy}{|\hat{n}\cdot\hat{k}|}$$

where R is the triangle (projection of S) in xy-plane

$$\Rightarrow \qquad \iint_S \vec{F}\cdot\hat{n}\, dS = \iint_R (6z\hat{i} + 6\hat{j} + 3y\hat{k}) \cdot \frac{(2\hat{i} + 3\hat{j} + 4\hat{k})}{\sqrt{29}} \frac{dx\, dy}{\left|\frac{4}{\sqrt{29}}\right|}$$

$$= \frac{1}{4}\iint_R (12z + 18 + 12y)\, dx\, dy$$

$$= \frac{1}{4} \iint_R (36 - 6x - 9y + 18 + 12y)\, dx\, dy$$

$$= \frac{1}{4} \int_0^6 \int_0^{\frac{12-2x}{3}} (54 - 6x + 3y)\, dx\, dy$$

$$= \frac{1}{4} \int_0^6 \left[(54 - 6x)\, y + \frac{3}{2} y^2 \right]_0^{\frac{12-2x}{3}} dx$$

$$= \frac{1}{4} \int_0^6 \left[(54 - 6x) \frac{1}{3} (12 - 2x) + \frac{1}{6} (12 - 2x)^2 \right] dx$$

$$= \frac{1}{6} \int_0^6 (360 - 102x + 7x^2)\, dx$$

$$= \frac{1}{6} \left[360x - 51x^2 + \frac{7}{3} x^3 \right]_0^6 = 138$$

Example 4.65 **Evaluate $\iint_S F \cdot \hat{n}\, dS$, where $\vec{F} = x^2 i + y^2 j + z^2 k$ and S is the surface of the solid cut off by the plane $x + y + z = a$ from the first octant.**

Solution : We have $\vec{F} = x^2 \hat{i} + y^2 \hat{j} + z^2 \hat{k}$

Let $f(x, y, z) = x + y + z - a$ be the surface.

Then $\text{grad} f = \hat{i} + \hat{j} + \hat{k}$ and $\hat{n} = \dfrac{\text{grad} f}{|\text{grad} f|} = \dfrac{\hat{i} + \hat{j} + \hat{k}}{\sqrt{3}}$

Let R be the projection of S in xy plane. Then R be the triangle bounded by $x = , y = 0$ and $x + y = a$.

$$\therefore \quad \iint_S \vec{F} \cdot \hat{n}\, d\vec{S} = \iint_R (x^2\hat{i} + y^2\hat{j} + z^2\hat{k}) \cdot \frac{(\hat{i} + \hat{j} + \hat{k})}{\sqrt{3}} \frac{dxdy}{\left| \frac{1}{\sqrt{3}} \right|}$$

$$= \iint_R (x^2 + y^2 + z^2)\, dxdy$$

$$= \int_0^a \int_0^{a-x} [x^2 + y^2 + (a - x - y)^2] dxdy$$

$$= \int_0^a \int_0^{a-x} [x^2 + y^2 + a^2 + x + y^2 - 2ax + 2xy - 2ay] dxdy$$

$$= \int_0^a \int_0^{a-x} [a^2 + 2x^2 + 2y^2 - 2ax - 2ay + 2xy] dxdy$$

$$= \int_0^a \left[(a^2 + 2x^2 - 2ax)y + \frac{2}{3} y^3 - ay^2 + xy^2 \right]_0^{a-x} dx$$

$$= \int_0^a \left[-\frac{5}{3} x^3 + 3ax^2 - 2a^2 x + \frac{2}{3} a^3 \right] dx$$

$$= \left[-\frac{5}{12} x^4 + ax^3 - a^2x^2 + \frac{2}{3} a^3 x \right]_0^a$$

$$= \frac{a^4}{4}$$

4.28 VOLUME INTEGRAL

As the name suggests the integration of a function over a given volume is termed as volume integral. Let V be a volume bounded by a surface S and $f(x, y, z)$ be a single valued function of a position defined over V divide the volume V into n small volumes $\delta V_1, \delta V_2, \ldots \delta V_n$. Let $P_k(x, y, z)$ be an arbitrary point on volume δV_k.

Let we define $f(P_k) = f(x_k, y_k, z_k)$

Here, we obtain sum $\sum_{k=1}^{n} f(P_k) \cdot \delta V_k$ and if the limit taken for this sum as $n \to \infty$ in such a way that the largest of the volumes δV_k approaches zero. This limit is called volume integral of $f(x, y, z)$ over V and denoted by $\iiint_V f(x, y)\, dV$. But if the function $f(x, y, z)$ is continuous over the volume V, then the above limit always exists.

Example 4.66 **Evaluate $\iiint_V \vec{F} \cdot dV$ for $\vec{F} = 2z\hat{i} - x\hat{j} + y\hat{k}$, where V is the region bounded by the srufaces $x = 0, y = 0, x = 2, y = 4, z = x^2, z = 2$.**

Solution : It is given that volume of integration is bounded by the surfaces $x = 0, y = 0, x = 2, y = 4, z = x^2, z = 2$.

Thus, $$\iiint_V \vec{F} \cdot dV = \iiint_V (2z\hat{i} - x\hat{j} + y\hat{k})\, dx\, dy\, dz$$

$$= \int_0^2 \int_0^4 \int_{x^2}^2 (2z\hat{i} - x\hat{j} + y\hat{k})\, dx\, dy\, dz$$

$$= \int_0^2 dx \int_0^4 dy\, [z^2\hat{i} - xz\hat{j} + yz\hat{k}]_{x^2}^2$$

$$= \int_0^2 dx \int_0^4 dy\, [4\hat{i} - 2x\hat{j} + 2y\hat{k} - x^4\hat{i} + x^3\hat{j} - x^2 y\hat{k}]$$

$$= \int_0^2 dx \left[4y\hat{i} - 2xy\hat{j} + y^2\hat{k} - x^4 y\hat{i} + x^3 y\hat{j} - \frac{x^2y^2}{2}\hat{k} \right]_0^4$$

$$= \int_0^2 [16\hat{i} - 8x\hat{j} + 16\hat{k} - 4x^4\hat{i} + 4x^3\hat{j} - 8x^2\hat{k}]\, dx$$

$$= \left[16x\hat{i} - 4x^2\hat{j} + 16x\hat{k} - \frac{4x^5}{5}\hat{i} + x^4\hat{j} - \frac{8x^3}{3}\hat{k} \right]_0^2$$

$$= 32\hat{i} - 16\hat{j} + 32\hat{k} - \frac{128}{5}\hat{i} + 16\hat{j} - \frac{64}{3}\hat{k}$$

$$= \frac{32}{5}\hat{i} + \frac{32}{3}\hat{k} = \frac{32}{15}(3\hat{i} + 5\hat{j})$$

Example 4.67 **If $\vec{F} = (2x^2 - 3z)\hat{i} - 2xy\hat{j} - 4x\hat{k}$, then evaluate $\iiint_V \nabla\cdot\vec{F}\, dV$ where V is the closed region bounded by the planes $x = 0$, $y = 0$, $z = 0$ and $2x + 2y + z = 4$.**

Solution : The region of integration is bounded by the planes $x = 0, y = 0, z = 0$ and $2x + 2y + z = 4$. If we take a thin column parallel to z-axis as the elementary volume it cuts the boundary of $z = 0$ and $z = 4 - 2x - 2y$. Also the projection of the plane $2x + 2y + z = 4$ on the xy-plane is the area bounded by $x = 0, y = 0$ and the line $x + y = 2$. Thus, the given volume integral would be

$$\iiint_V \nabla\cdot\vec{F}\, dV.$$

Also, we are given $\vec{F} = (2x^2 - 3z)\hat{i} - 2xy\hat{j} - 4x\hat{k}$

$$\nabla\cdot\vec{F} = \left(\hat{i}\frac{\partial}{\partial x} + \hat{j}\frac{\partial}{\partial y} + \hat{k}\frac{\partial}{\partial z}\right)\cdot[(2x^2 - 3z)\hat{i} - 2xy\hat{j} - 4x\hat{k}]$$

$$= \frac{\partial}{\partial x}(2x^2 - 3z) + \frac{\partial}{\partial y}(-2xy) + \frac{\partial}{\partial z}(-4x)$$

$$= 4x - 2x = 2x$$

Thus,

$$\iiint_V \nabla\cdot\vec{F} = \iiint_V 2x\, dx\, dy\, dz$$

$$= 2\int_0^2\int_0^{2-x}\int_0^{4-2x-2y} x\, dx\, dy\, dz$$

$$= 2\int_0^2 x\int_0^{2-x} [z]_0^{4-2x-2y}\, dx\, dy$$

$$= 2\int_0^2 x\int_0^{2-x} [4 - 2x - 2y]\, dx\, dy$$

$$= 2\int_0^2 [4xy - 2x^2y - xy^2]_0^{2-x}\, dx$$

$$= 2\int_0^2 [4x(2-x) - 2x^2(2-x) - x(2-x)^2]\, dx$$

$$= 2\int_0^2 [x^3 - 4x^2 + 4x]\, dx$$

$$= 2\left[\frac{x^4}{4} - \frac{4}{3}x^3 = 2x^2\right]_0^2$$

$$= 2\left[4 - \frac{32}{3} + 8\right]$$

$$= \frac{8}{3}$$

PROBLEM SET 4.5

1. Evaluate $\iint_S \vec{F}\cdot\hat{n}\,dS$, where $\vec{F} = y\hat{i} + (x - 2xy)\hat{j} - xy\hat{k}$ and S is the surface of the sphere $x^2 + y^2 + z^2 = a^2$, $0 \le x, y, z < a$. Also $\hat{n}$ is the unit vector along the outward drawn normal to the surface S.

   Ans. $\frac{2}{3}a^3 - \frac{3}{8}a^4$

2. Evaluate the surface integral $\iint_S \vec{F}\cdot\hat{n}\,dS$, where $\vec{F} = 6z\hat{i} - 4\hat{j} + y\hat{k}$ and S is the portion of the plane of $2x + 3y + 6z = 12$ in the first octant.

   Ans. 8.

3. Evaluate $\iint_S [(xy^2 + 2xz^2 - 1)\,dx + 2xy\,dy + 2x^2zdz]$, where S is the ellipse $\frac{x^2}{a^2} + \frac{y^2}{b^2} = 1$, $z = 0$.

   Ans. Zero.

4. Evaluate $\iint_S \vec{F}\cdot d\vec{S}$, where $\vec{F} = x\hat{i} + (2^2 - zx)\hat{j} - xy\hat{k}$ and S is the triangular surface with vertices (2, 0, 0), (0, 2, 0), (0, 0, 4).

   Ans. $-\frac{22}{3}$.

5. Evaluate $\iint_S \vec{F}\cdot\hat{n}\,dS$, where $\vec{F} = 2y\hat{i} - 3\hat{j} + x^2\hat{k}$ and S is the surface of the parabolic cylinder $y^2 = 8x$ in the first octant bounded by the planes $y = 4$ and $z = 6$.

   Ans. 132.

6. Evaluate $\iint_S \vec{F}\cdot\hat{n}\,dS$, where $\vec{F} = (x + y^2)\hat{i} - 2x\hat{j} + 2yz\hat{k}$ and S is the surface of the plane $2x + y + 2z = 6$ in the first octant.

   Ans. 81.

7. Evaluate $\iint_S \vec{F}\cdot\hat{n}\,dS$, where $\vec{F} = 18z\hat{i} - 12\hat{j} + 3y\hat{k}$ and S is the surface of the plane $2x + 3y + 6z = 12$ in the first octant.

   Ans. 24.

8. If $\vec{F} = 4x\hat{i} - 2y^2\hat{j} + z^2\hat{k}$ and S is the closed surface of the cylinder $x^2 + y^2 = 4$ and circular discs $z = 0$ and $z = 3$ then evaluate $\iint_S \vec{F}\cdot\hat{n}\,dS$.

   Ans. 84π

9. Evalute $\iiint_V \vec{F}\,dV$, where $\vec{F} = x\hat{i} + y\hat{j} + z\hat{k}$ and V is the region bounded by the surface $x = 0$, $x = 2$, $y = 0$, $y = 6$, $z = 4$ and $z = x^2$.

   Ans. $24\hat{i} + 96\hat{j} + \frac{384}{5}\hat{k}$.

10. Evalute $\iiint_V (2x+y)\,dV$ where V is the closed region bounded by the cylinder $z = 4 - x^2$ and the planes $x = 0, y = 0, y = 2$ and $z = 0$.

Ans. $\frac{80}{3}$

11. Evaluate $\iiint_V \nabla \cdot \vec{F}\,dV$, where $\vec{F} = (2x^2 - 3z)\,\hat{i} - 2xy\hat{j} - 4x\hat{k}$ and V is bounded by planes $x = 0, y = 0, z = 0$ and $2x + 2y + z = 4$.

Ans. $\frac{8}{3}$.

12. If $\vec{F} = 2xz\hat{i} - x\hat{j} + y^2\hat{k}$, evaluate $\iiint_V \vec{F}\,dV$, where V is the region bounded by the surfaces $x = 0, y = 0, y = 6$ and $z = x^2, z = 4$.

Ans. $16\hat{i} - 3\hat{j} + 48\hat{k}$.

4.29 GREEN'S THEOREM (In Plane)

Statement : If $M(x, y), N(x, y), \frac{\partial M}{\partial y}$ and $\frac{\partial N}{\partial x}$ are continuous function over a region R bounded by closed curve C in x-y plane, then

$$\int_C (M\,dx + N\,dy) = \iint_R \left(\frac{\partial N}{\partial x} - \frac{\partial M}{\partial y}\right) dx\,dy$$

The line integral is to be taken along the entire boundary C of R.

Example 4.68 **Verify Green's theorem in the plane $\int_C (xy + y^2)\,dx + x^2\,dy$ where C is the closed curve of the region bounded by $y = x$ and $y = x^2$.**

Solution : By Green's theorem in plane, we get

$$\iint_R \left(\frac{\partial N}{\partial x} - \frac{\partial M}{\partial y}\right) dx\,dy = \int_C (M\,dx + N\,dy)$$

Here, it is observed that $M = xy + y^2, N = x^2$.

The closed curve consists of a straight line $y = x$ and parabola $y = x^2$ between (0, 0) and (1, 1) as shown in figure 4.9.

Y
A (1, 1)
O
B (1, 0)
X

Fig. 4.9

Here, we have

$$\iint_R \left(\frac{\partial N}{\partial x} - \frac{\partial M}{\partial y}\right) dx\,dy = \iint_R \left[\frac{\partial}{\partial x}(x^2) - \frac{\partial}{\partial y}(xy + y^2)\right] dx\,dy$$

$$= \iint_R (2x - x - 2y)\,dx\,dy = \iint_R (x - 2y)\,dx\,dy$$

$$= \int_0^1 \int_{x^2}^{x} (x - 2y)\,dx\,dy$$

$$= \int_0^1 [xy - y^2]_{x^2}^{x}\, dx$$

$$= \int_0^1 [x^2 - x^2 - x^3 + x^4]\, dx$$

$$= \int_0^1 (x^4 - x^3)\, dx$$

$$= \left[\frac{x^5}{5} - \frac{x^4}{4}\right]_0^1 = \frac{1}{5} - \frac{1}{4} = -\frac{1}{20} \qquad \ldots(1)$$

Now, first we try to evaluate the line integral along $y = x^2$. Thus,

$$\int_0^1 [x(x^2) + x^4]\, dx + x^2\,(2x)\, dx] \qquad (\because y = x^2 \text{ and } dy = 2x\, dx)$$

$$= \int_0^1 (3x^3 + x^4)\, dx = \left[3\frac{x^4}{4} + \frac{x^5}{5}\right]_0^1$$

$$= \frac{3}{4} + \frac{1}{5} = \frac{19}{20}$$

Now, we evaluate the line integral along $y = x$. Thus,

$$\int_1^0 [x(x) + x^2]\, dx + x^2\, dx \qquad (\because\ y = x \text{ and } dy = dx)$$

$$= \int_1^0 3x^2\, dx = [x^3]_1^0 = -1$$

Therefore, the required line integral is

$$\int_C (M\, dx + N\, dy) = \frac{19}{20} - 1 = \frac{-1}{20} \qquad \ldots(2)$$

Hence, We find that

$$\iint_R \left(\frac{\partial N}{\partial x} - \frac{\partial M}{\partial y}\right) dx\, dy = -\frac{1}{20} = \int_C (M\, dx + N\, dy)$$

Example 4.69 **Evaluate $\int_C [(y - \sin x)\, dx + \cos x\, dy]$, by Green's theorem where C is the triangle formed by the lines $y = 0$, $x = \frac{\pi}{2}$ and $y = \frac{2}{\pi}x$.**

Solution : By Green's theorem, we know that

$$\int_C P dx + Q dy = \iint_S \left(\frac{\partial Q}{\partial x} - \frac{\partial P}{\partial y}\right) dx\, dy$$

Here $P = y - \sin x$, $Q = \cos x$

$$\therefore \quad \frac{\partial P}{\partial y} = 1, \quad \frac{\partial Q}{\partial x} = -\sin x$$

$\therefore \int_C [(y - \sin x)\, dx + \cos x\, dy] = \iint_S (-\sin x - 1)\, dx\, dy$ where S is the area of the given triangle.

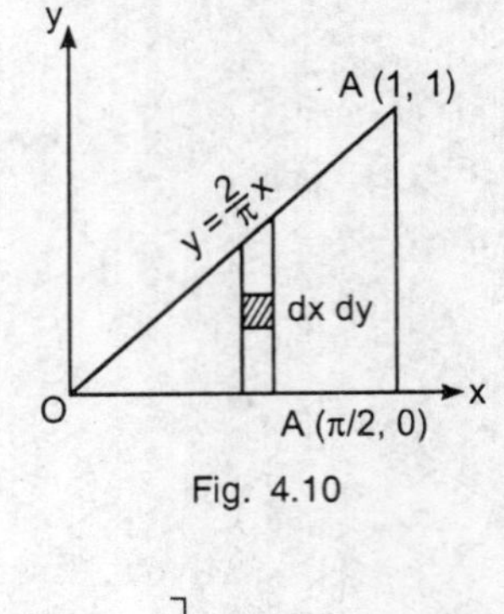

Fig. 4.10

$$= -\int_0^{\pi/2}\int_0^{2x/\pi} (1 + \sin x)\, dx\, dy$$

$$= -\int_0^{\pi/2} (1 + \sin x)\, [y]_0^{2x/\pi}\, dx$$

$$= -\int_0^{\pi/2} (1 + \sin x)\, \frac{2x}{\pi}\, dx$$

$$= -\frac{2}{\pi}\int_0^{\pi/2} (x + x \sin x)\, dx$$

$$= -\frac{2}{\pi}\left[\left\{\frac{x^2}{2}\right\}_0^{\pi/2} + \{x\,(-\cos x)\}_0^{\pi/2} - \int_0^{\pi/2} 1 \cdot (-\cos x)\, dx\right]$$

$$= -\frac{2}{\pi}\left[\frac{1}{2}\,\frac{\pi^2}{4} + 0 + \{\sin x\}_0^{\pi/2}\right]$$

$$= -\frac{2}{\pi}\left[\frac{1}{8}\pi^2 + 1\right] = -\left(\frac{\pi}{4} + \frac{2}{\pi}\right)$$

Example 4.70 **Apply Green's theorem to evaluate $\int_C [(2x^2 - y^2)\, dx + (x^2 + y^2)\, dy]$, where C is the boundary of the area enclosed by x-axis and the upper half of the circle $x^2 + y^2 = a^2$.**

Solution : Using Green's theorem we may write

$$\int_C [(2x^2 - y^2)\, dx + (x^2 + y^2)]\, dy = \iint_A \left[\frac{\partial}{\partial x}(x^2 + y^2) - \frac{\partial}{\partial y}(2x^2 - y^2)\right] dx\, dy$$

$$= 2\int_0^a \int_0^{\sqrt{a^2 - x^2}} (x + y)\, dx\, dy \qquad (\because A \text{ is symmetric about } y\text{-axis})$$

$$= 2\int_0^a \int_0^\pi r(\cos\theta + \sin\theta)\, r\, d\theta\, dr \qquad \text{(By charging the variables)}$$

$$= 2\int_0^a r^2\, dr \int_0^\pi (\cos\theta + \sin\theta)\, d\theta$$

$$= 2\int_0^a r^2\, dr\, [\sin\theta - \cos\theta]_0^\pi$$

$$= 2\int_0^a r^2\, dr\, [2] = 4\left[\frac{r^3}{3}\right]_0^a = 4\,\frac{a^3}{3}$$

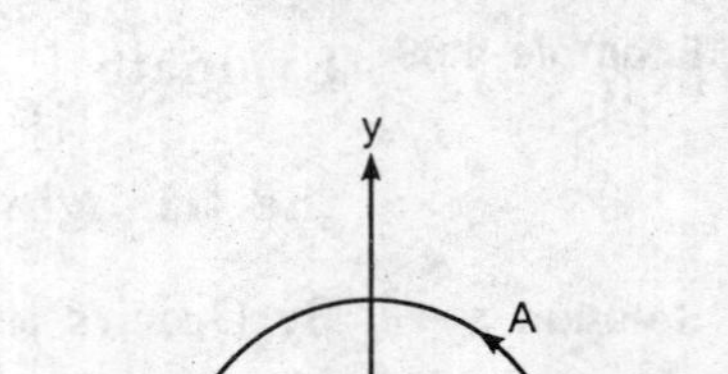

Fig. 4.11

4.30 STOKE'S THEOREM (Transformation between surface integrals and line integrals)

Suppose S be a piecewise smooth oriented surface in space and the boundary of S be a piecewise smooth simple closed curve C. Also, it is assumed that $\vec{F}(x, y, z)$ be a continuous vector function, which has continuous first order partial derivatives in a domain in space cotnaining S. Then

$$\iint_S (curl\ \vec{F}) \cdot \hat{n}\, dS = \int_C F \cdot dr$$

where $\hat{n} = \cos\alpha\hat{i} + \cos\beta\hat{j} + \cos\gamma\hat{k}$ is a unit vector normal to the surface dS.

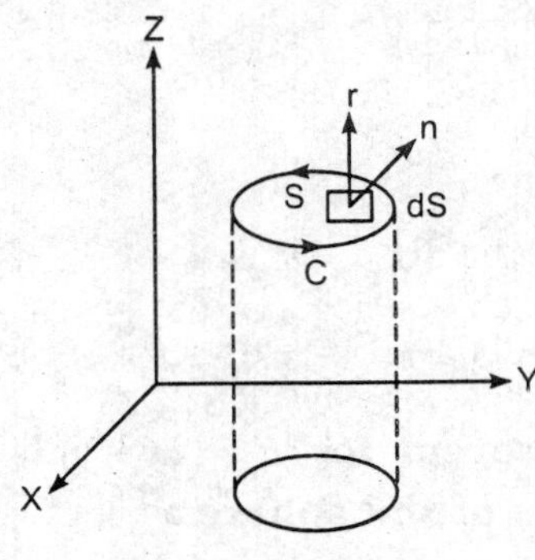

Fig. 4.12

Example 4.71 **Show that the area bounded by a simple closed curve C is given by $(xdy - ydx)$. Hence, find the area of the loop of the folium $x^2 + y^3 = 3axy,\ a > 0$.**

Solution : By Green's theorem, we have

$$\int_C (P\,dx + Q\,dy) = \iint_S \left(\frac{\partial Q}{\partial x} - \frac{\partial P}{\partial y}\right) dx\,dy$$

Let $P = -y$ and $Q = x$.

Then $\frac{\partial Q}{\partial x} = 1$ and $3\frac{\partial P}{\partial y} = -1$

$$\therefore \quad \int_C (-y\,dx + x\,dy) = \iint_S [1 - (-1)]\,dx\,dy$$

or $$2\iint_S dx\,dy = \int_C x\,dy - y\,dx$$

or $$\iint_S dx\,dy = \frac{1}{2}\int_C x\,dy - y\,dx.$$

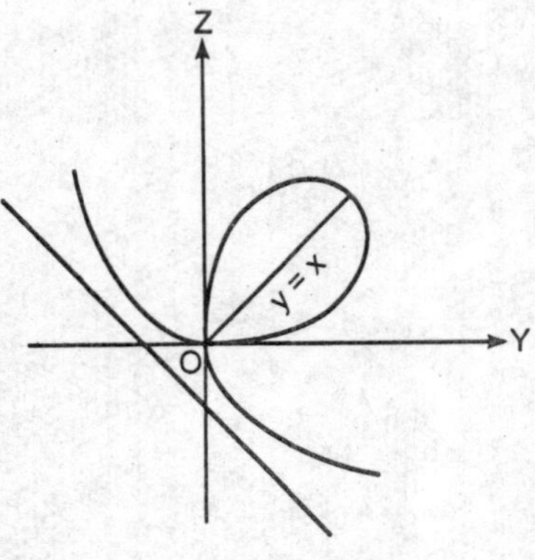

Fig. 4.13

$\therefore$ The area bounded by a simple closed curve C

$$= \frac{1}{2}\int_C x\,dy - y\,dx$$

Now required area (A) $= \frac{1}{2}\int_C x^2\left[\frac{x\,dy - y\,dx}{x^2}\right]$

$$= \frac{1}{2}\int_C x^2\, d\left(\frac{y}{x}\right)$$

Putting $y = tx$

$$= \frac{1}{2} \int_C x^2 \, dt$$

If we put $y = tx$ in the given folium, we get

$$x^3 + t^3 x^3 = 3axtx \text{ or } x = \frac{3at}{1 + t^3}$$

$$A = \frac{1}{2} \int_C \left(\frac{3at}{1 + t^3} \right)^2 dt$$

$$= \frac{1}{2} \cdot 2 \int_0^1 \frac{9a^2 t^2}{(1 + t^3)^2} \, dt \qquad [\because \text{ loop is symmerical about } y = x]$$

$$= 3a^2 \int_0^1 \frac{3t^2}{(1 + t^3)^2} \, dt$$

$$= 3a^2 \, [-(1 + t^3)^{-1}]_0^1$$

$$= 3a^2 \, [-(1 + 1)^{-1} + 1] = \frac{3}{2} a^2$$

Example 4.72 **Verify Stoke's theorem for $F = (2x - y)\,\hat{i} - yz^2\hat{j} - y^2z\hat{k}$, where S is the upper half surface of the sphere $x^2 + y^2 + z^2 = 1$ and C is its boundary.**
(UPTU 2002)

Solution : The boundary C of the surface S is a circle in the xy-plane of radius unity with centre origin. Suppose $x = \cos t$, $y = \sin t$, $z = 0$, $0 \le t \le 2\pi$ are parameteric equations of C. Then

$$\int_C \vec{F} \cdot dr = \int_C [(2x - y)\,\hat{i} - yz^2\hat{j} - y^2z\hat{k}] \cdot (dx\hat{i} + dy\hat{j} + dz\hat{k})$$

$$= \int_C [(2x - y\, dx - yz^2\, dy - y^2 z\, dz)]$$

$$= \int_C (2x - y)\, dx \qquad (\because z = 0 \text{ and } dz = 0)$$

$$= \int_0^{2\pi} (2 \cos t - \sin t) \frac{dx}{dt} \, dt$$

$$= -\int_0^{2\pi} (2 \cos t - \sin t) \sin t \, dt$$

$$= -\int_0^{2\pi} \left[\sin 2t - \frac{1}{2} (1 - \cos 2t) \right] dt$$

$$= -\left[-\frac{\cos 2t}{2} - \frac{t}{2} + \frac{1}{2} \frac{\sin 2t}{2} \right]_0^{2\pi}$$

$$= \left[\left(-\frac{1}{2} + \frac{1}{2} \right) - \frac{1}{2} (2\pi - 0) + \frac{1}{2} (0 - 0) \right] = \pi$$

Also $$\nabla \times \vec{F} = \begin{vmatrix} \hat{i} & \hat{j} & \hat{k} \\ \frac{\partial}{\partial x} & \frac{\partial}{\partial y} & \frac{\partial}{\partial z} \\ 2x - y & -yx^2 & -y^2 z \end{vmatrix}$$

$$= (-2yz + 2yz)\hat{i} - (0 -)\hat{j} + (0 + 1)\hat{k} = \hat{k}$$

Let S_1 be the plane region bounded by the circle C. If S_2 is the surface containing the surface S and S_1; then S_2 is closed surface. Since S_2 is a closed surface, therefore

$$\iint_{S_1} curl\ \vec{F} \,.\, \hat{n}\, dS = 0$$

or $$\iint_{S} curl\ \vec{F} \,.\, \hat{n}\, dS + \iint_{S_1} curl\ \vec{F} \,.\, \hat{n} = 0 \qquad (\because \text{ on surface } S_1, n = -k)$$

or $$\iint_{S} curl\ \vec{F} \,.\, \hat{n}\, dS - \iint_{S_1} curl\ \vec{F} \,.\, \hat{k}\, dS = 0$$

or $$\iint_{S} curl\ \vec{F} \,.\, \hat{n}\, dS + \iint_{S_1} curl\ \vec{F} \,.\, \hat{k} dS = 0$$

$$= \iint_{S_1} k \,.\, k\, dS = \iint_{S_1} dS$$

($\because$ S_1 is the surface bounded by the circle of radius 1)

$$= S_1 = \pi \qquad ...(2)$$

Hence, from equation (1) and (2) Stoke's theorem is verified.

Example 4.73 **Evaluate $\oint_C \vec{F} \,.\, dr$ by Stoke's theorem, where $\vec{F} = y^2\hat{i} + x^2\hat{j} - (x + z)\hat{k}$ and C is the boundary of the triangle with vertices at (0, 0, 0), (1, 0, 0) and (1, 1, 0).** *(UPTU 2001)*

Solution : It is obnserved that the z-coordinates of each vertex of the given triangle is zero, therefore, the triangle lies in the xy-plane and $\hat{n} = \hat{k}$

$$Curl\ \vec{F} = \begin{vmatrix} \hat{i} & \hat{j} & \hat{k} \\ \frac{\partial}{\partial x} & \frac{\partial}{\partial y} & \frac{\partial}{\partial z} \\ y^2 & x^2 & -(x + z) \end{vmatrix}$$

$$= \hat{j} + 2(x - y)\, \hat{k}$$

$$\therefore \quad Curl\ \vec{F} \,.\, \hat{n} = [\hat{j} + 2(x - y)\hat{k}\,] \,.\, \hat{k}$$
$$= 2(x - y)$$

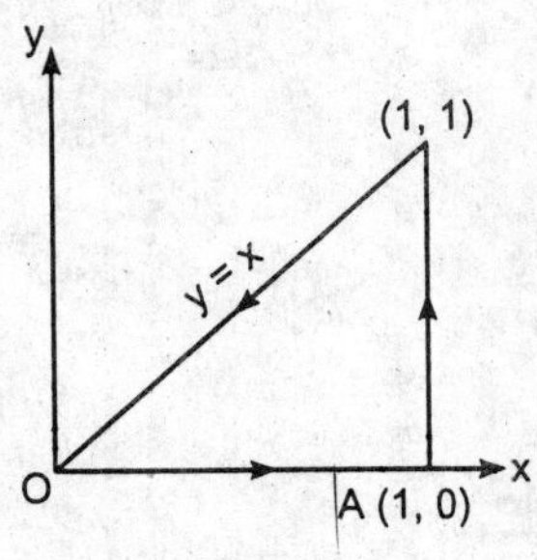

Fig. 4.14

So by Stoke's theorem

$$\oint_C \vec{F} \,.\, dr = \iint_S curl\ \vec{F} \,.\, \hat{n}\, dS$$

$$= \int_0^1 \int_0^x 2(x-y)\, dydx$$

($\because$ S is the surface bounded by triangle OAB)

$$= 2\int_0^1 \left[xy - \frac{y^2}{2} \right]_0^x dx$$

$$= 2\int_0^1 \left(x^2 - \frac{x^2}{2} \right) dx = \int_0^1 x^2\, dx = \left[\frac{x^3}{3} \right]_0^1 = \frac{1}{3}.$$

Example 4.74 **Verify Stoke's theorem $\vec{F} = (x^2 + y^2)\hat{i} - 2xy\,\hat{j}$ taken round the rectangle bounded by the lines $x = \pm a, y = 0, y = b$.** *(UPTU 2003)*

Solution : Let C be the closed curve comprised of the lines $x = \pm a, y = 0, y = b$ to form the rectangle $PQRS$.

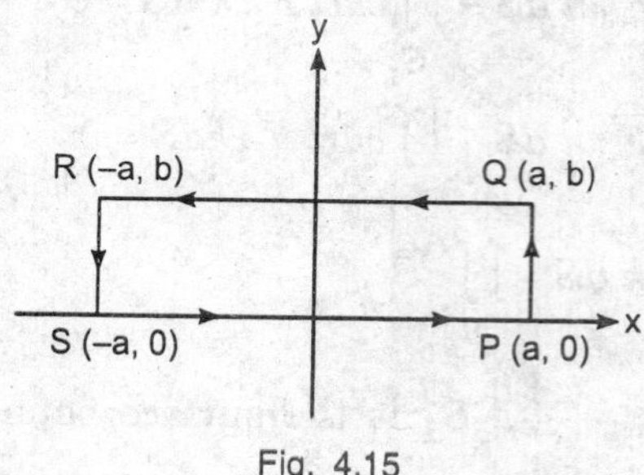

Fig. 4.15

$$\oint_C \vec{F} . dr = \int_C [(x^2 + y^2)\hat{i} - 2xy\hat{j}]\ (\hat{i}dx + \hat{j}dy)$$

$$= \oint_C [(x^2 + y^2)dx - 2xydy]$$

$$= \int_{PQ} [(x^2 + y^2)dx - 2xydy] + \int_{QR} [(x^2 + y^2)dx - 2xydy]$$

$$+ \int_{RS} [(x^2 + y^2)dx - 2xydy] + \int_{SP} [(x^2 + y^2)dx - 2xy]dy$$

Now, (i) $\int_{PQ} [(x^2 + y^2)dx - 2xydy] = \int_0^b - 2aydy$

($\because$ along PQ, $x = a$, $dx = 0$ and y varies from 0 to b)

$$= - 2a \left[\frac{y^2}{2} \right]_0^b$$

$$= -ab^2$$

(ii) $\int_{QR} [(x^2 + y^2)dx - 2xydy] = \int_a^{-a} (x^2 + b^2)dx$

($\because$ along QR, $y = b$, $dy = 0$ and x varies from a to $-a$)

$$= \left[\frac{x^3}{3} + b^2 x\right]_{a}^{-a}$$

$$= \frac{-2a^3}{3} - 2ab^2$$

(iii) $\int_{RS} [(x^2 + y^2)dx - 2xydy] = \int_b^0 -2(-a)ydy$

($\because$ along RS, $x = -a$, $dx = 0$ and y varies from b to 0)

$$= \int_b^0 2aydy = a\left[y^2\right]_b^0$$

$$= -ab^2$$

(iv) $\int_{SP} [(x^2 + y^2)dx - 2xydy] = \int_{-a}^{a} x^2\, dx$

($\because$ along SP, $y = 0$, $dy = 0$ and x varies from $-a$ to a)

$$= \left[\frac{x^3}{3}\right]_{-a}^{a} = \frac{2a^3}{3}$$

Thus, $$\int_C \vec{F} . d\vec{r} = -ab^2 - \frac{2a^3}{3} - 2ab^2 - ab^2 + \frac{2a^3}{3}$$

$$= -4ab^2 \qquad ...(1)$$

Now, $$Curl\, \vec{F} = \begin{vmatrix} i & j & k \\ \frac{\partial}{\partial x} & \frac{\partial}{\partial y} & \frac{\partial}{\partial z} \\ x^2 + y^2 & -2xy & 0 \end{vmatrix} = (-2y - 2y)k = -4y\hat{k}$$

For the given surface S; $\hat{n} = \hat{k}$

$\therefore \quad Curl\, \vec{F} . \hat{n} = -4y\hat{k} . \hat{k} = -4y$

$$\therefore \quad \iint_S curl\, \vec{F} . \hat{n}\, dS = \int_{-a}^{a} \int_0^b 4ydxdy$$

$$= -\int_{-a}^{a} [2y^2]_0^b\, dx$$

$$= -\int_{-a}^{a} 2b^2\, dx = -2b^2\, [x]_{-a}^{a}$$

$$= -2b^2[2a] = -4ab^2 \qquad ...(2)$$

Thus, equation (1) and (2) verify Stoke's theorem.

Example 4.75 **Verify Stoke's theorem for $\vec{F} = y\hat{i} + z\hat{j} + x\hat{k}$ where S is the portion of the surface $x^2 + y^2 - 2ax + az = 0$ above the plane $z = 0$.**

Solution : The given equation of the surface can be written as

$$(x - a)^2 + y^2 = -a(z - 0) \qquad ...(1)$$

Equation (1) is a paraboloid with vertex at $(a, 0, 0)$.

The paraboloid (1) meets the plane $z = 0$, *i.e.*, $(x - a)^2 + y^2 = a^2, z = 0$ which is a circle with centre $(a, 0, 0)$ and radius a.

Now, $\int_C \vec{F} \cdot \vec{dr} = \int_C (y\hat{i} + z\hat{j} + x\hat{k}) \cdot (\hat{i}dx + \hat{j}dy + \hat{k}dz)$

where C is the circle $(x - a)^2 + y^2 = a^2, z = 0$ and any point on the circle can be taken $(a + a\cos\theta, a\sin\theta, 0)$

$$\therefore \quad \int_C \vec{F} \cdot \vec{dr} = \int_C ydx + xdy + xdz$$

$$= \int_C ydx \qquad [\because z = 0 \Rightarrow dz = 0]$$

$$= \int_0^{2\pi} a\sin\theta\,(-a\sin\theta)\,d\theta$$

$$= -a^2 \int_0^{2\pi} \frac{(1 - \cos 2\theta)}{2} = -\frac{a^2}{2}\left[\theta - \frac{\sin 2\theta}{2}\right]_0^{2\pi}$$

$$= -a^2\pi \qquad ...(1)$$

$$Carl\ \vec{F} = \begin{vmatrix} \hat{i} & \hat{j} & \hat{k} \\ \frac{\partial}{\partial x} & \frac{\partial}{\partial y} & \frac{\partial}{\partial z} \\ y & z & x \end{vmatrix} = -(\hat{i} + \hat{j} + \hat{k})$$

and $\quad \hat{n} = \hat{k}$

$$\therefore \quad \hat{n} \cdot curl\ \vec{F} = \hat{k} \cdot (-1)(\hat{i} + \hat{j} + \hat{k}) = -1$$

Now $\quad \int_S \hat{n} \cdot curl\ \vec{F}\, dS = \int_0^{2\pi}\int_0^{a} (-1)\, rd\theta dr$

$$= -\left[\frac{1}{2}r^2\right]_0^a [\theta]_0^{2\pi} = -\pi a^2 \qquad ...(2)$$

From (1) and (2),

$$\int_C \vec{F} \cdot \vec{dr} = \int\int_S \hat{n} \cdot curl\ \vec{F}\, dS$$

Example 4.76 **Use Stoke's theorem of evaluate $\int curl\ \vec{F} \cdot \hat{n}\, dS$ over the open hemi-spherical surface $x^2 + y^2 + z^2 = a^2, z > 0$, where $\vec{F} = y\hat{i} + zx\hat{j} + y\hat{k}$.**

Solution : By stoke's theorem

$$\int_S curl\ \vec{F} \cdot \hat{n}\, dS = \int_C \vec{F} \cdot \vec{dr}$$

where C is the circle $x^2 + y^2 = a^2, z = 0$

$$= \int_C (y\hat{i} + zx\hat{j} + y\hat{k}) \cdot (\hat{i}dx + \hat{j}dy + \hat{k}dz)$$

$$= \int_C (ydx + zxdy + ydz)$$

On the circle C, $x = a\cos\theta, y = a\sin\theta, z = 0$ and θ varies 0 to 2π.

$$\therefore \quad \int_C \vec{F}.\,d\vec{r} = \int_C y\,dx \qquad [\because \; z = 0, dz = 0]$$

$$= \int_0^{2\pi} a\sin\theta\,(-a\sin\theta)\,d\theta$$

$$= -a^2\int_0^{2\pi} \sin^2\theta\,d\theta$$

$$= -a^2\int_0^{2\pi}\left[\frac{1-\cos 2\theta}{2}\right]d\theta = -\frac{a^2}{2}\left[\theta - \frac{\sin 2\theta}{2}\right]_0^{2\pi}$$

$$= -\frac{a^2}{2}[2\pi - 0] = -\pi a^2$$

Example 4.77 **Prove that**

$$\oint_C \vec{r}.\,d\vec{r} = 0.$$

Solution : By stoke's theorem, we know that

$$\oint_C \vec{F}.\,d\vec{r} = \int_S curl\,\vec{F}.\,\hat{n}\,dS$$

Let $\vec{F} = \vec{r}$

Now $$Carl\,\vec{r} = \begin{vmatrix} \hat{i} & \hat{j} & \hat{k} \\ \frac{\partial}{\partial x} & \frac{\partial}{\partial y} & \frac{\partial}{\partial z} \\ x & y & z \end{vmatrix} = i(0) + j(0) + k(0) = 0$$

$$\oint_C \vec{r}.\,d\vec{r} = curl\,\vec{r}.\,\hat{n}\,dS = 0 \qquad [\because\; curl\,\vec{r} = 0]$$

Example 4.78 **Verify Stoke' theorem if $\vec{F} = xy\hat{i} + yz\hat{i} + xz\hat{k}$ for the surface S be the part of the cylinder $z = 1 - x^2$ for $0 \le x \le 1, -2 \le y \le 2$.**

Solution : Stoke's theorem is given by

$$\oint_C \vec{F}.\,d\vec{r} = \iint_S (curl\,\vec{F}).\,\hat{n}\,dS$$

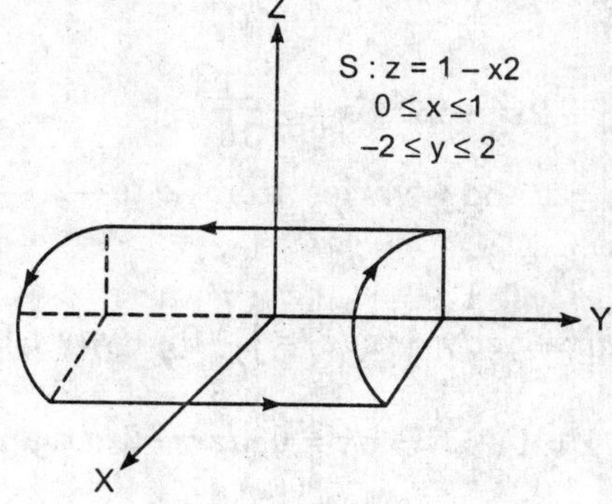

Fig. 4.16

Surface integral :

$\because \quad \overrightarrow{F} = xy\hat{i} + yz\hat{i} + xz\hat{k}$

$$\therefore \quad Carl\ \overrightarrow{F} = \nabla \times \overrightarrow{F} = \begin{vmatrix} \hat{i} & \hat{j} & \hat{k} \\ \frac{\partial}{\partial x} & \frac{\partial}{\partial y} & \frac{\partial}{\partial z} \\ xy & yz & xz \end{vmatrix}$$

$$= \hat{i}\left[\frac{\partial}{\partial y}(xz) - \frac{\partial}{\partial y}(xy)\right] + \hat{j}\left[\frac{\partial}{\partial z}(xy) - \frac{\partial}{\partial x}(xz)\right] + \hat{k}\left[\frac{\partial}{\partial x}(yz) - \frac{\partial}{\partial y}(xy)\right]$$

$$= -y\hat{i} - z\hat{j} - x\hat{k}$$

Now, the upper unit normal is

$$\hat{n} = \frac{\nabla f}{||\nabla f||} = \frac{2x\hat{i} + \hat{k}}{\sqrt{4x^2 + 1}}$$

$$\therefore \quad Curl\ \overrightarrow{F} . \hat{n}\, dS = \frac{-(y\hat{i} + 3\hat{j} + x\hat{k}) . (2x\hat{i} + \hat{k})}{\sqrt{rx^2 + 1}}$$

$$\iint_S \frac{-(2xy + x)}{\sqrt{4x^2 + 1}}\, dS = \iint_a -(2xy + x)\, dA$$

$$= \int_0^1 \int_{-2}^2 -(2xy + x) dx dy = -\int_0^1 [xy^2 + xy]_{-2}^2\, dx$$

$$= -\int_0^1 4x dx = -2 \qquad \text{...(1)}$$

Line integral :

On $C_1 : x = 1$ and $z = 0$

$\therefore \quad dx = 0, dz = 0$

$$\int_{C_1} \overrightarrow{F} . d\overrightarrow{r} = \int_{C_1} (xydx + yzdy + xzdz)$$

$$= \int_{C_1} . y(0) + y(0)dy + 1(0)dz = 0$$

On $C_2 : y = 2, z = 1 - x^2$ and x varies from 1 to 0.

$\therefore \quad dy = 0, dz = -2xd$

$$\therefore \quad \int_{C_2} (xydx + yzdy + xzdz) = \int_{C_2} 2xdx + 2(1 - x^2)0 + x(1 - x^2)(-2xdx)$$

$$= \int_1^0 (2x - 2x^2 + 2x^4)dx = -\frac{11}{15}$$

On $C_3 : x = 0, z = 1$ and y varies from 2 to -2.

$\therefore \quad dx = 0, dz = 0$

$$\therefore \quad \int_{C_3} (xydx + yzdy + xzdz) = \int_{C_3} 0y + ydy + 0 = \int_{-2}^2 ydy = 0$$

On $C_4 : y = -2, z = 1 - x^2 \Rightarrow dy = 0, dz = -2xdx$ and x varies from 0 to 1.

$$\therefore \int_{C_4} (xydx + yzdy + xzdz) = \int_C -2xdx - 2(1-x^2)0 + x(1-x^2)(-2xdx)$$

$$= \int_1^0 (-2x - 2x^2 + 2x^4)dx = -\frac{19}{15}$$

$\therefore$ Required line integral $\int_C xydx + yzdy + xzdz$

$$= 0 - \frac{11}{15} + 0 - \frac{19}{15} = -2 \qquad ...(2)$$

From (1) and (2), Stoke's theorem verified.

4.31 GAUSS DIVERGENCE THEOREM (Transformation between volume integrals and surface integrals)

Suppose T be a closed bounded region in space, whose boundary is piecewise smooth orientable surface S. Also if $F(x, y, z)$ be a vector function, which is continuous and has continuous first order partial derivatives in some domain containing Γ. Then

$$\iiint_\Gamma div\,\vec{F}\,dV = \iint_S \vec{F}\,.\,\hat{n}\,dS$$

where $\hat{n}$ is the outer unit normal vector of S.

Example 4.79 **For any closed surface S, prove that**

$$\iint_S \boldsymbol{curl\,\vec{F}\,.\,\hat{n}\,dS = 0.}$$

Solution : Using $\iint_S curl\,\vec{F}\,.\,\hat{n}\,dS = \iiint_K (div\,curl\,\vec{F})dv$, where V is the volume enclosed by S.

$$= 0 \qquad (\because\ div\ curl\ F = 0)$$

Example 4.80 **Apply the divergence theorem to compute $\iint u\,.\,\hat{n}\,dS$, where S is the surface of the cylinder $x^2 + y^2 = a^2$ bounded by the planes $z = 0, z = b$ and $u = x\hat{i} - y\hat{j} + z\hat{k}$.**

Solution : Using divergence theorem, we get

$$\iint \vec{u}\,.\,\hat{n}\,dS = \iiint_V (\nabla\,.\,\hat{u})dV$$

where V is the volume enclosed by the surface S.

$$= \iiint_V \left(\hat{i}\frac{\partial}{\partial x} + \hat{j}\frac{\partial}{\partial y} + \hat{k}\frac{\partial}{\partial z}\right) . (x\hat{i} - y\hat{j} + z\hat{k})dV$$

$$\iiint_V \left(\frac{\partial x}{\partial x} - \frac{\partial y}{\partial y} + \frac{\partial z}{\partial z}\right) dV = \iiint_V (1 - 1 + 1)\,dV$$

$$= \iiint_V dV$$

$$= \iiint_V dx\,dy\,dz = \text{Volume of the given cylinder}$$

Example 4.81 **The vector field $F = x^2\hat{i} + z\hat{j} + yz\hat{k}$ is defined over the volume of the cuboid given by $0 \le x \le a$, $0 \le y \le b, 0 \le z < c$ enclosing the surface S, evaluate the surface integral $\iint_S \vec{F}.\vec{ds}$.**

(UPTU 2002)

Solution : Using Gauss theorem, we may write

$$\iint_S \vec{F}.\vec{ds} = \iiint_V div\,\vec{F}\,dV$$

$$= \iiint_V \left[\frac{\partial}{\partial x}(x^2) + \frac{\partial}{\partial y}(z) + \frac{\partial}{\partial z}(yz)\right] dV$$

$$= \iiint_V (2x + y)\,dV = \int_0^a \int_0^b \int_0^c (2x + y)dxdydz$$

$$= \int_0^a \int_0^b (2x + y)\,(z)_0^c\,dxdy$$

$$= c\int_0^a \left[2xy + \frac{y^2}{2}\right]_0^b dx$$

$$= c\int_0^a \left(2xb + \frac{b^2}{2}\right)_0^b dx$$

$$= c\left[bx^2 + \frac{b^2}{2}x\right]_0^a = \left[a^2 b + \frac{b^2 a}{2}\right]$$

$$= abc\left[a + \frac{b}{2}\right]$$

Example 4.82 **Use divergence to evaluate the surface integral**

$$\iint_S (x\,dy\,dz + y\,dz\,dx + z dx\,dy)$$

where S is the portion of the plane, $x + 2y + 3z = 6$ which lies in the first octant.

(UPTU 2004)

Solution : Gauss Divergence theorem states that

$$\iint_S \vec{F}.\hat{n}\,\vec{dS} = \iiint_V div\,\vec{F}\,dV$$

We have $\vec{F} = x\hat{i} + y\hat{j} + z\hat{k}$

and $\hat{n} = dy\,dz\,\hat{i} + dz\,dx\,\hat{j} + dx\,dy\,\hat{k}$

Also,
$$div\,\vec{F} = \left(\hat{i}\frac{\partial}{\partial x} + \hat{j}\frac{\partial}{\partial y} + \hat{k}\frac{\partial}{\partial z}\right)\cdot(x\hat{i} + y\hat{j} + z\hat{k})$$

$$= \frac{\partial}{\partial x}(x) + \frac{\partial}{\partial y}(y) + \frac{\partial}{\partial z}(z)$$

$$= 3$$

Thus, $\iint_S (x\,dy\,dz + y\,dz\,dx + z\,dx\,dy) = \iiint_V 3dV$

$$= 3\int_{x=0}^{6}\int_{y=0}^{\frac{6-x}{2}}\int_{0}^{\frac{6-x-2y}{3}} dx\,dy\,dz$$

$$= 3\int_{0}^{6}\int_{0}^{\frac{6-x}{2}} [z]_0^{6-x-x\,2y/3}\,dx\,dy$$

$$= 3\int_{0}^{6}\int_{0}^{\frac{6-x}{2}} \left(\frac{6-x-2y}{3}\right) dx\,dy$$

$$= \int_{0}^{6}\Big[(6-x)y - y^2\Big]_0^{\frac{6-x}{2}} dx$$

$$= \int_{0}^{6}\left[\frac{(6-x)^2}{2} - \frac{(6-x)^2}{4}\right] dx$$

$$= \int_{0}^{6}\frac{(6-x)^2}{4}\,dx = \left[\frac{(6-x^3}{4(-3)}\right]_0^6$$

$$= \frac{1}{12}(216) = 18$$

Example 4.83 **Verify divergence theorem for $\vec{F} = 4xz\hat{i} - y^2\hat{j} + yz\hat{k}$ taken over the cube bounded by the planes $x = 0, y = 0, y = 1, z = 0, z = 1$.**

Solution : It is given that $\vec{F} = 4xz\hat{i} - y^2\hat{j} + yz\hat{k}$

$$div\,F = \left(\hat{i}\frac{\partial}{\partial x} + \hat{j}\frac{\partial}{\partial y} + \hat{k}\frac{\partial}{\partial z}\right)\cdot(4xz\hat{i} - y^2\hat{j} + yz\hat{k})$$

$$= \frac{\partial}{\partial x}(4xz) - \frac{\partial}{\partial y}y^2 + \frac{\partial}{\partial z}yz$$

$$= 4z - 2y + y$$

$$= 4z - y$$

So, $\iiint_V div\,\vec{F}\,dV = \int_{x=0}^{1}\int_{y=0}^{1}\int_{z=0}^{1}(4z - y)\,dx\,dy\,dz$

$$= \int_{0}^{1}\int_{0}^{1}\left[\frac{4z^2}{2} - zy\right]_0^1 dx\,dy$$

$$= \int_{0}^{1}\int_{0}^{1}(2 - y)\,dx\,dy$$

$$= \int_{0}^{1}\left[2y - \frac{y^2}{2}\right]_0^1 dx$$

$$\iiint_V div\,\vec{F}\,dV = \int_0^1 \left[\frac{3}{2}\right] dx = \left[\frac{3}{2}x\right]_0^1 = \frac{3}{2} \quad ...(1)$$

Now, we have to evaluate $\iint_S \vec{F}\,.\,\hat{n}\,dS$ over the surface of the given volume.

Hence, the surface S is consisting of six faces of a cuboid

(i) Over the face $OABC$, $z = 0$, $dz = 0$, $\hat{n} = -\hat{k}$, $dS = 0$

$$\therefore \quad \iint_{OABC} \vec{F}\,.\,\hat{n}\,dS = \int_0^1\int_0^1 (-y^2\hat{j})\,.\,(-\hat{k})dx\,dy = 0$$

(ii) Over the face $BCDE$, $y = 1$, $dy = 0$, $\hat{n} = \hat{j}$, $dS = dx\,dz$

$$\therefore \quad \iint_{BCDE} \vec{F}\,.\,\hat{n}\,dS = \int_0^1\int_0^1 (4xz\hat{i} - \hat{j} + z\hat{k})\,.\,\hat{j}dx\,dz$$

$$= \int_0^1\int_0^1 dx\,dz = \int_0^1 [z]_0^1\,dx$$

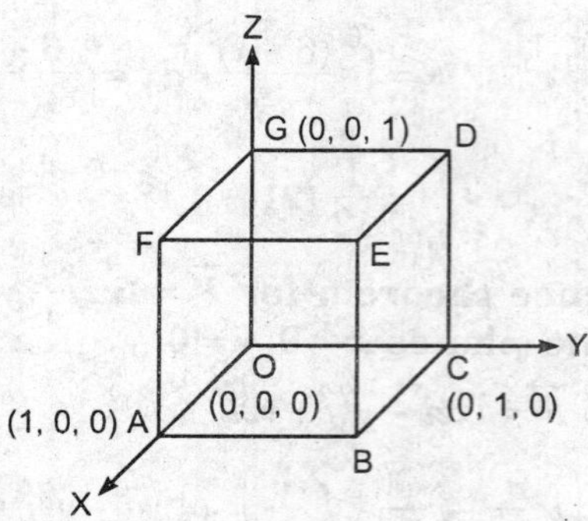

Fig. 4.17

$$= \int_0^1 dx = [x]_0^1 = -1$$

(iii) Over the face $DEFG$, $z = 1$, $dz = 0$, $\hat{n} = \hat{k}$, $dS = dxdy$

$$\therefore \quad \iint_{DEFG} \vec{F}\,.\,\hat{n}\,dS = \int_0^1\int_0^1 (4x\hat{i} - y^2)\hat{j} + y\hat{k})\,.\,\hat{k}\,dx\,dy$$

$$= \int_0^1\int_0^1 y\,dx\,dy$$

$$= \int_0^1 \left[\frac{y^2}{2}\right]_0^1 dx$$

$$= \int_0^1 \frac{1}{2}\,dx = \frac{1}{2}\,[x]_0^1 = \frac{1}{2}$$

(iv) Over the face $AOGF$, $y = 0$, $dy = 0$, $\hat{n} = -\hat{j}$, $dS = dx\,dz$

$$\therefore \quad \iint_{AOGF} \vec{F}\,.\,\hat{n}\,dS = \int_0^1\int_0^1 (4xz\hat{i})\,.\,(-\hat{j})dx\,dz = 0$$

(v) Over the face $OCDG$, $x = 0$, $dx = 0$, $\hat{n} = -\hat{i}$, $dS = dy\,dz$

$$\therefore \quad \iint_{OCDG} \vec{F}\cdot\hat{n}\,dS = \int_0^1\int_0^1 (-y^2\hat{j} + yz\hat{k})\cdot(-\hat{i})dy\,dz = 0$$

(vi) Over the face $ABEF$, $x = 1$, $dx = 0$, $\hat{n} = \hat{i}$, $dS = dy\,dz$

$$\therefore \quad \iint_{ABEF} \vec{F}\cdot\hat{n}\,dS = \int_0^1\int_0^1 (4z\hat{i} - y^2\hat{j} + yz\hat{k})\cdot\hat{i}\,dy\,dx$$

$$= \int_0^1\int_0^1 4z\,dy\,dz = \int_0^1 [2z^2]_0^1\,dy = 2\int_0^1 dy = 2[y]_0^1 = 2$$

Thus, the surface integral over the given surface, is

$$\iint_S \vec{F}\cdot\hat{n}\,dS = 0 - 1 + \frac{1}{2} + 0 + 0 + 2 = \frac{3}{2}$$

From equations (1) and (2) we find

$$\iint_S \vec{F}\cdot\hat{n}\,dS = \iiint_V div\,\vec{F}\,dV$$

Hence, the divergence theorem is verified.

Example 4.84 **Verify Gauss divergence theorem for $\vec{F} = [(x^3 - yz)\hat{i} - 2x^2y\hat{j} + 2\hat{k}]$, taken over the region S, here S denotes the surface of the cube bounded by the planes $x = 0$, $x = a$, $y = 0$, $y = a$, $z = 0$ and $z = a$.**

Solution : Given

$$\vec{F} = [(x^3 - yz)\hat{i} - 2x^2y\hat{j} + 2\hat{k}]$$

First, we shall calculate

$\int_S \vec{F}\cdot\hat{n}\,dS$ over the six faces of the cubes separately according to figure.

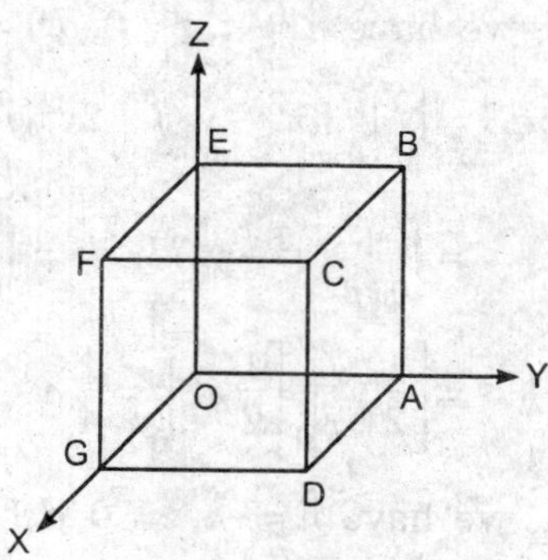

Fig. 4.18

For the face $GDCF$, we have $\hat{n} = \hat{i}$, $x = a$ and $dS = zdydz$

$$\int_{S_1} \vec{F}\cdot\hat{n}\,dS = \int_0^a\int_0^1 [(x^3 - yz)\hat{i} - 2x^2y\hat{j} + 2\hat{k}]\cdot\hat{i}dy\,dz$$

$$= \int_0^a\int_0^a (x^3 - yz)\,dy\,dz = \int_0^a\int_0^a (a^3 - yz)\,dy\,dz \qquad (\because\ x = a)$$

$$= \int_0^a \left[a^3 z - \frac{yz^2}{2} \right]_0^a dy$$

$$= \int_0^a \left(a^4 - \frac{a^2 y}{2} \right) dy$$

$$= \left[a^4 y - \frac{a^2 y^2}{4} \right]_0^a = a^5 - \frac{1}{4} a^4 \qquad \text{...(1)}$$

For the face $CBEF$, we have $\hat{n} = \hat{k}$, $z = a$ and $dS = dx\, dy$

$$\therefore \quad \int_{S_2} \vec{F} \cdot \hat{n}\, dS = \int_0^a \int_0^a [(x^3 - yz)\hat{i} - 2x^2 y \hat{j} + 2\hat{k}] \cdot \hat{k}\, dx\, dy$$

$$= \int_0^a \int_0^a 2 dx\, dy = 2[x]_0^a [y]_0^a = 2a^2 \qquad \text{...(2)}$$

For the face $ABCD$, we have $\hat{n} = \hat{j}$, $y = a$, $dS = dx\, dz$

$$\because \quad \int_{S_2} \vec{F} \cdot \hat{n}\, dS = \int_0^a \int_0^a [(x^3 - yz)\hat{i} - 2x^2 y \hat{j} + 2\hat{k}] \cdot \hat{j}\, dx\, dz$$

$$= \int_0^a \int_0^a (-2x^2 y)\, dx\, dz = \int_0^a \int_0^a (-2x^2 a)\, dx\, dz \qquad (\because y = a)$$

$$= -2a \int_0^a x^2 [z]_0^a\, dx = -2a \int_0^a x^2\, a dx$$

$$= -2a^2 \left[\frac{x^3}{3} \right]_0^a = -\frac{2}{3} a^5 \qquad \text{...(3)}$$

For the face $OABE$, we have $\hat{n} = -\hat{i}$, $x = 0$, $dS = dx\, dy$

$$\therefore \quad \int_{S_2} \vec{F} \cdot \hat{n}\, dS = \int_0^a \int_0^a [(x^3 - yz)\hat{i} - 2x^2 y \hat{j} + 2\hat{k}] \cdot (-\hat{i})\, dy\, dz$$

$$= \int_0^a \int_0^a (x^3 - yz) dy\, dz = \int_0^a \int_0^a yz\, dy\, dz \qquad (\because x = 0)$$

$$= \left[\frac{1}{2} y^2 \right]_0^a \left[\frac{1}{2} z^2 \right]_0^a = \frac{1}{4} a^4 \qquad \text{...(4)}$$

For the face $OGDA$, we have $\hat{n} = -\hat{k}$, $z = 0$, $dS = dx\, dy$

$$\therefore \quad \int_{S_2} \vec{F} \cdot \hat{n}\, dS = \int_0^a \int_0^a [(x^3 - yz)\hat{i} - 2x^2 y \hat{j} + 2\hat{k}] \cdot (-\hat{k})\, dx\, dy$$

$$= -2 \int_0^a \int_0^a dx\, dy = -2\, [x]_0^a [y]_0^a = -2a^2 \qquad \text{...(5)}$$

For the face $OGFE$, we have $\hat{n} = -\hat{j}$, $y = 0$, $dS = dx\, dz$

$$\therefore \quad \int_{S_2} \vec{F} \cdot \hat{n}\, dS = \int_0^a \int_0^a [(x^3 - yz)\hat{i} - 2x^2 y \hat{j} + 2\hat{k}] \cdot (\hat{i} - \hat{k})\, dx\, dz$$

$$= 2\int_0^a \int_0^a x^2 y \, dx \, dz = 0 \qquad \text{...(6)} \qquad (\because y = 0)$$

$$\therefore \quad \int_S \vec{F} \cdot \hat{n} \, dS = \int_{S_1} + \int_{S_2} + \int_{S_3} + \int_{S_4} + \int_{S_5} + \int_{S_6}$$

$$= a^5 - \frac{1}{4} a^4 + 2a^2 - \frac{2}{5} a^5 + \frac{1}{4} a^4 - 2a^2 + 0 = \frac{1}{3} a^5 \qquad \text{...(7)}$$

$$\because \quad \vec{F} = (x^3 - yz)\hat{i} - 2x^2 y\hat{j} + 2\hat{k}$$

$$\therefore \quad div\, \vec{F} = \left(\hat{i}\frac{\partial}{\partial x} + \hat{j}\frac{\partial}{\partial y} + \hat{k}\frac{\partial}{\partial z}\right) \cdot [(x^3 - yz)\hat{i} - 2x^2 y\hat{j} + 2\hat{k}\,]$$

$$= \frac{\partial}{\partial x}(x^3 - yz) + \frac{\partial}{\partial y} + \frac{\partial}{\partial y}(-2x^2 y) + \frac{\partial}{\partial z}(2)$$

$$= 3x^2 - 2x^2 + 0$$

$$= x^2$$

Now $$\iiint_V \nabla \cdot \vec{F} dV = \iiint_V x^2 \, dV$$

$$= \int_0^a \int_0^a \int_0^a x^2 \, dxdy \, dx$$

$$= \left[\frac{x^3}{3}\right]_0^a [y]_0^a \, [z]_0^a = \frac{a^5}{3} \qquad \text{...(8)}$$

Hence from equations (7) and (8), we get

$$\int_S \vec{F} \cdot \hat{n} \, dS = \int_V \nabla \cdot \vec{F} dV$$

Example 4.85 **Verify the Gauss divergence theorem for $\vec{F} = 4x\hat{i} - 2y^2\hat{j} + z^2\hat{k}$ taken over the region S bounded by $x^2 + y^2 = 4$, $z = 0$ and $z = 3$.**

Solution : Surface S is the given cylinder consisting curved surface S_1 and two equal circular ends S_2 and S_3 on the planes $z = 0$ and $z = 3$, respectively,

$$\int_S \vec{F} \cdot \hat{n} \, dS = \int_{S_1} \vec{F} \cdot \hat{n} + \int_{S_2} \vec{F} \cdot \hat{n} + \int_{S_3} \vec{F} \cdot \hat{n}$$

For the curved surface S_1, we have

$$\hat{n} = \frac{grad\, f}{|\, grad\, f\,|}$$

where $f = x^2 + y^2 = 4$

$$= \frac{2x\hat{i} + 2y\hat{j}}{\sqrt{4x^2 + 4y^2}} = \frac{x\hat{i} + y\hat{j}}{\sqrt{x^2 + y^2}} = \frac{x\hat{i} + y\hat{i}}{2}$$

Now $$\iint S_1 \vec{F} \cdot \hat{n} \, dS = \iint_{S_1} (4x\hat{i} - 2y^2\hat{j} + zz^2 k) \cdot \frac{1}{2}(x\hat{i} + y\hat{j}) \, dS$$

$$= \iint_{S_1} (2x^2 - y^2) dS$$

For S_1, $x = 2\cos\theta$, $y = 2\sin\theta$, θ varies from 0 to 2π and z varies from $z = 0$ to $z = 3$

$$\therefore \quad \iint_{S_1} \vec{F}\cdot\hat{n}\, dS = \int_{\theta=0}^{2\pi}\int_{z=0}^{3} [2(2\cos\theta)^2] - (2\sin\theta)^2]\, 2d\theta dz$$

$$= 2\times 8\int_0^{2\pi}\int_0^3 (\cos^2\theta - \sin^3\theta)\, d\theta dz$$

$$= 16\int_0^{2\pi} (\cos^2\theta - \sin^3\theta)\, [z]_0^3\, d\theta$$

$$= 48\int_0^{2\pi} \cos^2\theta\, d\theta - 48\int_0^{2\pi} \sin^3\theta\, d\theta$$

$$= 2\times 48\int_0^{\pi} \cos^2\theta\, d\theta - 0$$

$[\because\ (\sin(2\pi-\theta)]^3 = -\sin^3\theta$ and $[\cos(2\pi-\theta)^2 = \cos^2\theta]$

$$= 192\frac{\Gamma(3/2)\Gamma(1/2)}{2\Gamma\left(\frac{2+0+2}{2}\right)} = \frac{192\,\frac{1}{2}\sqrt{\pi}\sqrt{\pi}}{2.1} = 48\pi \qquad \text{...(1)}$$

For the lower circular end S_2 on the plane $z = 0$, $\hat{n} = -\hat{k}$

$$\iint_{S_2} \vec{F}\cdot\hat{n}\, dS = \iint_{S_2} (4x\hat{i} - 2y^2\hat{j} + z^2\hat{k})\cdot(-\hat{k})\, dS$$

$$= -\iint_{S_2} z^2\, dS = 0 \qquad \text{...(2)} \qquad (\because\ z = 0)$$

For upper circular end S_3, as the plane $z = 3$, $\hat{n} = \hat{k}$

$$\therefore \quad \iint_{S_3} \vec{F}\hat{n}\, dS = \iint_{S_3} (4\hat{i} - 2y^2\hat{j} + z^2\hat{k})\cdot\hat{k}\, dS$$

$$= \iint_{S_3} z^2\, dS$$

$$= 9\iint_{S_3} = dS$$

$\because$ Area of S_3 is $\pi(2)^2 = 9\,[\pi(2)^2] = 36\pi$

$$\therefore \quad \iint_S \vec{F}\cdot\hat{n}\, dS = \iint_{S_1} \vec{F}\cdot\hat{n} + \iint_{S_2} \vec{F}\cdot\hat{n} + \iint_{S_3} \vec{F}\cdot\hat{n}$$

$$= 48\pi + 0 + 36\pi$$

$$= 84\pi \qquad \text{...(4)}$$

$$\text{Now div } \vec{F} = \left(\hat{i}\frac{\partial}{\partial x} + \hat{j}\frac{\partial}{\partial y} + \hat{k}\frac{\partial}{\partial z}\right)\cdot(4x\hat{i} - 2y^2\hat{j} + 3z^2\hat{k})$$

$$= 4 - 4y + 2z$$

$$= 2(2 - 2y + z)$$

$$\because \quad \int_V \nabla\cdot\vec{F}\,dV = \int_{x=-2}^{2}\int_{y=-\sqrt{1-x^2}}^{\sqrt{4-x^2}}\int_{z=0}^{3} 2\,(2-2y+3)\,dx\,dy\,dz$$

$$= 2\int_{-2}^{2}\int_{-\sqrt{4-x^2}}^{\sqrt{4-x^2}}\left[2z - 2yz + \frac{z^2}{2}\right]_0^3 dx\,dy$$

$$= 2\int_{-2}^{2}\int_{\sqrt{4-x^2}}^{\sqrt{4-x^3}}\left(6 - 6y + \frac{9}{2}\right) dx\,dy$$

$$= 2\int_{-2}^{2}\int_{-\sqrt{4-x^2}}^{\sqrt{4-x^2}}\frac{21}{2}\,dx\,dy - 12\int_{-2}^{2}\int_{-\sqrt{1-x^2}}^{\sqrt{1-x^2}} y\,dx\,dy$$

$$= 21\times\int_{-2}^{2}[y]_0^{\sqrt{4-x^2}}\,dx - 12\times 0$$

[$\because$ y is an odd function]

$$= 42\int_{-2}^{2}\sqrt{4-x^2}\,dx = 84\int_0^2\sqrt{(2)^2-x^2}\,dx$$

$$= 84\left[\frac{x}{2}\sqrt{4-x^2} + \frac{1}{2}\cdot 4\sin^{-1}\left(\frac{x}{2}\right)\right]_0^2$$

$$= 84\,[2\sin^{-1}1]$$

$$= 84.2\,\frac{\pi}{2} = 84\,\pi \qquad \ldots(5)$$

From (4) and (5), we have

$$\int_S \vec{F}\cdot\hat{n}\,dS = \int_V div\,\vec{F}\,dv$$

PROBLEM SET 4.6

1. Verify Green's theorem in the plane for

$$\int_C [(xy+y^2)dx + x^2dy]$$

where C is closed curve of the region bounded by $y = x$ and $y = x^2$.

2. Verify Green's theorem in plane for

$$\int [3x^3 - 8y^2]dx + (4y - 6xy)dy]$$

where C is the region bounded by the parabola $y = \sqrt{x}$ and $y = x^2$.

3. Evaluate by Green's theorem

$$\int_C (x^2 - \cosh y)dx + (y + \sin x)dy$$

where C is the rectangle with vertices (0, 0), (π, 0), (π, 1), (0, 1).
Ans. $\pi(\cosh 1 - 1)$

4. Verify Green's theorem in plane for

$$\int_C (x^2 - xy^3)dx + (y^2 - 2xy)dy$$

where C is the square with vertices (0, 0), (2, 0), (2, 2), (0, 2).

5. Evaluate by Green's theorem

$$\int_C (e^{-x} \cos y dx + e^{-x} \sin y dy$$

where C is the rectangle with vertices (0, 0), (π, 0), $\left(\pi, \frac{\pi}{2}\right), \left(0, \frac{\pi}{2}\right)$.

Ans. $2(1 - e^{-\pi})$

6. Evaluate by Green's theorem

$$\int_C (\cos x \sin y - xy)dx + \sin x \cos y dy$$

where C is the circle $x^2 + y^2 = 1$.

Ans. 0

7. Verify Green's theorem in plane for

$$\int_C [(x^2 - 2xy)dx + x^2 y + 3)dy]$$

where C is the boundary of the region defined by $y^2 = 8x$ and $x = 2$.

8. Verify Stoke's theorem for the vector field $\vec{F} = (x - y^2)\hat{i} + 2xy\hat{j}$ integrated around the rectangle in the plane $z = 0$ and bounded by the lines $x = 0, y = 0, x = a, y = b$.

9. Verify Stoke's theorem for the vector field $\vec{F} = z\hat{i} + x\hat{j} + z\hat{k}$, integrated over the surface S, where S is the portion of sphere $x^2 + y^2 + z^2 = 9$ above the xy-plane.

10. Verify Stoke's theorem for the vector field $\vec{v} = z\hat{i} + (zx + z)\hat{j} + z\hat{k}$, C is the boundary of the triangle with vertices at (1, 0, 0), (0, 2, 0) and (0, 0, 3).

11. Use Stoke's theorem to evaluate $\iint_S (\nabla \times \vec{F}) \,.\, \hat{n}\, dS$, where $\vec{F} = y\hat{i} + (x - 2xz)\hat{j} - xy\,\hat{k}$ and S is the surface of the sphere $x^2 + y^2 + z^2 = a^2$ above the xy-plane.

Ans. 0

12. Prove by using Stoke's theorem that

$$\iint (e^x\, dx + 2y\, dy - dz) = 0$$

where C is the curve $x^2 + y^2 = 4, z = 2$.

13. Evaluate by Stoke's theorem

$$\oint_C (yz\, dx + zx\, dy + xy\, dz)$$

where C is the curve of intersection of $x^2 + y^2 = 1, y = z^2$.

Ans. 0

14. Evaluate the surface integral

$$\iint_S (\nabla \times \vec{v}) \cdot \hat{n}\, dS, \text{ where}$$

(i) $\vec{v} = (x+y)\hat{i} + (y+z)\hat{j} + (z+x)\hat{k}$, S is the portion of the cone

$$z = \sqrt{x^2+y^2} \text{ for } x^2+y^2 \le 4.$$

(ii) $\vec{v} = 2yz\hat{i} + 3zx\hat{j} + xy\hat{k}$, S is the paraboid $z = x^2+y^2$ for $x^2+y^2 \le 4$.

Ans. (i) -4π,

(ii) 16π

15. Evaluate

$$\iint_S xz^2\, dydz + (x^2y - z^3)dzdx + (2xy + y^2z)\, dxdy$$

where S is the surface of hemispherical region bounded by $z = \sqrt{a^2-x^2-y^2}$ and $z = 0$.

Ans. $\frac{2}{5}\pi a^5$

16. Evaluate

$$\iint_S (y^2z^2\hat{i} + z^2x^2\hat{j} + x^2y^2\hat{k}) \cdot \hat{n}\, dS$$

where S is the part of the sphere $x^2+y^2+z^2 = 1$ above the xy-plane and bounded by the plane.

17. Evaluate

$$\iint_S (\nabla \times A) \cdot \hat{n}\, dS$$

where S is the surface of the cone $z = 2 - \sqrt{x^2+y^2}$ above the xy-plane and $A = (x-z)\hat{i} + (x^3+yz)\hat{j} - 3xy^2\hat{k}$.

Ans. 12π

18. Use divergence theorem to evaluate

$$\iint_S \vec{v} \cdot \hat{n}\, dS$$

where $\vec{v} = x^2y\,\hat{i} + y^3\,\hat{j} + xz^2\,\hat{k}$ and S is the surface of parallelopiped $0 \le x \le 2$, $0 \le y \le 3$, $0 \le z \le 4$.

Ans. 384

19. Evaluate

$$\iint_S \vec{F} \cdot \hat{n} dS$$

using Gauss divergence theorem, where $\vec{F} = 2xy^2\,\hat{i} + x^2y\,\hat{j} + x^3\,\hat{k}$ and S is the surface of the region bounded by the cone $z = \sqrt{x^2+y^2}$ and the plane $z = 4$.

Ans. $\frac{3072\pi}{5}$

20. Using Gauss divergence theorem, evaluate

$$\iint_S \vec{F} \cdot \hat{n} dS \text{ for } \vec{F} = 5xy\,\hat{i} + 3y\,\hat{j} + x^2\,\hat{k}$$

where S is the surface of the region bounded by $x + y = 4, x = 0, y = 0, z = 0$ and $z = 4$.

Ans. $\frac{928}{3}$

21. Use divergence theorem to evaluate

$$\iint_S \vec{r} \cdot \hat{n}\, dS$$

where S is the surface of the sphere $x^2 + y^2 + z^2 = 9$.

Ans. 108π

22. Verify divergence theorem for the function $F = 2x^2y\hat{i} - y^2\hat{j} + 4xz^2\hat{k}$ taken over the region in first octant by $y^2 + z^2 = 9$ and $x = 2$.

23. Verify Gasuss divergence theorem for $\vec{v} = x^2\,\hat{i} + 2y^2\,\hat{j} + 3z^2\,\hat{k}$ and S be the surface of the region bounded by $x^2 + y^2 = 9$, $z = 0$ and $z = 3$.

24. Verify Gauss divergence theorem for $\vec{F} = 2xy\,\hat{i} + 6yz\,\hat{j} + 3zx\,\hat{k}$ taken over the region bounded by $x + y + z = 2$ in first octant.

25. Verify Gauss divergence theorem for $\vec{v} + 6x^2y\,\hat{i} - y^2\,\hat{j} + 8xz^2\,\hat{k}$ taken over the region bounded by the cylinder $y^2 + z^2 = 16$ and $x = 3$, lying in first octant.

TUTORIALS

AMITY SCHOOL OF ENGINEERING & TECHNOLOGY
AMITY UNIVERSITY UTTAR PRADESH

B.TECH (Ist Semester)

APPLIED MATHEMATICS - I

TUTORIAL SHEET-1

Successive differentiation, Leibnitz's theorem (without proof), Expansion of Functions of Several Variables, Asymptote & Curvature.

1. If $p^2 = a^2 \cos^2 \theta + b^2 \sin^2 \theta$, prove that

$$p + \frac{d^2p}{d\theta^2} = \frac{a^2b^2}{p^3}$$

2. Find the nth derivative of

(i) $\sin^3 x$ (ii) $\sin x \cos 3x$ (iii) $\frac{x+1}{x^2-4}$

(iv) $\frac{1}{1-5x+6x^2}$ (v) $\tan^{-1}\left(\frac{x}{a}\right)$ (vi) $\sin^2 x \cos^3 x$

3. Find the nth differential coefficient of $x^3 \log x$.

4. If $y = x^2 e^x$, show that

$$\frac{d^ny}{dx^n} = \frac{1}{2} n(n-1) \frac{d^2y}{dx^2} - n(n-2) \frac{dy}{dx} + \frac{1}{2} (n-1)(n-2)y.$$

5. If $y = e^{m \cos^{-1} x}$, show that $(1-x^2)y_{n+2} - (2n+1)xy_{n+1} - (n^2+m^2)y_n = 0$

6. If $y = \log (x + \sqrt{1+x^2})$

$$(1+x^2)y_{n+2} + (2n+1)xy_{n+1} + n^2y_n = 0$$

7. If $\cos^{-1}\left(\frac{y}{b}\right) = \log\left(\frac{x}{n}\right)^n$, prove that

$$x^2y_{n+2} + (2n+1)xy_{n+1} + 2n^2y_n = 0.$$

8. If $y = e^{m \sin^{-1} x}$, show that

$(1-x^2)y_{n+2} - (2n+1)xy_{n+1} - (n^2+m^2)y_n = 0$. Find $y_n(0)$

9. If $y = (\sin^{-1} x)^2$, prove that $y_n(0) = 0$ for n odd and

$$y_n(0) = 2.2^2 . 4^2 . 6^2 ..... (n-2)^2, n \neq 2 \text{ for } n \text{ even.}$$

10. Find $(y_n)_0$, if $y = \sin(a \sin^{-1} x)$.

11. (a) Prove that :

$$e^{x \cos x} = 1 + x + \frac{x^2}{2} - \frac{x^3}{3} - \frac{11}{24}x^4 - \frac{x^5}{5} - \ldots$$

(b) Use Maclaurin theorem to find the expansion of $\log(1 + e^x)$ in ascending powers of x so as to contain terms up to x^4.

12. Calculate the approximate value of $\sqrt{10}$ to four decimal places by taking the first four terms of an appropriate Taylor's expansion.

13. Apply Taylor's theorem to calculate the value of $f\left(\frac{11}{10}\right)$, where

$$f(x) = x^3 + 3x^2 + 5x - 10$$

14. Find the asymptotes of $y^3 - x^2y + 2y^2 + 4y + 1 = 0$.

Find the asymptotes of :

15. $x^3 - 2y + xy(2x - y) + y(x - 1) + 1 = 0$

16. $(x^2 - y^2)(y^2 - 4x^2) - 6x^3 + 5x^2y + 3xy^2 - 2y^3 - x^2 + 3xy - 1 = 0$

17. $y^3 - x^2y - 2xy^2 + 2x^3 - 7xy + 3y^2 + 2x^2 + 2x + 2y + 1 = 0$

18. $x^3 + y^3 - 3axy = 0$

19. $2x^3 + 3x^2y - 3xy^2 - 2y^3 + 3x^2 + y - 3 = 0$

20. $x^3 + 3x^2y - 4y^3 - x + y + 3 = 0$

21. Find the radius of curvature at the point $\left(\frac{a}{4}, \frac{a}{4}\right)$ on the curve $\sqrt{x} + \sqrt{y} = \sqrt{a}$.

22. Prove that for the ellipse $\frac{x^2}{a^2} + \frac{y^2}{b^2} = 1$, $\rho = \frac{a^2b^2}{p^3}$, p being the perpendicular distance from the centre upon the tangent at any point (x, y).

23. Show that the radius of curvature at a point of the curve $x = ae^\theta(\sin\theta - \cos\theta)$, $y = ae^\theta(\sin\theta + \cos\theta)$ is twice the distance of the tangent at the point from the origin.

24. Show that the radius of curvature at a point $(a\cos^3\theta, a\sin^3\theta)$ on the curve $x^{2/3} + y^{2/3} = a^{2/3}$ is $3a \sin\theta \cos\theta$.

25. Show that the ratio of the radii of curvature of the curve $xy = a^2$, $x^3 = 3a^2y$, at points which have same abscissae varies as the square root of the ratio of the ordinates.

TUTORIAL SHEET-2

Partial derivatives, Chain rule, Differentiation of Implicit functions, Total Differential.

1. If $u = \sin^{-1}\left(\frac{x}{y}\right) + \tan^{-1}\left(\frac{y}{x}\right)$, show that

$$x\frac{\partial u}{\partial x} + y\frac{\partial u}{\partial y} = 0$$

2. If $u(x+y) = x^2 + y^2$, prove that

$$\left(\frac{\partial u}{\partial x} - \frac{\partial u}{\partial y}\right)^2 = 4\left(1 - \frac{\partial u}{\partial x} - \frac{\partial u}{\partial y}\right)$$

3. If $z = e^{ax+by} f(ax - by)$, prove that

$$b\frac{\partial z}{\partial x} + a\frac{\partial z}{\partial y} = 2abz$$

4. If $x^x y^y z^z = c$, show that at $x = y = z$

$$\frac{\partial^2 z}{\partial x \partial y} = -[x \log (ex)]^{-1}$$

5. If $u = (1 - 2xy + y^2)^{-1/2}$, prove that

$$\frac{\partial}{\partial x}\left[(1 - x^2)\frac{\partial u}{\partial x}\right] + \frac{\partial}{\partial y}\left(y^2\frac{\partial u}{\partial y}\right) = 0$$

6. If $u = \tan^{-1}\left(\frac{x^3 + y^3}{x - y}\right)$, prove that

$$x^2\frac{\partial^2 u}{\partial x^2} + 2xy\frac{\partial^2 u}{\partial x \partial y} + y^2\frac{\partial^2 u}{\partial y^2} = \sin 4u - \sin 2u$$

7. If $u = \operatorname{cosec}^{-1}\left(\frac{x^{1/2} + y^{1/2}}{x^{1/3} + y^{1/3}}\right)^{1/2}$, prove that

$$x^2\frac{\partial^2 u}{\partial x^2} + 2xy\frac{\partial^2 u}{\partial x \partial y} + y^2\frac{\partial^2 u}{\partial y^2} = \frac{\tan u}{12}\left[\frac{13}{12} + \frac{\tan^2 u}{12}\right]$$

8. If $u = \log\frac{x^2 + y^2}{x + y}$, find the value of

$$x^2\frac{\partial^2 u}{\partial x^2} + 2xy\frac{\partial^2 u}{\partial x \partial y} + y^2\frac{\partial^2 u}{\partial y^2}$$

9. If $u = \sin^{-1}(x - y)$, $x = 3t$, $y = 4t^3$, show that

$$\frac{du}{dt} = \frac{3}{\sqrt{1 - t^2}}$$

10. If $z = \tan^{-1}\left(\frac{x}{y}\right)$, $x = u + v$, $y = u - v$, show that

$$\frac{\partial z}{\partial u} + \frac{\partial z}{\partial v} = \frac{u - v}{u^2 + v^2}$$

11. If $z = f(x, y)$ and $u = xy$, $v = \frac{y}{x}$, prove that

$$x\frac{\partial z}{\partial x} + y\frac{\partial z}{\partial y} = 2xy\frac{\partial z}{\partial u}$$

12. If $x + y = 2e^{\theta}\cos\phi$ and $x - y = 2ie^{\theta}\sin\phi$, prove that

$$\frac{\partial^2 v}{\partial \theta^2} + \frac{\partial^2 v}{\partial \phi^2} = 4xy\frac{\partial^2 v}{\partial x \partial y}$$

13. If $x^2 + y^2 + z^2 - 2xyz = 1$, show that

$$\frac{dx}{\sqrt{1 - x^2}} + \frac{dy}{\sqrt{1 - y^2}} + \frac{dz}{\sqrt{1 - z^2}} = 0$$

14. If $u = x^2 - y^2 + \sin yz$, where $y = e^x$ and $z = \log x$; find $\frac{du}{dx}$.

15. If $u = f(y - z, z - x, x - y)$, show that

$$\frac{\partial u}{\partial x} + \frac{\partial u}{\partial y} + \frac{\partial u}{\partial z} = 0$$

TUTORIAL SHEET-3

Jacobians, Maxima and Minima of two variables, Curve Tracing :

1. If $x = r \sin\theta \cos\phi, y = r \sin\theta \sin\phi, z = r\cos\theta$ then show that

$$\frac{\partial(x, y, z)}{\partial(r, \theta, \phi)} = r^2 \sin\theta.$$

2. If $u^3 + v^3 + w^3 = x + y + z, u^2 + v^2 + w^2 = x^3 + y^3 + z^3, u + v + w = x^2 + y^2 + z^2$ then prove that

$$\frac{\partial(u, v, w)}{\partial(x, y, z)} = \frac{(y-z)(z-x)(x-y)}{(u-v)(v-w)(w-u)}$$

3. If $u = x + y + z + t, v = x + y - z - t, w = xy - zt, r = x^2 + y^2 - z^2 - t^2$ show that

$$\frac{\partial(u, v, w, r)}{\partial(x, t, z, t)} = 0.$$

4. Verify $JJ' = 1$ if $x = \sin\theta \cos\phi, y = \sin\theta \sin\phi$, where $J = \dfrac{\partial(x, y)}{\partial(\theta, \phi)}$ and $J' = \dfrac{\partial(\theta, \phi)}{\partial(x, y)}$.

5. Find the maximum and minimum values of $x^3 + y^3 - 3axy$.

6. Find the points at which the absolute extreme values of the function $f(x, y) = 2 + 2x + 2y - x^2 - y^2$ on the triangular plate in the first quadrant bounded by the lines $x = 0, y = 0, y = 9 - x$ exist. Also write values of the function at these points

7. Use the method of undertermined multipliers to solve the problem given below:
   (i) Find the point $P(x, y, z)$ closest to the origin on :
   (a) the plane $2x + y - z - 5 = 0$ and the
   (b) hyperbolic cylinder $x^2 - z^2 - 1 = 0$.
   (ii) Find the points on the curve of intersection of the plane $x + y + z = 1$ and the cylinder $x^2 + y^2 = 1$ closest or farthest from the origin.
   (iii) The temperature at a point (x, y) on a metal plate is $T(x, y) = 4x^2 - 4xy + y^2$. An ant on the plate walks around the circle of radius 5 centred at the origin. What are the highest and lowest temperature encountered by the ant?

8. Trace the curve given below:
   (i) The Hypocoloid :

$$x = a\cos^3 t, y = b\sin^3 t \text{ (Parametric form)}$$

or

$$\left(\frac{x}{a}\right)^{2/3} + \left(\frac{y}{b}\right)^{2/3} = 1 \ (a, b >) \text{ (Cartesian form)}$$

   (ii) The Cissoid :

$$x = a\sin^2 t, y = b\frac{\sin^3 t}{\cos t} \text{ (Parametric form)}$$

or

$$y^2 = \frac{x^3}{(a-x)} \text{ (Cartesian form)}$$

   (iii) The Cycloid :

$$x = a(\theta + \sin\theta), \ y = a(1 + \cos\theta) \qquad (a > 0)$$

(iv) The Folium of Descartes :

$$x^3 + y^3 = 3axy \qquad (a > 0)$$

(v) The Catenary :

$$y = c \cosh \frac{x}{c} \qquad (c > 0)$$

(vi) The Circle :

$$x^2 + y^2 + 2gx + 2fy + c = 0 \qquad (f, g, c < 0)$$

(vii) The Cardiode :

$$r = a(1 + \cos \theta) \qquad (a > 0)$$

(viii) The Lemniscate of Bernoulli :

$$r^2 = a^2 \cos 2\theta \qquad (a > 0)$$

(ix) The Limacon :

$$r = a + b \cos \theta \qquad (a > b > 0) \text{ also for } (a < b < 0)$$

(x) The equiangular Spiral :

$$r = ae^{\theta \cot \alpha} \qquad (a > 0) \left(0 < \alpha < \frac{\pi}{2}\right)$$

(xi) The three leaved Rose :

$$r = a \sin 3\theta \qquad (a > 0)$$

TUTORIAL SHEET-4

1. Show that

$$\int_0^{\pi/2} \cos^m x \cos nx dx = \frac{m}{m+n} \int_0^{\pi/2} \cos^{m-1} x \cos (n-1) x dx$$

Hence deduce that

$$\int_0^{\pi/2} \cos^n x \cos nx dx = \frac{\pi}{2^{n+1}}$$

2. Obtain the reduction formula for $\int e^{ax} \sin^n x \, dx$ and hence evaluate $\int e^x \sin^3 x dx$.

3. Find the reduction formula for

$$I_{p,q} = \int_0^{\pi/2} \sin^p x \cos^q x dx$$

where p and q are +ve integers and hence evaluate $I_{p,q}$ when p and q are both even integers.

4. Obtain the reduction formula for :

(i) $\int \cos^n x dx$

(ii) $\int \tan^n x dx$

(iii) $\int \cot^n x dx$

5. Obtain the reduction formula for $\int x^n e^{ax} dx$.

6. Evaluate

(a) $\dfrac{\Gamma(7)}{2\Gamma(4)\,\Gamma(3)}$

(b) $\Gamma\left(\dfrac{-7}{2}\right)$

(c) $\Gamma\left(\dfrac{-1}{3}\right)$

7. (a) $\int_0^{\pi/2} \sin^6 \theta d\theta$

(b) $\int_0^{\pi/2} \sin^4 \theta \cos^5 \theta d\theta$

8. Show that :

$$\int_0^1 x^m (1-x^n)^p dx = \frac{1}{n} \beta\left(\frac{m+1}{n}, p+1\right)$$

9. Prove that

$$\Gamma n = \int_0^1 \left(\log \frac{1}{x} \right) dx, \; n > 0$$

10. Show that

$$\int_0^{\pi/2} \sqrt{\tan \theta} = \frac{1}{2} \Gamma\left(\frac{1}{4}\right) \Gamma\left(\frac{3}{4}\right) = \frac{\pi}{\sqrt{2}}$$

11. Show that

$$\beta\left(m, \frac{1}{2}\right) = 2^{2m-1} \beta(m, m)$$

12. Given that $\int_0^{\infty} \frac{x^{p-1}}{1+x} dx = \frac{\pi}{\sin p\pi}$. Show that

$$\Gamma(p)\, \Gamma(1-p) = \frac{\pi}{\sin p\pi}, \quad 0 < p < 1.$$

13. Show that

$$\int_0^1 x^m (\log x)^n \, dx = \frac{(-1)^n \, n!}{(m+1)^{n+1}}, \quad n \in z^+ \text{ and } m > -1.$$

14. Show that

$$\int_0^{\infty} x^{y-1} e^{-x} (\log x)^n \, dx = \frac{d^n}{dy^n} (\Gamma(y))$$

ANSWERS

2. $I_n = \dfrac{e^{ax} \sin^{n-1} x(a \sin x - n \cos x)}{n^2 - a^2}$

4. $I_n = \dfrac{\cot^{n-1} x}{n-1} - I_{n-2}$

6. (a) 30 (b) $\dfrac{16\sqrt{\pi}}{105}$ (c) $-3\,\Gamma\left(\dfrac{2}{3}\right)$

7. (a) $\dfrac{5\pi}{32}$ (b) $\dfrac{8}{315}$

TUTORIAL SHEET-5

1. Prove that the area of the region bounded by the curve $a^4y^2 = x^5(2a - x)$ is to that of the circle whose radius is, a, is 5 to 4.
2. Find the length of the arc of parabola $x^2 = 4ay$ measured from the vertex to one extremity of the latus rectum.
3. Find the perimeter of the loop of the curve $9ay^2 = (x - 2a)(x - 5a)^2$.
4. Find the volume of the solid obtained by the revolution of the cissoid $y^2(2a - x) = x^3$ about its asymptote.
5. Find the whole area of the astroid.
6. Find the area bounded by $y^2 = 5x + 6$ and $x^2 = y$.
7. Find the area of the segment cut off from the parabola $x^2 = 8y$ by the line $x - 2y + 8 = 0$
8. Find the volume of the solid generated by the revolution of the curve $xy^2 = 4a^2(2a - x)$ about as asymptote.
9. Find the area common to parabola $y^2 = ax$ and the circle $x^2 + y^2 = 4ax$.
10. Find the length of one arc of cycloid $x = a(t - \sin t), y = a(1 - \cos t)$.
11. Find the volume of a sphere of radius a.
12. Find the volume of the solid generated by revolving the ellipse $\frac{x^2}{a^2} + \frac{y^2}{b^2} = 1$.

    (i) about major axis.
    (ii) about minor axis.
13. Find the perimeter of loop of the curve $3ay^2 = x(x - a)^2$.
14. Find the length of the curve $y = \log\left[\frac{e^x - 1}{e^x + 1}\right]$ from $x = 1$ to $x = 2$
15. Find the length of the loop of the curve $x = t^2, y = t - \frac{t^3}{3}$.

ANSWERS

2. $a[\sqrt{2} + \log(1 + \sqrt{2})]$
3. $4\sqrt{3}a$
4. $2\pi^2a^3$
5. $\frac{3\pi a^2}{8}$ sq units

6. $\frac{27}{5}$

7. 36 sq units

8. $4\pi^2 a^3$ cubic units

9. $\left(3\sqrt{3}+\frac{4}{3}\pi\right)a^2$

10. $8a$

11. $\frac{4}{3}\pi a^3$

12. (i) $\frac{4}{3}\pi ab^2$ (ii) $\frac{4}{3}\pi a^2 b$

13. $\frac{4a}{\sqrt{3}}$

14. $\log\left(e+\frac{1}{3}\right)$

15. $4\sqrt{3}$

TUTORIAL SHEET-6

1. Evaluate $\iint (x+y)^2\, dxdy$ over the area bounded by the ellipse $\frac{x^2}{a^2}+\frac{y^2}{b^2}=1$.

2. Evaluate $\iint xy(x+y)dxdy$ over the area between $y=x^2$ and $y=x$.

3. Find the volume bounded by xy plane the cylinder $x^2+y^2=1$ and the plane $x+y+z=3$.

4. Find the volume bounded by the cylinder $x^2+y^2=4$ and the planes $y+z=4$ and $z=0$.

5. Change the order of integration in

$$I=\int_0^1 \int_{x^2}^{2-x} \frac{x}{y}\, dxdy.$$

6. Change the order of integration in

$$I=\int_0^{2a} \int_{\sqrt{2ax-x^2}}^{\sqrt{2ax}} f(x,y)\, dxdy.$$

7. Change the order of integration in

$$I=\int_0^{2a} \int_{\frac{x^2}{4a}}^{3a-x} f(x,y)dxdy.$$

8. Evaluate

$$\int_0^1\int_0^1 \frac{dxdy}{\sqrt{(1-x^2)(1-y^2)}}.$$

9. Evaluate $\iint e^{2x+3y}\, dxdy$ over the triangle bounded by $x=0$, $y=0$ and $x+y=1$.

10. By double integration find the area of region bounded by $y^2=4ax$ and $x^2=4ay$.

11. Evaluate

$$\int_0^1 \int_0^{\sqrt{1-x^2}} \int_0^{\sqrt{1-x^2-y^2}} xyzdxdydz.$$

12. Evaluate

$$\int_1^e \int_1^{\log y} \int_1^{e^x} \log zdzdxdy.$$

13. Evaluate $\iint_A xydxdy$, where A is the domain bold by x-axis, ordinate $x = 2a$ and the curve $x^2 = 4ay$.

14. Evaluate the following integrals by changing the order of integration

$$\int_0^a \int_{\frac{x}{a}}^{\sqrt{\frac{x}{a}}} (x^2 + y^2)dxdy.$$

15. Find $\int_0^4 \int_0^{x^2} e^{y/x}\, dydx.$

16. Evaluate $\iint xydxdy$ over the +ve quadrant of the circle $x^2 + y^2 = a^2$.

ANSWERS

1. $\frac{\pi}{4} ab(a^2 + b^2)$ **2.** $\frac{3}{56}$ **3.** $3\pi - \frac{4}{3}$ **4.** 16π

5. $I = \int_0^1 \int_0^{\sqrt{y}} \frac{x}{y} dxdy + \int_1^2 \int_0^{2-y} \frac{x}{y} dxdy$

6. $I = \int_0^a \int_{\frac{y^2}{2a}}^{\sqrt{a^2-y^2}} f(x, y)dxdy + \int_a^{2a} \int_{\frac{y^2}{2a}}^{2a} f(x, y)dxdy + \int_0^a \int_{\sqrt{a^2-y^2}+a}^{2a} f(x, y)dxdy.$

7. $I = \int_0^a \int_0^{2\sqrt{ay}} f(x, y)dxdy + \int_0^{3a} \int_0^{3a-y} f(x, y)dxdy.$

8. $\frac{\pi^2}{4}$ **9.** $\frac{1}{6}(e - 1)^2 (2e + 1)$ **10.** $\frac{16}{3} a^2$

11. $\frac{1}{18}$ **12.** $\frac{1}{2}(e^2 - 8e + 13)$ **13.** $\frac{a^3}{3}$

14. $\frac{a^2}{289} + \frac{a}{20}$ **15.** $\frac{1}{2}$ **16.** $\frac{a^4}{8}$

TUTORIAL SHEET-7

Solve the given differential equations:

1. $y(1-x)dx - x(1+y)dy = 0$
2. $\sqrt{(1-x^2)(1+y^2)}\,dx + xydy = 0$
3. $y - x\dfrac{dy}{dx} = a\left(y^2 + \dfrac{dy}{dx}\right)$
4. $\operatorname{cosec} x \log y dy + x^2y^2dx = 0$
5. $(xy^2 - x)dx + (yx^3 + y)dy = 0$
6. $(x^3 - y^3)dx = (x^2y + xy^2)dy$
7. $2\dfrac{dy}{dx} = \dfrac{y}{x} + \dfrac{y^2}{x^2}$
8. $x\dfrac{dy}{dx} = y - \sqrt{x^2 + y^2}$
9. $(x^2 - xy)dy = (x^2 + y^2)dx$
10. $(2x + y + 1)dx + (4x + 2y - 1)dy = 0$
11. $(4x + 6y + 5)dx = (2x + 3y + 4)dy$
12. $\dfrac{dy}{dx} = \dfrac{(ax + by - a)}{(bx + ay - b)}$
13. $x \log x \dfrac{dy}{dx} + y = 2 \log x$
14. $(x^2 - 1)\dfrac{dy}{dx} + 2xy = 1$
15. $\sin x \dfrac{dy}{dx} + y \cos x = x \sin x$
16. $(1 - x - xy^2)dy + (y + y^2)dx = 0$
17. $x(x^2 + 1)\dfrac{dy}{dx} = y(1 - x^2) + x^2 \log x$
18. $x\dfrac{dy}{dx} + y = y^2x^3 \cos x$
19. $\dfrac{dy}{dx} = y \tan x - y^2 \sec x$
20. $x\dfrac{dy}{dx} + y = x^3y^4$

ANSWERS

1. $x + y + \log(xy) = a$
2. $\sqrt{1+x^2} + \sqrt{1+y^2} - \log\left[\frac{\sqrt{1+x^2}+1}{xc}\right] = c$
3. $(1 - ay)(a + x) = cy$
4. $\left(\frac{1+\log y}{y}\right) x^2 \cos x - 2x \sin x - 2 \cos x = c$
5. $(x^2 + 1)(y^2 + 1) = c$
6. $y + x \log(c(x - y)) = 0$
7. $y - x = c\sqrt{xy}$
8. $c - y = \sqrt{x^2 + y^2}$
9. $(x - y)^2 = cxee^{-y/x}$
10. $\log(2x + y - 1) + x + 2y = c$
11. $\frac{1}{2}(2x + y^3) + \frac{9}{64}\log(16x + 24y + 23) = x + c$
12. $(y - x + 1)^{a+b}(y + x - 1)^{a-b} = c$
13. $y \log x = c + (\log x)^2$
14. $y(1 - x^2) = -x + c$
15. $(y - 1)\sin x + x \cos x = c$
16. $xy + \tan^{-1} y = c$
17. $4(x^2 + 1)y + x^3(1 - 2\log x) - cx$
18. $xy(x \sin x + \cos x + c) = 0$
19. $\sec x = (c + \tan x)y$
20. $(-3 \log x + c)x^3y^3 = 1$

TUTORIAL SHEET-8

Solve the given differential equations :

1. $x^2 - ay)dx = ax - y^2)dy$
2. $(x^2 - y^2 - a^2)xdx + (x^2 - y^2 - b^2)ydy = 0$
3. $(x^2 - 4xy - 2ys)dx + (y^2 - 4xy - 2x^2)dy = 0$
4. $(x^4 - 2xy^2 + y^4(dx - (2x^2y - 4xy^3 + \sin y)dy = 0$
5. $(2x^3 - xy^2 - 2y + 3)dx - (x^2y + 2x)dy = 0$
6. $ye^{xy}dx + (xe^{xy} + 2y)dy = 0$
7. $\dfrac{2x}{x^3}dx + \dfrac{y^2 + 3x^2}{y^4}dy = 0$
8. $\left[y\left(1 + \dfrac{1}{x}\right) + \cos y\right]dx + (x + \log x - x\sin y)dy = 0$
9. $(1 - xy)ydx + (1 - xy)xdy = 0$
10. $(2xy + y - \tan y)dx + (x^2 - x\sec^2 y + x + \sec^2 y)dy = 0$
11. $xdy - ydx + a(x^2 + y^2)dx = 0$

Solve the following equations using method of variation of parameters :

12. $\dfrac{d^2y}{dx^2} - 4y = e^{2x}$
13. $\dfrac{d^2y}{dx^2} + 4y = \tan 2x$
14. $\dfrac{d^2y}{dx^2} - 6\dfrac{dy}{dx} + 9y = \dfrac{e^{3x}}{x^2}$
15. $\dfrac{d^2y}{dx^2} + y = \sec x \tan x$

ANSWERS

1. $x^3 - y^3 - 3axy = c$
2. $x^4 + 2x^2y^2 - y^4 - 2a^2x^2 - 2b^2y^2 = c$
3. $x^3 - 6x^2y - 6xy^2 + y^3 = c$

4. $\frac{x^3}{5} - x^3y^2 + xy^4 + \cos y = c$
5. $x^4 - x^2y^2 - 4xy + 6x = c$
6. $e^{xy} + y^2 = c$
7. $x^2 - y^2 = cy^3$
8. $y(x + \log x) + x \cos y = c$
9. $\log \frac{x}{y} - \frac{1}{xy} = c$
10. $x^2y + xy - x \tan y + \tan y = c$
11. $ax + \tan^{-1}\left(\frac{y}{x}\right) = c$
12. $c_1e^{2x} + c_2e^{-2x} + \frac{x}{4}e^{2x} - \frac{e^{2x}}{16}$
13. $c_1 \cos 2x + c_2 \sin 2x - \frac{\cos 2x \log(\sec 2x + \tan 2x)}{4}$
14. $(c_1 + c_2 x)e^{3x} - e^{3x} \log x - e^{3x}$
15. $c_1 \cos x + c_2 \sin x + x \cos x + \sin x \log \sec x - \sin x$

TUTORIAL SHEET-9

Solve the following differential equations :

1. $(D^2 + 1)y = 0$

2. $\dfrac{d^2x}{dt^2} + 3a\dfrac{dx}{dt} - 4a^2 x = 0$

3. $(D^2 + D - 1)y = 3e^x$

4. $(D^2 + 4D + 3)y = \sin 3x \cos 2x$

5. $(D^2 + 9)y = 2 \cos 3x + e^{-3x}$

6. $\dfrac{d^2y}{dx^2} + 4y = x^2 + \cos 2x$

7. $\dfrac{d^2y}{dx^2} - \dfrac{dy}{dx} + y = e^{2x} + e^x$

8. $(D + 3)^2y = 2e^{3x} + e^x$

9. $(D^2 - 8D \_ 9)y = 40 \sin 5x$

10. $2\dfrac{d^2y}{dx^2} - 3\dfrac{dy}{dx} + 4y = 3 \cos \dfrac{x}{2}$

11. $\dfrac{d^2y}{dx^2} + 2\dfrac{dy}{dx} + y = x + x^2$

12. $\dfrac{d^2y}{dx^2} + 4y = x^2$

13. $\dfrac{d^2y}{dx^2} + 2y = x^2e^{3x} + e^x \cos 2x$

14. $(D^2 + 3D + 2)y = e^{2x} \sin x$

15. $x^2\dfrac{d^2y}{dx^2} + x\dfrac{dy}{dx} - 4y = x^3$

16. $x^2\dfrac{d^2y}{dx^2} + 3x\dfrac{dy}{dx} + y = (1 - x)^2$

17. $x^2\dfrac{d^2y}{dx^2} - 2x\dfrac{dy}{dx} + 2y = x + x^2 \log x + x^3$

18. $x^2\dfrac{d^2y}{dx^2} + 2x\dfrac{dy}{dx} - 20y = (x + 1)^2$

19. Solve the following simultaneous differential equations :

(a) $(D+1)x + 2y = e^t$

$(D+1)y + 2x = 3e^t$

(b) $Dx + y = \sin t$

$Dy + x = \cos t$ given that $x = 2$ and $y = 0$ when $t = 0$.

ANSWERS

1. $y = c_1 \cos x + c_2 \sin x$

2. $x = c_1 e^{-4at} + c_2 e^{at}$

3. $y = c_2 e^{xc} + c_2 e^{-2x} + xe^x$

4. $y = c_1 e^x + c_2 e^{3x} + \frac{1}{884}(10 \cos 5x - 11 \sin 5x) + \frac{1}{20}(\sin x + 2 \cos c)$

5. $y = c_1 \cos 2x + 3x + c_2 \sin 3x + \frac{x}{3} \sin 3x + \frac{1}{18} e^{-3x}$

6. $y = c_1 \cos 2x + c_2 \sin 2x + \frac{1}{8}(2x^2 - 1) + \frac{1}{4} x \sin 2x$

7. $y = \left(a \cos \frac{\sqrt{3}}{2} x + b \sin \frac{\sqrt{3}}{2} x\right) e^{x/2} + e^x + \frac{e^{2x}}{4}$

8. $y = (c_1 + c_2 x)e^{3x} + \frac{2}{14} e^{-x} + x^2 e^{3x}$

9. $y = c_1 e^{(4+\sqrt{7})x} + c_2 e^{(4-\sqrt{7})x} + \frac{5}{29}(5 \cos 5x - 2 \sin 5x)$

10. $y = \left(c_1 \cos \frac{\sqrt{23}}{4} x + c_2 \sin \frac{\sqrt{23}}{4} x\right) e^{3x/4} + \frac{3}{29}\left[7 \cos \frac{x}{x} - 3 \sin \frac{x}{2}\right]$

11. $y = (c_1 + c_2 x)e^{-x} + x^2 - 2x + 2$

12. $y = c_1 e^{2x} + c_2 e^{-2x} = \frac{1}{4}\left(x^2 + \frac{1}{2}\right)$

13. $y = c_1 \cos \sqrt{2x} + c_2 \sin \sqrt{2x} - \frac{1}{17(}\cos 2x - 4 \sin 2x)e^x + \frac{1}{11}\left(x^2 - \frac{12x}{11} + \frac{50}{121}\right) e^{3x}$

14. $y = c_1 e^{-2x} + c_2 e^{-x} + \frac{1}{170}(11 \sin x - 7 \cos x)e^{2x}$

15. $y = (c_1 x^2 + c_2 x^{-2} + \frac{1}{5} x^3$

16. $y = (c_1 + c_2 \log x)x^{-1} + 1 - \frac{1}{2} x + \frac{1}{9} x^3$

17. $y = c_1 x^2 + c_2 x + \frac{1}{2} x^3 - x \log x + x^2 \left[\frac{1}{2} (\log x)^2 - \log x\right]$

18. $y = c_1 x^4 + c_2 x^{-5} - \frac{1}{14} x^2 - \frac{1}{9} x - \frac{1}{20}$

19. (a) $x = c_1 e^t + c_2 e^{-3t} - te^t$; $y = e^t - c_1 e^t + c_2 e^{-3t} + te^t$

(b) $x = e^t + e^{-t}$; $y = e^{-t} - e^t + \sin t$.

TUTORIAL SHEET-10

1. If $\phi = \log |\vec{r}|$, show that $\nabla\phi = \frac{\vec{r}}{r^2}$, where $\vec{r} = x\hat{i} + y\hat{j} + z\hat{k}$ and $r = |\vec{r}|$.

2. Show that $\nabla.(\nabla\phi) = \nabla^2\phi$, where $\nabla^2 = \frac{\partial^2}{\partial x^2} + \frac{\partial^2}{\partial y^2} + \frac{\partial^2}{\partial z^2}$. ($\nabla^2$ is called Laplacian Operator).

3. If $\vec{A} = x^2 z\hat{i} + yz^3\hat{j} - 3xy\hat{k}$, $\vec{B} = y^2\hat{i} - yz^3\hat{j} + 2x\hat{k}$, $\phi = 2x^2 + yz$, find
   (i) $\vec{A} . (\nabla\phi)$
   (ii) $(\vec{A} . \nabla)\phi$
   (iii) $(\vec{A} \times \nabla)\phi$

4. Show that :
   (i) $\nabla\left(\frac{1}{4}\right) = -\frac{\vec{r}}{r^3}$
   (ii) $\nabla r^n = mr^{n-2}\vec{r}$

5. Show that :
   (i) $\nabla^2 (\log r) = \frac{1}{r^2}$
   (ii) $\nabla^2 (r^n) = n(n+1)r^{n-2}$

6. If $\vec{A} = xz^3\hat{i} - 2x^2 z^4 \hat{k}$, find $\nabla \times \vec{A}$ at the point (–1, –1, –2).

7. (a) Prove that the vector $\vec{A} = 3y^4z^2\hat{i} + 4x^3z^2\hat{j} - 3x^2y^2\hat{k}$ is solenoidal.
   (b) Find the constant a so that the vector $\overset{\mu}{V} = (2x + y)\hat{i} - (3y + 2z)\hat{j} + (x + ax)\hat{k}$ is solenoidal.

8. Show that the vector $\vec{V} = (6xy + z^3)\hat{i} + (3x^2 - z)\hat{j} - (3xz^2 - y)\hat{k}$ is irrotational.

9. If $\vec{A}$ is a constant vector, prove that $\nabla(\vec{r}.\vec{A}) = \vec{A}$.

10. If $\vec{a}$ is a differentiable vector function and u is a differentiable scalar function, prove that

$$div(u\vec{a}) = u div\vec{a} + (\nabla u) . \vec{a}$$
$$cur(u\vec{a}) = \nabla u \times \vec{a} + u curl\, \vec{a}$$

11. If u and $\vec{a}$ have continuous second order partial derivatives, then

$$curl\ grad\ u = 0$$
$$curl\ (u\vec{a}) = 0$$

12. Show that :

$$\nabla . \left[r\nabla\left(\frac{1}{r^3}\right)\right] = 3r^{-4}$$

13. Show that $curl(r^2\,\vec{a}) = 0$, where $\vec{r} = x\hat{i} + y\hat{j} + z\hat{k}$ and $r = |\vec{r}|$.

14. Show that :

(i) $\nabla \times \left(\frac{\vec{r}}{r^2} \right) = 0$

(ii) $\nabla \cdot \left(\frac{\vec{r}}{r^2} \right) = r^{-2}$

15. Prove that :

$$\nabla \times (\nabla \times \vec{A}) = \nabla(\nabla \cdot \vec{A}) - \nabla^2 \vec{A}.$$

16. If a is a constant vector, prove that :

(i) $div\ (\vec{a} \times \vec{r}) = 0$

(ii) $\nabla\ (\vec{r} \cdot \vec{a}) = -2\vec{a}$

(iii) $curl\ (\vec{r} \times \vec{a}) = 0$

(iv) $div\ [r^n\ (\vec{a} \cdot \vec{r})] = 0$

(v) $\nabla(\vec{r} \cdot \vec{a}) = \vec{a}$

17. Prove that :

$$\nabla \cdot \left(\frac{\vec{r}}{r} \right) = \frac{2}{r}.$$

TUTORIAL SHEET-11

1. Show that $\nabla\phi$ is a vector normal to the surface $\phi(x, y, z) = c$, where c is a constant.

2. Find the tangent plane and normal line to the cone $z = \sqrt{x^2 + y^2}$ at the point $(1, 1, \sqrt{2})$.

3. Find a unit vector which is normal to the surface $z = x^2 + y^2$ at the point (1, 2, 5).

4. Find the equation for the tangent plane to the surface $xz^2 + x^2y = z - 1$ at the point (1, –3, 2).

5. Find a unit normal to the surface $xy^2 + 2xz = 4$ at the point (2, –2, 3).

6. Find the unit outward drawn normal to the surface $(x - 1)^2 + y^2 + (z - 2)^2 = 9$ at the point (3, 1, –4).

7. Find the directional derivative of $\phi = xy + yz + zx$ at (1, 2, 0) in the direction of $\hat{i} + 2\hat{j} + 2\hat{k}$.

8. Find the directional derivative of $\phi = xy + yz + zx$ at the point $P(1, 2, 3)$ in the direction of the line PQ, where Q has coordinate (5, 0, 4).

9. In what direction from the point (1, 3, 2) is the directional derivative of $\phi = 2xz - y^2$ a maximum? What is the magnitude of this maximum?

10. Find the directional derivative of $p = 4e^{2x - y + z}$ at the point (1, 1, –1) in a direction toward the point (–3, 5, 6).

ANSWERS

2. Tangent plane : $x + y - 2z = 0$, Normal line : $\dfrac{x-1}{1} = \dfrac{y-1}{1} = \dfrac{z-\sqrt{2}}{-\sqrt{2}}$.

3. $-\dfrac{1}{21}(2\hat{i} + 4\hat{j} - \hat{k})$

4. $2x - y - 3z + 1 = 0$

5. $\dfrac{1}{3}(-\hat{i} + 2\hat{j} + \hat{k})$

6. $\dfrac{1}{3}(2\hat{i} + \hat{j} - 2\hat{k})$

7. $\dfrac{10}{3}$

8. $\dfrac{4\sqrt{21}}{3}$

9. $2\sqrt{14}$

10. $\dfrac{20}{9}$

TUTORIAL SHEET-12

1. If $f(t) = (2t^2 - 5)\vec{i} + (1 - et)\vec{j} + t^3\hat{k}$, evaluate $\int_1^2 f(t)dt$.

2. If $\vec{A} = (3x - 2y)\hat{i} + (y + 2z)\hat{j} - x^2\hat{k}$, evaluate $\int_C \vec{A} . dr$ from (0, 0, 0) to (1, 1, 1) along the curve $C : x = t, y = t^2, z = t^3$.

3. If $\vec{F} = (5xy - 6x^2)\hat{i} + (2y - 4x)\hat{j}$, evaluate $\int_c \vec{F} . dr$ along the curve C in the plane, $y = x^3$ from the point (1, 1) to (2, 8).

4. If $\vec{F} = (xy\hat{i} - z\hat{j} + x^2\hat{k})$ and C is the curve $x = t^2, y = 2t, z = t^3$ from $t = 0$ to $t = 1$, evaluate the line integral $\int_C \vec{F} \times dr$.

5. If $\vec{F} = (2y + 3)\hat{i} + xz\hat{j} + (yz - x)\hat{k}$, evaluate $\int_C \vec{A} . dr$ along the following paths
   $C : x = 2t^2,\ y = t,\ z = t^3$ from $t = 0$ to $t = 1$.
   C : The straight line from (0, 0, 0) to (0, 0, 1) then to (0, 1, 1) and then to (2, 1, 1).
   C : The straight line joining (0, 0, 0) and (2, 1, 1).

6. Evaluate $\iint \vec{A} . \hat{n}dS$, where $\vec{A} = y\hat{i} + 2x\hat{j} - z\hat{k}$ and S is the surface of the plane $2x + y = 6$ in the first octant cut off by the plane $z = 4$.

7. Evaluate $\iint \vec{A} . \hat{n}dS$, where $\vec{A} = (x + y^2)\hat{i} - 2x\hat{j} + 2yz\hat{k}$ and S is the surface of the plane $2x + y + 2z = 6$ in the first octant.

8. If $\vec{F} = 2y\hat{i} - z\hat{j} + x^2\hat{k}$ and S is the surface of the parabolic cylinder $y^2 = 8x$ in the first octant bounded by the planes $y = 4$ and $z = 6$, evaluate $\iint_S \vec{F} . \hat{n}dS$.

9. Evaluate $\iint \vec{A} . \hat{n}dS$ where $\vec{A} = z\hat{i} + x\hat{j} - 3y^2z\hat{k}$ and S is the surface of the cylinder $z = x^2 + y^2 = 16$ included in the first octant between $z = 0$ and $z = 5$.

10. Evaluate $\iint_S \vec{A} . \hat{n}dS$, over the entire surface of the region above the xy-plane boundedby the cone $z^2 = x^2 + y^2$ and the plane $z = 4$, if $\vec{A} = 4xz\hat{i} + xyz^2\hat{j} + 3z\hat{k}$.

11. Evaluate $\iiint_V (2x + y)dV$, where V is the closed region bounded by the cylinder $z = 4 - x^2$ and the planes $x = 0, y = 0, y = 2$ and $z = 0$.

12. If $\vec{F} = (2x^2 - 3z)\hat{i} - 2xy\hat{j} - 4x\hat{k}$, evaluate $\iiint_V \nabla \times \vec{F} dV$ where V is the closed region bounded by the planes $x = 0, y = 0, z = 0$ and $2x + 2y + z = 4$.

13. Evaluate $\iiint_V V dV, F = 2xz\hat{i} - x\hat{j} + y^2\hat{k}$ and V is the region bounded by the surfaces $x = 0, x = 2, y = 0, y = 6, z = x^2, z = 4$.

ANSWERS

1. $\frac{19}{6}\hat{i} - \frac{7}{2}\hat{j} + \frac{15}{4}\hat{k}$

2. $\frac{23}{15}$

3. 35 [illegible]

4. $\frac{-9}{10}\hat{i} + \frac{2}{3}\hat{j} + \frac{7}{5}\hat{k}$

5. (a) $\frac{288}{35}$, (b) 10, (c) 8

6. 108

7. 81

8. 132

9. 90

10. 320*pi*

11. $\frac{80}{3}$

12. $\frac{8}{3}(\hat{j} - \hat{k})$

13. $128\hat{i} - 124\hat{j} + 384\hat{k}$

TUTORIAL SHEET-13

1. For any closed surface S, prove that $\iint_S curl\,\vec{A}\,.\,\hat{n}\,dS = 0.$

2. Show that $\iint_S \vec{r}\,.\,\hat{n}dS = 3V$, where V is the volume enclosed by S.

3. If $\hat{n}$ is the unit outward drawn normal to any closed surface S, show that
$$\iiint_V div\,\hat{n}dV = S.$$

4. If S is any closed surface enclosing a volume V and $\vec{A} = ax\hat{i} + by\hat{j} + cz\hat{k}$, prove that
$$\iint \vec{A}\,.\,\hat{n}dS = (a + b + c)V.$$

5. Prove that :
$$\iiint_V \frac{dV}{r^2} = \iint_S \frac{\vec{r}\,.\,\hat{n}}{r^2}\,dS.$$

6. Apply Divergence Theorem to evaluate $\iint_S \vec{F}\,.\,\hat{n}dS$, where $\vec{F} = 2xy\hat{i} + yz^2\hat{j} + xz\hat{k}$ and S is the surface of the parallelopiped bounded by $x = 0, y = 0, z = 0, x = 2, y = 1$ and $z = 3$.

7. Evaluate $\iint_S \vec{r}\,.\,\hat{n}dS$, where S is the sphere of radius 2 with centre at (0, 0, 0).

8. Apply Divergence Theorem to evaluate $\iint_S (x + z)dydz + (y + z)dzdx + (x + y)dxdy$ where S is the surface of the sphere $x^2 + y^2 + z^2 = 4$.

9. Verify Divergence theorem for $\vec{F} = x^2\hat{i} + y^2\hat{j} + z^2\hat{k}$ taken over the cube $0 \le x, y, z \le 1$.

10. Verify Divergence Theorem for $\vec{F} = 2x^2y\hat{i} - y^2\hat{j} + 4xz^2\hat{k}$ taken over the region in the first octant bounded by $y^2 + z^2 = 9$ and $x = 2$.

11. Verify Stokes Theorem for $F = y\hat{i} + z\hat{j} + x\hat{k}$ where S is the upper half surface of the sphere $x^2 + y^2 + z^2 = 1$ and C is its boundary.

12. Verify Stokes Theorem for $\vec{A} = (y - z \mid 2)\hat{i} + (yz + 4)\hat{j} - xz\hat{k}$ where S is the surface of the cube $x = 0, y = 0, z = 0, x = 2, y = 2, z = 2$ above the xy-plane.

13. Show that $\iint_S curl\,\vec{F}\,.\,\hat{n}dS = 12\pi$ where $F = (x - z)\hat{i} + (x^3 + yz)\hat{j} - 3xy^2\hat{k}$ and S is the surface of the cone $z = 2 - \sqrt{x^2 + y^2}$ above the xy-plane.

14. Verify Green's Theorem in the plane for $\int (3x^2 - 8y^2)dx + 4y - 6xy)dy$ where C is the boundary of the region bounded by $y = \sqrt{x}, y = x^2$.

15. Evaluate $\int_C (3x + 4y)dx + (2x - 3y)dy$, where C, a circle of radius 2 with centre at the origin of the plane, is traversed in the positive sense.

16. Evaluate by Green's Theorem $\int_C (x^2 + y^2)dx + 3xy^2dy$, where $C : x^2 + y^2 = 4$.

17. Verify Green's theorem in the plane for $\int x^2ydx + xy^2dy$ taken along the closed path formed by $y = x$ and $x^2 = y^3$ in the first octant.

ANSWERS

6. 30

7. 32π

8. $\frac{64\pi}{3}$

15. -8π

16. 12π